THE ULTIMATE BOOK OF EXPLOSIVES, BOMBS AND IEDs

Dr. N.C. Asthana, IPS, Inspector General, is a nuclear physicist by training with a doctorate. He is, however, a veritable polymath, the very last of its kind, possessing an unprecedented depth of knowledge in the most diverse fields; a fact all too evident from the numerous outstanding and critically acclaimed reference works he has produced earlier on metaphysics, the science of intelligence, terrorism, military science, and weapon systems etc. On any subject he chooses to expound upon, his is the final word with the authority of the ultimate resource person on the subject. Explosives are one such field.

Dr. Anjali Nirmal is a scholar and researcher of outstanding merit. With a Master's in political science, she holds a doctorate in the field of police administration. She has been associated with Dr. Asthana in the works on intelligence, terrorism, military science, and weapon systems and has independently authored valuable reference works on police atrocities, constitutional law, the Kashmir problem, and police administration. This is their eleventh work.

For more details on the authors and their other works, please visit www.thelastpolymath.com.

THE ULTIMATE BOOK OF EXPLOSIVES, BOMBS AND IEDs

Dr. N.C. Asthana, IPS
M.Sc., Ph.D.
Inspector General of Police
Dr. Anjali Nirmal
M.A., Ph.D.

Pointer Publishers
Jaipur 302 003 (Raj), India

Distributed by
Prem C. Bakliwal for
Aavishkar Publishers, Distributors
807, Vyas Building, Chaura Rasta
Jaipur 302 003 (Raj.) India
Phone : 0141-2578159
e-mail : aavishkarbooks@hotmail.com

ISBN 978-81-7132-552-8

First Published in 2008
Reprinted in 2016 by
Mrs. Shashi Jain for
Pointer Publishers
807, Vyas Building, Chaura Rasta
Jaipur 302 003 (Raj.) India
Phones : 0141-2578159, 2708286
e-mail : pointerpub@hotmail.com
Visit us at : www.pointerpublishers.com

Printed at
Sheetal Printers
Jaipur (Raj.) India

Dedicated to
the very fond and loving memory of
our dearest,
the one and the only one,
Col. Mickey.

From the Big Bang to cars, explosives drive the world!
—Priyamvada

ACKNOWLEDGEMENT

An unprecedented work of this sort could be accomplished only by the grace of the Almighty Bhagwan and nothing else. Man can only aspire and strive to acquire knowledge; the intellect to comprehend it is given by Him. Equally important is the capability to present knowledge in such a form so that it becomes comprehensible to others also—that's how knowledge is preserved and perpetuated, one of the sacred duties that have been cast upon the human race. This capability too is given by Him. We thank Bhagwan that in our lives He gave us the opportunities and the resources to acquire knowledge and also the intelligence to comprehend it. We have made an all-out effort to present the subject in such a way that it becomes a source of knowledge and inspiration for posterity. If this work reflects genius and scholarship, the source is none but Him; if some mistakes are left, they are entirely ours. We humbly offer this treatise on explosives at His feet for all the kindness he has bestowed upon us.

PREFACE

Why did we write this treatise? There are three reasons. The first is what Goswami Tulsidas had said about the reason of his having written the immortal 'Ramcharitmanas': 'swantah sukhay'. If a sublime work like the 'Ramcharitmanas' could avowedly be written for the exalted author's own pleasure, we are fully justified in writing on a subject like explosives that has fascinated us all our lives and on which we have done research for well over three decades. It is but natural that we must distill the essence of all that we have learnt over the years and preserve it in the form of a treatise for the sake of posterity. The second reason is loftier. One of us has been a scientist by training and has also been in the Indian Police Service for long. As such we have had extensive interaction with the paramilitary forces and the armed forces also besides the police forces. We have also been observing media reports on any matter pertaining to explosives very keenly for several decades now. We have very strong reasons to believe that there is terrible ignorance regarding explosives; in fact to such an extent that it has become an arcane subject.

Ignorance is found in many spheres. It is pandemic; it has become particularly acute after the TV has started molding the intellectual sensibilities of the people and reading has been relegated to an activity that is undertaken only when the TV is out of order or when one is waiting at the barber's. With regard to a subject like explosives which is highly specialized in any case, the hopelessness of ignorance can only be imagined. All sorts of misconceptions regarding explosives, bombs and IEDs abound. Through this book, we have undertaken to eradicate the scourge of ignorance regarding explosives. Why? Because, as Martin Luther King, Jr. had said: "Nothing in the world is more dangerous than sincere ignorance and conscientious stupidity." Ignorance is dangerous because it distorts perceptions. Distorted perceptions result in defective policy-making. Ignorance sends you on a wild-goose chase. Ignorance makes you spend billions on useless things which any intelligent terrorist can easily counter. Ignorance makes you see enemies where they are not. Ignorance makes you perceive threats which do not exist. At the same time, it keeps you oblivious of the real threats which may be far more serious. The common man, in whose name everything is done, has a right to know the sources of real threat to him as against imaginary threat; what things can harm him and to what extent, and what he should know about explosives to defend him and his family from such threats. It is also imperative that he should not be misled or misinformed by ignoramuses in the media or amongst the officers in the intelligence agencies and the security forces. In an era when 24-hour TV channels and weekly magazines were not competing with each other in turning out garbage, a cripple never taught anyone to run; unfortunately this is what is happening now.

Examples of ignorance regarding explosives are legion. We will cite but a few to illustrate the point.

- That the terrorists have invented such mysterious explosives which they can make on board aircrafts using innocuous ingredients. The ingredients will be smuggled in bottles of cold drink etc. and the explosive will be made in the toilets of the aircrafts. After that they will blow up the aircrafts in mid-air. (*And hence liquids must be banned on flights. Have you forgotten? They had indeed banned all liquids in cabin baggage on aircrafts! They could have at least considered a Nobel Prize in chemistry for such a wonderful discovery!*)
- That terrorists are coming by the sea route in dinghies and with just a few pounds of explosives carried in their backpacks they will be able to breach our dams, and destroy vital installations like power stations, etc. (*Yeah, you know the Arabian sea is their backwater anyway and these old dams are really fragile! A few pounds are all that they will need and you will have a veritable deluge! Strike off dams from the list of picnic spots!*)
- That the Naxalites are using Claymore mines (used in the US Army) which they have made by themselves in their jungle hideouts. (*Why do the Americans bother to produce the Claymore mines—they could as well order it from the Naxalites! It will take care of their financial problems and could perhaps wean them from the path of terrorism!*)
- That the terrorists are able to make 'deadly explosives' from things as simple as hair bleach, nail polish remover and battery acid. (*How ingenious! Smart guys, isn't it? They should have banned such things immediately! Find some herbal solution quickly, please.... otherwise what will happen to the fashion industry and the millions of cars?*)
- That if there is a bomb blast, RDX must have been used. The only explosive terrorists now use is RDX. (*Incidentally, it was Semtex in the days of terrorism in Punjab. The Al Qaeda has decreed that any terrorist using any other explosive shall be excommunicated from the fraternity and his license to practice terrorism shall be revoked!*)
- Forensic experts have identified the explosive material used in the 'landmine' as ammonium nitrate. (*Yeah, just ammonium nitrate and you did not know all these years that it was a bloody explosive? These fertilizers must be banned! We support organic farming!*)
- That the terrorists are able to make 'high intensity' explosives using substances as simple as aluminum powder. (*Looks like even paint shops will have to be closed! Next they will make explosives out of thin air!*)

- That the terrorists will simply shoot at our lakhs of LPG and petrol tankers across the country and blow them up. Or they will strap small charges to them and they will have 'huge bombs on wheels'. Have they not been showing this in films all along? (*What do we do now, momma? Never knew bombs could be soo... big! Stay away from them, baby! Shall we suggest underground gas pipelines all over the country?*)
- That the terrorists have made explosives which also release poisonous fumes after explosion, thereby invoking a chemical warfare aspect to it. (*Now these guys are really getting nasty, isn't it? Adding insult to injury—as if the blast itself was not enough?*)
- That they use a timer circuit called ABCD timer in their time bombs. (*Ever heard of it; even if you have a Ph.D. in electronics? And you thought the ISI was teaching them English alphabet? That's the wonder timer, baby!*)
- That the terrorists have only to carry a couple of LPG cylinders in a car and crash it—they will be able to blow up entire airport buildings. (*They do invent such brilliant things. You know, the Glasgow bomber Kafeel was doing Ph.D. in aeronautical engineering in UK? And your stupid son-in-law? The bugger did not learn a thing all these years with NASA!*)
- That the IED (which was no bigger than a cricket ball), if it had succeeded in exploding, would have demolished a three-storey building. A major tragedy was thus averted. (*Thank God! Had it been of the size of a football it would have taken the entire town down!*)
- That when the bomb disposal squad detonated the IED (once again no bigger than a school geometry box), it shook houses as much as three miles away and shattered widow glasses. (*Yeah, some people thought the Krakatoa had erupted again! Must take insurance against bomb blasts in the town.*)
- That the explosive used in the pressure cooker bomb was a high explosive as well as incendiary and that's why it had set the train compartment ablaze. (*You know, what these American scientists will ever invent? The ISI knows how to invent explosives! They are the real master blasters! And you thought a master blaster is a batsman?*)

Do not be surprised, dear readers. Einstein had said once: "Only two things are infinite, the universe and human stupidity, and I'm not sure about the former." We have not cooked up the list. If you have been monitoring the media closely, you too would have seen such gems some time or the other. Such outrageous notions are held not only by your own intelligence and security agencies but, as we have discussed in the book, also by premier agencies across the world including the FBI, the British police and the German police. For some it is pure ignorance; for some it is prejudice that has closed the mind to reason; and for some it is both. Ignorance about explosives amongst the laymen and the media people is understandable; it is particularly painful in respect of intelligence officers and officers of the security forces. That they do not know is one thing; that they maintain that they need not know because they happen to wear the uniform is quite another. Being ignorant is not so much a shame as being unwilling to learn or closing one's mind to reasonis. We have nothing but utter contempt for ignorance for there is no sin but ignorance.

This treatise is a fierce counterassault against the universal onslaught of ignorance. It is intended to educate and enlighten officers of the police forces, paramilitary forces, armed forces, the policy-makers, the general public and the media people all over the world and to dispel their ignorance. This is not a pretentious claim. As you go through the book, you will realize it for yourself. We are very much aware that it is a technical subject and not all readers are expected to have a science education. Still, we promise that we will make you understand. We have explained the most complex things in the most lucid way and that too without compromising on rigor anywhere. It is easy to write in jargon, any expert can do that as medical scientists and management gurus know it too well; it will nevertheless be absolutely incomprehensible to the layman—the challenge lies in making a layman understand a technical thing. Glory be to God for this gift. We also take this opportunity to remember George Gamow, a Nobel Laureate in Physics, with great reverence for he has been a source of great inspiration since our childhood. His wonderful books like 'Matter, Earth And Sky', 'Mr. Tompkins In Wonderland' and 'The Atom And Its Nucleus' had shown us in our childhood as to what a rare gift it is to be able to explain technical things to laymen. We guarantee that any reader who has studied science up to high school will be able to understand this book.

The third reason of writing this book is that explosives are extremely important. People don't realize it but the fact is that explosions and explosives pervade every sphere of life. The universe itself started with an unimaginably gigantic explosion: the Big Bang! Life exists on this planet because of the heat of the sun. How does the sun get its heat? It is from the explosive generation of heat in thermonuclear reactions. Explosives are, of course, virtually synonymous with weapon systems. Every single implement of the military depends on explosives. Bullets, shells and missiles are propelled by explosives. From grenades to the biggest bombs, you will have a disreputable dud if you do not have an explosive filling in it. High explosives lie at the heart of even nuclear weapons. High explosive 'lenses' compress the fissile material in a plutonium bomb to super criticality. And every hydrogen bomb has the trigger of a plutonium bomb. Even futuristic weapons like the E-bombs (EMP or Electromagnetic Pulse Bombs) will be using high explosives. As far weapon systems go, anything that moves, from a tiny bullet to the largest Inter-Continental Ballistic Missile, moves by the power of some explosive only.

Even your car runs because of a series of explosions that take place inside it. The explosions are produced when the compressed mixture of air and gasoline burns explosively in the cylinder and pushes the piston down. And remember there are over 600 million cars in the world today, each one of them being driven by explosions. The airbag of your car, which saves you in the eventuality of a crash, uses an explosive. Ever wondered how does the rocket bearing the Space Shuttle free itself from the gantry which supports it on the launch pad? Well, it is done with the help of explosive bolts. Similar explosive bolts are used to jettison the boosters also. The ejector seats on military aircrafts are shot up in the sky by explosives. By the way, construction of aircrafts encounters a strange problem. Owing to their unique design, there are a very large number of places which are inaccessible for normal riveting—they are riveted using explosive driven rivets. But for that the construction of modern aircrafts would have been very difficult.

Then come to mining. Everything that we use on this planet comes from some mining activity. We may change the form of millions of things drastically in the factories or call something synthetic—there is nothing really synthetic. The starting materials for everything come from the Mother Earth only; we do not create anything out of thin air. Now the entire mining industry survives on explosives and they are the biggest consumer of non-military explosives. In the USA, coal mining consumes 67% of the explosives; non-metal mining and quarrying consume 14%; mining of metals takes up 10%, and construction accounts for 7%. Explosives are used for things as diverse as seismic exploration, controlling avalanches, oil-well perforations, pile driving, controlled demolition of buildings, hardening of metals and railroad construction. Great tunnels, dams and canals have been made only with the help of explosives—otherwise it would have been impossible to make them. It is therefore imperative that we must learn about explosives. This book is also a tribute to the tremendous contribution which explosives have had on our lives.

It takes all sorts of people to make this world. Most readers would be delighted to find the wealth of information in this book. There could, however, be some ill-informed, conceited souls who, instead of admiring a work of such phenomenal labor or learning from it, might express a doubt that the contents of this book could possibly be abused by terrorists in making explosives or bombs themselves. For their kind information, it is just not possible. In this book, we have taken great care to purposefully withhold all the crucial information which could be of help to a potential terrorist in actually making explosives, bombs or IEDs. The reader will learn all about the science of explosives, bombs and IEDs from this book but will still not be able to make a terrorist device. This is a reference work on the science of explosives and explosive devices intended to impart technical knowledge to all those for whom it had hitherto remained esoteric or mysterious. But it is NOT a manual on 'how to make explosives, bombs and IEDs'. Any fool who tries making explosives or IEDs after reading this book even for his amusement if not for breaking the law, runs the risk of losing his life and limb, and of course, the heavy legal punishment that he will get. The authors, publisher, and distributors of this book disclaim any liability for any damages or injuries of any type that an irresponsible reader may cause or suffer from the use or misuse of the academic information in this book. The information in this book, as is the case with all books on science, is presented purely for academic interest and study only and not for starting a laboratory in the reader's kitchen. If you really want to spend your life with explosives, become a scientist first and then specialize in the field.

One of us is a very senior IPS officer in a very responsible position. We have already written ten reference works on various subjects including terrorism, counter-insurgency operations, and intelligence. We are also experts in the Conduct Rules of the Indian Police Service. We know what can be written and what cannot be written. Moreover, we know what ought to be written and what ought not to be written. The content of the book has therefore been extremely carefully regulated. We will give an example. Suppose you want to make a super heterodyne radio receiver. Take a textbook on electronics. It will explain you the principle of a super heterodyne receiver with the help of block diagrams. But you won't be able to make the receiver with the block diagrams. For making it you need a detailed circuit diagram with the values of all the electronic components. This is what we have done in this book.

Still, for the information of such doubting ignoramuses, this is not the first book dealing with the subject of explosives, bombs and IEDs. We have, for example, given a selected but still a very sizeable bibliography of 232 books. All of them deal with the subject of explosives, bombs and IEDs. There are thousands of other books which we have referred to but not cited. A google search for books on explosives and explosive devices would be enlightening. The knowledge presented in this book has not been invented by us or is being presented before the world for the first time. All the knowledge has been available in the world since long in the public domain and it is not a military secret. We have only distilled the knowledge contained in myriad sources and presented the essence in a highly readable format.

We must inform that scores of people like William Powell, David Harber, Ragnar Benson, The Jolly Roger, Kurt Saxon, J. Flores, Dirk Goldman, Jo Jo Gonzales and Uncle Fester, to name just a few, have already written a number of manuals on bomb-making for terrorists—all of them are very much available in print as well as on the Internet free of cost. In 1997 the US Department of Justice published a special report on the subject of the public availability of bomb-making information. There are hundreds of publications, particularly of the Paladin Press and the Desert Publications of USA which are devoted to making explosives at home, bomb-making and improvised explosive devices (IEDs). There are numerous publications, field manuals, and technical manuals of the US Army and the CIA also which deal with explosives for sabotage and improvised sabotage devices. In fact the CIA manuals and the technical and field manuals of the US Army could be far more useful to a terrorist than the Anarchist Cookbook. They are all available in the public domain. The Internet has thousands of forums like www.roguesci.org and millions of websites which deal with the subject of improvised explosives and bomb-making and have thousands of books to download. Some of them have posted the information straightaway; some have very elaborate discussion forums where members can exchange information.

It is only some ignorant people who think that such books or Internet resources have or could have contributed to terrorist activities. The fact is that such resources have never been of help to a terrorist. They are neither accurate nor adequate in terms of content. The books of the type mentioned above are written by people who are not scientists themselves. They had acquired their knowledge from dubious sources. Hence the methods which they have described are not standard. Their target readership was people who were not likely to have even high school education. They could not expect their readers to bother themselves with chemistry or the skills required in a chemical laboratory. As a result their books carried erroneous facts, and eventually became source of amusement to fanciful adolescents. No terror strike in the world has ever been made with the help of information gained from such books. The fact is that they cannot. That's why governments of the West have not bothered to ban them or the millions of websites on the subject. They know that what these books and websites teach is possible but impractical. Anyone who tries making explosives or bombs using those methods is most likely to get himself blown off in the process.

On the other hand we have great books on the technology of explosives written by outstanding scientists like Tadeusz Urbanski. They have described explosive manufacture in great technical detail. Had the doubting ignoramuses known about such great works, probably they would have made similar observations regarding them also. But they too have not been of any help to the terrorists. No terrorist has ever been able to make any explosive from books like Urbanski's though it has been around since 40 years! Why? The reason is that while scientists like Urbanski describe the methods of manufacturing practically all explosives, the methods are not suitable for the terrorist—they require elaborate apparatus, extremely high purity chemicals (most of them rare), control of temperature, pressure and other parameters, and very costly procedures. In short, they are suited for nations wanting to make explosives and not for terrorist groups with limited funds, space and technical expertise. You do not expect the terrorists to be running huge factories, do you?

There is one serious limitation of the books by scientists like Urbanski, however. Their books are beyond the comprehension of anybody who is not a scientist himself. We have great regards for men like Urbanski but the fact remains that their works did nothing to lift the veil of darkness as far as the layman was concerned. For the vast majority of law-abiding citizens who, out of sheer curiosity and a sincere desire to learn, would like to educate themselves about explosives and who would not like to be left entirely to the knowledge doled out by half-baked journalists or even worse security or intelligence officers, the field had remained as arcane as ever. There was nothing which could clear their misconceptions. Don't they have a right to learn? Don't they deserve to learn?

We have tried to fill this yawning gap. This is a book that will tell you everything that Urbanski does and more, and that too in a manner that even a layman will be able to understand them. Book written by chemists are, for example, woefully poor on the applications of explosives. You will learn the chemistry of explosives but will not come to know even as much as the charging of a bore hole in mining, not to speak of military ordnance. Are not they important? If you do not know the applications of explosives, you will remain in the dark and will only keep on indulging in fantasies about the mysterious powers of explosives. Such a book was therefore an absolute necessity in a world which is increasingly being driven by myths, misconceptions, pre-conceived notions and pre-digested information doled out by the TV. In the absence of such knowledge, even responsible people have been forming grossly wrong perceptions which have a terribly deleterious effect on policy-making and even execution of policies. Unless you have this knowledge, you will keep on wallowing in the misconceptions of the type we listed in the beginning. There is no end to the stupidities that can be committed out of sheer ignorance. We know that everyone is entitled to be stupid but some abuse the privilege. We have written this book so that the common man does not become the victim of this abuse of privilege. Remember, what Socrates had said: "The only good is knowledge and the only evil is ignorance."

Let us put the clincher now and move ahead. As one cannot make a nuclear weapon simply by reading books on nuclear physics or even doing a Ph.D. in nuclear physics; similarly, one cannot make explosive devices fit for use by terrorists by reading this book. Those who have made nuclear weapons across the world have been Ph.D.s in nuclear physics. It does not mean that to check the proliferation of nuclear weapons, we should restrict Ph.D.s in nuclear physics or proscribe all the books on nuclear physics. Making nuclear weapons requires something more than a Ph.D. in nuclear physics. Do you think that most nations except a few could not make nuclear weapons simply because they did not have enough Ph.D.s in nuclear physics? No! You need much more both in terms of knowledge as well as resources. So is the matter with bombs and explosive devices; they require knowledge more than the chemistry of explosives or the functioning of electronic circuits. Period.

This book tells you not only all about explosives, their invention, and their inventors but also their place in history and industry, and the beauty of the technology involved therein. Explosives and bombs are thus, for the first time in the world, discussed in the proper context of technology and history together. This is a fascinating journey affording you a panoramic view of the entire history of the development of explosives and military ordnance of all kinds, from ancient to modern, from simple to sophisticated, from individual combat to mass destruction, all seen in the backdrop of the men who made them, the men who used them and the impact they made on the course of history. This is a most unique effort in the sense that it is a technical compendium, a collection of human stories, a historical survey, and a tribute to the conquests of technology in different ages, all blended into a highly readable treatise. This is as much a very high-standard reference book for those having some knowledge of explosives as it is a lively introduction to the subject for the layman.

In short, it is a single-window solution to all questions and doubts related to the field of explosives. All that is essential, all that is important, all that needs to be known is to be found here. You will have to consult a thousand reference books if you want all the information that is there in this book. We maintain that while modesty is a virtue, false modesty is not. It is an open challenge—if anyone can find any book in the whole world written before this and which has more wealth of information, we promise to withdraw this book from circulation.

Welcome to the fascinating world of explosives—from gunpowder to polymer-bonded explosives used in nuclear weapons; from dynamites to RDX; from ANFO to Kinepak; from binary explosives to liquid explosives; from safety fuses to electronic detonators; from single-base propellants to composite propellants; from door breaching charges used in urban combat to the fascinating spectacle of controlled demolitions; from shaped charges to explosive lenses; from dumb bombs to smart bombs; from grenades to the Daisy Cutter; from landmines to the bunker buster bombs; from napalm to fuel-air explosives; from the alarm clock time bombs to integrated circuit chip controlled IEDs, from trip-wire booby traps to sound-actuated devices; from wired devices to remote control devices based on cell phones and VHF/UHF sets; from improvised flame throwers to IEDs made by cannibalizing unexploded ordnance; from suicide bomb jackets to car bombs, everything is here. This is the only affordable and complete encyclopedia of explosives in the whole world—exhaustive yet attractive, informative yet interesting, it is the **ultimate encyclopedia of explosives, bombs and IEDs.**

Dr. N. C. Asthana, IPS

Dr. Anjali Nirmal

drncasthana@gmail.com

CONTENTS

PART I
EXPLOSIVES: THE POWER BEHIND EVERYTHING

PART II
BOMBS: TO BREAK THE BACK OF NATIONS

PART III

IEDs: INGENUITY UNLIMITED

PART I
EXPLOSIVES : THE POWER BEHIND EVERYTHING

Chapter 1

EXPLOSIVES AND EXPLOSIONS: BASICS YOU MUST KNOW

(1.1) WHAT ARE EXPLOSIVE SUBSTANCES?

To put it very simply, an explosive is a chemical substance that can be made to explode under certain conditions. But is that a precise definition? Steam boilers are also known to have exploded. Does that make steam an explosive? No! A chemical explosive is better defined as a compound or a mixture of compounds which, when subjected to heat, shock or friction, undergoes a very rapid, self-propagating, heat-producing decomposition. In terms of chemistry, we could describe explosive materials as **energetic materials** also because they convert their chemical energy into kinetic energy of the gaseous products of their decomposition.

(1.1.1) How Do We Create Explosive Substances?

Explosives are essentially unstable compounds which seek to decompose into more stable compounds. That is the law of nature. Everything tends to acquire a state of higher stability. Energetic materials can be made in two ways. The first is by physically mixing solid oxidizers and fuels, a process that, in its basics, has remained virtually unchanged for centuries. The idea is to make something burn faster than usual. Such a process results in a composite energetic material like black powder, a low explosive. The second process involves creating a monomolecular energetic material, such as TNT (trinitrotoluene), in which each molecule contains an oxidizing component and a fuel component. For composite energetic materials, the total energy can be much greater than that of monomolecular materials. However, the rate at which this energy is released is relatively slow when compared to the release rate of monomolecular materials. Monomolecular energetic materials such as TNT work fast and thus have greater power than composites, but they have only moderate energy densities—commonly half of those of composites. Greater energy density versus greater power has been the traditional trade-off in the business of explosives. Your choice is dictated by your intended purpose.

(1.1.2) What Do We Understand By The Explosion of A Chemical Substance?

Decomposition and production of heat are fine but not enough to produce an explosion. A chemical explosion must also produce gases that must exert a high pressure as they expand at the high temperature of the reaction. For the gases to exert a high pressure it is necessary that the rise in temperature should be a quick one. That is to say, they must become hot very quickly. This means that the chemical reaction should involve not only release of a considerable quantity of heat but this heat should be released in a very short time. If the reaction produces a great deal of gaseous products but at a low temperature or slowly then you do not get an explosion. In terms of chemistry you require an exothermic reaction, liberation of gaseous products and a high rate of reaction for an explosion to take place. An explosion results when you get all the three conditions satisfied simultaneously. You get a powerful explosion if the heat released, the volume of gases produced and the reaction rate are all high. If they have relatively lower values you get a milder explosion. If only two of the three conditions are satisfied, you will not get an explosion. For example, the burning of coal produces a great deal of heat and gaseous products both. But the burning of coal does not produce an explosion because the rate at which heat and gases are liberated is very low. Some other chemical reactions, on the other hand, could be very rapid but may fail to generate enough heat or gases.

Fig. 1 The Raw Power Unleashed By An Explosion: A Large Volume Of Gases Released At A High Temperature In A Very Short Time

(1.1. 3) What Other Factors Help In The Making Of Explosions?

An external factor also helps in the making of an explosion. This relates to how the explosive material has been confined and what is the breaking strength of the container. For an explosion to result the confinement of the explosive must offer a certain resistance before it gives way and the gases are released. The gases will develop a high pressure only if a considerable quantity of them is forced to remain for a while in a small volume, that is, the volume of the container. If the confinement gives way very easily then the gases will not be able to develop a high pressure. On the other hand if the confinement is too strong then the gases will waste a considerable part of their energy in breaking the container itself and little would be left for damage outside. For a solid explosive, it means how it has been packed and how strong is the packing material. Small firecrackers are wrapped simply in a tight wadding of paper. The so-called cherry bombs require packing with twine. For liquid explosives, it relates to the strength of the container.

The importance of packing can be understood better with an example. A burning matchstick satisfies all the three conditions of an explosion—it produces heat and gases, and both of them fairly quickly too. Yet you don't get an explosion with a matchstick. Why? The reason is that it is made to burn in the open and is not confined properly! Scrape a considerable quantity of match heads, pulverize them and pack it tight—it will explode if properly ignited. Fill half of a room with matchboxes, close all the windows and lock the room. If you could set fire to the whole thing simultaneously, an explosion would result. People have been able to make improvised .22 rifle cartridges also with 'strike anywhere' match heads—you will have to remove the primary tip, of course! In fact the US Army's Technical Manual TM 31-210 on improvised munitions states that one could make a .222 Remington cartridge propellant with 26 match heads packed tightly into the cartridge!

(1.2) DIFFERENT WAYS IN WHICH EXPLOSIONS CAN TAKE PLACE

(1.2.1) Deflagration

The decomposition of a chemical explosive as described above can take place in two ways. It can be a very rapid burning called deflagration. Burning or combustion is a simple process we are all familiar with. Deflagration is a process of subsonic combustion that usually propagates through thermal conductivity, that is, one layer of hot burning material heats the next layer of cold material and ignites it and so on. Or it can be a phenomenon called detonation discussed below. Deflagration is different from detonation which is supersonic and propagates through shock compression. Deflagration is the process by which low explosives explode and propellants propel things like bullets and rockets.

Fig. 2 Deflagration

Deflagration proceeds in a direction normal to the surface of the burning material. Material is consumed in parallel layers, so the geometry of the grain does not change as the burning takes place. The material is volatized by heat transfer from the flame zone which is in the gas phase above the surface of the material. An increase in the ambient pressure causes the flame zone to move closer to the surface. This increases the rate of heat transfer. The more rapid heat transfer causes an increase in the rate of volatilization of the material which correspondingly increases the rate of deflagration. If the flow pattern of the hot combustion gases is perturbed and penetrates the flame zone, an increase in the rate of heat transfer may occur. An increase in the flame temperature also causes an increase in the rate of heat transfer. The flame temperature is a function of the composition of the burning material. Propellants tend to burn more smoothly at high pressure than at low pressure.

(1.2.2) Detonation

Detonation is a more complex process. Burning requires oxygen either from the atmosphere or from within the explosive mixture. Gases are produced as a result of burning. Detonation does not need oxygen. In detonation, gases are produced when the large molecule of the solid or liquid explosive simply break into smaller molecules that are gaseous.

For detonation to take place, the explosive has to be given a thermo-mechanical shock of a critical value. Any shock less than that would not trigger it. Such shocks are usually given by devices known as detonators. Detonators contain explosives which can be triggered by electrical current or the flame of a burning fuse and hence are easy to use in a 'command' mode. A very significant part of the research on high explosives has gone in developing explosives that are more and more insensitive to moderate shocks. That makes them safer to handle and chances of accident are reduced. Nitroglycerine in its pure form was a great explosive but was notorious for its high sensitivity. The genius of Alfred Nobel converted it into a form which was less sensitive to shock and thus the famous dynamite was born. It is quite possible for one to shoot a bullet through a block of a good high explosive—it will not explode. Throw it in fire, it will simply burn but not explode. For any artillery shell or bomb that is expected to penetrate armor or concrete before it explodes, it is necessary that the charge in the shell or bomb should not go off with the shock of the impact of the shell or bomb with the armor or concrete—it should go off only when its own detonator triggers it.

(1.2.3) The Physics Of Detonation

Detonation is a process of supersonic combustion that involves a shock wave and a reaction zone behind it. A **shock wave** is a very rapid, almost discontinuous change in pressure. The manner in which shock waves propagate through a medium is different from the manner in which sound waves move through the medium though both of them involve periodic changes in pressure in the medium. Shock waves are formed by a pressure pulse of very high amplitude, much higher than the pressures involved in propagation of sound waves. The propagation velocity of a shock wave is not constant and increases as the pressure increases. The shock velocity is the velocity at which the shock wave moves through the material. The shock velocity is greater than the velocity of sound in the material at a given density. In a non-explosive material even a shock wave would eventually dampen and die due to normal attenuation. In an explosive material, however, pressure, density, internal energy, and temperature all increase as a result of the passage of the shock wave through the material because the shock wave itself produces further decomposition of the material. The shock wave compresses the material thus increasing the temperature to the point of decomposition or ignition as the case could be. The compressed material decomposes or gets ignited behind the shock and releases energy that supports the shock propagation. A detonation shock wave therefore becomes self-sustaining because the energy added by the chemical reaction taking place behind the shock wave balances the losses due to normal attenuation.

A self-sustained detonation wave is different from a deflagration that propagates with a subsonic speed and without a shock. To qualify as detonation, the speed of the chemical reaction which constitutes the decomposition propagating in the explosive material should be higher than the speed of sound in the same explosive material in the un-exploded state. In deflagration the decomposition is propagated by the flame front which necessarily moves at a slower speed; in detonation the decomposition is propagated by a shock wave traversing through the explosive material at a very high speed.

Detonations too can be of two types. If you use a weak detonator or a high explosive that is packed very loose or is degraded otherwise, you get a **'low order' detonation**. The resulting explosion is weaker and there is considerable smoke also. A proper detonation is a **'high order' detonation** with all the expected explosive power released and little smoke.

Fig. 3 Detonation Of High Explosives: The Extent Of Damage In The Oklahoma City Bombing

(1.2.4) Deflagration-To-Detonation Transfer (DDT)

In some energetic materials deflagration can eventually lead to detonation under certain conditions of porosity of the material, confinement and initiation. In a deflagration-to-detonation transfer situation, local ignition forms a convective flame front. A convective flame front is an ignition wave which is propagated by the passage of hot combustion gases through the pores of the explosive material or the charge. The hot gases are forced into the pores of the charge by the pressure built up as a result of charge confinement. A continuous flow of hot gases is required to raise the temperature of the pore walls to the point of ignition. Assuming that the pores are interconnected, ignition of the pores causes the convective flame front to advance and as the convective flame front advances, the pressure increases linearly with time. The pressure increase leads to the formation of a post-convective or compressive wave, thereby marking the beginning of an accelerated pressure buildup. Subsequently the compressive waves may coalesce into a shock wave and that would result in a deflagration-to-detonation transfer. The degree of compaction of the explosive determines how much material will be consumed before the deflagration-to-detonation transfer occurs. The run-up distance is defined as the length of material which will deflagrate before detonation occurs. A deflagration-to-detonation transfer can therefore occur only if the explosive is relatively sensitive and there is a rapid pressure rise in the ignition region that produces a sufficiently strong shock wave.

This explains the phenomenon of some explosives simply burning when set on fire. They burn when they are unconfined. But they may detonate if they are suitably confined.

This also means that if you have an explosive material which is not porous, such as a cast charge, then deflagration-to-detonation transfer becomes difficult. On the other hand, excessive porosity would mean that there will be smaller rates of pressure increase near the ignition point. Burning for a longer period of time would therefore be required before the rate of pressure increase begins to accelerate. The deflagration-to-detonation transfer occurs only after the acceleration occurs. A sharp rate of pressure increase is required to produce compressive waves which can coalesce into shock waves. The run-up distance is therefore correspondingly increased.

❑❑❑

Chapter 2

Low But Not Lowly Explosives

(2.1) WHAT ARE LOW EXPLOSIVES?

Explosive materials which undergo deflagration are called **low explosives**; explosive materials which undergo detonation are called **high explosives**. This is a rule of thumb and exceptions do exist. As we saw above, deflagration-to-detonation transfer is possible. Several explosives can both deflagrate and detonate depending upon the conditions of initiation or confinement—nitrocellulose for example, will deflagrate if simply ignited but would detonate if initiated by a detonator. Thus we could say that low explosives generally deflagrate and high explosives generally detonate though the behavior might change under certain conditions of confinement or initiation.

During the Vietnam War many American troops had discovered a simple way of boiling water or heating food. Explosives were to be found aplenty in those days and a small piece of the plastic explosive C-4 was used as a fuel. When lit, it could boil a canteen cup of water very quickly thus making cooking fun and easy. It so happened that once a GI who was cooking his food in this manner was suddenly ordered to move. This guy thought of being a good guy and tried to stomp the fire out before leaving the place. The burning C-4 blew up and took a better part of his foot with it. Never stomp on a high explosive which is burning or try to smother the fire in any way. Don't even throw water on it—it does not help. Doing that amounts to a sort of confinement and may result in detonation.

Low explosives create sub-sonic explosions, that is, in which the speed is below 1089 fps. For high explosives, the most important single property in rating an explosive is **detonation velocity**, which may be expressed for either confined or un-confined conditions. It is the speed at which the detonation wave travels through the explosive. The detonation velocity of an explosive is dependent on the density, ingredients, particle size, charge diameter, and degree of confinement. Decreased particle size, increased charge diameter, and increased confinement all tend to increase the detonation velocity. Similarly, denser explosives usually give higher detonation velocities and pressures. Unconfined detonation velocities are generally 70 to 80 percent of confined detonation velocities. The confined detonation velocity of commercial high explosives varies from 4,000 to

25,000 fps or more. Manufacturers typically measure detonation velocities in an unconfined column of explosive 1.25 inches in diameter.

(2.1.1) The Primary Uses Of Low Explosives

Low explosives are primarily used as propellants whether in guns or in rockets. They produce gases at relatively lower pressures. Thus they have a heaving or pushing effect on the object in front or the confinement. That's how they propel things. High explosives, on the other hand, produce such high pressures that they tend to shatter things. For military warheads or any useful bomb we must necessarily use high explosives. On the other hand, if you use high explosives in a gun or rocket, they will burst instead of flying. Brig. John Masters recalls with great nostalgia in his lovely book 'Bugle and a Tiger' how the British had successfully tried a covert operation on the unruly Afghans during the colonial wars. The Afghans were always in short supply of ammunition. The British 'doctored' some ammunition in the sense that the propellant was replaced by high explosive of equal weight and then the bullet was re-crimped. This doctored ammunition was then arranged to be supplied to the Afghans through some agents. When these cartridges were used, the cartridge exploded in the chamber blowing off the chamber—the firer was sure to lose his arms, eyes and parts of his face if not his life altogether. Imagine the terror in the hearts of the Afghans, not knowing which cartridge was going to explode in the gun! Many accidents in the early days of rocketry, particularly with solid fuel rockets had taken place because the packing density became high or due to some other defect in the design of the combustion surfaces, the rate of burning could not be kept controlled.

(2.2) THE OLD FAITHFUL BLACK POWDER, A GREAT LOW EXPLOSIVE

(2.2.1) The Invention Of Gunpowder In China

The earliest clear, certain references to saltpeter explosives come from China. In 220 BC an accident was reported when some **Chinese alchemists** accidentally seem to have made black powder while separating gold from silver during a low-temperature reaction. They had reportedly added saltpeter (potassium nitrate) and sulfur to the gold ore in the alchemists' furnace but forgot to add charcoal in the first step of the reaction. Trying to rectify their error they added charcoal in the last step. Unknown to them they had just made black powder which resulted in an explosion. Joseph Needham, on the other hand, argues that ancient Chinese alchemists got to an early form of gunpowder, as part of their search for elixirs of immortality. In any case, it is generally agreed that the ancient Chinese made some form of gunpowder and it could well be called one of the Four Great Inventions of ancient China.

The word 'gunpowder', widely defined, should include all mixtures of saltpeter, sulfur and carbonaceous material; but any composition not containing charcoal, for example those which incorporated honey, may be termed 'proto-gunpowder'. The word gunpowder arises from the fact that Europe knew it only for cannon or handguns. In China, however prototype mixtures were known to alchemists, physicians and perhaps fireworks technicians, for their deflagrative properties, some time before they began to be used in weapons. Hence the Chinese name for gunpowder, 'huo yao', literally meaning 'fire-chemical' or 'fire drug'. In Europe, although the earliest clear, written reference to black powder by itself, without other ingredients, was made by Roger Bacon in England in 1267, he implies that he did not invent it himself, and that the technology was already widespread in his time.

(2.2.2) Gunpowder In Medieval Europe

There is no direct record of how the modern formula for black powder was invented, or how it came to be known in Europe, but most scholars believe that this technology spread west from China to the Middle East and then Europe, possibly via the Silk Road. In the mid-12th century, John Arderne, an Englishman, gave the proportions of saltpeter, charcoal, and sulfur as 6:2:1, the same as recommended by Marcus Graecus around 1275 and by Albertus Magnus in 1300. The Germans who, as some scholars claim, invented gunpowder and guns around 1313 (that is, the legendary Berchtold Schwarz), were using a mixture of 4:1:1 in 1350.

The earliest mention of black powder on military supply lists is seen in 1326 AD. Also at that time, a Latin manuscript called "On the Duties of Kings," written by Walter de Milemete, King Edward III's chaplain, included the first known illustration of cannon. In 1320 a German monk called **Berthold Schwartz** studied the writings of Bacon and began to make black powder and study its properties. The results of Schwartz's research probably speeded up the adoption of black powder in central Europe. By 1400, iron cannons, bound by iron hoops to keep them from bursting, and iron cannon balls were coming into use. At the end of the 14th century, a handgun was invented that weighed 10 lbs and fired lead shot. One man carried the gun, which was mounted on a stand and aimed at the enemy. This weapon evolved into the arquebus or harqebus, a handgun with a bent stock and a forked staff to support the barrel while firing. A lit match or fuse of serpentine black powder was used to ignite the propellant powder charge through a touch hole in the breech (closed end) of the gun barrel. By the end of the 13th century many countries were using black powder as a military aid to breach the walls of castles and cities. Depending on the weapon or the use, some countries varied the proportions of saltpeter, charcoal, and sulfur. For example, the French at one time made six different mixes one strong and one weak each for pistols, muskets, and cannons. Different mixes have been used for blasting, mining, and sporting weapons. In 1895, most countries, including the USA, adopted the English standard 75:15:10 (saltpeter: charcoal: sulfur), which had actually been in effect since 1781. This standard is generally used for all arms, the only difference being in the size of the grains.

The records of towns and armies during the early days of gunpowder contain many references to accidental explosions. Other mostly useless ingredients, such as amber, alcohol, camphor, wine, urine, arsenic, and sal ammoniac, were added to the recipe to improve stability, reduce absorption of moisture, and to prevent crumbling.

The ingredients were ground, mixed, and used in a fine powder or meal. At first, the materials were mixed in a mortar using a pestle operated by hand and were later mixed by horse power. Later, improvements in machinery mixed gunpowder with rollers on a marble slab and with wooden stamps. Mixing black powder was considered a highly technical art and a dangerous one too. Sometimes the final mixing was done at the gun site to reduce the hazard of stores exploding. In 1425, **corning**, a new process for making black powder into grains instead of the finer serpentine gunpowder, was invented. Corning consisted of wetting black powder to form a paste, which was then pounded into a cake. The cake was broken into small lumps by hand and sifted. The resulting pieces, which were called corned powder, absorbed less moisture and were more suitable for use in cannons than the serpentine powder. Flame from the priming powder was distributed more efficiently by the larger grains. This resulted in a more complete explosion and reduced fouling.

(2.2.3) Some Historic Military Exploits Of Black Powder In Guns

Do not deride this great explosive as a relic of the past or something insignificant; till nearly the end of the 19th century it took part in the creation of history in the course of numerous battles as propellant for small arms and guns and also as explosive used in mines against fortifications. Black Powder reigned as the king of explosives for some 600 years! Production in 1860 was about 27 million pounds and it peaked around the turn of the century with nearly 100 million pounds produced. Black Powder consumption in the U.S. peaked at 277 million pounds in 1917. It is used even now in the manufacture of safety fuses and fireworks.

In 1453, gunpowder signaled the end of an era in more than one way. The Turks had besieged the city of Constantinople. The fortification of the Byzantine city was the greatest and the most complex fortification system hitherto built in the world and the 'Walls of Constantinople' were legendary. Sultan Muhammad II was attacking the so-called 'Theodosian Walls'. This was a chain of double walls about three and a half miles long. The walls were made of alternating layers of stone and brick. The inner wall was about 16.5 feet thick and 39.4 feet high. The outer wall, separated from the inner wall by a distance of about 57 feet was about 6.6 feet thick and 27.8 feet high. The Turks got 68 cannons made in Hungary to batter these formidable walls. The cannons were made by a genius of the era called Urban. This man was a Christian and had first offered his services to the Byzantines. When they refused, he joined the Turks. The siege guns which he made were up to 27 feet long and weighed between 17 to 20 tons each. They were capable of throwing stone balls weighing 700 to 1200 lbs. The bronze gun became famous as the **Dardanelles gun** and one of them still survives, having been presented to Queen Victoria by Sultan Abdul Aziz in 1867. They needed a crew of nearly 200 men each and 70 to 100 oxen were required to drag them. The effect of the guns is described by the Greek chronicler Kritovoulos in the following words: "The cannon balls hit the wall, which it immediately shook and knocked down and was itself broken into many fragments and scattered, hurling the pieces everywhere and killing those who happened to be nearby." Constantinople fell in less than two months of battering by these monsters whereas it was common for sieges to last several years!

Fig. 4 Theodosian Walls: The Most Formidable Fortification Of The 15th Century In The World In Constantinople

The Dardanelles guns showed what gunpowder could do and dramatically ended the hitherto unchallenged belief in the security offered by fortifications. Encouraged by the exploits of the Dardanelles guns people elsewhere in the world made even bigger and heavier guns, but they were not destined to fire in combat. In 1586 Andrey Chokhov in Russia made the Tsar Pushka, the world's biggest howitzer weighing 18 tons. In India, Sawai Jai Singh II of Jaipur got the world's biggest cannon on wheel mounts made in 1720. Called 'Jaivaan' (meaning 'arrow of victory'), the gun weighs about 50 tons. None of them ever fired in combat, however. The guns were too unwieldy and the rate of fire was excruciatingly slow to be of any combat effectiveness—the Jaivaan required four elephants just to swivel it around its axis.

Fig. 5 And The Gun That Humbled The Theodosian Walls: The Dardanelles Gun: A Most Emphatic Demonstration Of The Awesome Power Of Gunpowder

Gunpowder was the explosive that propelled the shots from the famous **twelve-pounder** and bigger guns that decided the course of many a decisive battle in European history and the colonial wars. Later it also formed the bursting charge when exploding shells were invented. The 12-pounder or the **'Napoleon'** as it came to be called, was light enough to be moved rapidly on the battlefield by horses, heavy enough to destroy field fortifications almost a mile away, and versatile enough to fire solid shot, shell, spherical case, grapeshot and canister. The gun could be used for ranges as low as 200 yards or even less for canister shot to as high as 2000 yards for round shot. Till breach-loading guns were invented, the 'Napoleon' was the mainstay of the European armies and the 'lord of the battlefield'. No gun in history has been so famous. The 12-pounder needed a charge of 2.5 lbs of gunpowder for one round shot. The charges were contained in cartridge bags, made of woolen material free from any mixture of cotton, and of sufficiently close texture as to prevent the powder from sifting through.

Guns consumed huge quantities of gunpowder. A 12-pounder battery of six guns carried at least 200 rounds per gun in battle and typically 1344 rounds were carried for the entire battery. You can see that a battery of six guns only needed about 3360 lbs of gunpowder in the field! In major battles like Waterloo where Napoleon had about 250 guns and Wellington about 150 guns, this means that up to 224,000 lbs of gunpowder could have been present on the battlefield for artillery alone, not to speak of the powder carried on their persons by soldiers wielding the flintlocks! This gives some idea of the huge quantities of gunpowder that were consumed in battles and eventually decided the fate of the nations. This also shows how important it was for warring nations to produce good quality gunpowder in great quantities. The industrial capacity of a nation and the logistic capability of an army were both taxed in producing such huge quantities and making them available on the spot. Even in the wars of the Orient where the armies were notorious for their poor logistic arrangements, those wishing to be victorious could not afford to ignore this vital aspect. According to the contemporary source Zawabit-i-Alamgiri, the great Mughal emperor Aurangzeb carried about 478,000 lbs of gunpowder and 102,000 lbs of shot for his 106 field pieces of diverse sizes for the siege of Golconda that lasted more than seven months (224 days).

(2.2.4) Powder Charged Mines That Humbled Many Fortifications

People used to apply a great deal of mind in making fortifications and choosing their locations in such a manner that attacking cannon could not hit their walls at proper angles. The solution for the attackers was once again black powder, though used in a different manner. It was this explosive which, when used in **mines** sunk under fortifications, breached them and enabled the besiegers to storm the fort. The history of the world is replete with instances where a fortification could be taken only after it was breached this way. The amount of powder needed varied with the job in hand, but it was common for about 1250 to 8000 lbs of powder to be used in a mine. In the siege of Petersburgh, Virginia in July 1864, it took 34 days for nearly 400 men to dig a 511 feet long main gallery excavating about 18000 cubic feet of earth. The mine was charged with 8000 lbs of powder in 320 kegs of powder of 25 lbs each. This gives some idea of tremendous effort it took to fight wars in those days. The explosion created a crater 200 feet long, 50 feet wide and 25 feet deep. The Mughals in India depended heavily on mining operations in their sieges and persisted with them even as neglecting safe practices resulted in many accidents and failures.

In his work on siege warfare published in 1740, Sebastien Le Prestre de Vauban (French Marshal, 1630-1707) codified principles of military mining that remained valid well into the 19th century. Vauban, in what could be considered the first scientifically based demolitions manual, described a method of charge calculation and placement based on characteristics of the target fortress and the desired effect. He defined these mines by the depth and size of the charge:

- For depths less than 3 meters, it was called a **fougasse** (or contact mine).
- For depths greater than 3 meters, it was called a **mine.**
- When used as a "countermine" against an enemy mine, it was called a **camouflet.**
- When intended to destroy an entire fortification (using 5,500 lbs of powder or more), it was called **globes de compression** (pressure balls).

Incidentally, mining operations in the battlefields continued even when high explosives were invented. The Somme Offensive was an example where mining operations were conceived and executed as an integral part of the battle plan for destruction of enemy works and other things. In 1916 in the Bluff, nine mines totaling 200, 000 lbs of HE were used, the largest of them carrying 40,600 lbs of Ammonal sunk 75 feet deep. Incidentally, the biggest mines of history were used in Messines in 1917. Twenty mines totaling 1000,000 lbs were deployed, the largest carrying 95,600 lbs of ammonal at 125 feet depth. You can imagine the quantity of gunpowder required had the HE not been invented!

Storage of vast quantities of gunpowder on the field meant that one had to be very careful with it. If gunpowder could inflict damage on the enemy, accidental explosions in powder magazines could inflict great damage on one's own side also. Sher Shah Suri, the great Afghan ruler in India and a very able administrator, was killed in the explosion of a powder magazine only in 1545 when he had besieged the fort of Kalinjar. But for his sudden demise due to that explosion, the history of India would have taken a different course as he was a great hindrance to the Mughal dynasty. In Europe, however, great thought was devoted to designing powder magazines and elaborate rules were evolved with regard to proper storage of gunpowder.

(2.2.5) Some Great Constructive Exploits Of Gunpowder

Black powder was used for the first time for a civil work during 1548-1572 for the dredging of the River Niemen in Northern Europe. In 1627 black powder was used as a blasting aid for recovering ore in Hungary. Soon, black powder was being used for blasting in Germany, Sweden and other countries. In England, the first use of black powder for blasting was in the Cornish copper mines in 1670. Bofors Industries of Sweden was established in 1646 and became the main manufacturer of commercial black powder in Europe.

Gunpowder was used extensively in civil construction works also. The first time gunpowder was used on a large scale in civil engineering was in the construction of the Languedoc canal in Southern France. It was completed in 1681 and linked the Mediterranean Sea with the Bay of Biscay with 240 km of canal and 100 locks. Another noteworthy consumer of black powder was the Erie canal in New York. It was 585 km long and took 8 years to complete starting in 1817. Canal construction led to a frenzy of activity among American gunpowder manufacturers.

Black powder was also extensively used in railway construction. At first tracks ran on level ground, but later railways made extensive use of cuttings and tunnels. One 800 meter stretch of the notorious Box Tunnel on the Great Western Railway line between London and Bristol consumed gunpowder at the rate of a ton a week for over two years! Over 100 lives were lost construction the 3.3 km tunnel. The 12.9 km long Mont Cenis tunnel was completed in 13 years starting in 1857.

(2.2.6) Landmarks In The History Of Gunpowder

- 1540. Standard gunpowder grain sizes for pistols, muskets, and cannons were adopted by the French. They discovered that less powder was needed to propel a projectile if the grain size was controlled. By using uniform, slow-burning grains they were able to control the pressure of the explosion in the gun, and thereby reduce the hazard of blowing up the gun. This discovery also led to improvements in gun design.
- 1560. Linen or paper bags containing gunpowder were used for charging cannons.
- 1578. An instrument for testing the strength of gunpowder in laboratory was devised by Bourne of England. Until then, the only way was to burn a small quantity to see how much residue would be left. The testing device consisted of a small metal cylinder with a heavy metal lid on a hinge. The lid was prevented from falling by a ratchet, and the angle to which the lid rose when powder was fired inside the cylinder measured the strength.
- 1588. Explosive shells were used at the siege of Bergen op-Zoom and Wachtendonck, Holland by the Dutch. The shells were made from hollow iron balls filled with gunpowder. The gunpowder was ignited by a fuze consisting of a hollow tube filled with slow burning serpentine powder, and timed to burn 14 to 20 seconds, corresponding to ranges of 1000 to 2000 yards for mortar fire. The knowledge for making this fuze was brought to the Dutch by an Italian deserter from Parma.
- 1590. Cartridges with ball and powder combined were introduced for small arms.

- 1602-1604. Fulminating gold, later used as a priming explosive, was invented by Johann Tholden, a Dutch chemist in the employ of the British Navy.
- 1613. Blasting powder, a variant of the military propellant gunpowder is reported to have been used in German mines to blast rocks. It used less saltpeter and more charcoal and sulfur than was used in the black powder for guns.
- 1627. An improved instrument for testing black powder was devised by Curtenbach.
- 1627. Black powder was used to blast ore in the royal mines at Schemnitz, Hungary following experiments by a Hungarian engineer, Kaspar Weindl.
- 1628. Gold fulminate priming explosives were invented by J. Thollen for the British Navy.
- 1647. Master-Gunner Nye in his "Art of Gunnery" proposed the so-called mortar test for measuring the strength of gunpowder.
- 1654. Ammonium nitrate was first prepared by J. R. Glauber, a German chemist.
- 1679. Blasting powder was used on a large scale in the construction of the Mal pas Tunnel, Canal de Midi, France.
- 1684. Glazing process for gunpowder was introduced. Rough gunpowder corns were glazed by tumbling them in large wooden drums until the grain surfaces were smooth. Graphite was added to coat the grains. Glazed gunpowder was less hygroscopic and flowed more easily than the unglazed gunpowder.
- 1738. d'Antony of Italy devised the so-called 'closed bomb' for measuring the pressure generated by gunpowder fired in a closed vessel.
- 1742. The ballistic pendulum, which determined the velocity of a projectile's impact by measuring the recoil of a struck pendulum, was invented by Benjamin Robins, an English mathematician.
- 1784-1788. Potassium chlorate was prepared in its pure state by Claude L. Berthollet, a French chemist who discovered its detonating properties. However, his idea of using potassium chlorate instead of potassium nitrate in black powder was abandoned after a disastrous explosion during manufacture in 1788.
- Berthollet also prepared fulminating silver in 1788.
- 1784-1789. Compact black powder grains were obtained by pressing the cakes with screw presses before corning at Faversham, England.

(2.3) MANUFACTURING GUNPOWDER: THE ELABORATE METHOD EVOLVED BY THE EUROPEANS

Over the centuries the Europeans perfected the art of manufacturing this simple explosive so as to render it relatively safe in use and dependable in storage. It is not that one could simply mix potassium nitrate, sulfur and charcoal and get black powder fit for military use. It requires potassium nitrate of at least 99% purity. The sulfur should be the one distilled from retorts heated to a temperature of 400 degrees Celsius and the receiver at above 115 degrees, that is, above the melting point of sulfur. The distillate must be cast into sticks or blocks. The point to be noted is that if the receiver temperature is lower, one gets flowers of sulfur which have impurities of sulfur dioxide and sulfuric acid. It is hygroscopic also and cannot be used for making gun powder. The charcoal should be made from soft, but resinous wood, and preferably from the white wood of trees like alder, poplar, willow, hazel, etc. For production of good black powder, one has to go through several stages. The ingredients are to be milled first in an **Iron Ball mill** or a **Disk mill** which must be non-sparking.

Then they are to be mixed in a wooden drum with humps, or in an **Edge Runner**. Then the mass needs to be pressed to form press cakes, which are then dried. The mixing process was improved tremendously in 1425 when the corning, or granulating, process was developed. **Corning** is a process which first compresses the fine black powder "meal" into blocks with a fixed density (1.7 g/cc). There were two types of corning mills, **Roll corning mills** or **Drum corning mills**. Heavy wheels were used to grind and press the fuels and oxidizer into a solid mass, which was subsequently broken down into smaller grains. These grains contained an intimate mixture of the fuels and oxidizer, resulting in a black powder which was physically and ballistically superior. These grains were then sorted by size to give the various grades of black powder with different rates of burning. Corned black powder gradually came into use for small guns and hand grenades during the 15th century and for big guns in the 16th century. Black powder mills (using the corning process) were erected at Rotherhithe and Waltham Abbey in England between 1554 and 1603.

Fig. 6 Scene From An Old Gunpowder Mill: Double Runners Mix The Ingredients

The powder from the corning mills is polished, dried and graded according to grain size in the process called **glazing**. Glazing process for gunpowder was introduced in 1648. Rough gunpowder corns were glazed by tumbling them in large wooden drums until the grain surfaces were smooth. Graphite was added to coat the grains. **Glazed gunpowder** was less hygroscopic and flowed more easily than the unglazed gunpowder. It is vital to coat the grains with graphite so that they become electrically conducting and do not go off by sparks produced by static electricity. All these stages required very elaborate precautions and special conditions. The slightest error or variation could result in a most unpleasant accident or, at best, a highly unsatisfactory product.

Fig. 7 That's How They Stored Gunpowder For The Great Battles: Gunpowder Barrels Kept in a Magazine

The optimum proportions for gunpowder are: 74.64% potassium nitrate, 13.51% charcoal, and 11.85% sulfur (by weight). The current standard for black powder manufactured by pyrotechnicians today is 75% potassium nitrate, 15% softwood charcoal and 10% sulfur.

Commercial gunpowder or black powder is graded according to the grain size. The smaller the grain the faster is the burning rate. In one of the **grading systems**, in decreasing size of grain the nomenclature is Fg, FFg, FFg, FFFFg, and FFFFFg. The FFFFFg type was used in flintlock guns as the priming powder. Fg was used in large bore rifles and small cannon; FFg in medium and small bore rifles and FFFg was used in pistols.

(2.4) FLASH POWDERS

Flash powders are generally mixtures of an oxidizer and a metallic fuel. They burn extremely fast and the finer the powder, the faster is the rate of burning. They are called flash powders because once upon a time some of them were used in flashes for cameras. They are also used in pyrotechnics. Numerous combinations have been produced. One of the combinations is ammonium perchlorate for oxidizer and aluminum powder for fuel which, with a rubber binder, is used as a rocket propellant.

□□□

Chapter 3

HIGH EXPLOSIVES: PROPERTIES THAT MATTER TO KEEP THEM ON HIGH

(3.1) HOW WOULD YOU SELECT A HIGH EXPLOSIVE?

How do you go about selecting high explosives? On what basis would you select a high explosive? Is detonation velocity the only criterion? Would you select a high explosive simply because it has a detonation velocity higher than others? It is not so. Merely having a high explosive compound with a good detonation velocity is not enough. The high explosive must satisfy many other criteria before it becomes the explosive of choice for a particular job. Just about any high explosive would not do everywhere. You must get the best explosive or the explosive which is most suited for a particular purpose.

For example, to enable a high explosive to be stored and used safely, one requires several other things. In the first place it should be reasonably insensitive to thermal or mechanical shock, friction and impact. That makes it safer to use as chances of accidental detonation are reduced. However, unless you are going to use the explosive for a highly specialized purpose of say, making bunker buster bombs or for nuclear weapons, extreme insensitivity would mean that you would require a rather powerful detonator to set it off. This may not always be practicable in the field. That's why we used the phrase reasonably insensitive. Then it should be stable in storage. Militaries cannot accept explosives which have got a low **shelf life**—they may have to store their shells and bombs for years! Then whether the explosive can be melt cast or not. That makes a great deal of difference. From the point of view of large scale industrial production of shells and bombs running into hundreds of thousands or even more, an explosive that can be melted and poured down the bomb where it sets later is preferable to an explosive that has to be pressed into the cavity which is necessarily a slower process.

The whole science of explosives has developed out of the desire to optimize as many of such factors simultaneously as possible. This is what has led to the invention of numerous explosives.

In the following we will have a look at the various parameters that are used to define the properties of explosives.

(3.2) PARAMETERS USED TO DEFINE THE PROPERTIES OF HIGH EXPLOSIVES

(3.2.1) Strength Or Power Of Explosives

The strength or power of explosives refers to their total ability to do mechanical work. It is related to the volume of gases produced and the heat liberated. It is possible to theoretically calculate it but it is more accurate to measure it experimentally by standard procedures like the **Ballistic Pendulum test**, the **Cylinder-Expansion test**, the **Trauzl Lead Block test** or in very large Calorimetric bombs.

In practice, the term power or performance of an explosive is defined as its ability to accomplish what is intended in the way of energy delivery (i.e., fragments, air blast, high-velocity jets, underwater bubble energy, etc.). Explosive power or performance is evaluated by a tailored series of tests to assess the material for its intended use. Of the test listed below, cylinder expansion and air-blast tests are common to most testing programs, and the others support specific uses.

(1) Cylinder Expansion Test—A standard amount of explosive is loaded in a cylinder usually manufactured of copper. Data is collected concerning the rate of radial expansion of the cylinder and maximum cylinder wall velocity.

(2) Cylinder Fragmentation Test—A standard steel cylinder is charged with explosive and fired in a sawdust pit. The fragments are collected and the size distribution analyzed.

(3) Detonation Pressure (Chapman-Jouget)—Detonation pressure data are derived from measurements of shock waves transmitted into water by the detonation of cylindrical explosive charges of a standard size.

(4) Determination Of Critical Diameter—This test establishes the minimum physical size a charge of a specific explosive must be to sustain its own detonation wave. The procedure involves the detonation of a series of charges of different diameters until difficulty in detonation wave propagation is observed.

(5) Infinite Diameter Detonation Velocity—The detonation velocity is dependent on landing density, charge diameter, and grain size. The hydrodynamic theory of detonation used in predicting explosive phenomena does not include diameter of the charge, and therefore a detonation velocity, for an imaginary charge of infinite diameter. This procedure requires a series of charges of the same density and physical structure, but different diameters, to be fired and the resulting detonation velocities interpolated to predict the detonation velocity of a charge of infinite diameter.

(6) Pressure Versus Scaled Distance—A charge of a specific size is detonated and its pressure effects measured at a standard distance. The values obtained are compared with that for TNT.

(7) Impulse Versus Scaled Distance—A charge of a specific size is detonated and its impulse (the area under the pressure-time curve) measured versus distance. The results are tabulated and expressed in TNT equivalent.

(8) Relative Bubble Energy (RBE)—A 5- to 50-kg charge is detonated in water and piezoelectric gauges are used to measure peak pressure, time constant, impulse, and energy.

(3.2.2) Brisance Of Explosives

Brisance refers to the shattering effect produced by an explosive. The rapidity with which an explosive reaches its peak pressure is a measure of its brisance. It is obviously related to the rate of the chemical reactions leading to decomposition and also on the nature of confinement. Two explosives could have the same strength but one in which the gaseous products reach their peak pressure more quickly will have more brisance. This characteristic is of practical importance in determining the effectiveness of an explosion in fragmenting shells, bomb casings, grenades, and the like. Experimentally, it is measured by the Sand Crush test to give the relative brisance in comparison to a reference explosive like the TNT. No single test is capable of directly comparing the explosive properties of two or more compounds; it is important to examine the data from several such tests (Sand Crush, Trauzl, and so forth) in order to gauge relative brisance. True values for comparison will require field experiments.

(3.2.3) Sensitivity Or Insensitivity Of Explosives

Secondary explosives must be quite insensitive. This is required not only for safe handling but also for practical problems involved in their use in military objects like projectiles. Chamber pressures in excess of 40,000 psi are produced when a gun is fired. The explosive which is inside the shell should sustain this shock and not go off inside the barrel itself. When an armor piercing shell strikes the armor at speeds in excess of 3600 fps, the explosive must sustain the shock of this terrible impact and yet not go off till the shell has penetrated to the point desired. The explosive in an armor-piercing projectile must be relatively insensitive, or the shock of impact would cause it to detonate before it penetrated to the point desired.

We speak of three types of sensitivities; **sensitivity to impact, sensitivity to heat,** and **sensitivity to friction**. Sensitivity to impact is determined by the so-called **Impact test** in which a 2 kg weight is made to fall on a small quantity of confined explosive. Sensitivity to heat refers to the temperature at which the material would explode. This is determined by the temperature required to detonate an explosive in less than 0.1 second. Sensitivity to friction is measured by the so-called **Pendulum Friction test** in which a standard weighted pendulum is made to scrape against the explosive material.

Desensitizing a pure chemical which could be highly sensitive by adding other things to it or doing something else to it makes up half of the practical science of explosives. You may recall how Alfred Nobel tamed nitroglycerin (a very sensitive oily liquid and hence difficult to handle) by absorbing it in kieselguhr (a porous mineral earth) and produced dynamite, a great explosive.

It is customary to speak in terms of the so called **Figure of Insensitivity**. This is a relative scale with Picrite given the value of 100. On this scale Tetryl gets a figure of 70-75, TNT 115, RDX 60-65 and PETN 85. You can see how insensitive RDX is.

(3.2.4) Relative Effectiveness Factor

The relative effectiveness factor (R. E. factor) is a measurement of an explosive's power for military demolitions purposes. It measures the detonation velocity relative to that of TNT, which is taken to have an R. E. factor of 1.00.

(3.2.5) Stability

Stability refers to the ability of an explosive to be stored without deterioration. When we speak of stability we mean stability in terms of chemical degradation of the explosive compound in storage, exposure to temperature and sun etc. You are aware of the date of expiry for medicines. Explosives behave in a similar manner. The date of expiry does not mean that the drug will become absolutely useless the very next day after the date of expiry while it would have been perfect till the last evening. There is really no such sharp transition. Molecules of the drug keep on degrading or decomposing even before the period of expiry. The date of expiry only means that by that time, so many molecules of the drug are likely to have degraded or decomposed into other compounds that in some cases the efficacy of the drug could be in doubt. The same concept applies to explosives also. In their case, however, the products of degradation may make them behave in an undesirable fashion. While for some explosives, the degraded products may no longer be powerful explosives, the products of degradation for some others may make them highly sensitive and thus very dangerous to use. Dynamite sticks which have started oozing become notoriously sensitive.

Coming to explosive substances, the very fact that some chemical compounds can undergo explosion when heated indicates that there is something unstable in their structures. That is intrinsic instability. For example, certain groups like azide are intrinsically in a condition of internal strain. Increased strain through heating can cause a sudden disruption of the molecule and consequent explosion. In some cases, this condition of molecular instability is so great that decomposition takes place at ordinary temperatures also. We don't require such compounds as military high explosives. We are concerned about compounds that do not have any problem with regard to intrinsic stability but are also stable against extrinsic factors as far as possible. In any case, the rate of decomposition of all explosives increases at higher temperatures. All of the standard military explosives may be considered to be of a high order of stability in the temperature range of -10 to +35 degrees Celsius, but each has a high temperature at which the rate of decomposition becomes rapidly accelerated and stability is reduced. As a rule of thumb, most explosives are treated as dangerously unstable at temperatures exceeding 70 degrees Celsius. Exposure to the ultraviolet rays of the sun also affects stability adversely. Addition of antacids to explosives is found to improve the stability of many explosives.

(3.2.6) Density Or Density Of Loading

Density of loading refers to the unit weight of an explosive per unit volume. A high density is obviously desirable as an explosive that permits high density of loading allows you to use a higher quantity of explosive in a given volume thereby increasing the power of the warhead. Several methods of loading are available, and the one used is determined by the characteristics of the explosive. The methods available include **pellet loading**, **cast loading**, or **press loading**. Depending upon the method employed, an average density of the loaded charge can be obtained that is within 80-95% of the theoretical maximum density of the explosive. High load density can reduce sensitivity by making the mass more resistant to internal friction. But if the density is increased to the extent that individual crystals are crushed, the explosive becomes more sensitive.

(3.2.7) Volatility

Volatility refers to the tendency of evaporation. The explosive must have as low a **vapor pressure** as possible for prolonged storage and resistance to detection—the lower the vapor pressure the more difficult it is to detect the explosive. Volatility, or the readiness with which a substance vaporizes, is an undesirable characteristic in military explosives. Explosives must be no more than slightly volatile at the temperature at which they are loaded or at their highest storage temperature. Excessive volatility often results in the development of pressure within rounds of ammunition and separation of mixtures into their constituents. Volatility also affects the chemical composition of the explosive such that a marked reduction in stability may occur, which results in an increase in the danger of handling. The maximum allowable volatility is 2 ml of gas evolved in 48 hours.

(3.2.8) Hygroscopicity

Hygroscopicity refers to the tendency of the material to absorb **moisture** and degrade as a result of that. Obviously we require the least hygroscopic material. Moisture affects explosives adversely by acting as an inert material that absorbs heat when vaporized and thus reduces the temperature of the reaction. Moisture also acts as a solvent medium that can cause undesired chemical reactions. The introduction of moisture into an explosive is highly undesirable since it reduces the sensitivity, strength, and velocity of detonation of the explosive. Moisture has an adverse effect on the stability of explosives also. It promotes decomposition of the explosive and, in addition, causes corrosion of the explosive's metal container. For all of these reasons, hygroscopicity must be negligible in military explosives.

(3.2.9) Oxygen Balance

Oxygen balance is used to indicate the degree to which an explosive can be oxidized. If an explosive molecule contains just enough oxygen within itself to convert its entire carbon to carbon dioxide, all of its hydrogen to water, and all of its metal to metal oxide with no excess, the molecule is said to have a zero oxygen balance. The molecule is said to have a positive oxygen balance if it contains more oxygen than is needed and a negative oxygen balance if it contains less oxygen than is needed. The sensitivity, strength, and brisance of an explosive are all somewhat dependent upon oxygen balance and tend to approach their maximum values as oxygen balance approaches zero.

However, it must be kept in mind that sensitivity, brisance, and strength are properties resulting from a complex explosive chemical reaction, and hence a simple relationship such as oxygen balance cannot be depended upon to yield universally consistent results. Still, this property is highly useful in calculating the percentage of various components if you are making an explosive composition or mixture—the percentages are so calculated that you tend to get the oxygen balance as close to zero as possible. When using oxygen balance to predict properties of one explosive in comparison to another, it can be expected that one with an oxygen balance closer to zero will be the more brisant, powerful, and sensitive; however, many exceptions to this rule do exist.

For example, consider the family of explosives called amatols. They are mixtures of ammonium nitrate and TNT. Ammonium nitrate has an oxygen balance of +20% and TNT has an oxygen balance of -74%. Theoretically, it would appear that a mixture yielding an oxygen balance of zero (i.e., 78.74% ammonium nitrate and 21.26% TNT) should result in the best explosive properties. In practice some fine tuning is, however, required. In actual practice a mixture of 80% ammonium nitrate and 20% TNT by weight with oxygen balance of +1% yields the best explosive properties of all mixtures, and an increase in strength of 30% over TNT.

(3.2.10) Whether The Explosive Can Be Melt-Cast Or Not?

As mentioned above, it is highly desirable to have an explosive which can be melt- cast. It is a highly useful property from the point of view of large scale manufacture of bombs and shells. An explosive which can be melted can be simply poured into the cavities of the bomb or shell and will automatically acquire the desired shape. The next desirable property which we seek in a melt cast explosive is that it should neither shrink upon setting nor develop air cavities in a random manner. Shrinkage will make the amount of explosive less than what was intended to be used and will thus affect the performance of the warhead. Air cavities in high explosives act as 'hot spots'. A random distribution can make the behavior of the charge rather unpredictable.

(3.2.11) Toxicity

Due to their chemical structure, most explosives are toxic to some extent. Since the effect of toxicity may vary from a mild headache to serious damage of internal organs, care must be taken to limit toxicity in military explosives to a minimum. Any explosive of high toxicity is unacceptable for military use.

Chapter 4

PRIMARY HIGH EXPLOSIVES: THEY KICK OFF EVERYTHING

(4.1) WHAT ARE PRIMARY HIGH EXPLOSIVES AND WHAT IS THEIR MAIN USE?

High explosives are classified as primary or secondary based on their susceptibility to initiation. **Primary high exploslves** are highly susceptible to initiation. In other words they are highly sensitive to heat, shock or friction. Primary explosives often are referred to as initiating explosives because they can be used to ignite secondary explosives. **Secondary high explosives** are relatively insensitive to heat, shock and friction. In fact most of them, if ignited in small unconfined quantities, would simply burn and not detonate. Detonating them requires a critical value of heat or mechanical shock. We get high explosives with a wide range of sensitivity. Then where do we draw the line? **PETN** is often considered a benchmark compound, with materials that are more sensitive than PETN being classified as primary explosivcs and those which are less sensitive, classified as secondary explosives.

Primary explosives may have detonation velocities in the same range as secondary explosives. Yet because they are so sensitive, it is difficult to use them safely in bulk or in the field. You cannot expect a soldier in battle to handle an explosive as a scientist handles it in the lab! You can use only secondary explosives in bulk and in the field. Because they are formulated to detonate only under specific circumstances, secondary explosives are very useful as main charges for military purposes. Primary explosives in small quantities are used to initiate them. Thus a combination of primary and secondary explosives does the trick.

If you have several explosives with different degrees of sensitivity, the way to explode the most insensitive of them is to make an **explosive train**. You use the most sensitive explosive to initiate the explosive which is not so sensitive and yet sensitive enough to be initiated by the first explosive. Then the explosion of the second explosive which will be a little more powerful than the first explosion can be used to initiate a third explosive and so on. This is precisely what detonators and booster charges do.

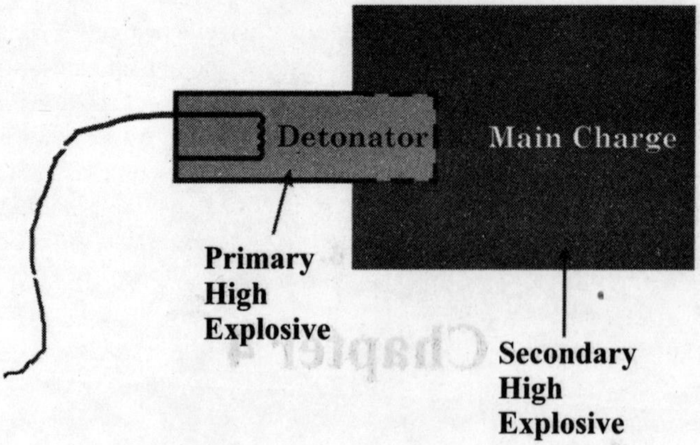

Fig. 8 A Two-Step Explosive Train

Also called an **initiation sequence** or a **firing train**, this is the sequence of events that progresses from relatively low levels of energy to cause a chain reaction to initiate the final explosive material or main charge. They can be either low or high explosive trains. Low explosive trains are as simple as a rifle cartridge, including a primer and a propellant charge. High explosives trains can be more complex, either two-step (e.g., detonator and dynamite) or three-step (e.g., detonator, booster of some primary explosive or a relatively sensitive secondary high explosive like tetryl, and the main charge of relatively insensitive secondary explosive).

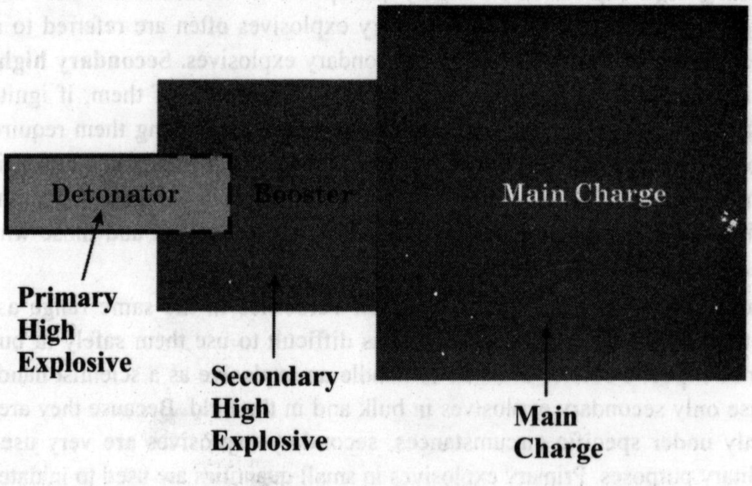

Fig. 9 A Three-Step Explosive Train

(4.2) FAMOUS PRIMARY EXPLOSIVES

(4.2.1) Mercury Fulminate: The Grand Daddy Of Them All

Mercury fulminate is the oldest primary explosive. It is a salt of fulminic or paracyanic acid. Mercury Fulminate appears to have been prepared for the first time by **Hohann Kunckel von**

Lowenstern (1630-1703). He obtained this dangerous explosive by treating mercury with nitric acid and alcohol. At that time, Kunckel and other alchemists could not find a use for the explosive and the compound was forgotten until **Edward Howard** of England rediscovered it between 1799 and 1800. The preparation and properties of mercury fulminate were described in much detail by Edward Howard in 1800 in a paper presented to the Royal Society of London. Howard had made it by dissolving mercury in nitric acid and adding ethyl alcohol to the solution. Howard examined the properties of mercury fulminate and proposed its use as a percussion initiator for black powder.

In 1807 a Scottish clergyman, **Alexander Forsyth** patented the device called **percussion cap** which made firing a much simpler and reliable affair. It was a great invention made possible only with the help of this wonderful compound. The percussion cap contained a small amount of mercury fulminate in a cap-shaped device made of thin ribbed copper. A blow by the hammer set off the fulminate and the heat of explosion was transmitted to the propellant charge in the chamber through a tube leading into the barrel. Earlier, firing even a flintlock musket with a flash pan was a pain in the neck. The priming powder used to be kept loose in the flash pan and was always open to the elements like wind, rain and even moisture in the air. In fact one could not even move the musket violently in battle lest the powder fall out from the pan! The musket would become useless if the powder trail got even as much as broken. One could not even run with a primed musket. Once primed, the musket had to be handled very delicately. The percussion cap solved all these problems in one stroke. It made the weapon easier to load, weather-resistant and generally more reliable. In a way, the concept of the percussion cap paved the way for the development of the self-contained cartridge which revolutionized warfare. Preparation of Mercury Fulminate is even now carried out by a process essentially the same as described by Howard. The compound served for a long time in spite of the accidents that took place due to it.

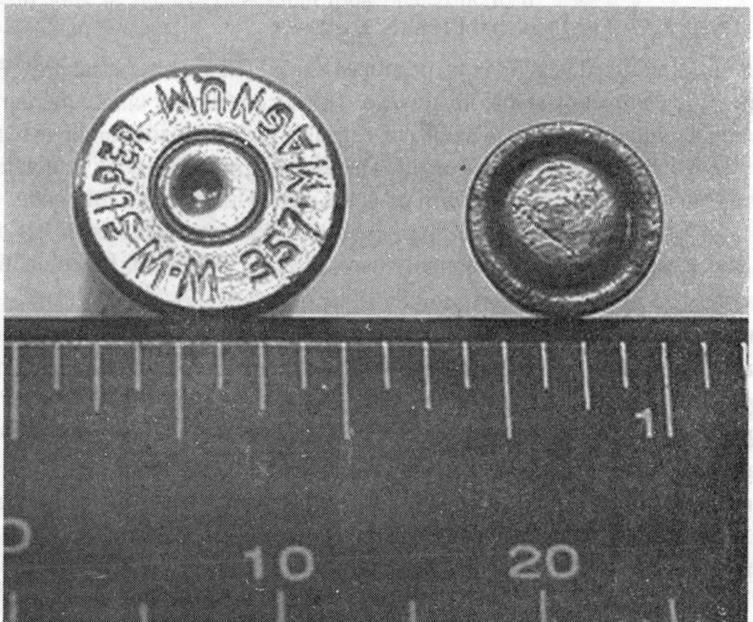

Fig. 10 Primers (Percussion Caps) Of A Pistol Cartridge (Fired) And A Shotgun: The Invention Of The Modern Cartridge Owes A Lot To Mercury Fulminate

Making mercury fulminate was not a problem. Storing it in bulk was a problem in view of its notorious sensitivity. It is only slightly soluble in both water and ethyl alcohol. Because of these slight solubilities, mercury fulminate can be stored underwater or, if there is danger of freezing, in a mixture of equal volumes of water and ethanol or methanol. The compound can be subsequently dried easily whenever desired. There is one more way of storing it safely. Aqueous ammonium hydroxide, aqueous potassium cyanide, and pyridine dissolve mercury fulminate. The compound can subsequently be recovered by treating the ammonium or potassium solution with acid and the pyridine solution with water.

Fulminates are friction-sensitive due to the instability of the anion. It is largely the presence of the weak single nitrogen-oxygen bond which leads to its instability. Mercury fulminate is more sensitive to impact than lead azide and lead styphnate. Being of the same sensitivity to impact as diazodinitrophenol and tetrazene, these three compounds are the most sensitive initiating explosives used in military ammunition. Mercury fulminate is more sensitive to friction than lead azide and lead styphnate. It is also more sensitive to heat than lead azide or lead styphnate, but less sensitive than diazodinitrophenol and tetrazene. It is also highly sensitive to electrical sparks.

With a detonation velocity of 13940 fps, mercury fulminate was used extensively as an initiator for quite some time; however, poor stability has eventually prompted replacement of this compound by such initiators as lead azide. There was one more problem with it. When dry, mercury fulminate reacts rapidly with aluminum and magnesium and reacts slowly with copper, zinc, brass, and bronze. When wet, mercury fulminate reacts immediately with aluminum and magnesium; and the reaction is rapid with copper, zinc, brass, or bronze. Dry or wet, the compound does not affect iron or steel. It meant that if you are using mercury fulminate in the detonator or blasting cap, you could not make the body of aluminum!

(4.2.2) Lead Azide: Now The Prinicpal Primary Explosive

Two things weighed heavily in favor of lead azide. The first is its stability. When protected from humidity, it is completely stable in stowage. The second is that unlike mercury fulminate, it does not react with aluminum. Hence detonator capsules for lead azide can be made of aluminum, a much cheaper and lighter metal than copper. Then the hygroscopicity of lead azide is very low. Water does not reduce its impact sensitivity, as is the case with mercury fulminate. It is therefore widely used as an initiating explosive in high-explosive detonator devices. It has a high ignition temperature and is today the most commonly used primary explosive. Its detonation velocity is 17,500 fps.

Lead azide is an excellent initiating agent for high explosives. While not superior to mercury fulminate for detonating the less sensitive explosives such as TNT, lead azide is markedly superior as an initiator for the more sensitive explosives such as tetryl, RDX, and PETN. Unlike diazodinitrophenol, lead azide cannot initiate the detonation of ammonium picrate or cast TNT.

There are several forms of lead azide. **Dextrinated Lead Azide (DLA)** is also known as type I lead azide. Dextrin is used as a colloiding agent which prevents the formation of large, sensitive crystals and regulates crystal shape during manufacture. **Service Lead Azide (SLA)** consists of lead azide crystals each containing a nucleus of lead carbonate. This form of lead azide is used extensively in the United Kingdom. The method of manufacture of this compound is confidential. SLA is practically nonhygroscopic and is superior to DLA in functioning characteristics. Storage under water is considered hazardous due to the possibility of growth of the crystals and formation of agglomerates which detonate spontaneously. Other long term storage tests dispute this result. **Colloidal Lead Azide (CLA)** is also known as type II lead azide. CLA is non-dextrinated lead azide in which the particle size is very small. The size of the particles is in the range of three to four microns. CLA is not suited for uses requiring good flow characteristics but, because of the very fine particle size, is ideal as a spot charge and a priming charge in low energy, electric initiators. CLA successfully replaced the milled DLA formerly used for this purpose, thus eliminating the milling operation which was always considered dangerous, even under carefully controlled conditions. In preparing a spot charge for a low energy, electric detonator, dry CLA is mixed with a concentrated solution of nitrocellulose in ether, alcohol, or other organic solvent, and a small quantity of the resulting paste is placed on the bridge wire to form a droplet called a spot. For a type of detonator in which the bridge wire is located inside a cavity, the charge of CLA can be made in the form of a pellet by pressing wet CLA into the cavity. **Polyvinylalcohol Lead Azide (PVA-LA)** consists of lead azide crystals coated with polyvinylalcohol. PVA-LA possesses practically the same sensitivity to impact as DLA, but is much more efficient in detonators and is practically non-hygroscopic.The ignitability is about the same as for straight lead azide and better than for DLA. The normal lead azide content is 93 to 96 percent. Other substances, such as polyethylene glycols and urea-formaldehyde polymer, can be used to produce mixtures with properties very similar to PVA-LA. **RD-1333 lead azide** is an insensitive form of lead azide. The methods of manufacture as well as some properties of RD-1333 are military secrets. **Dextrinated Colloidal Lead Azide (DCLA)** is essentially DLA with a very small particle size. The particle size is in the range of one to two microns.

(4.2.3) Lead Styphnate

The stability of lead styphnate is the same as lead azide and better than that of mercury fulminate or diazodinitrophenol. Lead styphnate is less sensitive to shock and fiction than lead azide. It does not react with metals also. Its detonation velocity is 17,000 fps. It is used as an initiating explosive in propellant primer and high-explosive detonator devices. The problem is that when used alone, despite a favorable rate of detonation and good power characteristics, lead styphnate is a relatively poor initiator of detonation as its brisance is much less than that of lead azide, being only about 62.9% of the brisance of lead azide. Detonation of 60 percent dynamite and sometimes gelatin dynamite by lead styphnate is possible. However, of the military high explosives, only un-pressed PETN can be detonated by it. Hence one of the practices has been to use it as a cover charge for lead azide in priming compositions.

Lead styphnate can be prepared by the reaction of lead nitrate or acetate with either sodium or magnesium styphnate in the presence of an acid. In commercial production, the magnesium salt is used. Normal lead styphnate is slightly less sensitive to impact than mercury fulminate or diazodinitrophenol, but is more sensitive than lead azide. The friction sensitivity is less than that for mercury fulminate or lead azide. Normal lead styphnate is much more easily ignited by flame or electric spark than mercury fulminate, lead azide, and diazodinitrophenol. When the styphnate is dry, it can readily ignite by static discharges from the human body also! That should give some idea of how dangerous it is. The longer and narrower the crystals, the more susceptible the material is to static electricity.

(4.2.4) Diazodinitrophenol (DDNP)

DDNP is considerably more brisant than mercury fulminate, lead azide, lead styphnate and tetrazene. It is often used as an initiating explosive in propellant primer devices. DDNP is a better initiator of detonation than mercury fulminate or lead azide for less sensitive high explosives. The most marked evidence of this is the ability of DDNP to initiate the detonation of ammonium picate and cast TNT; these explosives, as we saw above, are not detonated by lead azide. For initiation of the more sensitive high explosives, however, DDNP is not superior to lead azide.

DDNP is not as stable as lead azide, but is markedly more stable than mercury fulminate. Storage tests have shown dry DDNP to withstand storage at 50 degrees Celsius for at least 30 months, as compared with nine months for mercury fulminate. When stored under water, DDNP is of unimpaired brisance for 24 months at ordinary temperature and for 12 months at 50 degrees Celsius. The stability of DDNP, therefore, is considered satisfactory for commercial and military use, and DDNP and lead azide have replaced mercury fulminate in blasting caps to a large extent.

The process used for the industrial manufacture of DDNP is a military secret, but the compound can be prepared in the laboratory by the diazotization of picramic acid by means of sodium nitrite and hydrochloric acid.

It is less sensitive to impact but more powerful than lead azide or mercury fulminate. The sensitivity of DDNP to friction is approximately the same as that of lead azide but less than that of mercury fulminate. DDNP is desensitized by immersion in water and does not react with it at normal temperatures. A mixture of DDNP and water is desensitized to the extent that a number eight blasting cap cannot cause detonation. DDNP detonates when struck with a sharp blow. An unconfined sample burns with a flash if ignited but even the slightest confinement causes a transition from deflagration to detonation. A charge of DDNP undergoes detonation when ignited if pressed into a blasting cap shell with a reinforcing cap and a piece of black powder safety fuse crimped in the shell.

(4.2.5) Tetrazene

Ever heard of something called **explosive rivets**? An explosive rivet is a rivet used in a place which is rather inaccessible and where you cannot have space to hammer. This was a major problem encountered in manufacturing airplanes. Most of the planes have riveting points which can be reached from only one side. So troublesome have been these inaccessible points that plane designs have often been modified to avoid them, but there are still about 10,000 in a large bomber. It was a difficult engineering problem, and solved only by the invention of the explosive rivet. Explosive rivets were developed by Du Pont after their basic invention in 1937 by two employees of the famed German aircraft builder Ernst Heinkel. In an explosive rivet a small amount of high explosive nests in a cavity at the headless end of an aluminum-alloy rivet. When heat is applied to the head by an electric riveting gun, the charge explodes at the other end forming a "blind" head, and thus sets the rivet. Explosive charges can be controlled to adjust the size and shape of the head to within .02 in.

Why are we making a reference to explosive rivets here? Because the explosive used in them is tetrazene and that alone is sufficient to secure its place in history! Tetrazene is also known as guanyldiazoguanyl tetrazene and guanyl nitrosoaminoguanyl tetrazene. It is made by the reacting the solution of aminoguanidine carbonate in water and glacial acetic acid with sodium nitrite.

Tetrazene is of the same order of sensitivity as mercury fulminate and diazodinitrophenol. This ease of ignition along with a relatively high heat of explosion and large gas volume produced makes tetrazene useful in priming compositions and with lead azide in explosive rivets. Tetrazene can be used in detonators also when initiated by another primary explosive. In such cases, tetrazene functions as an intermediate booster or as a sensitizer to flame or heat. In some cases, tetrazene is used in primers where as little as two percent tetrazene in the composition results in uniformity of percussion sensitivity.

Tetrazene is practically insoluble in water and ethanol and hence can be stored wet with water or a mixture of water and ethanol. The compound is also insoluble in ether, benzene, acetone, carbon tetrachloride, and ethylene dichloride. Tetrazene is soluble in dilute nitric acid or strong hydrochloric acid. In a solution with hydrochloric acid, the hydrochloride is precipitated by the addition of ether. Tetrazene may then be recovered by treatment with sodium acetate or ammonium hydroxide.

(4.2.6) Lead Mononitroresorcinate (LMNR)

LMNR has slow burning properties and a low combustion temperature. The compound is used in electric detonators with DLA as the spot charge to initiate a PETN base charge, as an upper charge, and as an ingredient in primary compositions. LMNR is prepared by mono-nitrating resorcinol then allowing the slurry of mononitroresorcinol to react with lead hydroxide at elevated temperatures.

(4.2.7) Potassium Dinitrobenzofuroxane (KDNBF)

The sensitivity of KDNBF is between that of mercury fulminate and lead azide.

(4.2.8) A Comparison Of The Relative Efficiencies Of Some Initiating Compounds

Minimum detonating charge (in grams) of initiator required to to detonate

	Tetryl	TNT	Ammonium picrate
Lead azide	0.10	0.26	No detonation
Mercury fulminate	0.19	0.24	No detonation
Diazodinitrophenol	0.12	0.15	0.28

(4.3) VARIOUS TYPES OF PRIMARY COMPOSITIONS

Primary compositions are mixtures of primary explosives, fuels, oxidizers, and other ingredients used to initiate detonation in high explosive charges or ignite propellants and pyrotechnics. The ingredients and the portions of the ingredients for individual priming compositions are determined empirically from the use the composition is intended for. Fuels commonly used in priming compositions are lead thiocyanate, antimony sulfide, and calcium silicide. The last two also serve to sensitize the composition to friction or percussion. Oxidizing agents include potassium chlorate and barium nitrate. Other ingredients include primary explosives and binders. The major determining factor in ingredient selection is the impetus which is to detonate the priming composition. The types of impetus commonly used are percussion and electrical.

(4.3.1) Percussion Type Priming Compositions

Percussion type priming compositions are used extensively for the ignition of propellants. In the early 1900's this type of priming composition contained mercury fulminates. Now, of course, as we saw, mercury fulminate has been replaced by more stable compounds. Potassium chlorate was used in this type of priming composition from about 1917 through World War II. Except for special applications, the use of this compound has been discontinued because of the corrosive action of potassium chloride, a decomposition product of potassium chlorate, on gun barrels. Potassium chlorate is still used in high temperature priming compositions. Two of them are described below.

	G-11	G-16
Potassium chlorate %	53	53
Antimony trisulfide %	25	30
Calcium silicide %	12	17
Temperature resistant explosive %	10	-

The ingredients of some of the modern percussion priming compositions are:

	FA959	Composition FA982	FA956
Normal lead styphnate, %	35	36	37
Tetrazene, %	3.1	12	4
Barium nitrate, %	31	22	32
Antimony sulfide, %	10.3	7	15
Powdered zirconium, %	10.3	9	-
Lead dioxide, %	10.3	9	-
PETN, %	-	5	5
Aluminum, %	-	-	7
Gum Arabic, %	-	-	0.2 (added)

The output of percussion type primers includes hot gases, hot particles, thermal radiation, and a pressure pulse. In some cases the pressure pulse may be a shock wave.

(4.3.2) Stab Detonator Priming Compositions

Stab detonators are used to directly initiate secondary explosives. As the name implies, a stab detonator is initiated by the penetration of a firing pin. The mechanical energy of the firing pin is converted to heat, which causes detonation. Rather than using one homogenous priming mixture as percussion type primers do, stab detonators contain three layers of charges; primary, intermediate, and base. Priming compositions used in the primary charge are shown in the table below. The intermediate charge is an energetic material such as lead azide or black powder. The base charge may be lead azide, black powder, or a more brisant explosive such as RDX or PETN. The base charge initiates the secondary explosive.

	Composition		
	NOL 130	PA101	NOL 60
Lead azide, %	20	-	-
Basic lead styphnate, %	40	53	60
Tetrazene, %	5	5	5
Barium nitrate, %	20	22	25
Antimony sulfide, %	15	10	10
Powdered aluminum, %	-	10	-

(4.3.3) Electric Priming Compositions

Electrical primers are of several types: exploding bridge wire; hot wire bridge; film bridge; conductive mixture; and spark gap. It must be kept in mind that in an exploding bridge wire detonator, a large current is passed through a wire which causes the wire to burst. The bursting of the wire causes a shock wave that is of sufficient strength to cause detonation. With this type of primer no priming composition is used; the wire is placed directly in a charge of RDX or PETN. Hot wire bridges and film bridges use priming compositions.

Some of the famous electric priming compositions are given below.

Composition

	I	II	III	IV	V	VI
Potassium chlorate, %	8.5	55	25	60	66.7	66.7
Lead mononitroresorcinate, %	76.5	-	-	-	-	-
Nitrocellulose, %	15.0-	-	-	-	-	-
Lead thiocynate, %	-	45	-	-	-	-
DDNP, %	-	-	75	20	-	-
Charcoal, %	-	-	-	15	-	-
Nitrostarch, %	-	-	-	5	-	-
Titanium, %	-	-	-	-	-	33.3
Aluminum, %	-	-	-	-	-	33.3

❑❑❑

Chapter 5

High Explosives: The Illustrious Nitric Esters

(5.1) ALKYL ESTERS OF NITRIC ACID

Alkyl esters of nitric acid are very strong explosives. It is so because their oxygen balance is higher than nitroaromatic compounds having the same number of carbon atoms and nitro groups and hence they are more powerful explosives than nitro compounds. Their disadvantage is that they are more sensitive to shock and friction than nitro compounds. Furthermore, they are prone to hydrolytic decomposition. Usually they are made by the so-called "O-nitration" reaction of alcohols with nitric acid in the presence of sulfuric acid.

(5.2) THE MOST ILLUSTRIOUS OF THEM ALL: GLYCEROL TRINITRATE OR NITROGLYCERIN

(5.2.1) Nitroglycerin: A Great Invention

Nitroglycerin, undoubtedly the most illustrious explosive, is the basic material used in **dynamite** which has got everlasting fame as one of the most widely used explosives in history—if you include non-military uses as well then it is the most widely used explosive. Who does not recognize the eight inches long and one-and-a-quarter inch thick sticks wrapped in brown wax paper and weighing about half a pound each? They are the most universal symbol of explosives in the world. If you have to depict a bomb pictorially in such a manner that even dummies don't mistakenly take it for something else, show them dynamite sticks.

Nitroglycerin was first prepared late in the year 1846 or early in 1847 by the Italian chemist **Ascanio Sobrero** (1812 - 1888) who was at that time professor of Applied Chemistry at the University of Torino. Sobrero was authorized to practice medicine and had studied in Paris and Giessen, returning to Torino in 1845 where he equipped a laboratory. The earliest printed account of nitroglycerin appears in a letter which Sobrero wrote to Pelouze and which Pelouze caused to be published in L'Institut of February 15, 1847. In the same month Sobrero presented to the Academy of Torino a paper in which he described nitroglycerin, nitromannite, and nitrated lactose. Later in the year he presented another paper before the chemistry section of the Ninth Italian Scientific Congress at Venice. Sobrero had found that if concentrated nitric acid or strong mixed acid is added to glycerin, a violent reaction ensues and red fumes are evolved, but that, if syrupy glycerin is added to a mixture of sulfuric acid and nitric acid under strict controls the results are entirely different. The glycerin dissolves, and the solution when poured into water gives an oily precipitate of nitroglycerin.

For many years Sobrero kept in his laboratory and guarded jealously a sample of the original nitroglycerin which he had prepared in 1847. In 1886 he washed this material with a dilute solution of sodium bicarbonate and took it to the Nobel-Avigliana factory, of which he was a consultant, where he gave verbal testimony of its authenticity and where it has since been stored in one of the magazines.

In its pure form nitroglycerin is a colorless heavier-than-water oily liquid; the commercial product could be yellowish to pale brownish. The pure liquid also tends to decompose due to acidic pH and turns yellow. It has got a bitter-sweet 'burning' taste. The detonation velocity is about 25,000 fps. In its pure form it is very sensitive—in fact so sensitive as if it has a mind of its own to sense if a stupid or clumsy person is around and then it tries to kill him. Yet the new explosive, being liquid, could be poured into crevices in rocks, and was so convenient as a blasting agent that its manufacture went on until a large vessel carrying cases of the explosive from Hamburg to Chili blew up at sea. The ship was blown to bits and her crew killed, and the disaster caused so great a sensation that the manufacture of nitroglycerin was prohibited for the time being in Sweden, Belgium, and England.

It is so sensitive that to avoid dangers, internal transport within the factories is made by water injection (Water-driven Injector Transport). The Picatinny Arsenal impact test shows nitroglycerin is more sensitive than mercury fulminate itself!

Incidentally, this great explosive also finds therapeutic use as a drug—being a vasodilator, it has been used to alleviate angina pectoris.

(5.2.2) Dynamite: A Greater Invention

The high sensitivity of nitroglycerin to shock and friction was a matter of serious concern. Numerous ways had been tried to make it safe for handling before a determined **Alfred Nobel** hit upon the winning formula. Alfred Nobel, his father Immanuel, and younger brother Emil began experimenting with nitroglycerin near Stockholm in 1862. It was highly unstable and difficult to handle, and accidental explosions were not uncommon. One such accident killed Emil, among others, at a plant in 1864. Despite the personal tragedy, Alfred continued his work with this dangerous liquid, working on a boat in the middle of a lake before conducting his experiments in a factory! We salute his single-minded devotion to nitroglycerin!

Previous attempts at reducing the sensitivity had used magnesium carbonate, saw dust and gun powder as absorbents, and methyl alcohol as a solvent. They were not satisfactory, however. Nobel's perseverance eventually paid and he got the patent for his formula in 1867. His original composition used 75% nitroglycerin mixed in 25% **kieselguhr**, a diatomaceous earth to form a doughy mass which was safer to handle. Nobel controlled the patent tightly and made a great deal of money out of it—he eventually established the corpus for the Nobel Prize out of those earnings only. In all he held 355 patents!

However, other people were also smart and got around the patent by making many other variants. That's how we speak of '**dynamites**' now and not simply dynamite. In fact some people use the term dynamite generically to refer to any high explosive used in mining, quarries, tunnel construction and other engineering blasting operations; many compositions had in fact replaced nitroglycerin with nitrostarch!

Nitroglycerin has the problem of freezing in cold climates and this necessitates some careful thawing. It has therefore become customary to add a little '**anti-freeze**' material like nitrotoluene, nitrated diglycerin, nitrated sugars, or nitroglycols to obviate the necessity of thawing nitroglycerin explosives when used in cold climates.

Dynamite quickly caught the fancy of the mining industry, mainly because it was powerful. Until 1870, black powder was the only explosive used in coal mining, and several disastrous explosions had occurred. Many attempts were made to modify black powder; these included mixing black powder with cooling agents such as ammonium sulfate, starch, paraffin, etc., and placing a cylinder filled with water into the bore hole containing the black powder. None of these methods proved to be successful. When nitrocellulose and nitroglycerine were invented, attempts were made to use these as ingredients for coal mining explosives instead of black powder but they were found not to be suitable for use in gaseous coal mines. It was not until the development of dynamite and blasting gelatin by Nobel that nitroglycerine-based explosives began to dominate the commercial blasting and mining industries.

Dynamite was, however, used without such hazards in construction work. The first large-scale use of dynamite for construction purposes was in the creation of the Hoosac Tunnel, completed in 1876.

Fig. 11 A Bundle Of Dynamite Sticks: The Most Enduring Symbol Of Explosives Since 140 Years

(5.2.3) Various Types Of Dynamites

(a) Straight Dynamites

Various compositions of the so-called straight dynamites exist. Usually they contain 20-60% nitroglycerin absorbed in 13-18% wood pulp of good grade as the carbonaceous material, 22-

60% of sodium nitrate as oxidizer, and calcium carbonate, zinc oxide or some antacid (0.8-1.3%) to neutralize any traces of acidity that might be produced by the decomposition of nitroglycerin. Antacids are necessary because the presence of more than a trace of free acid renders nitroglycerin quite unstable; decomposition with the appearance of red fumes takes place within a few days. A little sulfur (2-3%) is also sometimes added. Note that in the straight dynamites, the wood pulp is combustible whereas in the original dynamite, the kieselguhr is inert and does not contribute to the explosion. Straight dynamites are therefore more powerful. The original dynamite is called '**inert-dope**' whereas the straight dynamites are called '**active-dope**' dynamite. They are the costliest of the lot and also the most sensitive and inflammable.

(b) Ammonia Dynamites

In the so-called Ammonia dynamites (also known as '**extra dynamites**'), a part of the nitroglycerin or other organic nitrate is replaced with ammonium nitrate. Thus they contain 12-23% nitroglycerin, 11-50% ammonium nitrate, 8-10% carbonaceous material, 15-57% sodium nitrate, antacid (0.8-1.2%) and a little sulfur (1-7%). They have lower brisance and are more susceptible to adverse effects of water, but ammonium nitrate produces more gas and thus the pushing or heaving action is substantial. Ammonia dynamites are also cheaper and hence very popular in mining and quarrying. It can be used effectively where the rock is not extremely hard and water conditions are not severe. It is widely used in quarrying, stripping, and in well-ventilated mines for smaller diameter holes of small blasting operations.

(c) Gelatin Dynamites

In 1875 Alfred Nobel made another important discovery. This time he used the colloiding action of nitrocellulose to solidify nitroglycerin. He found that if one uses only little quantities of nitrocellulose then the effect is merely to increase the viscosity of nitroglycerin but if you increase the percentage of nitrocellulose to 8-12, a tough product with rubber-like texture results. This he called **blasting gelatin** and since both the components are explosive, it is a particularly strong blasting explosive.

Fig. 12 Miners Ignite Safety Fuse: A Scene From The Early Days Of The Use Of Dynamite In Mining

An antacid is added for stability in storage. Wood meal is usually added to improve sensitivity. Blasting gelatin attains a very high detonation velocity and has excellent water resistance, but it emits large volumes of noxious fumes upon detonation. It is the most powerful of all commercial non-military explosives. Blasting gelatin is also known as **'oil well explosive'**. Blasting gelatin was used to pierce the great St. Gothard Railway tunnel through rock so hard that without it the task could never have been accomplished.

Fig. 13 Dynamite Tore The Heart Of Rock Hard Mountains To Make Tunnels Like This

Blasting gelatin was tried in guns, but burst them, so Nobel set himself to discover an explosive less violent, yet equally clear and smokeless. By mixing nitroglycerin and guncotton he found a comparatively slow-burning powder which he called **ballistite**, and this, when he gave it to the world in 1888, caused a very great sensation. We have talked more about ballistite later in connection with guncotton.

Commercially, gelatin dynamites are now made by adding a little nitrocellulose to the ordinary 'active dope' (i.e., straight) dynamites. Thus they could have 20-91% nitroglycerin, 0.4-8% nitrocellulose, 19-60% sodium nitrate, 8-11% carbonaceous matter, 0.8-1.5% antacid and 1-8% sulfur. They are characterized by high density and excellent resistance to water. For this reason they are the explosives of choice for submarine operations or in any work where the explosive may have to remain submerged in water for prolonged periods.

Variations in the concentration of gelatinized nitroglycerin (that is, the thickened product) by the addition of sodium nitrate and wood dust or cereal meal yielded **'gel dynamites'**, which are known as **'gelignites'** in English-speaking countries.

(d) Ammonia Gelatin Dynamites

They have generally replaced straight gelatin dynamites because they are less sensitive to detonation. They are formulated to combine, as far as possible, the best characteristics of gelatin dynamites and ammonia dynamites. They are also known as '**special gelatin**' or '**gelatin extra**'. They consist essentially of ammonia dynamite in which the nitroglycerin is present in the form of a nitrocellulose colloid. Their water resistance is obviously not as good as that of the gelatin dynamites but still better than that of the ammonia dynamites. In terms of gas volume and heaving action they are superior to the gelatin dynamites but inferior to the ammonia dynamites. They contain 22-38% nitroglycerin with 0.3-0.9% nitrocellulose, 19-55% sodium nitrate, 4-34% ammonium nitrate, 0.7-0.9% antacid and 3-7% sulfur.

(e) Semi-Gelatin Dynamites

They are intermediate in composition to ammonia gelatin dynamites and ammonia dynamites. Essentially they are ammonia dynamites to which has been added insufficient nitrocellulose to from a stiff gel. They are less sensitive than gelatin dynamites.

(f) Things Which Go By The Name Of Dynamites But Do Not Have Any Nitroglycerin In Them

Du Pont markets certain water-gel based explosives containing mono methylamine nitrate and ammonium nitrate under the trade name of **Tovex series dynamites**. Sensitizing additives like TNT and PETN are also added in small amounts. They also have thickeners and cross-linking agents such as gum and borax. Fuels like aluminum powder and water soluble fuels like glycol are also added.

The US Army uses something called **Military Dynamite** for general military work like quarrying and service demolitions. It is actually 75% RDX, 15% TNT, 5% SAE 10 motor oil and plasticizer, and 5% cornstarch. It is much more stable in storage and can be used over a wider range of temperature.

(5.2.4) Storage Of Dynamites: Problems And Correct Methods

Dynamite sticks require careful storage. Over a period of time, the nitroglycerin tends to settle to the bottom of the stick if they have been left standing on one end. It is called '**leaking**'. In fact any stick having oily stains would indicate either an old stick or improper storage. One simple way of preventing this is not to keep the sticks standing and also to keep on turning dynamite cases every month or so. If you land up with a stick that has become mushy instead of feeling firm and resilient like modeling clay, the dynamite has gone bad and has become very sensitive. Fool around with it at your risk. If you have no other option, you could wet the sticks liberally with acetone and that would make it somewhat safer, though acetone itself is highly inflammable. In cold climates if you find a dynamite stick that has frozen, it would have become very insensitive and would require considerable effort to detonate it. To ascertain whether a stick has frozen, insert a nail in it; if it does not go easily, you will have to thaw the dynamite. Thaw it in a warm room but never by a fire. Further, dynamite that is undergoing thawing becomes very sensitive due to the crystalline changes that are taking place in it. Leave the stick alone and do not fumble with it till it has completely thawed.

(5.3) NITROCELLULOSE AND GUNCOTTON

Nitrocellulose was the key to the development of the **smokeless powder** and therefore deserves great respect. The smokeless powder revolutionized firearms and thus warfare itself. Chemists had known it all along that most nitrate compounds are strong oxidizing agents and some can react violently with oxidizable substances and may explode if exposed to heat or shock.

In 1832 itself Bracon had shown that wood fibers could be turned into explosive by the action of nitric acid. Then in 1845 **Schonbein** treated cotton wool with nitric and sulfuric acids and produced the famous **guncotton. Paul Vieille** used this guncotton to produce his **Poudre B**.

In 1847 an English firm, Messrs. Hall and Son of Faversham, began to manufacture guncotton, and military experts hailed it as the new explosive which would take the place of gunpowder. But this explosive was so terribly powerful that, when used in a gun or rifle, it blew the barrel to pieces. Worse than that, it was most dangerous to manufacture.

Two main problems had to be solved before it could be used as a gun propellant. First, the velocity of the explosion had to be reduced so that the charge weight required to propel the projectile would not shatter the gun tube. Second, the density had to be increased so that a given charge weight would pack into a reasonable space. The first problem was solved in part by igniting nitrocellulose instead of firing it with a detonator. The solution to the second problem actually solved both. In 1886, Vielle first colloided or **gelatinized nitrocellulose** with alcohol and ether and, thus reduced the burning rate to acceptable levels. The procedure significantly increased the loading density of nitrocellulose, establishing it as the foundational element in gun propellants used through the present day. Further developments resulted in materials that could be added to improve stowage qualities, reduce or eliminate flash, reduce hygroscopicity, reduce flame temperature, and even increase the propellant force or impetus.

Almost pure cellulose is found in the pith of certain plants, in absorbent cotton, and in some filter papers. Pure cellulose is most readily obtained from cotton by treating it with a dilute acid or base solution and then thoroughly washing it with water. At the present time most of the cellulose for nitrocellulose preparation is obtained from coniferous wood, which is 50 to 60 percent cellulose. Another source is straw, which is 30 to 40 percent cellulose.

Different products result depending on the degree of nitration. Thus you can get what are known as **pyroxylin** or **collodion** (8-12.3% nitrogen) or **pyrocellulose** (12.6$\pm$0.10% nitrogen) or **high nitrogen nitrocellulose** (13.75 to 14.14%). Guncotton is the grade of nitrocellulose that contains the highest percentage of nitrogen obtainable by nitrating cellulose with mixed nitric and sulfuric acids. The percentage of nitrogen is 13.35 to 13.45 percent.

A point should be made here regarding the mechanism of solubility of nitrocellulose. As a general rule, solvents for polymers like nitrocellulose act at first as swelling agents and it is only later that they break the fibrous structure of nitrocellulose. When nitrocellulose is treated with a solvent a gel is obtained and the phenomenon known as **gelatinization** occurs. A volatile solvent can be evaporated in order to obtain a plastic-like colloided material. You get the smokeless nitrocellulose propellant when nitrocellulose is processed with the aid of an ether-alcohol mixture. If the solvent is nonvolatile or only slightly volatile, the nitrocellulose forms a gel, but usually heat is required for completion of the operation. If heating is undesirable because of danger, gelatinization can be achieved by blending nitrocellulose at room temperature with a gelatinizer dissolved or suspended in a liquid, such as alcohol, which is not necessarily a solvent for nitrocellulose. If a solid solvent such as camphor is used in the preparation of colloid, the water-wet nitrocellulose is mixed with powdered camphor, some alcohol is added, and the mass is kneaded at room temperature for several hours. The process is usually called **plasticization**. Both gelatinization and plasticization yield similar materials which are flexible and crack resistant. The difference is that a gelatinizer tends to draw the molecules together and a plasticizer spreads them out.

Alfred Nobel used guncotton to make his famous propellant **ballistite** (made by 10% camphor and equal parts of nitroglycerin and collodion which was a solution of nitrocellulose in ether or acetone). We will talk more about that when we discus propellants.

(5.4) PETN

PETN, as we have noted above, is a benchmark compound. Explosives more sensitive than this are classified as primary explosives, whereas explosives less than sensitive than this would be called secondary explosives. PETN itself is taken to be a secondary explosive. PETN or Pentaerythritol Tetranitrate was synthesized in 1894 by the German firm **Rheinisch-Westfalische Sprengstoff A.G**. Chemically it is the ester of a polyhydric alcohol, made by nitrating pentaerythritol.

PETN has a R.E. factor of 1.66 and this should tell you how powerful an explosive it is. It has had an illustrious history. During World War II it made up the warhead of the legendary Bazooka anti-tank rocket. The warhead consisted of 8 oz of pentolite and could penetrate up to 5 inches of armor. Pentolite, as we shall see later, is a mixture of PETN and TNT. The famous demolition charge, M118 of the US Army, commonly called Flex-X or sheet explosive, in its original form consisted of 4 half-pound sheets of flexible explosive packed in a plastic envelope. Each sheet is approximately 3 inches wide, 12 inches long, and 1/4 inch thick. Originally, the explosive used in M118 was made from PETN only. Later other explosives were also used under the same name.

While PETN can be manufactured by treating pentaerythritol with nitric acid and adding concentrated sulfuric acid to complete the separation of the PETN, manufacture in the United States has been with nitric acid alone. PETN is not as sensitive to impact as nitroglycerin or nitrocellulose, but is slightly more sensitive than RDX and distinctly more so than tetryl.

It is a white crystalline substance melting at 141 degrees Celsius. It is insoluble in water but soluble in acetone. Its detonation velocity is 27,500 fps. It is also known by the names of Penthrite, Penta, Pentrit, Nitropenta, NP, and Ten. PETN is highly sensitive to shock and initiation by explosion though not very sensitive to friction. It is desensitized by adding 10% of montan (lignit) wax to it. Since it has high explosive power and appreciable sensitiveness, it is an excellent explosive for transmitting a detonation from an initiating explosive to a high and more insensitive explosive charge, that is, as a secondary filling of detonating caps or as a filling for boosters.

PETN is very sensitive to the presence of acid or alkali in storage. Even 0.01 percent of free acid or alkali would make storage difficult beyond 15 months at 65 degrees as it results in rapid acceleration of the rate of decomposition. Since RDX, tetryl, and TNT are more resistant to storage at 65°C, PETN is not as suitable for storage and use under tropical conditions.

Even lasers can be used to initiate PETN. The energy required for initiation increases with increased packing density. PETN five millimeters thick with a density of one gram per cubic centimeter can be detonated by a neodymium glass laser with an output at 10600 Angstroms when the beam is focused to produce a power density in excess of 0.08 megawatts per square millimeter. Initiation appears to be a thermal process. The laser energy is absorbed rapidly in a thin layer of explosive which produces a shock wave that causes the initiation of the entire sample.

PETN is there to stay forever as a great explosive. One of the most famous uses of PETN is in forming the explosive core of a **detonating cord**. It is incomparable in this role and as long as detonating cords will be in existence, PETN has its place assured in history. It also finds therapeutic use as a drug to alleviate angina pectoris and as a long-acting coronary vasodilator.

Fig. 14 Destroying An Unexploded Old Bazooka Warhead With Detonating Cord: Both Of Them Have PETN In Them

(5.5) EGDN

EGDN or ethylene glycol dinitrate is obtained by nitrating ethylene glycol. It is also an oily liquid like nitroglycerin (NG) and hence mixed with it in some dynamites. It has the effect of lowering their freezing point. The EGDN/NG ratio is about 8/2 or 9/1. This is the only commercial use of EGDN. Since its vapor pressure is high, it is also used as a taggant for explosives with very low vapor pressure so that their detection becomes easier.

□□□

Chapter 6

HIGH EXPLOSIVES: THE CELEBRATED NITROAROMATICS

(6.1) NITRO DERIVATIVES OF AROMATIC HYDROCARBONS

A very large number of nitro derivatives of various aromatic hydrocarbons are found to be explosives. They include nitro derivatives of benzene, toluene, higher benzene homologues (like xylene), naphthalene, phenol, aniline, and aliphatic nitro compounds. We will discuss the most important ones among them.

(6.2) TNT: THE MOST CELEBRATED HIGH EXPLOSIVE

TNT or trinitrotoluene is the most celebrated of them all; it is in fact so hugely famous that it is taken as a reference for many things including the power of explosives. You may recall, even the power of nuclear explosions is described in terms of tons of TNT! Made by the nitration of toluene, it is the most widely produced military explosive in history. The compound was synthesized in 1863 itself by a German chemist **Joseph Wilbrand**. The first detailed study of the preparation of trinitrotoluene was undertaken by Beilstein and Kuhlberh in 1870. Pure trinitrotoluene (i.e., separated from its isomer) was prepared in 1880 by Hepp and its structure was established in 1883 by Claus and Becker. The manufacture of TNT began in Germany in 1891. In 1899 somebody tried mixing aluminum with it to produce an explosive composition. However, the Germans could use pure TNT in artillery shells in 1902 only and since then it has never looked back.

In 1902, TNT was adopted for use by the German Army replacing picric acid, and in 1912 the US Army also started to use TNT. The British gradually started using it as replacement for lyddite in 1907. It took some time for the armies of the world to reconcile to the relative insensitivity of TNT mentally attuned as they were to high explosives which used to be quite sensitive. In fact in 1910 it was exempted from the British 1875 Explosives Act from actually being considered an explosive for the purposes of manufacture and storage! How governments work! By 1914, TNT became the standard explosive for all armies during World War I. The advantages of TNT include low cost, safety in handling, fairly high explosive power, good chemical and thermal stability, compatibility with other explosives, a low melting point favorable for melt casting operations and moderate toxicity. It is very stable, neutral, and does not attack metals—all making it convenient for storage. It is so insensitive that you can shoot a rifle bullet through it and 96% of the time it will not explode; you can take a bomb containing TNT and drop it from a height of 4000 feet on a concrete floor and 92% of the time it will not explode. In direct flame it ignites only at temperatures greater than 300 degrees Celsius. TNT is one of the least sensitive of military explosives; only ammonium picrate, nitroguanidine, and ammonium nitrate are less sensitive. TNT shows no deterioration after even 20 years' storage in a magazine.

Fig. 15 TNT Makes It way Into The Heart Of Artillery Shells As Soon As Its Importance Is Recognized: A Scene From The Early Days

It is also known commercially as Trilite, Tolite, Trinol, Trotyl, Tritolol, Tritone, Trotol, and Triton. It lends itself eminently to mixing with many other explosives and is a constituent of many explosives, such as **amatol, pentolite, tetrytol, torpex, tritonal, picratol, ednatol, and Composition B**. When TNT is pure, it is known as grade A TNT and varies from white to pale yellow crystals in the form of small columns or needles. When the proportion of impurities is much greater, the color is darker, often brown, and the chemical is known as grade B TNT. The velocity of detonation is approximately 22,300 fps.

Fig. 16 TNT Filled Up Great Shells Like These Too

It can be melted using steam or hot water and poured into shells or wherever you want. The ability to melt cast was one of the strongest points in favor of TNT as the standard bursting charge. Cast charges of TNT are insensitive to blasting caps and need a booster for safe initiation. Pressed TNT is, however, cap-sensitive.

It is made by the nitration of toluene, which is produced from light oil, the gases obtained from the dry distillation of coal, aromatic fractions of petroleum or by aromatization of heavy petroleum hydrocarbons by cracking processes. After nitration, the crude TNT requires purification. The ultimate objective of TNT purification is to remove all those products produced during nitration which are more reactive than TNT or which can easily give oily exudation products during storage. In the early years large-scale production of TNT was limited by the availability of toluene from coal tar and it failed to meet demand for the filling of munitions. Use of a mixture of TNT and ammonium nitrate, called amatol, became wide spread to relieve the shortage of TNT. Underwater explosives used the same formulation with the addition of aluminum, and were called **aminal**.

Exudate has been known to separate from cast TNT. It may appear pale yellow to brown and may vary in consistency from an oily liquid to a sticky substance. The amount and rate of separation depend primarily upon the purity of the TNT and, secondarily, upon the temperature of the stowage place. Grade B (low-melting point) TNT may exude considerable liquid and generate some gas. This exudation is accelerated with an increase in temperature. Pure TNT will not exude since the exudate consists of impurities that have not been extracted in the refining process. The exudate is a mixture of lower melting isomers of TNT, nitro compounds of toluene of lower nitration, and possible nitro compounds of other aromatic hydrocarbons and alcohols. It is flammable and has high sensitivity to percussion when mixed with absorbents. TNT exudate, when mixed with a combustible material, such as wood chips, sawdust, or cotton waste, will form a low explosive that is highly flammable and ignites easily from a small flame. Its presence does no appreciable harm to the stability but somewhat reduces the explosive force of the main charge. Accumulation of the exudate is considered a great risk of explosion and fire. Its accumulation should always be avoided by continual removal and disposal as it occurs. The exudate is soluble in acetone or alcohol. One of these solvents (requiring adequate ventilation) or clean, hot water should be used to facilitate removal and disposal of the exudate. Under no circumstances should soap or other alkaline preparations be used to remove this exudate. The addition of a small amount of hydroxide, caustic soda, or potash will sensitize TNT and cause it to explode if heated to 74.4 degrees Celsius.

Oil exudation from TNT is an undesirable characteristic for the following reasons:

(a) Exudation produces porosity in the TNT charge and the corresponding reduction in density. This reduces the explosive effect. In artillery shells this can produce a dislocation of the charge on firing, compression of the air included in the cavities, and premature explosion.

(b) The oily products can penetrate into the threaded parts of the shell and form "fire channels" through which the ignition of propellant can ignite the charge.

(c) The oily products can penetrate into the detonating fuze if the explosive of the fuze is not protected by a metal envelope, and reduce the detonating power of the fuze.

(6.3) PICRIC ACID OR LYDDITE: THE EXPLOSIVE THAT CAUSED THE BIGGEST NON-NUCLEAR EXPLOSION IN HISTORY

Till the First World War picric acid was almost universally used as the bursting charge in shells, grenades and mines of all description; it does not matter that it has since then been preceded by better explosives—its universality of use is enough to secure its place in history. Picric acid will be remembered for one more reason; as the explosive that caused the biggest man-made non-nuclear explosion in the world, albeit accidental. Chemically, Picric acid is trinitrophenol, a nitro derivative of phenol. Picric acid is also a constituent of some over-the counter antiseptic creams like Burnol used for treating minor burn injuries.

Picric acid is probably the earliest known nitrophenol; it is mentioned in the alchemical writings of Glauber as early as 1742. In the second half of the 19th century, picric acid was widely used as a fast dye for silk and wool. It was not until 1830 that the possibility of using picric acid as an explosive was explored by Welter.

Designolle and Bruge suggested that picrate salts could be used as a propellant, while in 1871; Abel proposed the use of ammonium picrate as an explosive. In 1873, **Sprengel** showed that picric acid could be detonated to an explosion and **Turpin**, utilizing these results, replaced black powder with picric acid for the filling of munition shells. In Russia, Panpushko prepared picric acid in 1894 and soon realized its potential as an explosive. Eventually, picric acid was accepted all over the world as the basic explosive for military uses.

The British called it **Lyddite** also because they used it as a shell filling at Lydd, Kent. It was extensively used as a shell-filling, in grenades and mines by almost every nation till it was supplanted by TNT. It was noticed during the World War I that Lyddite-filled shells of the Royal Navy tended to explode right after hitting the armor of battleships whereas TNT-filled armor piercing shells of the Germans were capable of penetrating the armor and then exploding. The effectivity of the German shells was therefore higher and they were found causing greater damage. This was an observation which was instrumental in convincing the British that they must switch over to TNT. Still, picric acid and ammonium picrate were used as booster explosive for TNT for quite some time.

Picric acid did have its problems: in the presence of water it caused corrosion of the shells, its salts were quite sensitive and prone to accidental initiation, and picric acid required prolonged heating at high temperatures in order for it to melt. Before TNT came in use, people managed with Picric acid in spite of the high pouring temperature needed. It reacts with metals to form metal picrates, which like picric acid itself are highly sensitive explosives that can be detonated by heat, flame, shock, or friction. TNT does not suffer from this defect. The use of picric acid in spite of these difficulties is an illustration of the indomitable spirit of mankind—if they had to fight and if it was the only suitable explosive, they made millions of shells with picric acid and used them successfully, the problems be damned.

(6.3.1) The Halifax Explosion Of 1917

Picric acid was responsible for the biggest man-made non-nuclear explosion in the history of mankind! Yes, we are talking of the explosion of the ship Mont Blanc off Halifax on December 6, 1917 which destroyed practically the entire city of Halifax in Canada.

At 8:45 that bright morning, a French munitions freighter, the Mont Blanc, was coming through the Narrows to wait for a convoy to accompany her across the Atlantic. This small, barely seaworthy vessel was carrying a full cargo of explosives. Stored in the holds, or stacked on deck, were 35 tons of high octane benzol gasoline, 300 rounds of ammunition, 10 tons of gun cotton, 200 tons of TNT, and most importantly, 2,300 tons of picric acid! The Mont Blanc was not flying the regulation red flag to indicate she was carrying explosives.

Coming into the harbor was a Belgium relief vessel Imo, on its way to New York City. The Imo was a much larger and faster ship. She was traveling fast and a bit too close to Dartmouth when the Mont Blanc first spotted her. She signaled that she was in her correct channel, but Imo signaled that she was intending to bear even further to port and into the Mont Blanc's channel. The Mont Blanc signaled again that she was still intending to pass to starboard. By this time she was traveling 'dead slow' close to the Dartmouth shore, expecting the Imo to swing towards Halifax, but she didn't, and the Mont Blanc made the fateful decision to swing to port, towards Halifax, across the bows of the Imo. Just as this happened; the captain of Imo panicked, and signaled 'full speed astern'. So did the Mont Blanc, but it was too late—the Imo cut into the Mont Blanc's side, striking deadly sparks, igniting the benzol gasoline, and propelling the blazing ship towards the piers of Halifax.

The terrified crew took to the lifeboats, screaming warnings that no one heeded, and rowed for their lives to Dartmouth, where most of them survived. Soon the drifting ship brushed by Pier 6, setting the wooden pilings on fire, while spectators, including many children, ran down to the waterfront to watch the billowing smoke and sparks from the deck of the vessel. Many just watched the fireworks from their windows, while crews from nearby ships raced to put out the blaze. The Halifax Fire Department responded quickly, and they were just positioning their engine up to the nearest hydrant when the Mont Blanc exploded at 9:04:35 a.m. in a blinding white flash. The boom was big enough to ring church bells 100 km away. The shock wave shattered windows at Truro, and the sound could be heard in Charlottetown and Sydney.

Water around the ship vaporized, and a huge wave flooded up the streets of Halifax and Dartmouth, sweeping many people back into the harbor where they drowned. The blast leveled downtown Halifax, flattening everything within 800 meters, destroying 1600 buildings and killing over 1,900 people instantly. Hundreds more died of their wounds in the weeks and months that followed.

Fig. 17 Aftermath Of The Halifax Explosion: A Scene Of The Destruction Caused By Picric Acid

The explosion leveled much of the poorer North End of the city, while the bulk of Citadel Hill absorbed or deflected the blast, protecting the wealthier South End neighborhoods. The hospitals soon were packed with wounded and dazed citizens. A total of 9,000 were left injured, many permanently. Over 1000 fire watchers sustained eye damage from broken glass shards. What was not immediately leveled by the blast eventually burned to the ground, aided by winter stockpiles of coal in cellars. In all, 12,000 houses were badly damaged, leaving 10,000 people homeless and 25,000 without adequate housing. The damage amounted to over $50 million.

(6.4) AMMONIUM PICRATE

Ammonium picrate is distinctly less sensitive to impact than TNT. The low degree of sensitivity of ammonium picrate to impact is paralleled by a relative insensitivity to initiation. Add to this the combination of brisance and power, almost equal to those of TNT and you get a wonderful explosive. For these reasons, ammonium picrate is used in armor piercing projectiles. The compound is also used in rocket assisted takeoff units for aircraft.

The manufacture of ammonium picrate (also called **Yellow Explosive** and **Explosive D**) is relatively simple. Picric acid, which is only partially soluble in water, is suspended in hot water and neutralized by the addition of gaseous or aqueous ammonia. The picrate, which is more soluble in water, is formed and immediately goes into solution. The specifications for ammonium picrate cover one grade of material representing two classes with respect to granulation. Class 1 material, which is the coarser of the two classes, is intended for use in the press-loading of shells, while Class 2 is used for the manufacture of picratol and other compositions. However, the impracticality of melt-loading ammonium picrate has led to its partial replacement by picratol, which can be melt-loaded.

(6.5) SOME OTHER NITROAROMATIC EXPLOSIVES

In this section we will merely list some of the not-so-famous nitroaromatic explosives and the sources from which they are derived so that if you happen to hear the name somewhere you can have some idea of where it comes from.

Dinitrobenzene, Trinitrobenzene (TNB): Benzene

TNB has close structural relationship with TNT and is, incidentally, a manufacturing by-product of TNT.

TNX (trinitroxylene): Xylene

Trinitronaphthalene: Naphthalene

Hexyl (hexanitrodiphenylamine): Aniline

Tetranitromethane: Methane

❑❑❑

Chapter 7

HIGH EXPLOSIVES: THE LEGENDARY NITRAMINES

(7.1) NITRAMINES

Nitramines are amines substituted at N with a nitro group. They are the most recent class of explosives. They have in general higher densities and bigger molecules packing more realizable energy in the same space.

(7.2) RDX: MORE THAN A LEGEND IN ITSELF

Georg Henning of Germany had obtained the patent on its manufacture by nitration of hexamethylenetetramine (hexamine) with nitric acid in 1899. Incidentally, he did not cite it as an explosive and recommended its use in medicine! Hexamine is indeed used as a medicine for the prophylaxis of urinary tract infections. Its explosive properties were recognized only during the 1920s by **George Hale**, chief chemist at the Picatinny Arsenal, USA.

It came to the frontline of explosives during the Second World War and was second only to TNT. Since then it has been used so much in formulating the new generation of explosives that it has easily and most deservedly acquired the crown of the King of Explosives. The most illustrious member of the nitramine family, RDX is also known by the names of **cyclonite, hexogen**, and **T4**. The name cyclonite was given to this explosive by Clarence J. Bain because of its cyclic structure and its cyclonic nature. Chemically, it is a heterocyclic nitramine called cyclo-trimethylenetrinitramine. The acronym RDX is an abbreviation of either **Research Department Formula X** or **Research Department Explosive** or **Royal Demolition Explosive**. The British gave this acronym to it as it was made in Britain at the Research Department of the Royal Arsenal at Woolwich. The Americans had developed a secret process to manufacture it on a large scale and they adopted the British name. Many people regard it as the ultimate explosive. Better explosives might have been developed but nothing comes close to the tremendous reputation which RDX enjoys.

The 'classical' method of production invented by Henning is the nitration of hexamethylenetetramine to 'hexogen' using concentrated nitric acid; the concentrated reaction mixture is poured into iced water, and the product precipitates out. As such the process was both expensive and unattractive for large scale production. The structural formula showed that three methylene groups must be destroyed or split off by oxidation. As soon as this problem and the attendant dangers had been mastered, industrial-scale production became possible, and during the Second World War Cyclonite was manufactured in large quantities on both sides, using several mutually independent chemical methods. No doubt that TNT ruled the roost during the Second World War but RDX was used to boost the power of TNT.

Now there are several industrial processes for the manufacture of RDX which are known by the names of their inventors, such as S-H process of Schnurr, K process of Knoffler, KA process of Knoffler and Apel, E process of Ebel and W process of Wolfram. In the USA and Canada the so-called **Bachmann's process** is popular. The product resulting from Bachmann's process was known as Type B RDX and this RDX necessarily comes with an impurity of 8 to 12%. The explosive properties of this impurity were later utilized and the explosive **HMX**, also known as Octogen, was developed. This manufacturing process was more economical and also led to the discovery of several new explosives. A manufacturing route for the synthesis of pure RDX (no impurity of HMX) was developed by Brockman, and this became known as Type A RDX. Type A RDX melts between 202 and 203 degrees Celsius; Type B RDX melts between 192 and 193 degrees. Type A and type B RDX are equally sensitive to impact, friction, heat, and initiation. This is due to the fact that HMX, the chief impurity in Type B RDX, has sensitivity characteristics that are very similar to those of RDX.

In the USA and England, RDX is stored wetted with water. In Germany and Japan it is stored dry. RDX does not cause corrosion of aluminum or stainless steel, even in the presence of moisture. Dry RDX causes no corrosion of and RDX with 0.5 percent moisture causes only slight corrosion of nickel or zinc. Dry or damp RDX causes very slight corrosion of copper, brass, mild steel, or cadmium.

In its pure form it is a white crystalline solid. The detonation velocity is about 28,000 fps. PETN matches it in terms of power and detonation velocity but RDX beats it in terms of stability. It is generally regarded as one of the most powerful and brisant military explosives. Reported sand test results indicate RDX is from 125 to 145 percent as brisant as TNT. Fragmentation tests indicate a brisance of 141 percent of TNT. Tests against earthworks, masonry, and in underwater explosions indicate a power of close to 200 percent that of TNT! In one test a 25 kilogram charge of RDX in a shell produced a crater in packed earth 12.60 cubic meters in volume. The same amount of TNT under the same conditions produced a crater 6.75 cubic meters in volume. This indicates a power of 186.5 percent of TNT.

However, it is rarely used alone. Usually it is mixed with other explosives, oils or waxes to reduce its sensitivity. Pure RDX is used in press-loaded projectiles but not in cast loaded projectiles because of extensive decomposition at the melting point. Cast loading is accomplished by blending RDX with a relatively low melting point substance. Straight RDX is used as a base charge in detonators and in some blasting caps, and as an oxidizer in specialized gun propellant.

Compositions in which the RDX particles are coated with wax are called Composition A; mixtures with TNT are called Composition B; and blends with a non-explosive plasticizer are known as Composition C. RDX is insoluble in water and very stable in storage. With plasticizers it forms the basis of many plastic explosives. In fact there are innumerous ways in which this brilliant explosive may be used. The compositions have been used extensively as charges for aerial bombs since the Second World War.

Fig. 18 The Compositions Of RDX Are Important Ingredients Of Bomb Fillers Since The WWII

Phlegmatized and pressed cyclonite is used as a highly brisant material for the manufacture of W Booster and W Hollow Charges. Non-phlegmatized Cyclonite in combination with TNT is also used as a pourable mixture for hollow charges and brisant explosive charges (W Compositions B); mixtures of cyclonite with aluminum powder are used as torpedo charges (Hexotonal, Torpex, Trialen). Cyclonite may also be used as an additive in the manufacture of smokeless powders. In manufacturing explosive charges which are required to have a certain mechanical strength or rubber-elastic toughness, Cyclonite is incorporated into curable plastic materials such as polyurethanes, polybutadiene or polysulfide and is poured into molds (W Plastic Explosives).

RDX is made in different granule sizes and they are put to different uses. The granule size is easily determined by the number of the standard sieves (US numbers 8 to 325—size 8 corresponds to aperture size 2.36 mm, size 20 to 0.85 mm, size 50 to 0.30 mm, size 100 to 0.15 mm and size 325 to 0.045 mm) which a certain percentage of granules are able to pass. Granules are not exactly of the same size and that's why we speak of the percentage of granules of a certain size. RDX of different granule sizes is named from Class A to Class H. As a rule of thumb, granules of Class A are bigger than granules of Class H.

The different uses of RDX of different class are enumerated below:

(a) Class A, for use in Composition A3, Composition B, Composition C4, and cyclotols

(b) Class B, for use in Composition A3 and Composition C4

(c) Class C, for use in the preparation of RDX booster and detonator pellets

(d) Class D, for use in cyclotols

(e) Class E, for use in Composition C4 and plastic bonded explosives (PBX's)

(f) Class F, for use in Composition B3

(g) Class G, for use in PBX's

(h) Class H, for use in Composition C-4

Fig. 19 A Demolition Charge Of C-4 Being Prepared: RDX Is An Important Constituent Of C-4

(7.3) HMX: AN EXPLOSIVE THAT BEATS THE RDX

This is one explosive which clearly beats RDX in terms of detonation velocity. Its detonation velocity is 29,880 fps! It also has more density than RDX. Its chemical name is cyclo-tetramethylenetetramine; the relationship to RDX should be obvious. In fact it was discovered as a byproduct during the production of RDX. RDX produced by the Bachmann Process usually contains about 8-12% HMX as an acceptable byproduct. A modification of the process would yield about 55-60% HMX with RDX as an impurity! It may also be produced by nitration of hexamine in the presence of acetic anhydride, paraformaldehyde and ammonium nitrate. There are four polymorphs of HMX: alpha, beta, gamma, and delta. Military grade HMX consists only of the beta polymorph.

It is also known as **Octogen**. The abbreviation HMX could have come from High Melting Explosive (its melting point is about 280 degrees Celsius in comparison to 202 degrees of RDX), High Molecular Weight RDX, or High Velocity Military Explosive. Like RDX it is also a white crystalline substance and is used along with other explosives, and also lends itself to plasticizing. It is used in making explosive lenses for implosion type nuclear devices also. When mixed with TNT it becomes a melt-cast explosive called Octol.

HMX and RDX are chemically very similar except that HMX is not easily decomposed by alkaline hydroxide. HMX is used as an explosive charge when desensitized, as a booster charge in admixtures with TNT called octols, and as an oxidizer in solid rocket and gun propellants.

(7.4) TETRYL

Tetryl is an aromatic nitramine known since 1877 and we must remember it for its historical significance as it was used as a **booster explosive** for a long time. Tetryl was first prepared in 1877 by **Mertens** and its structure established by Romburgh in 1883. It is rather sensitive and hence was once used extensively as a booster explosive in blasting caps for detonating TNT. The sensitivity of Tetryl is roughly the same as that of picric acid. It is sensitive to friction, shock or spark. Its use as a booster was eventually foreshadowed by PETN and RDX. It was also the standard bursting charge for small-caliber projectiles. The Americans have discontinued the use of tetryl on the argument that it is relatively unstable after storage at elevated temperatures. Various NATO nations have, however, not discontinued its use in boosters and lead charges.

Its chemical name is trinitrophenylmethylnitramine. It is also known as **Tetrylit**, **Tetralite**, **Tetralita** and **Pyronite**. In its pure form it forms yellow crystals. Its detonation velocity is 23,750 fps. Two methods have been used extensively for the manufacture of tetryl. The first method uses methylamine, a cheap commercial bulk chemical, in the production of dinitrophenylmethylamine from dinitrochlorobenzene. This can be nitrated to tetryl with relative ease. Tetryl does not react with copper, tin, nickel, lead, copper plated steel, bronze, stainless steel, cadmium, aluminum, silver, and titanium. Tetryl, however, corrodes steel heavily. Slight corrosion is reported for iron, zinc, zinc-plated steel, tin-plated steel, parkerized steel, brass, and Monel metal.

(7.5) HNIW OR CL-20: ONE THAT BEATS EVEN HMX AND PERHAPS THE MOST SIGNIFICANT EXPLOSIVE DISCOVERED IN THE PAST 50 YEARS!

HNIW or hexanitrohexaazaisowurtzitane is a new nitramine explosive that is 20 percent more powerful that HMX. It is also known as **CL-20**. It beats HMX in both detonation velocity and density. CL-20 was a breakthrough in energetic materials with higher performance, minimum signature, and reduced-hazard characteristics. CL-20 has numerous military and commercial applications. The trend today is to explore the possibilities that HNIW can provide to munitions, from high performance gun propellants to shaped charges etc. Its only limitation is the high cost of manufacture.

HNIW is a polyazapolycyclic caged polynitramine. **Arnold Nielson**'s original discovery in 1987 had astonished the scientific community because he had constructed the CL-20 cage using a single chemical reaction and, in the process, established a new type of amine glyoxal chemistry. The starting materials are derivatives of acetyl isowurtzitane. Nitration is not the problem. The cost goes up on account of the necessity of using Palladium as a catalyst in one stage of manufacture.

Many people have described CL-20 as the most significant explosive discovered in the past 50 years. It offers great potential for insensitive munitions to be used in future weapons systems. CL-20-based shaped charges are already being used in the oil well industry. Other commercial applications we expect to see include specialty demolition, because CL-20 has good properties for use as a '**cutter**'. And it will likely be used in high-rate detonating cords also.

❑❑❑

Chapter 8

Composite Explosives Take the Best of Many Things and Make the Very Best

(8.1) WHY DO WE NEED COMPOSITE EXPLOSIVES?

It is difficult to find a single compound which would have all the desired characteristics. Therefore, most modern explosives used for practical military purposes are not single chemical substances, however exotic they be individually. Rather they are composites, their ingredients selected in such a manner that required properties are obtained. Long back when picric acid was used in shells, the French scientists had found that adding 40% dinitrophenol to picric acid improved its melt-loading. When people felt the need of decreasing the sensitivity of picric acid to mechanical impact, they added paraffin to it. Then as we mentioned earlier, many ingredients are added and in such proportions as to bring the oxygen balance close to zero. Addition of a high melting point material such as ammonium nitrate also regulates the fusibility of a mixture so that they may become fusible, semi-fusible or infusible, free-pouring. Similarly, addition of a plastic substratum results in explosives with plastic consistency. That's how the making of composite explosives has become a science in itself. When the components are two in number, we call them **binary mixtures or binary compositions**— when they are three in number; we call them **ternary mixtures** and so on.

(8.2) BINARY COMPOSITIONS

Most of the binary compositions have TNT as an important ingredient and that's a testimony to how great an explosive TNT is.

(8.2.1) Amatols

Discovered during the World War I, amatols were expedient solutions to a war-time shortage of TNT. It was found that 50% ammonium nitrate could be added to TNT to yield a composition that could still be melt-loaded. Its brisance was certainly less than that of TNT but it had 120% more strength than TNT. The detonation velocity was about 21,000 fps. People tried even 80% ammonium nitrate and 20% TNT—it did give about 75% of the brisance of TNT but could not be melt-loaded. Amatols were used during World War II also, particularly in mines as there was a shortage of both TNT and RDX on account of the tremendous consumption in the Great War. The 80/20 mixture was used in the Bangalore Torpedo. Amatols were also used in large caliber shelis.

We would, however, remember the 50/50 amatol for one more reason—the warhead of the legendary **V-2 rocket** contained 2000 lbs of amatol. It is enough to immortalize it. The **V-1 flying bomb** had also carried a warhead of 1870 lbs of amatol. These two weapons were trail blazers. The V-1 was the forerunner of the modern cruise missiles and the V-2 was the father of all modern liquid fuel rockets.

Fig. 20 V-1 Flying Bomb, The Forerunner Of Modern Cruise Missiles, Carried A Warhead Of 1870 Lbs Amatol

All the great accomplishments of the NASA and the Soviet Union in the field of rocketry owe heavily to the V-2 rockets that were captured intact at the end of the War and also the talent of the German scientists who were involved in the V-2 project. After the end of the World War II,

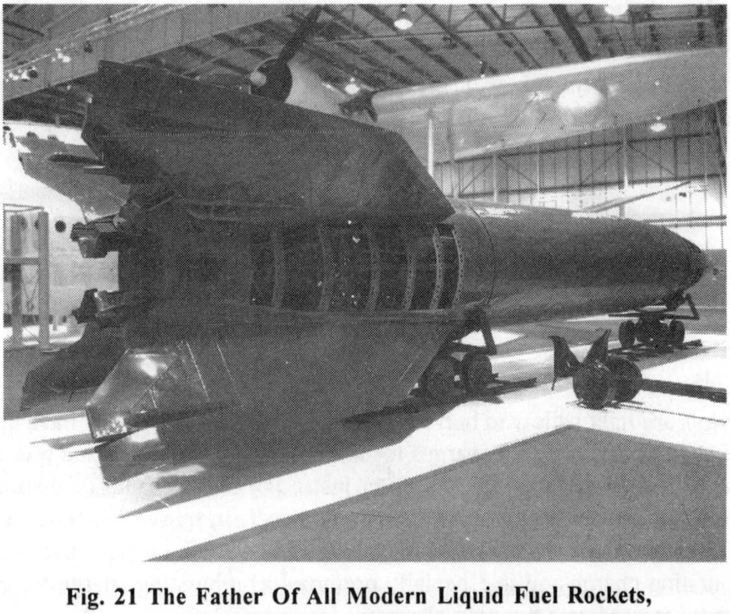

Fig. 21 The Father Of All Modern Liquid Fuel Rockets, The Legendary V-2 Rocket Also Carried A Warhead Of 2000 Lbs Amatol

under Operation Paperclip, nearly three hundred train loads of captured V-2s and their parts and about 500 German scientists including von Braun himself (many of them were to become naturalized American citizens) were to become the basis of all subsequent rocket research in the USA. No one can deny the tremendous help which the Americans received through the V-2s and the German scientists. The American rocket Redstone launched in 1953, which helped them greatly in their space program, was a direct descendent of the V-2. Werner von Braun remained employed in the US space program till 1972 and was associated with all the great things that happened there over the years including the Jupiter, Pershing and Saturn rockets—the Saturn V eventually landing man on the moon in 1969. The Soviets had also got some V-2s and German scientists and the first Soviet missile, the R-1 was almost an exact copy of the V-2. Something which was used in the V-1 and the V-2 deserves to be remembered.

(8.2.2) Tritonal

Tritonal is a silvery solid that contains 80% TNT and 20% flaked aluminum. The beauty of this composition is that it can be melt-cast without the separation of aluminum. Metallic aluminum is insoluble in TNT so tritonal begins to melt at the melting point of TNT. The TNT and aluminum are run slowly through separate chutes into a steam-heated melting kettle equipped with a stirrer which is kept in motion while the ingredients are being added. Heating and mixing are continued until all the TNT is melted and the fluidity of the mixture is considered satisfactory. The tritonal is then loaded by pouring into bombs, using at least the same precautions and techniques used in the melt-loading of TNT.

It has been used extensively in bombs and continues to be used even now. The detonation velocity is 21,650 fps and the strength is 124% that of TNT. Tritonal is good for blast bombs but is slightly less effective (91% than pure TNT) in fragmentation. The explosive will retain its place in history because it was used in the famous **bunker buster bombs** GBU-28 and BLU-109. It is also used in the warhead of the **MK-48 torpedo**.

Tritonal, like TNT, can undergo exudation if stored at elevated temperatures, but this tendency is less than that of TNT because of the presence of 20 per cent by weight of metallic particles that tend to hold the oily exudate by surface tension. This prevents the oily exudate from coalescing and exuding.

Fig. 22 The Famous Bunker Buster Bomb, The GBU-28 Uses Tritonal In Its Warhead

(8.2.3) Tetrytols

Tetrytols are light yellow to buff mixtures of tetryl and TNT. They have been found useful as demolition charges and bursting charges for mines because their brisance was greater than that of TNT. It was also used as an auxiliary booster instead of plain tetryl. The detonation velocity is over 24,000 fps and explosive power is 120% that of TNT. It, however, needed a heavier booster charge. The 75/25 tetrytol is used as a bursting charge and demolition explosive. The 70/30 mixture is used as a bursting charge and is especially preferred as a bursting charge for **chemical shells**. The 65/35 terytol is used as a bursting charge in landmines.

Tetrytols resemble tetryl more closely than they resemble TNT. They are more powerful but less sensitive than TNT. Tetrytols can be cast into munitions, which is an advantage over press-loading.

Tetrytol is manufactured by heating TNT in a melting kettle equipped with an agitator until all the TNT is melted. The proper amount of tetryl is then added with continued stirring. Part of the tetryl dissolves in the TNT. The temperature is then allowed to decrease until the mixture has thickened somewhat and is considered to be of the maximum viscosity suitable for pouring in the melt loading operation. The tetrytol is then poured into the ammunition component or a mold. As a cast explosive, tetrytol is a mixture of tetryl and the TNT-tetryl eutectic. The eutectic contains approximately 55 percent tetryl. There is no difference in the impact sensitivities of 65/35, 70/30, or 75/25 tetrytols. All are less sensitive than tetryl and more sensitive than TNT. Storage at 65 degrees or above causes some separation of an oily extrudate and distortion of blocks of tetrytol. This problem of storage has caused the United States to discontinue the use of tetrytol though it continues to be used by other nations including various NATO allies.

(8.2.4) Pentolites

Pentolites are compositions of PETN and TNT that can be cast. The 50/50 composition has been used in grenades and as a booster-surround charge because of its sensitivity to initiation and greater brisance. The detonation velocity is nearly the same (24,000-25,000 fps) as that of Tetrytol but they are significantly more powerful than Tetrytols. The warhead of the famous anti-tank rocket **bazooka** used in the Second World War had a charge of pentolite.

Fig. 23 The Famous WWII Anti-Tank Rocket Bazooka Used Pentolite In Its Warhead

The most commonly used blend consists of 50/50 PETN/TNT. Other blends such as 75/25, 40/60, 30/70, and 10/90 have been occasionally employed but the 50/50 blend is superior in the characteristics of sensitivity to initiation, brisance, and suitability for melt-loading. Pentolite is manufactured by the slurry method. In this the PETN is suspended by agitation in water heated above 80 degrees Celsius. TNT is then added. The temperature of the water causes the TNT to melt and coat the particles of PETN. The slurry is then cooled with rapid agitation, causing the TNT to solidify. The granules are then separated on a filter or in a centrifugal wringer and dried at a temperature below 75 degrees. Pentolite is gradually losing ground to Composition B. Bazooka's successor; the **M28 anti-tank rocket** used during the Korean War employed Composition B in its warhead. This is on account of the somewhat lower stability and greater sensitivity of PETN in comparison to RDX making pentolite also less stable and more sensitive than Composition B, and also the tendency of pentolite to undergo some exudation above 50 degrees Celsius.

(8.2.5) Picratol

Picratol is a composition of approximately same amounts of TNT and ammonium picrate. It is a mixture of 52 percent ammonium picrate and 48 percent TNT. Though it retains the same detonation velocity and fragmentation efficiency as TNT, it is remarkably less sensitive to impact. Hence it was used in armor-piercing shells.

Molten TNT has little or no solvent action on ammonium picrate, and consequently, cast picratol consists essentially of a physical mixture of crystals of the two explosives. Picratol is manufactured by heating TNT to above 90 degrees in a steam jacketed melting kettle equipped with a stirrer. The ammonium picrate is added slowly, without preheating, with continuous agitation and this is continued for a short time after all the ammonium picrate has been added. The thick slurry is allowed to cool to about 85 degrees before being loaded into ammunition components. Like molten TNT, picratol undergoes contraction in volume upon solidification. The contraction is, however, much less than that in the case of TNT.

(8.2.6) Cyclotol

Cyclotol is a 60/40 composition of RDX (cyclonite) and TNT. Because of its great brisance and blast effects it was used extensively in bombs during the Second World War. Its detonation velocity of 25,600 fps is greater than any other binary explosive used for melt-loading bombs. Its explosive power is 130% that of TNT and it scores above TNT in both blast and fragmentation effects.

(8.2.7) Octols

Octols are melt-cast mixtures of HMX and TNT. Type I Octol is 75 $\pm$ 2 % HMX and 25 $\pm$ 2 % TNT. Type II Octol is 70 $\pm$ 2% HMX and 30 $\pm$ 2% TNT. The detonation velocity for Type I Octol is 28350 fps; that for Type II is 27480 fps.

Octol is used as an oil well formation agent and in fragmentation and shaped charges. In fragmentation it is definitely better than Composition B. In fragmentation tests using a 105 millimeter M1 shell, 15 percent more fragments are produced and the average velocity of the fragments is 100 meters per second faster than with a similar shell loaded with Composition B. This improvement is attributed to both the higher rate of detonation of octol and the greater density of octol which permits a greater weight of explosive in the same volume.

To manufacture octols, TNT is melted in a steam jacketed kettle equipped with an agitator. The temperature of the TNT is maintained at about 100 degrees and water wet HMX is added slowly. Stirring is continued until the water is driven off. The kettle temperature is then reduced until a viscosity suitable for casting is obtained. The viscosity of the slurry has a strong dependence on the particle size distribution and polymorphic variety of HMX used. In order for the slurry to have an efflux time of less than 15 seconds the solid HMX must consist of the beta polymorph having particle diameters in the range of 500 to 800 microns. Even when octol is cast at a temperature of less than 90 degrees, there is a large amount of settling of HMX crystals while the charge is still molten. This can lead to erratic performance of the ammunition loaded. Several methods to prevent the settling are being considered. Casting temperature is kept as low as possible to prevent air entrapment in the cast.

(8.3) TERNARY MIXTURES

(8.3.1) Amatex 20

Amatex 20 consists of 40% RDX 40% TNT and 20% ammonium nitrate. Its detonation velocity is 22780 fps.

(8.3.2) Ammonal: The Explosive That Had Charged The Biggest Military Mines In History

Ammonal is an explosive made up of ammonium nitrate, TNT, and aluminum powder mixed in a ratio of roughly 22/67/11. Ammonium nitrate in ammonal functions as an oxidizer and aluminum as a power enhancer. Aluminum increases the temperature greatly, from 1700 degrees to nearly 4000 degrees! This results in an overall increase in power by about 20%. To some extent aluminum makes it less sensitive to detonation. The use of the relatively cheap ammonium nitrate and aluminum make it a good replacement for pure TNT and it has been used in vast quantities. It is still used as an industrial explosive. Storage of ammonal is a problem because of phase changes in ammonium nitrate and also its hygroscopicity. In the presence of moisture, ammonals react with the same metals as amatols: copper, bronze, lead, and copper plated steel.

The British employed ammonal for their mines during the First World War. The composition of the ammonal used was 65% ammonium nitrate, 15% TNT, 17% coarse aluminum and 3% charcoal. The basic motivation for devising this mixture was to cope with the shortage of TNT.

Ammonal has the distinction of having been used as the explosive in some of the **biggest military mines** in history. In 1916 in the Bluff, nine mines totaling 200,000 lbs of HE were used, the largest of them carrying 40,600 lbs of Ammonal sunk 75 feet deep. Incidentally, the biggest mines of history were used in Messines in 1917. Twenty mines totaling 1000,000 lbs were deployed, the largest carrying 95,600 lbs of ammonal at 125 feet depth. These mines blew the Germans off the ridge and the position was easily recaptured by the British. The mine craters are still there as large and deep pools marking the old German front line. Tunnellers of the Royal Engineers dug galleries over 2000 feet long from behind the British lines to reach the German strong points at various depths. The biggest and the deepest crater made by the largest mine of 95,600 lbs of ammonal charge is now private property and the pool has been named 'Pool of Peace'.

Fig. 24 A Crater Produced At Messines In The WWI By One Of The Biggest Tunnel Mines In Human History: The Mines Were Charged With Up To 96,500 Lbs Of Ammonal Buried At A Depth Of 125 Feet

Ammonals have been made with DNT and RDX also. Many different combinations of ingredients have been used, both militarily and commercially. The most recent United States military ammonal consists of 22 percent ammonium nitrate, 67 percent TNT, and 11 percent flaked aluminum. The major use of this composition is as projectile filler.

Ammonals are manufactured by a method similar to that used for the preparation of amatols. The calculated amount of TNT is placed in a kettle that is equipped with an agitator and steam jacket. The temperature of the kettle is raised to between 85 and 100 degrees. The calculated amount of ammonium nitrate, which was previously heated to the same temperature as the contents of the kettle, is then added. Finally aluminum powder is added and the mass cooled while continuing the agitation. If the ammonal mixture contains less than 40 percent TNT, press-loading becomes necessary. For compositions with more than 40 percent TNT, the mixture can be cast-loaded.

(8.3.3) Torpex

Torpex is RDX/TNT/Aluminum/Wax desensitizer. Torpex would live forever in history because it was the explosive chosen for the great **dam buster bombs** 'Upkeep' designed by the brilliant Sir Barnes Wallis, which breached great dams like the Mohne and the Eder in Germany in the WWII. The Upkeep bombs had a charge of 5047 lbs of torpex! Wallis chose torpex for the dam buster bombs because of its great underwater power.

Fig. 25 This Is The Legendary Dam Buster Bomb, The Upkeep: Containing 5047 Lbs Of Torpex Each, They Destroyed Several Dams In Germany During The WWII

Torpex had one problem. It was sensitive to bullet impact. Due to its sensitivity to bullet impact, the first weapons loaded were ones for which there would be the least possibility of rifle bullet and fragment attack, namely, submarine delivered mines and torpedoes. It was believed that the British had been able to handle it without incident for 2 years and that the risk was worth the advantage gained in its underwater power. It was replaced by HBX or HBX-1 later in WWII.

Torpex is a silvery white solid when cast. The composition of torpex is 41.6 percent RDX, 39.7 percent TNT, 18.0 percent aluminum powder, and 0.7 percent wax. Torpex is manufactured by melting TNT in a steam jacketed kettle equipped with a stirrer and heating the molten TNT to approximately 100 degrees. RDX that is slightly wetted with water is added slowly and mixing and heating are continued until all water has been driven off. Grained aluminum is added and the mixture is stirred until uniformity is obtained. The mixture is then cooled, with continued stirring, until the viscosity is suitable for pouring. The aluminum tends to settle out of the torpex in large castings if the liquid is allowed to set too long. To prevent this, a portion of the casting is poured and allowed to cool to the point of crusting over before another layer of torpex is poured. Torpex of slightly different composition can be manufactured by melting 12.5 parts of TNT, adding 69.5 parts of Composition B and, after heating and stirring, adding 18 parts of aluminum. This illustrates the fact that torpex is essentially aluminized 50/50 octol. Torpex is considerably more sensitive to impact than Composition B. Cast torpex is much more sensitive to initiation than cast TNT and pressed torpex is as sensitive as RDX to initiation.

(8.3.4) HBXs (High Blast Explosives): The Result Of The Port Chicago Naval Magazine Explosion

HBX or hexahydrotrinitrotriazine is a composition of RDX (40%), TNT (38%), Aluminum (17%), lecithin and wax (5%). It is less impact sensitive than Torpex and hence eventually replaced it. The development of the HBXs was spurred by an accident.

At 10:18 p.m. on Monday evening, July 17, 1944, a giant explosion rocked the town of Port Chicago, California located in the Suisun Bay. The blast killed 320 naval personnel and registered 3.4 on the Richter scale in parts of Nevada. This was the site of a U. S. Navy munitions depot which was later renamed the Concord Naval Weapons Station. Bombs, shells, mines, and other explosive ordnance devices were transferred from railcars to ships, whence they would be moved to locations in the Pacific Theatre. At that time two ships, namely, SS Quinault Victory and SS E. A. Bryan were being loaded. There was a small explosion on the pier that started a fire. Six seconds later the entire contents of E. A. Bryan exploded. The reported cargo of the ship was 4600 tons of ammunition, including 1780 tons of high explosives. The nighttime explosion was reported as a bright white light over the sky of the San Francisco Bay Area, followed by a mushroom cloud and a strong concussion. Windows in Vacaville, Concord, Vallejo, Benicia, Martinez, Napa, and San Francisco were all blown out. Heavy doors and locks in Yountville and ship hatches at Mare Island were blown off because of the resulting concussion from the explosion.

The actual cause of the Port Chicago disaster was never officially or publicly established. Three days after the explosion, Rear Admiral Carleton H. Wright, commander of the 12th Naval District, convened a board of inquiry in San Francisco. After hearing all the evidence, the board could not determine the exact cause or circumstance of the explosion, but did issue a list of seven circumstances that might have caused the disaster. One of them was the presence of some sensitive explosive that might have exploded in the process of handling.

After this accident research was devoted to developing safer explosives. HBX and H-6 explosives were developed in this process. They incorporated wax and other chemicals to desensitize the explosive and hot-melt liners were introduced for lining bombs and warheads to give some thermal protection and eliminate potential pinch points from cracks or fissures in the bomb or warhead case. Later, plastic-bonded explosives were developed for increased thermal protection and fragment impact resistance. The hot-melt liners coated and eliminated metal-to metal pinch points.

Three compositions of HBX are in use. The detonation velocities are roughly the same and they are in the range of 22700 fps to 23700 fps.

HBX explosives are prepared by melting TNT in a steam jacketed kettle equipped with a mechanical stirrer. Water wet RDX is added slowly with stirring and heating until all the water is evaporated. Powdered aluminum is added and the mixture is stirred until uniform. D2 wax and calcium chloride are then added and the mixture is cooled to a temperature suitable for casting. HBX can also be prepared by adding the calculated amount of molten TNT to molten Composition B to obtain the desired proportion of RDX and TNT. Then the appropriate weights of other ingredients are added to complete the composition. D2 wax, also referred to as Composition D2, consists of 84 percent paraffin wax, 14 percent nitrocellulose, and 2 percent lecithin.

	HBX-1	HBX-3	H-6
RDX (including nitrocellulose, calcium chloride, and calcium silicate)	40.4 ± 3.0%	31.3 ± 3.0%	45.1 ± 3.0%
TNT	37.8 ± 3.0%	29.0 ± 3.0%	29.2 ± 3.0%
Aluminum	17.1 ± 3.0%	34.8 ± 3.0%	21.0 ± 3.0%
Wax and lecithin	4.7 ± 1.0%	4.9 ± 1.0%	4.7 ± 1.0%

H-6 is an Australian composition. H-6 is used as the main charge filling for underwater blast weapons such as mines, depth charges, torpedoes and mine disposal charges.

(8.3.5) Minol

Since amatols were regarded as expedient solutions, some people bettered them by adding aluminum to that. Minol is 40% TNT, 40% ammonium nitrate and 20% aluminum. Minol-2 which is the type currently used, has the following composition: TNT 40 ± 3 % , Ammonium nitrate 40 ± 3 % , and Aluminum 20 ± 3 % .

Minol is used in four types of ordnance: underwater depth charges, block buster bombs, concrete fragmentation bombs, and general purpose bombs. In underwater ordnance the confinement effects of the water somewhat offset the effects of the relatively low detonation velocity. In block buster bombs a sustained and powerful impulse is more destructive than a high peak pressure. In concrete fragmentation bombs, the high impulse will impart a satisfactory velocity to the fragments but will not pulverize the concrete.

Minols are manufactured by adding appropriate quantities of dry ammonium nitrate and aluminum powder to molten TNT at about 90 degrees under agitation. Originally aluminum was used in the form of a fine powder, but it has been found to be unnecessary. The larger aluminum particles do not cause any loss of performance and have a smaller surface area per unit volume and so reduce the reactivity of the aluminum in the mixture and increase the stability.

(8.3.6) HMX, TNT And Aluminum Mixture-3 (HTA-3).

	Type I	Type II
Percent HMX	49	49
Percent TNT	29	28.65
Percent aluminum	22	22
Percent calcium silicate	-	0.35

To manufacture HTA-3, TNT is heated to about 100 degrees in a steam jacketed kettle equipped with an agitator. Water wet HMX is added slowly to the molten TNT. Stirring and heating are continued until all the water is evaporated. Aluminum powder is then added and the mixture is cooled with continued stirring. When a satisfactory viscosity is obtained the HTA-3 is cast.

(8.4) OTHER FAMOUS EXPLOSIVE COMPOSITIONS

(8.4.1) Composition A Explosives

Composition A explosives consist of a series of formulations of RDX and a desensitizer. It is a wax-coated, granular explosive consisting of RDX and plasticizing wax which also acts as desensitizer. Compositions A and A2 contain the same percentages of materials as Composition A3 but the type of wax used and the granulation requirements for the RDX are different. Composition A, for example, contains beeswax, while Composition A2 contains a synthetic wax. All of the Composition A explosives are press-loaded.

Compositions A and A2 are no longer used. **Composition A3** is used as a bursting charge for medium-size projectiles and in landmines. It contains 91.0 ± 0.7% RDX and 9.0 ± 0.7% desensitizing wax. Composition A-3 is appreciably more brisant and powerful than TNT. The detonation velocity is about 27,000 fps. It is non-hygroscopic and possesses satisfactory stowage properties.

The manufacture of composition A3 is carried out by heating a water slurry of RDX at nearly 100 degrees and agitating. The wax, which contains a wetting agent, is added during agitation and agitation is continued while the mixture is allowed to cool to a temperature less than the melting point of the wax. After being caught on a filter, the composition is air-dried at 77 degrees.

(8.4.2) Composition B Explosives

Composition B refers to mixtures of approximately 60 percent RDX and 40 percent TNT. Compositions resulting from other combinations of RDX and TNT are called **cyclotols**. TNT enables the composition to be cast-loaded and also reduces the sensitivity of RDX to some extent. It is non-hygroscopic and remains stable in stowage.

Composition B has its place assured in history because it was the '**fast explosive**' used in the '**explosive lenses**' of the **Fat Man nuclear bomb** dropped on Nagasaki and innumerous subsequent bombs of that type. At that time it was a composition that offered very high shaped-charge efficiency and could convert the theory of 'explosive lenses' to a reality. The detonation velocity is about 26,000 fps. We will discuss explosives lenses in greater detail in a later chapter.

Composition B grade A is formulated as a 60/40 RDX/TNT mixture, but high quality castings usually are higher in RDX content because a TNT-rich section is removed from the top of the casting. The grade A composition B is made from grade A RDX. Composition B is manufactured from TNT and water wet RDX. Sometimes a little wax is also added. The TNT is melted at approximately 100 degrees in a steam jacketed melting kettle equipped with a stirrer. The damp RDX is added slowly with stirring, and after addition of the RDX is completed, most of the water is poured off and heating and agitation are continued until all moisture has been driven off. The wax is then added and mixed thoroughly with the other ingredients by stirring. The nature of the wax is important since only certain waxes mix with the RDX and TNT and do not tend to segregate during cooling. The thoroughly mixed Composition B then is cooled with continued agitation until the fluidity is satisfactory for casting. The Composition B is either cast directly into bombs or solidified into chips approximately 4.8 square centimeters by 0.6 centimeters. Composition B is made in the form of chips when it is to be stored or shipped for use elsewhere.

Composition B comes in three varieties: Composition B, Composition B2 and Composition B3. Cyclotol also comes in three varieties: Cyclotol Type I, Cyclotol Type II A and Cyclotol Type II B.

	C-Type I	C-Type IIA	C-Type IIB	Comp-B	Comp-B2	Comp-B3
RDX	75±2	70±2	69.6±2.0	59.5±2.0	60	59.9±1.0
TNT	25±2	30±2	29.9±2.0	39.5±2.0	40	40.5±1.0
Wax	-	-	1±.03	-	-	-
Calcium Silicate	-	-	0.5±0.15	-	-	-
Cast - ability	Difficult to pour	Difficult to pour	-	Easily poured	Easily poured	-

(8.4.3) Baratol

Baratol would live in history because of its use as the **'slow explosive'** in the **'explosive lenses'** of the **Fat Man nuclear bomb** and subsequent bombs of that type. Baratol is a composition of barium nitrate and TNT. TNT is typically 25-33% of the mixture with 1% wax as a binder. The high density of barium nitrate gives Baratol a density of at least 2.5. Explosive lenses have been discussed at great length later in the work. The explosive lens effect occurs in a more pronounced manner when there is considerable difference in the detonation velocities of the tow explosives used. The detonation velocity of Baratol is just about 16,000 fps. The difference with the detonation velocity of Composition B was highly significant and hence made the lens action in the first nuclear device quite possible.

(8.5) QUATERNARY MIXTURES

Depth Bomb Explosive (DBX) is the only explosive covered under quaternary mixtures. DBX consists of: TNT 40 %, RDX 21 %, ammonium nitrate 21 %, and aluminum 18 %. DBX was developed to replace torpex which was, as we saw, more sensitive. As the name indicates, the primary use of DBX is in depth charges. The method of manufacture is similar to that of torpex except that casting is done at 90 to 95 degrees. The impact sensitivity of DBX is slightly less than that of TNT. The detonation velocity is 21750 fps.

(8.6) PLASTIC EXPLOSIVES OR PLASTIQUE: VERSATILITY UNLIMITED

As the name suggests, plastic explosives are compositions which are soft and hand malleable. Technically, plastic explosive means an explosive material in flexible or elastic sheet form formulated with one or more high explosives which in their pure form have a vapor pressure less than 10^{-4} Pa at a temperature of 25 degrees Celsius, is formulated with a binder material, and is as a mixture malleable or flexible at normal room temperature. This makes them especially suitable for controlled demolition work as they can be easily formed into the best shapes for cutting structural members. You would not like to waste plastic explosives for regular blasting work as they are costlier and do not offer any significant advantage. The origin of the US term plastique is due to the plastic explosive introduced to the US by the British in 1940. The samples of explosive brought to the USA by the Tizard Mission had been packaged by SOE (Special Operations Executive) ready for dropping to the French Resistance and were labeled in French, as Explosif Plastique. The Americans took the print as a product name rather than an explanatory label!

Many plastic compositions were made before the World War II also, such as mixtures comprising a solution of collodion cotton in liquid aromatic compounds as plasticizers. **Plastrotyl,** for example, was 86% TNT, 10% liquid dinitrotoluene (DNT), 0.3% collodion cotton and 3.7% of turpentine. During the Second World War, plastic explosives based on RDX took off. RDX was especially suitable for plastic explosives as its detonation velocity is very high. One of the early products was called **Nobel 808**. Nobel 808 contained RDX, mineral oil, and lecithin. It had a distinct smell of almonds. This was used by the British SOE (Special Operations Executive) for their sabotage missions and was also used in their HESH anti-tank shells.

(8.6.1) Composition C or C-1

During World War II, the British used a plastic demolition explosive that could be shaped by hand and had great shattering power. As standardized by the US, it was designated as Composition C (or C-1). It contained 88.3 % RDX and 11.7 % of a non-explosive oily plasticizer. Included in the plasticizer was 0.6 % lecithin, which helped to prevent the formation of large crystals of RDX which would increase the sensitivity of the composition. Composition C was plastic from 0 degrees Celsius to 40 degrees Celsius. It, however, became brittle and less sensitive below 0 degrees and tended to become gummy and exude oil at temperatures above 40 degrees.

(8.6.2) Composition C-2

Composition C was replaced by Composition C-2 which contained 80 % RDX and 20 % explosive plasticizer. This explosive plasticizer was composed of mononitrotoluene (MNT), a liquid mixture of dinitrotoluenes (DNT), TNT, nitrocellulose, and dimethylformamide. Composition C-2 remained plastic from -30 degrees to 52 degrees Celsius, but became less plastic in hot storage because of the evaporation of the volatile matter. MNT and DNT have two good properties. They are oily liquids and thus help towards plasticizing the mixture; at the same time they are explosives themselves.

(8.6.3) Composition C-3

Composition C-2 was replaced by Composition C-3, which contains 77 ± 2 % RDX and 23 ± 2 % explosive plasticizer. The plasticizer contains 5% mononitrotoluene (MNT), 10% liquid mixture of dinitrotoluenes (DNT) , 4% TNT, 3% tetryl, and 1% nitrocellulose or butyl phthalate and nitrocellulose. It is a yellowish, puttylike solid that has a density of 1.60 and is soluble in acetone. The detonation velocity is about 26,000 fps.

The essential difference between Composition C-3 and Composition C-2 is the substitution of 3% tetryl for 3% RDX, which improves the plastic qualities. The changes were made in an effort to obtain a plastic, puttylike composition to meet the requirements of an ideal explosive for molded and shaped charges that will maintain its plasticity over a wide range of temperatures and not exude oil.

In the manufacture of Composition C-3, the mixed plasticizing agent is placed in a steam jacketed, melting kettle equipped with a stirrer and heated to nearly 100 degrees. Water wet RDX is added slowly, and heating and stirring are continued until a uniform mixture has been obtained and all the water has been driven off. The mixture then is cooled while being agitated. Composition C-3 is less sensitive to impact and initiation than TNT. The chemical stability of Composition C-3 is acceptable, but the physical stability is not entirely satisfactory.

(8.6.4) The Celebrated Plastic Explosive Composition C-4

The world-famous plastic explosive C-4 followed the C-3 in the 1960s. For the American army it is immensely popular as a demolition explosive which can be filled even in cracks. In fact all their calculations for explosive charges for demolitions are now done for C-4 only and if you are obliged to use some other explosive then you must use the RE factor to get the correct amount. This great plastic explosive consists of 91% RDX, 2.1% polyisobutylene, 1.6% motor oil, and 5.3% diethylhexylsebacate. In this composition polyisobutylene acts as the plasticizer and diethylhexylsebacate as the binder. It is an off-white solid with the feel of soft clay. It is less volatile than C-3 and has less tendency to harden at low temperatures even if you go down to -55 degrees Celsius; it does not exude even if you go up to 77 degrees Celsius! Composition C-3 has a tendency to harden at -29 degrees and is somewhat hygroscopic also. C-4 is also less sensitive than C-3; another attractive quality.

Once upon a time the West valued it highly for operations in arctic climate because it retains its effectiveness even in the extremely cold weather of the arctic. Imagining that perhaps they would have to fight the Soviet Union in those inhospitable climates also, the US army recommended to its soldiers that a fox hole could be created in the frozen ground by using about 10 lbs of C-4! Subsequently it was to become a darling of their adversaries coming from a different region altogether. Now you mention C-4 and it reminds them of terrorists. They claim that C-4 was used in the attack on the USS Cole in October 2000 and that it has also been a favorite of the Palestinian suicide bombers against Israeli targets. That's why these days they are insisting that a taggant compound dimethyldinitrobutane (DMDNB) be added to C-4 in the manufacturing stage itself to help detect the explosive and locate its source.

Fig. 26 A 1.25 Lbs Block Of C-4 (Demolition Charge M112)

Composition C-4 is manufactured by placing the water wet RDX in a stainless steel mixing kettle and adding the plastic binder, dissolved in a solvent. The mass is blended by tumbling the kettle or with an agitator until a homogeneous mixture is obtained. The solvent is removed by distillation. The resulting dough is then dried in trays by forced air at 50 degrees to 60 degrees. About 16 hours of drying are required to reduce the moisture content to below 0.25 %.

There are three classes of Composition C-4:

Class I: 61 % RDX class A or B, 30 % RDX class E, 9 % polyisobutylene

Class II: 92 % RDX class H, 9 % polyisobutylene

Class IV: 89.9 $\pm$ 1 % RDX, 10 $\pm$ 1 % polyisobutylene, 0.2 $\pm$ 0.02 % dye composition which consists of 90 % lead chromate and 10 % lamp black

(8.6.5) Semtex: The Scourge Of The West

This famous plastic explosive was made by **Stanislav Brebera**, a scientist at the Czech firm VCHZ Synthesia and named after a place called Semtin where it was first made. The Western propaganda machine has branded it forever as an explosive synonymous with terrorism simply because quantities of it were exported to North Vietnam and Libya and perhaps because it was used in the bombing of the Pan-Am flight at Lockerbie, Scotland in 1988—as if the terrorists cannot use any other plastic! Semtex, on account of its extremely low vapor pressure, is very difficult to detect. That does make it an attractive option for the terrorists. Semtex offers one more advantage over other plastic explosives like C-4—it can be used over a much greater temperature range. Now the West insists that a high vapor pressure explosive EGDN (ethylene glycol dinitrate) be added to it as a taggant. Semtex H consists of 50.2% RDX and 49.8% PETN by way of explosive material, dioctylphthalate or tributylcitrate as plasticizer and styrene-butadiene as binder, and phenylnaphthylamine as anit-oxidant. Semtex for industrial use is colored orange-red with a dye and because of the taggant is easily detectable by explosive detectors. The West keeps on accusing that controls in both the Czech and Slovak republics are lax and there could be a black market trade in Semtex or quantities of it could be getting diverted from the two countries.

(8.6.6) Composition CH6

CH6 is an explosive mixture containing 97.50 $\pm$ 0.50 % RDX, 1.50 $\pm$ 0.15 % calcium stearate, 0.50 $\pm$ 0.10 % graphite, and 0.50 $\pm$ 0.10 % polyisobutylene. The primary use of CH6 is in boosters and leads. When compared with tetryl, the material withstands higher temperature before cook off and has higher output yield, yet matches the sensitivity of tetryl. The velocity of detonation is nearly 27,000 fps.

(8.6.7) Detasheet

Detasheet is a plastic explosive, manufactured by DuPont containing PETN with nitrocellulose and a binder. However, it should be noted that some manufacturers use RDX as the explosive material. This explosive is flexible over a wide range of temperatures. It is waterproof and available in a variety of extruded shapes and in sheets and cords. In use, it is typically cut to shape for precision engineering charges. Sheet explosives are designed for use as cutting, breaching, or cratering charges, and especially for use against steel targets. The sheets of explosives may be quickly applied to irregular and curved surfaces, and are easily cut to any desired dimension.

The thin flexible sheets have a rubbery texture, and are generally colored either reddish/ orange (commercial) or green (military). Detasheet 'A', the commercial form is red and contains 85% PETN. Detasheet 'B', the military variety is colored OD green for identification and contains 63% PETN.

(8.7) POLYMER BONDED EXPLOSIVES (PBXs) HOLD THE KEY TO FUTURE

(8.7.1) What Are PBXs And What Are Their Intrinsic Advantages?

PBX, that is, **polymer bonded explosive** or **plastic bonded explosive,** is a class of explosives in which the explosive material is bound together in a matrix using small quantities (typically 5-10% by weight) of a synthetic polymer. They are characterized by high mechanical strength, good explosive properties, excellent chemical stability, relative insensitivity to handling and shock, and high thermal input sensitivity. The detonation velocity of PBX's is usually about 26,000 fps, the average auto-ignition temperature is above 250 degrees Celsius and the shock sensitivity is usually

10 to 40 percent above that required for detonating the basic explosive. PBX's contain a large percentage of basic explosives such as RDX, HMX, HNS, or PETN in intimate mixture with a polymeric binder such as polyester, polyurethane, nylon, polystyrene, various types of rubbers, nitrocellulose, or teflon. In some instances a plasticizer such as dioctylphthalate (DOP), DPA, or butyldinitrophenylamine (BDNPA) is included in the ingredients as well as a fuel such as powdered aluminum or iron.

Polymer bonded explosives (PBXs) were developed to reduce the sensitivity of the newly-synthesized explosive crystals by embedding the explosive crystals in a rubber-like polymeric matrix. The first PBX composition was developed at the **Los Alamos Scientific Laboratory** in USA in 1952. The composition consisted of RDX crystals embedded in plasticized polystyrene. Since 1952, Lawrence Livermore Laboratory, the US Navy and many other organizations have developed a series of PBX formulations.

The great advantage of PBXs is that the material can be machined into three-dimensional shapes. This makes them ideal candidates for explosives used in nuclear weapons where one has to make **'explosive lenses'** by giving the explosives very intricate and precise shapes. PBX powders can be pressed into a particular shape at room temperature, when casting normally requires hazardous melting of the explosive. PBXs retain their precise engineering shapes even under high stress. Further, high pressure pressing can achieve density for the material very close to the theoretical crystal density of the base explosive material.

The plastic coating that binds the explosive granules, typically 5 to 20% of each formulation by weight, is what gives each PBX its distinctive characteristics. Pressing a PBX molding powder converts it into a solid mass, with the polymer binder providing both mechanical rigidity and reduced sensitivity to accidental detonation. The choice of binder affects hardness, safety, and stability. Too brittle a PBX can sustain damage in normal handling and succumb to extreme temperature swings or thermal shocks; while too soft a PBX may be susceptible to creep and may lack dimensional stability or strength.

Fig. 27 The Warhead Of The Famous Tomahawk Cruise Missile Uses PBXN-7 Polymer Bonded Explosive

(8.7.2) Pressed, Extruded, Cast, And Injection-Molded PBXs

Much of the advantage stemming from the use of PBX lies in the simplicity of the technique of end item manufacture. About half of the developed PBX composites are used to directly cast end items. For this procedure no elaborate melting equipment or controlled cooling cycles are required. All operations may be conducted at ambient temperatures or, in some instances, can even be carried out in the field. However, where extraordinary mechanical strength is required or where complicated shapes are desired, a pressing or injection technique is applied using a molding powder or slurry of the PBX. Thus we can have pressed PBXs, extruded PBXs, cast PBXs and injection-molded PBXs.

Pressed PBXs:

PBX-9010:	90% RDX, 10% Kel F*
PBX-9011:	90% HMX, 10% Estane
PBX-9404-03:	94% HMX, 3% NC, 3% chloroethylphosphate
PBX-9205:	92% RDX, 6% polystyrene, 2% ethylhexylphthalate
PBX-9501:	95% HMX, 2.5% dinitropropylacrylate-furmarate, 2.5% estane
PBXN-1:	68% RDX, 20% Al, 12% nylon
PBXN-2:	95% HMX, 5% nylon
PBXN-3:	86% HMX, 14% nylon
PBXN-4:	94% DATNB, 6% nylon
PBXN-5:	95% HMX, 5% Viton A
PBXN-6:	95% RDX, 5% Viton A

Extruded PBXs:

PBXN-201:	83% RDX, 12% Viton A, 5% Teflon

Cast PBXs:

PBXN-101:	82% HMX, 18% Laminac
PBXN-102:	59% HMX, 23% Al, 18% Laminac

Injection-molded PBXs:

PBXC-303:	80% PETN, 20% Sylgard 183*

(8.7.3) Some Great PBXs And Their Compositions

A large number of PBXs have been made. The first PBX made at Los Alamos National Laboratory in 1947 was designated **PBX-9205**. It was a RDX-polystyrene formulation. Most of the later ones are based on HMX, RDX or **TATB**. (TATB is triaminotrinitrobenzene. It is more powerful than TNT but less powerful than RDX. However, it is extremely insensitive and hence preferred for use in nuclear weapons so that any possibility of an accidental detonation is eliminated even in the extreme scenario of a bomber crashing or a missile going astray during launch.) The binders used include polyurethane rubber, Estane, Viton-A, polystyrene and Kel-F.

Polymer bonded explosives are generally designated by the prefix **PBX-**, **PBXN-**, or **LX,** the acronyms indicating the place where they were originally developed.

Name	Explosive Ingredients	Binder Ingredients	Use
PBXN-4	94% DATB	6% nylon	
PBXN-5	95% HMX	5% co-polymer of vinylidene fluoride and hexalfluoroprpylene	
PBXN-6	95% RDX	5% co-polymer of vinylidene fluoride and hexalfluoroprpylene	
PBXN-201	83% RDX	12% co-polymer of vinylidene fluoride and hexalfluoroprpylene and 5% polytetrafluoroethylene (Teflon)	
PBX-0280	95% RDX	5% polyethylene	
PBX Type I	90% RDX	8.5% polystyrene and 1.5% dioctylphthalate	
PBXC-116	86% RDX	14% elastomeric binder	
PBXAF-108	82% RDX	16% plasticizer and 2% binder	
X-0242	92% HMX	8% polymer	
EDC-37	91% HMX/NC	9% polyurethane rubber	
PBXN-5	95% HMX	5% fluoroelastomer	Naval shells
PBXN-7	60%TATB and 35% RDX	5% Viton	WDU-36 warhead of Tomahawk Block III cruise missile
PBXN-9	92% HMX		Hellfire ATGM
PBXN-106	RDX	polyurethane rubber	Naval shells
LX-14-0	95.5% HMX	Estane & 5702-Fl 4.5%	
LX-10-0	95% HMX	Viton-A 5%	
LX-10-1	94.5% HMX	Viton-A 5.5%	
PBX-9501	95% HMX	Estane 2.5%; BDNPA-F 2.5%	
PBX-9404	94% HMX	NC 3%; CEF 3%	Nuclear Weapons
LX-09-1	93.3% HMX	BDNPA 4.4%; FEFO 2.3%	
LX-09-0	93% HMX	BDNPA 4.6%; FEFO 2.4%	

contd. ...

... contd.

Name	Explosive Ingredients	Binder Ingredients	Use
LX-07-2	90% HMX	Viton-A 10%	
PBX-9011	90% HMX	Estane and 5703-Fl 10%	
LX-04-1	85% HMX	Viton-A 15%	
LX-11-0	80% HMX	Viton-A 20%	
LX-15	95% HNIS	Kel-F 800 5%	
LX-16	96% PETN	FPC461 6%	
PBX-9604	96% RDX	Kel-F 800 4%	
PBX-9407	94% RDX	FPC461 6%	
PBX-9205	92% RDX	Polystyrene 6%; DOP 2%	
PBX-9007	90% RDX	Polystyrene 9.1%; DOP 0.5%; rosin 0.4%	
PBX-9010	90% RDX	Kel-F 3700 10%	
PBX-9502	95% TATB	Kel-F 800 5%	Nuclear Weapons
LX-17-0	92.5% TATB	Kel-F 800 7.5%	Nuclear weapons
PBX-9503	80% TATB, 15% HMX	Kel-F 800 5%	

(8.7.4) Manufacturing PBXs

Several manufacturing procedures are currently used to produce PBX's. One of these techniques is that of **casting**. This procedure, at first glance, merely involves combining a dried explosive, such as HMX, with binder constituents and curing initiators in a mixing vessel, blending to desired homogeneity, then casting into a given warhead or other ordnance item. Unfortunately, there are hazards associated with the drying of large quantities of explosives such as HMX or RDX. Hence, a desensitizing procedure must be added for production-scale operations. This procedure involves coating the water wet HMX or RDX with the alkyd or polyester portion of the binder. The resulting lacquer is added slowly to an aqueous slurry of the HMX or RDX. Agitation at 250 revolutions per minute in the presence of water causes the resin to precipitate onto the surface of the HMX or RDX, producing an insensitive powder which may be safely dried, handled, shipped, and stored until ready for use in the final PBX composition. The pre-coated explosive is then combined in a mixing kettle with sufficient copolymer to constitute the final binder composition. At this point an accelerator such as cobalt naphthenate is added. The mixture is then stirred until homogeneous, at first under ambient pressure and finally in a partial vacuum to remove entrapped air. A curing agent such as methyl ethyl ketone peroxide is then added. After another short mixing period, the explosive composition is poured into the desired mold. **Vacuum casting** has not been found necessary to obtain good density provided that the mold design is not too intricate or the height-to-diameters ratio is not too great.

Another method of casting is called **injection-molding**. This method is used to produce intricate shaped or small diameter castings. The PBX usually employed with this technique has a high solid content of explosive such as PETN, which provides for steady state detonation in small cross-sectional channels or tubes. This method was developed by the US Navy to provide explosive logic links of less than .25 centimeters in diameter. The technique of injection-molding involves movement of explosive material at room temperature into a cavity or tubing from a reservoir. This is accomplished by applying about 82,740 kilopascals (12,000 pounds per square inch) to the piston. This forces the explosive into the cavity or tube. When the accepting device is full, pressure on the piston is relaxed and flow stops. The device is then removed and set aside to allow the explosive to polymerize into an elastic, rubber-like solid.

Another manufacturing technique produces a molding powder which is **pressed** into the desired end item shape under pressure in excess of 206,850 kilopascals (30,000 pounds per square inch). A slurry method can be used to prepare RDX type PBX's. In this method 42.5 grams of polystyrene and eight cubic centimeters of dioctylphthalate are dissolved in 200 cubic centimeters of toluene in a lacquer dissolver. Steam is introduced into the jacket until the temperature is 65 degrees. The lacquer is agitated constantly then added to a granulator. This lacquer contains a four to one ratio of plastic-plasticizer to toluene. The granulator which is agitated at 400 revolutions per minute, contains 450 grams of RDX and 4,500 grams of water which has been heated to 75 degrees. The lacquer solution is poured into the granulator followed by a solution of gelatin in water. Mixing is continued for five minutes to insure the RDX is well dispersed. Granulation takes place at this point. Steam is introduced into the jacket again to distill the solvent until the temperature reaches 98 degrees. Cooling water is then run into the jacket to cool the batch to 40 degrees. The coated material from the granulator is collected on a Buchner funnel and dried in a tray at 70 degrees for 24 hours.

Making polymer bonded explosives is not difficult but if you want to use them specifically in nuclear weapons then they must have the right combination of adequate energy content, mechanical properties, sensitivity, and chemical stability required for stockpile nuclear weapons.

(8.7.5) Energetic Polymers

Development is continuing to produce PBXs which contain polymers that are energetic and will contribute to the explosive performance of the PBX. Energetic plasticizers have also been developed for PBXs.

Examples of energetic polymers under investigation are:

Common name	**Chemical name**
GLYN (monomer)	Glycidyl nitrate
polyGLYN	Poly(glycidyl nitrate)
NIMMO (monomer)	Nitratomethyl-methyl oxetane
polyNIMMO	Poly(nitratomethyl-methyl oxetane)
GAP	Glycidyl azide polymer
AMMO(monomer)	Azidomethyl-methyl oxetane
PolyAMMO	Poly(azidomethyl-methyl oxetane)
BAMO (monomer)	Bis-azidomethyl oxetane
PolyBAMO (monomer)	Poly(bis-azidomethyl oxetane)

(8.7.6) Energetic Plasticizers

Examples of energetic plasticizers under investigation are:

Common name	Chemical name
NENAs	Alkyl nitratoethyl nitramines
EGDN	Ethylene glycol dinitrate
MTN	Metriol trinitrate
BTTN	Butane-triol trinitrate
K10	Mixture of di- and tri-nitroethylbenzene,
BDNPA/F	Mixture of bis-dinitropropylacetal and bis-dinitropropylformal

❑❑❑

Chapter 9

Insensitive High Explosives and Some Novel Explosives

(9.1) INSENSITIVE HIGH EXPLOSIVES AND INSENSITIVE MUNITIONS: WHAT THEY ARE AND WHY WE NEED THEM?

'Insensitive high explosives' are being developed for using them in the warheads of **penetrator bombs** and **anti-ship missiles**, etc. Such munitions encounter tremendous shock on impact and the explosive should be able to withstand the shock without exploding unless triggered by its specific primer. Penetrator bombs have to penetrate earth or concrete and the anti-ship missiles have to penetrate layers of steel of a ship. They are also useful in making **'insensitive munitions' (IM)** which will not be as prone to accidental explosions by fire or impact by bullets or fragments, etc. as the regular munitions are.

An **'insensitive munition'** is one that will not detonate under any conditions other than its intended mission to destroy a target. If it is struck by fragments from an explosion or hit by a bullet, it will not detonate. It will not detonate even if it is in close proximity to a target that is hit. In extreme temperatures, the missile will only burn (no detonation or explosion). This increased safety also allows greater numbers of missiles to be packaged, handled, stored, and transported in smaller containers.

The first insensitive high explosive of the world was Picatinny Arsenal's **PAX-21**. As we have seen, half the research in explosives had been directed towards desensitizing high explosives by suitable additives or by making suitable compositions; the thrust of this new research is to develop explosives which are insensitive in their pure form itself. One great advantage of insensitive explosives is that the performance and the shock-sensitivity can be tailored for a given system based on particle size and the percentage of ingredients.

Research into insensitive munitions got a fresh impetus in the context of the Gulf War. Making penetrating bombs was, of course a necessity as the Americans were consumed with the fear of Saddam Hussain hiding in impenetrable bunkers reminiscent of the Fuhrer's bunker in the

Second World War. Then they also found that most of the disabling damage to fighting vehicles was in fact caused by their own munition payloads, inadvertently triggered by unwanted stimuli.

Insensitive explosives pose a problem of their own. This pertains to reliable detonation. Existing booster explosives and fuses have insufficient energy output to reliably initiate the new insensitive main charge explosives. Increasing the amount of booster explosive will increase the weapon's sensitivity and the chance of an accidental detonation. What is needed is an inexpensive method of reliably initiating the new, more insensitive main charge explosive while at the same time reducing the chance of the accidental initiation of a fuse booster system.

(9.2) THE PAX FAMILY (PICATINNY ARSENAL EXPLOSIVE FORMULATIONS)

Since the mid-1980s the Picatinny Arsenal has developed about 24 formulations. They are melt-pour explosives and intended to gradually phase out TNT, Composition B, Tritonal, Octol and Cyclotol. Most of them are based on RDX or HMX, DNAN (dinitroanisole), AP (ammonium perchlorate), Al (aluminum) and MNA (methyl nitroaniline).

Interest in melt-pour explosives is maintained because most of the unitary warheads launched by artillery have to have them for cost and surge capability. The great advantage of these compositions is that performance and shock sensitivity of the explosive can be tailored for a given system based on particle size and the percentage of ingredients. For example, PAX-40 and -41 exceed Composition B's detonation velocity. PAX-40 and -41 are also less shock sensitive than Composition B.

The compositions of some of the Picatinny Arsenal Explosive Formulations are as follows:

- PAX-21 (Composition B replacement): RDX, DNAN, AP and trace amounts of MNA (for processability) currently in production
- PAX-24 (TNT replacement): DNAN, AP and MNA
- PAX-25 (Comp B replacement): RDX, DNAN, AP and MNA (different proportions for RDX, DNAN, and AP) better performance than PAX-21
- PAX-26 (Tritonal replacement): DNAN, Al, AP, MNA
- PAX-28 (Unitary warheads): RDX, DNAN, Al, AP, MNA
- PAX-40 (Octol replacement): HMX, DNAN, MNA
- PAX-41 (Cyclotol replacement): RDX, DNAN, MNA

PAX-21 was the US Army's first high performing IM (insensitive munition) explosive. It is less sensitive to initiation by outside stimuli, but retains all the requisite performance capabilities of the high explosive that was used in the past, particularly with regard to fragmentation. It has matured through extensive loading, performance testing and hazard threat analysis testing in various current and future warhead configurations of the Army, Navy, and Air Force munitions systems. PAX-21 is now considered to be a viable alternative to current HMX formulations and has been successfully demonstrated in Hellfire, Javelin, M830A1, HE-WAM, SADARM, and XM915 Dual Purpose Improved Conventional Munitions (DPICM) XM80 grenade sub-munitions.

(9.3) INSENSITIVE PBXs

(9.3.1) PBXW-122, PBXW-124 And PBXN-110

Two examples of these insensitive main charge explosives are PBXW-122 and PBXW-124. The composition of PBXW-122 by weight is 47% nitrotriazolone (NTO), 5%

cyclotrimethylenetrinitramine (RDX), 20% ammonium perchlorate (AP), 15% aluminum, and 13% binder. The composition of PBXW-124 by weight is 27% NTO, 20% RDX, 20% aluminum, 20% ammonium perchlorate, and 13% binder. The NTO used in PBXW-122 and PBXW-124 is an interesting thing. It is under consideration for use as a gas producer in automobile air bags.

One of the important criteria for insensitive high explosives is that they must make main charge explosives that have critical diameters greater than 1 inch. Future underwater and bombfill explosives will have critical diameters greater than one inch. The critical diameter for an explosive is the minimum diameter mass of that explosive that can be detonated without being heavily confined. PBXW-122 has a critical diameter of 7 inches. It means that it cannot be detonated in less than a 7 inch diameter mass unless heavily confined. PBXW-122 has a sensitivity of 130K bars (ELSGT). PBXW-124 has a critical diameter of between 3 and 4 inches.

The composition of PBXN-110 by weight is 86% cyclotetramethylenetetranitramine (HMX) and 14% Binder. A 3 inch high by 5 inch diameter cylindrical charge of PBXN-110 is required to initiate 65 pounds of PBXW-122 main charge explosive. That gives some idea of the insensitivity of the explosive. Whereas other HE can be initiated by a few grams of explosive in the detonator, this requires a charge that big!

(9.3.2) PBXIH-135

The US Naval Surface Warfare Center has developed the PBXIH-135. As the name suggests, it is a polymer bonded explosive. It is actually a **thermobaric explosive** or a fuel-air explosive in solid form. In a thermobaric weapon, the fuel consists of a monopropellant and energetic particles. The monopropellant detonates in a manner similar to TNT while the particles burn rapidly in the surrounding air later in time, resulting an intense fireball and high blast overpressure. The term 'thermobaric' is derived from the effects of temperature (the Greek word 'therme' means "heat") and pressure (the Greek word 'baros' means 'pressure') on the target. The idea is to use it against tunnels and other buried structures where simple explosives are not so effective. This offers sustained overpressure by prolonged reaction of explosive. The secret to PBXIH-135 is the addition

Fig. 28 The BLU-118/B Designed To Destroy Tunnels Uses PBXIH-135 Thermobaric Explosive

of a precise mixture of aluminum powder, which burns in the hot gases. Long after the initial shock wave, the burning aluminum sends heat and pressure bounding through corridors. The Americans became obsessed with destroying tunnels during the attack on Afghanistan and the Gulf war. A laser-guided bomb was developed with PBXIH-135 in the warhead and tested in the full-scale tunnel facilities of the Department of Energy's Nevada test site.

(9.3.3) PBXN-109

PBXN-109 is used in penetrator bombs like the BLU-109, Penguin anti-ship missile and Tomahawk cruise missile. It is 64% RDX, 20% aluminum and 16% binder plus stabilizers consisting of HTPB (hydroxy-terminated polybutadiene rubber R-45HT) 7.35%, IPDI (Isophorone diisocyanate) .95%, dioctyl adipate 7.35%, Dantocol DHE (a hydantoin glycol, dihydroxyethyl dimethylhydantoin), 0.26%, anti-oxidant 0.1%, and Triphenyl Bismuth 0.02%. The shock sensitivity of PBXN-109 can be further reduced by the use of Reduced Sensitivity RDX instead of conventional RDX.

(9.4) OTHER GREAT INSENSITIVE HIGH EXPLOSIVES

(9.4.1) AFX-757

The AFX-757, a very insensitive explosive, is developed by the US Air Force Research Laboratory. It is being used in the warhead of the JASSM (Joint Air-to-Surface Standoff Missile—a weapon under joint development by the US Air Force and Navy). The warhead contains 240 lbs of AFX-757. This new formulation boasts of higher blast characteristics, is less sensitive to many physical effects which can trigger unwanted explosions, uses less expensive ingredients, and is easier to process than current explosive fills like Tritonal and PBXN-109.

Incidentally, the fuze for this contains 150 grams of PBXN-9 as booster. It is a plastic bonded explosive intended to be used for other hard target penetrators also. It consists of RDX (19-21%), Ammonium Perchlorate (29-32%), Aluminum (32-35%), Poly BD 4.44%, Dioctyladipate 6.56%, Isophorone diisocyanate 0.45%, Lecithin 0.30-0.50%, Triphenyl Bismuth 0.01-0.50%. The specific characteristics of AFX-757 were achieved by having the proper fuel (aluminum) and oxidizer (ammonium perchlorate) ratios allowing for more complete combustion in the product reaction zone. Additionally, the formulation was tailored to reduce the detonation velocity and increase the reaction temperature. All these techniques are combined to ensure that more of the available chemical energy is released in the blast.

(9.4.2) DNI (Dinitroimidazoles)

Dinitroimidazoles, such as DNI (2,4-dinitroimidazole), are a group of very insensitive explosives.

(9.4.3) MNX-194

MNX-194 is a melt-castable, wax-binder explosive intended to replace TNT in US Army's 155 mm artillery rounds. The Munitions Directorate researchers, funded by the US Army Tank-Automotive and Armaments Command/Armament Research, Development and Engineering Center, have developed the MNX-194. MNX-194 is believed to be RDX plus wax and three varieties have been developed. The primary difference between the three varieties is the source and/or pre-treatment of the RDX component. Additional characterization and optimization work is currently under way. MNX-194 has a melting temperature similar to that of TNT.

The US Air Force is also considering this formulation as a fill for their MK series bombs. The explosive for the air force is believed to be an aluminized version of the MNX-194 binder matrix.

(9.4.4) Reduced Sensitivity RDX (RS-RDX)

Recently it has been shown that if RDX is manufactured in certain ways, it acquires such properties that it leads to reduced shock sensitivity in cast-cured polymer bonded explosives (PBX) formulations in comparison to equivalent formulations based on conventional RDX. The phenomenon was first discovered by a French company called SNPE. The process they used was a **re-crystallization technique** in a process known as **Woolwich synthesis**—conventional RDX is produced by a process called Bachmann synthesis. The process also seems to work for HMX. The exact phenomenon has not been clearly understood yet. The use of RS-RDX in AP shells could result in lowered standoff distances, for example. Dyno Nobel ASA have also claimed that they have produced RS-RDX by treating their Type II RDX.

(9.4.5) HNS

HNS is hexanitrostilbene and it was synthesized in 1966 by Shipp. It is also one of the explosives that can withstand relatively high temperatures.

(9.4.6) TATB: Presently The Mainstay Of Explosive Lenses Of Nuclear Weapons

TATB is triamino trinitrobenzene. TATB was first prepared in 1888 by **Jackson** and **Wing**, who also determined its solubility characteristics. In the 1950s, the USA Naval Ordnance Laboratories recognized TATB as a useful heat-resistant explosive, and successful small-scale preparations and synthetic routes for large-scale production were achieved to give high yields. **Adkins** and **Norris** developed a viable method for its production in 1978. The preferred method of preparation in the laboratory consists of the nitration of strichlorobenzene with a mixture of nitric acid and oleum at a temperature of 150 degrees for three hours. A toluene solution of the trinitrotrichlorobenzene (TCTNB) produced by the nitration is then aminated to yield TATB.

The present thrust of research is on to finding relatively cheaper ways of making TATB. The latest method of manufacturing TATB uses a process known as **VNS (Vicarious Nucleophile Substitution)** on picramide and TMHI (trimethyl hydrazinium iodide). The process can make use of two chemicals, ammonium picrate (an explosive) and UDMH (uns-dimethyl hydrazine—a rocket propellant), large stocks of which survive as a legacy of the Cold War era. Ammonium picrate is used to produce picramide and UDMH is used to produce TMHI.

TATB lends itself for the manufacture of plastic explosives and in explosive mixtures with TNT that can be cast. It is more powerful than TNT but less powerful than RDX. For an explosive like this, power is not the main criterion; insensitivity is. The major military use is in special applications in warheads of high-speed guided missiles and in the explosive lenses of nuclear weapons. The idea is that the charge in the warhead must remain unaffected by the tremendous impact of the high speed missile and should explode only by its detonator. When used in a nuclear warhead its insensitivity ensures that any possibility of an accidental detonation is eliminated even in the extreme scenario of a bomber crashing or a missile going astray during launch. Its resistance to heat and shock is greater than any other explosive of comparable energy. The possibility of using TATB as an insensitive booster material and as a main charge explosive for new weapons designed to penetrate hardened bunkers is also being explored. However, its most important use is as an explosive for nuclear weapons, so much so that the US Department of Energy currently maintains an estimated five-year supply of TATB for its Stockpile Stewardship and

Management Program, which is designed to ensure the safety, security, and reliability of the U.S. nuclear stockpile.

In addition to its military uses, TATB has been proposed for use as a reagent in the manufacturing of components for liquid crystal computer displays. There is also interest in employing the explosive in the civilian sector for deep oil well explorations where heat-insensitive explosives are required.

TATB is normally used as the explosive ingredient in Plastic bonded explosive compositions, such as PBX-9502, LX-17-0, and PBX-9503 (with 15% HMX). For use in the PBXs you require particle size greater than 20 micron in 96% of the particles and for that a different method is used for manufacture on industrial scale. The procedure is carried out in two steps; a nitration step and an amination.

Fig. 29 The Famous W80 Thermonuclear Warhead, Derived From The Highly Successful B61 Bomb And Used On Cruise Missiles: The Primary Of The Device Uses TATB Based PBX-9502: Nearly 5272 Of The W80 And B61 Have Been Produced

(9.4.7) DATB

DATB or diamino trinitrobenzene can be prepared by nitrating dichlorobenzene to trinitro-dichlorobenzene and aminating in methanol solutions. It is an insensitive explosive.

(9.4.8) HNAB

HNAB is hexanitroazobenzene. One method of obtaining HNAB is to treat picryl chloride, with hydrazine, to obtain hexanitrohydrazobenzene. The hexanitrohydrazobenzene is then oxidized by nitric acid or nitrogen oxide gas in a glacial acetic acid solution to yield HNAB. HNAB is a very powerful and brisant high explosive. At a density of 1.77 g/cc, its detonation velocity is 23800 fps. Its power, by the Trauzl test, is 123 percent of TNT.

(9.4.9) TNAZ

TNAZ is trinitroazetidine, a strained ring heterocyclic nitramine. It is 30-40% more powerful than TNT and roughly equal to HMX in power but less sensitive than both. It is melt-castable and could replace TNT one day. TNAZ is one of the few new energetic materials found to be thermally stable above its melting point. Once again the high cost of manufacture is a major problem at the moment. Researchers have evaluated five synthetic routes for producing TNAZ. The two most likely methods to manufacture TNAZ in a sustainable green manufacturing process are those due to **Axenrod**, **Coburn** and **Hiskey**. However, in some formulations studies, it was found that TNAZ has high volatility that could severely inhibit its utility in military explosive and propellant applications. Further limitations to its use could include the processing, polymorph, and material costs. In any case, more studies have to be carried out for perfecting melt-casting and sorting out sensitivity issues of various compositions.

(9.5) LLM-105 AND LLM-121: TWO EXTREMELY NOVEL EXPLOSIVES

Let us first understand the motive for undertaking the research into making some extremely novel explosives. Since the invention of black powder the technology for making solid energetic materials has remained either the physical mixing of solid oxidizers and fuels, referred to as composite energetic materials (e.g., black powder); or the incorporation of oxidizing and fuel moieties into one molecule, referred to as monomolecular energetic materials (say, for example, TNT). In composite systems, desired energy properties can be attained through readily varied ratios of oxidizer and fuels. A complete balance between the oxidizer and fuel may be reached to maximize energy density. Current composite energetic materials can store energy as densely as >23 kJ/cc. However, due to the granular nature of composite energetic materials, reaction kinetics is typically controlled by the mass transport rates between reactants. Hence, although composites may have extreme energy densities, the release rate of that energy is below that which may be attained in a chemical kinetics controlled process.

In monomolecular energetic materials the rate of energy release is, however, primarily controlled by chemical kinetics, and not by mass transport. Therefore, monomolecular materials can have much greater power than composite energetic materials. A major limitation with these monomolecular energetic materials, however, is the total energy density achievable. Currently, the highest energy density for monomolecular materials is approximately 12 kJ/cc, about half that is achievable in composite systems. The reason for this is that the requirement for a chemically stable material and the current state of the art synthetic procedures limit both the oxidizer-fuel balance and the physical density of the material.

If we want to make novel explosives with energy densities as high as that of composite explosives, we will have to make materials that have a very high degree of uniformity, that is, with very few or no defects, perfect crystalline structure, composites with molecularly engineered uniformity, laminated composites with structures built of molecularly controlled and selectable layers, etc. This is intended to be accomplished by manipulating individual atoms and molecules (i.e., nano-scale work) and control their assembly into a large-scale bulk energetic material.

The current emphasis in the research on **nano-scale energetic materials** is on the preparation and characterization of single nano-scale energetic particles. These particles are then utilized in an otherwise conventional composite formulation, incorporating the nano-particles (typically aluminum, the fuel) in a matrix with micron-sized oxidizer particles. What are the technical issues involved in accomplishing this? We need macro-scale formulations of energetic materials that preserve the intrinsic nano-scale structure of the individual components so as to realize the true potential of nano-scale energetic materials. An optimal material may be a three-dimensional macro-scale, ordered array of nano-scale constituents, with spacing and interstitial/bonding materials chosen to optimize both stability and reactivity. At a minimum, these macro-scale units would be of the order of millimeter size, which could then be processed into centimeter to meter sizes needed for practical propellants and explosives. A new bulk process for manufacturing nano-structured energetic materials using sol-gel chemistry has also been developed with the promise of precise control of material homogeneity, properties, and geometry.

The complexity does not end there. The chemistry, physics and materials science of nano-scale energetic material preparation need to be developed, focusing on those processes that lead to well-ordered structures, e.g. self-assembly, vapor deposition, etc. Computational methods are needed to assess the reactivity of candidate structures and to predict the stability of the energetic material structure, to both hazards (shock, spark, etc.) and to long-term degradation. Then experimental methods of characterizing nano-energetic structures are needed to verify structure and performance.

Eventually we may be able to develop a single smart material that can function as a structural material, embedded sensor, and have real-time selectable propellant, explosive, or non-lethal functionality within a precision munition!

One new explosive under development is **LLM-105**. LLM-105 was synthesized at the **Lawrence Livermore National Laboratory**. Chemically it is diamino-dinitropyrazine-oxide. It is dense and has 81% the energy of HMX and excellent thermal stability. It is made by the oxidation of diamino-dinitropyrazine with trifluoroaceticperacid. It has performance and insensitivity between those of HMX and TATB. Its calculated energy content is about 85% that of HMX and 15% more than that of TATB. It is insensitive to shock, spark and friction and has impact insensitivity level approaching that of TATB. These combined properties make it a realistic high-performance IHE material, attractive for applications that require moderate performance and insensitivity. With 30% more energy than TNT it has possible detonator and booster applications and is an alternative to TATB in special purpose weapons such as hard target penetrators that have to survive high shock loading. The synthesis, scale-up, and characterization of this material have been completed and its use as insensitive booster material for the US Navy weapons applications is now being evaluated.

A still better explosive named **LLM-121** is also round the corner. Its predicted energy density is greater than that of CL-20. To study it further it is necessary to make pure crystals of it. Efforts to crystallize the pure form are continuing. Characterization work is pending the crystallization of the material in pure form.

Extended **solid high energy density materials** are also under development. This work uses intense pressure and temperature to force elements into highly energetic bonding states that can be recovered to ambient conditions. Current synthesis techniques have produced CO-derived solids and a family of novel nitrogen materials, but in very small quantities. These materials are expected to be highly energetic, but characterizing them, and particularly verifying the energy content, has been difficult due to the microscopic quantities of material available.

(9.6) POLYNITROCUBANES: SUPER EXPLOSIVES OF THE FUTURE?

Polynitrocubanes are a very recent discovery. They are also very insensitive explosives. **Octanitrocubane** was synthesized in 1999 at the University of Chicago by Philip Eaton et al and is yet to be studied extensively. Cubane is a synthetic hydrocarbon in which eight carbon atoms are there at the corners of a cube with one hydrogen atom each attached to all of them. In Octanitrocubane the hydrogen atoms are replaced by nitro groups. It is believed to have a R. E. factor of 2.7 against 1.7 of HMX. The increased power of the explosive is believed to result from its greater molecular weight and the highly expansive breakdown into carbon dioxide and nitrogen molecules as well as to the presence of strained chemical bonds in the molecule. In addition, Octanitrocubane requires no external oxygen source to decompose and produces no water vapor which makes it less visible in use. It is supposed to have a density higher than any other explosive though in practice the theoretical densities are yet to be realized.

Heptanitrocubane has been studied a bit more than Octanitrocubane and since presently a very little quantity of the latter is available, is found to be an extremely powerful explosive, approximately equal to what we would expect from Octanitrocubane.

□□□

Chapter 10

Ammonium Nitrate Based Explosives Capture the World of Non-military Explosives

(10.1) THE RECOGNITION OF AMMONIUM NITRATE AS AN ALLY OF EXPLOSIVES: THE RESULT OF AN ACCIDENT AT TEXAS CITY

Ammonium nitrate was first prepared in 1654 by **Glauber** but it was not until the beginning of the 19th century when it was considered for use in explosives by Grindel and Robin as a replacement for potassium nitrate in blackpowder. Its explosive properties were also reported in 1849 by Reise and Millon when a mixture of powdered ammonium nitrate and charcoal exploded on heating. Ammonium nitrate was not considered to be an explosive although small explosions involving ammonium nitrate had occurred throughout the world. After the end of World War II, the USA Government began shipments to Europe of so-called **Fertilizer Grade Ammonium Nitrate (FGAN)**, which consisted of grained ammonium nitrate coated with about 0.75% wax and conditioned with about 3.5% clay. Since this material was not considered to be an explosive, no special precautions were taken during its handling and shipment; workmen even smoked during the loading of the material.

Numerous shipments were made without trouble prior to 16 and 17 April 1947, when a terrible explosion occurred. The **SS Grandchamp** and then the **SS Highflyer**, both moored in the harbor of Texas City and loaded with FGAN, blew up. In all 576 people were killed and a considerable part of the town was destroyed. As a consequence of this disaster, a series of investigations was started in the USA in an attempt to determine the possible causes of the explosions. At the same time a more thorough study of the explosive properties of ammonium nitrate and its mixtures with organic and inorganic materials was also conducted. The explosion at Texas City had barely taken place when a similar one aboard the **SS Ocean Liberty** shook the harbor of Brest in France on 28 July 1947.

As for the accident aboard the SS Grandchamp is concerned, it is believed that a fire broke out in the cargo hold of a ship full of ammonium nitrate. The ship was equipped with a newfangled steam fire fighting system which works pretty well on the average fire. They sealed up the hatches and started pumping in steam to smother the flames. The sealed, steam pressurized cargo hold made a very effective bomb casing and led to the explosion. Ammonium nitrate by itself is just an oxidizer and would not normally detonate. It can be detonated only under high pressure at high temperature. Looks like the conditions required for detonation got somehow created in the ill-fated ships.

In 1867, two Swedish chemists, **C. J. Ohlsson** and **J. H Norrbin**, patented an explosive, called **ammoniakrut**, which consisted of ammonium nitrate either alone or in mixture with charcoal, sawdust, naphthalene, picric acid, nitroglycerin, or nitrobenzene. Theoretical calculations had shown that large quantities of heat and gas were given off by the explosions of these mixtures. The proportions of the materials were selected in such a manner that all of the carbon should be converted to carbon dioxide and all the hydrogen to water. Some of these explosives were difficult to ignite and to initiate, but the trouble was remedied by including some nitroglycerin in their compositions and by firing them with fulminate detonators. They were used to some extent in Sweden. Nobel purchased the invention from his fellow-countrymen early in the 1870's, and soon afterwards took out another patent in connection with it, but still found that the hygroscopicity of the ammonium nitrate created several problems. He was not able to deal satisfactorily with the problems until after the invention of gelatin dynamite.

(10.1.1) How Dynamites Ran Into Trouble In The Mining Industry

The growing use of dynamites in coal mining brought a corresponding increase in the number of gas and dust explosions, with appalling casualties. Worried, some European governments went to the extent of considering prohibiting the use of explosives in coal mines and resorting to the use of hydraulic devices or compressed air! However, before resorting to such drastic measures, some governments decided to appoint scientists, or commissions headed by them, to investigate this problem. Between 1877 and 1880, commissions were created in France, Great Britain, Belgium and Germany. As a result of the work of the French Commission, maximum temperatures were set for explosions in rock blasting and gaseous coal mines. In Germany and England, however, it was recognized that regulating the temperature of the explosion was only one of the factors in making an explosive safe and that other factors should also be considered. Consequently, a testing gallery was constructed in 1880 at Gelsenkirchen in Germany in order to test the newly-developed explosives. The **testing gallery** was intended to imitate as closely as possible the conditions in the mines. A Committee was appointed in England in 1888 and a trial testing gallery at Hebburn Colliery was completed around 1890. After experimenting with various explosives, the use of several explosive materials was recommended, mostly based on ammonium nitrate. Explosives which passed the tests were called **permitted explosives**. Dynamite and blackpowder both failed the tests and were replaced by explosives based on ammonium nitrate. The results obtained by this Committee led to the Coal Mines Regulation Act of 1906. Following this Act, testing galleries were constructed at Woolwich Arsenal and Rotherham also in England.

(10.1.2) Ammonium Nitrate Makes Deep Inroads Into The Mining Industry

By 1913, British coal production reached an all-time peak of 287 million tons, consuming more than 5000 tons of explosives annually and by 1917, 92% of these explosives were based on ammonium nitrate! Ammonium nitrate virtually wrote the epitaph of dynamite as far as the mining industry was concerned.

In 1935, experimentation had begun on blasting with a mixture of prilled fertilizer grade ammonium nitrate (FGAN) and carbon black or coal dust at a surface coal mine in Indiana. The mixture was packaged in large tubes and was primed with 20 pound charges of dynamite. It is popularly believed that this product was developed because of publicity about the 1947 Texas City disaster in which two ships laden with ammonium nitrate fertilizer had blown up. In reality, that explosion had occurred with grained AN, not prilled. Still, the success of the Indiana experiments was that prills became commonly available (all fertilizer production in the US having converted to prilling by 1948), and that dry drilling of large diameter blastholes in the surface mines was becoming the norm. The person responsible for the product was **Bob Akre** of the Maumee Collieries, and the product was called **Akremite**.

In order to reduce the cost of explosive compositions the explosives industry added more of the cheaper compound ammonium nitrate to the formulations, but this had an unfortunate side effect of reducing the explosive's resistance to water. This was a significant problem because mines and quarries were often wet and the holes drilled to take the explosives got regularly filled with water. Chemists tried to overcome this problem by coating the ammonium nitrate with various inorganic powders before mixing it with dynamite, and by improving the packaging of the explosives to prevent water seepage. Accidental explosions still occurred involving mining explosives, and in 1950 manufacturers started to develop explosives which were waterproof and solely contained the less hazardous ammonium nitrate. Shortly after May 1955 many mines began using prills mixed with common No.2 diesel fuel, and the name **ANFO (Ammonium Nitrate Fuel Oil)** was soon applied to them.

Fig. 30 Machine To Prepare ANFO On The Site And Load It Into Bore Holes

Although ANFO became immensely popular, it has not prevented further research into explosives based on ammonium nitrate. In the 1970s, the US companies Ireco and DuPont began adding paint-grade aluminium and monomethylamine nitrate (MAN) to their formulations to produce **gelled explosives** which could detonate more easily. More recent developments concern the production of **emulsion explosives** which contain droplets of a solution of ammonium nitrate in oil. These emulsions are waterproof because the continuous phase is a layer of oil, and they can still readily detonate since the ammonium nitrate and oil are in close contact. Emulsion explosives are safer than dynamite, and are simple and cheap to manufacture.

(10.1.3) What Is ANFO And How To Make The Best Of It?

In ANFO the nitrate works as oxidizer and the oil as the fuel. The fuel oil used is diesel or kerosene. ANFO has turned out to be three times cheaper than dynamites and is easy to use; hence it has virtually replaced gelatins in coal mining, quarrying and metal mining. In North America they use about 6 billion tons of explosive in commercial activities every year and nearly 80% of that is ANFO! What more needs to be said about its popularity? It is safe to use, inexpensive and very easy to load. The miners love it because in the free-flowing form, it has a great advantage over cartridge explosives as it completely fills the borehole. This direct coupling to the walls assures efficient use of explosive energy. When properly oxygen- balanced, the fume qualities of dry blasting agents permit their use underground also.

ANFO is classified as a **blasting agent**. It means that for proper detonation it would require a booster charge of conventional high explosive and cannot be reliably detonated with a blasting cap or detonator alone. Detonation velocity for charge diameter of 6 inches or more is about 12,000 fps but in a diameter of 1.5 inch, the velocity is reduced to 60 percent. Pneumatic loading results in high detonation velocities and higher charge concentrations, particularly in holes smaller than 3 in. (otherwise such small holes are not usually recommended for ANFO blasting). However, when loading pneumatically, precautions should be taken to avoid a possible building up of static electrical charge. A number of materials have a strong catalytic effect on the thermal decomposition of AN. These include acids, chlorides, organic materials, chromates, dichromates, salts of manganese, copper and nickel and certain metals such as zinc, copper and lead. This should be kept in mind.

Best results are obtained with 94.5 percent ammonium nitrate and 5.5 percent fuel oil. In the field if they have poured ammonium nitrate directly into the bore hole, it is common to use one gallon of oil for 100 lbs of nitrate and let it soak thoroughly. That makes for about 6% oil. If the oil is mixed outside then they use a little more oil—about 8 to 10%. Drain it for about half an hour or so and you have the right proportions. If there is any doubt, it is better to err on the higher side. A little excess of oil will result in post-blast fumes but if it happens to be less than the critical value the explosion will be poorer. Diesel used for motor vehicles can be used for ANFO. Kerosene has also been used. Most miners make ANFO on the spot. The catch in using ANFO is to ensure proper mixing otherwise full explosive force is not developed. Then AN is highly hygroscopic; that is, it readily absorbs water from air. Care must be taken with its storage in humid environments, as any absorbed water interferes with its explosive function. Since ammonium nitrate is water soluble, its use in wet holes requires special precautions. Many miners first pump out the water, insert a plastic sleeve into the hole and then charge it.

The ammonium nitrate used in ANFO to make a good and proper blasting agent is somewhat different from that used in fertilizers. Here for best results you require prills that have a bulk density of approximately 0.84. A prill is a dry pellet form of a solid. Prills are formed by allowing drops of the prill substance to congeal in mid-air by being dripped from the top of a tall Prilling Tower. The density of individual prills is about 1.3, while the density of pure crystalline ammonium nitrate is 1.7. It is notable that AN prills used for explosive applications are physically different from fertilizer prills; the former contain approximately 20% air. **Fertilizer grade prills** have a bulk density of about 1.0. These voids in **blasting grade prills** are necessary to sensitize ANFO; they create so-called "**hot spots**" in which the interaction of the detonation front with a spherical void concentrates energy. Blasting-grade AN prills are typically between 0.9 and 3.0 mm in diameter.

Regular FGAN can also make passable ANFO but best results would be obtained if you use blasting grade prills.

A blasting agent based on ANFO may contain additional non-explosive substances such as powdered aluminum or ferrosilicon to increase its density. For proper detonation it requires a primer and also booster explosives spaced along the borehole to maintain propagation. However, when you add a little TNT to the blasting agents made as above, they become cap sensitive and are classified as explosives. Then you don't require boosters.

Fig. 31 Another Type Of ANFO Loader

(10.1.4) Emulsion Explosives

The so-called emulsion explosives are also based on the ANFO chemistry only. However, they differ from ANFO in the physical form the reactants take. An emulsion is made up of a mixture of two liquids that do not normally mix with one another as one liquid is oil-based and the other liquid is water-based. When the appropriate emulsifier is present and enough mechanical energy is exerted, the two phases can be forced to blend together. Emulsion explosives are, therefore, prepared in the form of **water-in-oil emulsions**. The internal phase is composed of an almost saturated solution of oxidizer salt suspended as microscopically fine droplets, which are surrounded by a continuous fuel phase. The fuel is in the form of lightly dispersed droplets. An emulsifying agent stabilizes the emulsion thus formed against separation of the liquids. Most of the commercial emulsion explosives thus made consist of small droplets of saturated ammonium nitrate solution surrounded by oil.

In terms of chemistry, in the emulsion we have the aqueous phase which is known as the disperse phase, the oil phase which is known as the continuous phase, and the surfactants. The surfactants could be anionic, cationic, nonionic, or amphoteric. You first form the emulsion by mixing the oil-surfactant phases in a mixer and then add the solids in blending.

The most notable properties of emulsions are water resistance for several days and higher bulk density. Since each micro-cell of the oxidizer is coated with an oily exterior, the emulsion has excellent water resistance and does not rely on a package for their ability to function efficiently in water. What makes a water-in-oil emulsion particularly efficient is that the major component by weight is mixed into the minor component. This not only results in a finely dispersed, homogenous mixture, but one in which the oil phases constitutes the outside or continuous phase. It is this feature that imparts excellent water resistance to emulsion type explosives.

The small uniform emulsion droplets provide high detonation velocity and a rapid release of energy. The base emulsion matrix is not classified as an explosive; therefore it provides safety advantages and also lower costs for shipping and storage. The composition of a typical emulsion matrix is ammonium nitrate (77.66%), sodium nitrate (4.68%), water (11.22%), wax and emulsifier (5.40%) by weight.

The detonation behavior can be widely controlled by the selection of the size and volume of voids contained in the emulsion explosives. These voids act as hot spots. Sensitized emulsion products are sensitized with glass or plastic micro-spheres (made of acrylonitrile, vinylidene chloride or polystyrene) which are uniform and have diameters measured in microns. They provide the voids. Alternatively, chemical gassing agents can also be used which generate gas bubbles chemically and which have the same effect of sensitization.

For manufacturing emulsion explosives, the ammonium nitrate solution is mixed with the oil phase mixture. The oil phase mixture is a blend of mineral oil and liquid emulsifier. The mineral oil is blended with the appropriate emulsifier. The second process step is the formation of un-sensitized base matrix emulsion. This is still a non-detonable product. The base emulsion contains all of the ingredients for the explosive but still requires a bulking agent to lower density and increase sensitivity in order to form explosive characteristics. The matrix is pumped to a paddle blender for mixing with a metered amount of a non-explosive dry ingredient sensitizing agents in order to modify or enhance the product's characteristics or performance. Glass micro-spheres are added as a physical sensitizer to lower product density and increase the explosives sensitivity and performance. As an option, another additive that may also be added may be aluminum or even a demilitarized propellant for increasing the total energy of the formula and to produce a high performance, detonator sensitive product.

(10.2) HOW THE BOMB USED IN THE OKLAHOMA CITY BOMBING WAS MADE?

A variant of the ANFO was used in the Oklahoma City bombing by **Timothy McVeigh** and **Terry Nichols** in April 1995. They used **nitromethane** instead of diesel. In effect he produced an improvised version of the commercial explosive **Kinepak**.

Nitromethane is used as a fuel in racing, particularly **drag racing**. The oxygen content of nitromethane enables it to burn with much less atmospheric oxygen in comparison to hydrocarbons such as gasoline. 14.6 kg of air are required to burn one kg of gasoline, but only 1.7 kg of air for one kg of nitromethane. Since an engine's cylinder can only contain a limited amount of air on each stroke, 8.7 times more nitromethane than gasoline can be burned in one stroke. However, since its energy density is low, nitromethane does not produce 8.7 times the energy of gasoline. Still it produces about 2.3 times more power than gasoline and hence its use in drag racing. A typical supercharged drag racing engine of 8.9 liters that develops 6000 hp can burn up to a gallon of nitromethane per second!

McVeigh's bomb consisted of over two tons of explosive mixture; all kept in a 20-foot Ryder truck. He used 108 bags of 50 lbs each of ammonium nitrate and 6 drums of 55 gallons each of nitromethane. Some Tovex was also used. Detonation was by Primadet shock tube detonator used in mining and at least two regular detonators. The regular detonators were fired by cannon fuse and they initiated the Primadet. The ammonium nitrate was not explosive grade but fertilizer grade and he had purchased it in two installments of one ton each from a farm co-op. The nitromethane which he had purchased disguised as a biker had cost him just $2775!

The ANFO mixture was placed in barrels and 11 barrels were arranged in the form of a 'V' to give it some sort of improvised shaped charge effect. Boards were nailed into the floor to hold the barrels in place. McVeigh put some Tovex in each barrel and also at the inside point of the 'V' shape of the barrels as booster charge. One Primadet was placed in each barrel and one in the inside point of the 'V'.

The bomb was exploded in front of the north side of the nine-story Alfred P. Murrah Federal Building. The effects of the blast were equivalent to 4,000 pounds of TNT and could be heard and felt up to fifty-five miles away. The blast destroyed a third of the building and created a thirty-foot wide and eight-foot deep crater on NW 5th Street next to the building. The bombing killed 169 and injured 853 people. Of the killed 163 were inside the building. The blast destroyed or damaged 324 buildings in a sixteen-block radius, destroyed or burned 86 cars around the site, and shattered glass in 258 nearby buildings. The broken glass alone had accounted for 5% of the total deaths and 69% of the injuries outside the Murrah Federal building. The destruction of the building left several hundred people homeless and shut down multiple offices in downtown Oklahoma City.

Fig. 32 Damage Caused To The Murrah Federal Building In The Oklahoma City Bombing

McVeigh was incidentally arrested within 90 minutes of the bombing, though not for bombing. He was arrested for driving his car without a license plate and possessing a concealed weapon! He could have been released but fortuitously they were able to link him with the explosion through the Vehicle Identification Number of an axle from the destroyed Ryder truck! Criminals do commit such elementary mistakes.

(10.3) A GREAT INVENTION CALLED SLURRIES OR WATER GELS

Water is popularly considered a great destroyer of explosives. To a layman give some explosive and ask him to neutralize it. Most people would dunk it in water! It is a collective hangover of the gunpowder era when dampness was a great enemy of gunpowder. Could anybody ever think that some day someone would make an explosive based on water? This is precisely what a genius called Dr. Mervin Cook did. Dr. **Mervin Cook** was a professor of metallurgy at the University of Utah and also a businessman. His greatest commercial explosives invention was formulated in 1956 with the creation of a new blasting agent using an unusual mixture of ammonium nitrate, aluminum powder and water. **Slurry explosives** were born. This invention converted the commercial explosives industry from "dangerous dynamite" to "safe slurry" and dry blasting agents (ANFO). Incorporating water in an explosive was nothing short of revolutionary. One great advantage of slurry is that it does not have any problem with wet holes.

Incidentally, Cook is also the designer of the biggest conventional bomb used in history, the famous 15,000 lbs heavy **Daisy Cutter** which uses **Gelled Slurry Explosive**.

Slurries, sometimes called **water gels**, contain ammonium nitrate partly in aqueous solution. In general, slurry explosives are based on concentrated solutions of mixed inorganic nitrates, usually those of calcium, ammonium, and sodium, which may be thickened or gelled with natural gums, such as guar. Furthermore, aluminum powder, organic fuels, and solid oxidants may be introduced in this matrix. **Detagel**, a proprietary product, for example, is a high strength hexamine nitrate, slurry water gel. It contains ammonium nitrate (<65%), sodium nitrate (<20%), sodium perchlorate (<7%), nitric acid (<5%), hexamine (<15%) and aluminum (<7%).

Several mixtures have been developed since then. There are as many types of slurries as there are manufacturers, so a general description is not easy to give. Three basic types could be distinguished, according to the method of sensitization: **metallized slurries, slurries based on emulsions/occluded gas** and **slurries based on other explosives**, such as TNT, ethylene glycol di- and mononitrates, RDX, HMX, PETN, etc. Common to all of them is that the most of the oxygen comes from an inorganic (usually, but not necessarily) salt dissolved in water. The fuel can be a metal powder, oil (as an emulsion) or another explosive or even an organic subtance dissolved in water. Essential for all slurries is a gelatinizing agent. The consistency varies from pourable to stone hard solid; this is adjusted with the gelatin (usually polyvinyl alcohol, guar gum, dextran gums, urea-formaldehyde resins).

Depending on the remainder of the ingredients, slurries can be classified as either blasting agents or explosives. **Slurry blasting agents** contain nonexplosive sensitizers or fuels such as carbon, sulfur, or aluminum, and are not cap sensitive; whereas **slurry explosives** contain cap-sensitive ingredients such as TNT and the mixture itself may be cap-sensitive.

Slurry blasting agents require adequate priming with a conventional high explosive to attain proper detonation velocities, and often require boosters of high explosive spaced along the borehole to as sure complete detonation. Slurry explosives may or may not require priming. The detonation velocities of slurries, between 12,000 and 18,000 fps, vary with ingredients used, charge diameter, degree of confinement, and density. The detonation velocity of a slurry, however, is not as dependent on charge diameter as that of a dry blasting agent.

The sensitivity of slurry explosives may be increased by the incorporation of small air or gas bubbles. Air bubbles may be incorporated during the mixing of the explosive, or gas bubbles may be added to a more permanent form as small gas-filled silicate spheres. It has been shown that bubbles provide "**hot-spots**", as the gas they contain is adiabatically compressed by the advancing detonation front and that these hot spots assist propagation of the detonation

Some slurries maintain their fluidity even at freezing temperatures. Slurries consequently give the same advantageous direct borehole coupling as dry blasting agents as well as a higher detonation velocity and a higher density. Thus, more energy can be loaded into a given volume of borehole. Saving in costs realized by drilling smaller holes or using larger burden and spacing will often more than offset the higher cost per pound of explosive. Adding powdered aluminum as a sensitizer to slurries greatly increases the heat of explosion or the energy release. Aluminized slurries have been used in extremely hard rock with excellent results.

A slurry and a dry blasting agent may be used in the same borehole in "**slurry boosting,**" with the bulk of the charge being dry blasting agent. **Boosters placed at regular intervals may improve fragmentation.** In another application of slurry boosting, the slurry is placed in a position where fragmentation is difficult, such as a hard toe or a zone of hard rock in the burden. The combination will often give better overall economy than straight slurry or dry blasting agent.

(10.4) GELLED SLURRY EXPLOSIVES (GSX) GO A STEP FURTHER

Gelled slurry explosives consist essentially of concentrated solutions of ammonium nitrate and alkali metal nitrates and an appropriate selection of sensitized energetic fuels and other modifying additives. In general they contain as an essential ingredient a thickener or gelling agent in order to prevent segregation or precipitation of the ingredients, to provide mixtures which may be more easily worked and packaged in firm, shape-retaining cartridges and to impart properties of water resistance which are required when the explosives are used in wet environments. People have also made GSX containing water, at least one selected from the group of a polysaccharide and synthetic polymer; water-soluble melamine resin and a water-soluble explosive salt. The GSX used in the famous **Daisy Cutter bomb** is believed to consist of ammonium nitrate, aluminum powder, and polystyrene. The gelled explosive requires a high explosive booster to detonate.

Fig. 33 The Biggest Conventional Bomb (15,000 Lbs) In Existence That Has Been Used In Combat, The Daisy Cutter Uses A Gelled Slurry Explosive (GSX)

There was no fundamental problem with the GSX as far as their properties as explosives are concerned. The problem arose on account of gel instability. Gel instability has been manifested by syneresis and bleeding of the gel after a relatively short period of time. Hence much of the research in this field is directed towards solving this problem.

The thickening or gelling agents generally include materials such as, for example, gum arabic, agar-agar, Irish moss, locust bean, tamarind, psyllum or guar gums, starches, hydroxyethylcellulose and hydrophilic vinyl polymers such as polyacrylamide. The most widely employed of the thickeners which have been used with aqueous slurry explosives has been **guar gum** which may be procured in self-complexing forms where a cross-linking agent is incorporated or in non-complexing forms which contain no cross-linking agent. When the non-complexing form is used, usually small amounts of a conventional cross-linking agent, such as borax or potassium pyroantimonate and the like are employed to produce a firmer or stronger gel. Generally the explosives industry has preferred to use the non-complexing form of thickener in order to more easily control the properties of the mixture, particularly with regard to the degree of firmness desired. Depending on the desired end use of the aqueous slurry explosive and the method of manufacture employed, its thickness or degree of gelatinization may range from a highly flowable consistency to that of a very firm, rubber-like gel. These properties result in large part from the amount and kind of gelling agent employed. In the preparation of slurry explosives containing soluble organic nitrate sensitizers, particular problems are encountered since the more conventional gelling agents employed in most explosive slurries are generally ineffective or unduly slow in their activity so as to be unsuitable for use. This unsuitability is attributed to the interfering action in solution of the dissolved organic nitrate sensitizer which tends to prevent or retard hydration of the thickener. It has been generally found that in the presence of soluble organic nitrate, the so-called modified guar gums, that is, hydroxyethyl- and hydroxypropyl-modified guars, are most effective as thickeners and are therefore preferred. Indeed, in some compositions these modified guars are the only thickeners which can be usefully employed.

When soluble organic nitrate sensitized aqueous slurries are thickened by means of hydroxyethyl guar or hydroxypropyl guar, it is found that very high viscosities are produced very rapidly. This, in turn, leads to difficulty in processing since the mixed slurry must be packaged relatively quickly after mixing. Any delay can result in a mixture which is too viscous to handle in the conventional packaging machines. When the amount of thickener employed is reduced in order to achieve suitable lower viscosities for easy packaging, the slurry explosive product generally remains unduly soft resulting in a limp cartridge or package which stores poorly and is difficult to handle in blasting operations. The elastic memory of the gel also often results in packages which burst after sealing and the borehole tampability of the cartridge is reduced. What the industry has been seeking in order to overcome the afore-mentioned difficulties is a thickener for use with organic nitrate sensitized aqueous slurry explosives which results in a product which is easily mixed and which retains flowability and pumpability for several hours to permit easy packaging yet which will result in a firm and tampable cartridge or borehole charge. It has now been found that these industry objectives can be achieved by employing as a thickener for organic nitrate sensitized aqueous slurry explosives, a proportioned mixture of **hydroxypropyl-modified guar and an unmodified guar.**

It has been discovered that the use of a blend of unmodified guar and hydroxypropyl-modified guar produces an unexpected synergistic effect with slurry systems containing organic nitrate sensitizers in that the final gel strength of the compositions after packaging is substantially greater than the gel strength of similar compositions when thickened either by the modified guar or the unmodified guar alone. Further, it has also been found that this synergism is accelerated when calcium nitrate is present as an oxidizing salt ingredient in the explosive mix.

(10.5) BOOSTER EXPLOSIVES IN MINING INDUSTRY

Booster explosives are those components of the explosive train that function to transmit and augment the force and flame from the initiating explosive for the convenience of the relatively insensitive explosive. They ensure the reliable detonation or burning of the main bursting charge or the propellant charge. The booster explosive thus acts as a bridge between the high sensitivity low energy primer and the low sensitivity high energy main charge. Some of the most popular high-explosive boosters use one of the following: Tetryl, Composition H-6, or Composition A-5. **Tetryl** we have already discussed. Tetryl had a major problem of being easily degradable to a toxic compound like methanol. **Composition H-6** or **CH-6** is a mixture of 97.5% RDX, 1.5% calcium stearate, 0.5% polyisobutylene, and 0.5% graphite. It is a finely divided gray powder that is less toxic and more available than tetryl. **Composition A-5** is a mixture of 98.5% RDX and 1.5% stearic acid.

The word '**boosters**' has a different meaning in mining industry, however. In mining industry boosters are used to initiate blasting agents like ANFO which are not cap sensitive, i.e., for which a cap is not adequate.

The development of non-cap sensitive and insensitive explosives for civilian and military use created the need for compact high detonation pressure boosters. NG based versions of boosters were available, but were in less-than-convenient packages and did not fit the overall trend toward bore holes containing non-nitroglycerin based explosives.

There are several forms of boosters available and they have varying properties and sensitivities. Three forms are available: **Cast, pliable** and **binary**. They are normally cylindrical in shape and designed to be initiated with a detonator or detonating cord.

Cast Boosters are cap-sensitive explosives that typically contain the high explosive trintotoluene (TNT) as the casting material. Different molecular explosives are mixed into the melted TNT and impart additional energy and/or sensitivity to the booster. Thus you can have **pentolite boosters, Composition B boosters, Torpex boosters, tetrytol boosters**, and **amatol/sodatol boosters** (when you replace ammonium nitrate with sodium nitrate, you get sodatol).

(10.6) OXYLIQUITS

We are mentioning it essentially for the sake of completeness from a historical point of view. An oxyliquit is an explosive material made of a mixture of liquid oxygen (LOX) with a suitable fuel, usually carbon (as lampblack) or some organic chemical (e.g. a mixture of soot and naphthalene), wood meal, or aluminium powder or sponge; the material is capable of absorbing several times its weight of LOX.

Oxyliquits have numerous advantages. They are cheap to make, can be initiated by a safety fuse, and in case of a misfire the oxygen evaporates quickly, rendering the charge quite safe in a short period of time. The first large scale deployment was in 1899 during building of the Simplon tunnel, in the form of cartridges filled with diatomaceous earth soaked with petroleum, or an absorbent cork charcoal, dipped in liquid oxygen immediately before use, or in another modification the cartridge is filled with liquid oxygen after placement in the borehole.

The problem with oxyliquits is that, once mixed, they are sensitive to sparks, shock and friction, and there were reported cases of spontaneous ignition also. The power relative to weight is high, but the density is low, and hence the brisance is low as well. Ignition by a fuse alone is sometimes unreliable. The charge should be detonated within 5 minutes of soaking, but even after 15 minutes it may be capable of exploding, even though the explosion would be weaker and would also result in the production of carbon monoxide. Further, the flame it produces has too long a duration to be safe in possible presence of explosive gases. As a result oxyliquits had found their use mostly in open quarries and strip mining.

Due to the complicated machinery required for the manufacture of liquid oxygen, oxyliquit explosives were used mostly only where their consumption was high. In the United States, some such locations were the strip mines in coal mining areas of the Midwest. Its consumption peaked in 1953 with 10,190 tons, but then decreased until zero in 1968, when it was totally replaced with even cheaper ANFO. ANFO sounded the death knell for oxyliquit.

❑❑❑

Chapter 11

The Deceptive Beauty of Binary Explosives

(11.1) WHAT ARE BINARY EXPLOSIVES AND WHAT MAKES THEM SO BEAUTIFUL?

Binary explosives are beautiful things. As the name suggests, they have two components. Individually they are not explosives. However, when they are mixed, the mixture becomes explosive! Wonderful, isn't it? You can make an explosive when you want it and that solves a million problems associated with the storage of explosives. The invention of binary explosives must be ranked as one of the best inventions in the field of explosives. In most binary systems one of the components is an oxidizer (ammonium nitrate, sodium perchlorate) and the other is a fuel (nitromethane, aluminum). As with all explosives, the potential uses and effects are determined by several properties such as detonation velocity, density, gas production, etc. Binary explosives are usually not as cost effective and convenient as traditional explosives. However, for the low volume, specialized user (such as a cartridged explosive in small-to-medium bore hole applications), they can save a great deal of time and expense in the area of transportation and logistics.

(11.2) SOME GREAT BINARY EXPLOSIVES

(11.2.1) The Magic Called Kinepak

Kinepak is a binary explosive invented by **Gerald L. Hurst**. The two components can be transported and stored without any problem as they do not require magazine storage. When mixed, KinePak can be initiated with a Number 8 detonator or 10 gram per meter detonating cord. The Kinepak containers incorporate a recess for the detonator or the detonating cord. When mixed, KinePak has a density of 1.20 g/cc and a detonation velocity of 20,700 fps.

The components that make up Kinepak remain a trade secret. There seem to be two schools of thought about the liquid component of Kinepak. Some think it is neat nitromethane; others hold that it contains additional material. We are inclined to accept the other school because even their official literature says that it is 95-100% nitromethane. As for the solid, the official literature says it is 96-100% ammonium nitrate. They are obviously deliberately being vague about the little amount of extra things that they might have added. The extra ingredients might be responsible for the extra properties of Kinepak but a basic binary explosive can be made out of ammonium nitrate and nitromethane (AN-NM)—a fact so loudly demonstrated by Timothy McVeigh in the Oklahoma bombing.

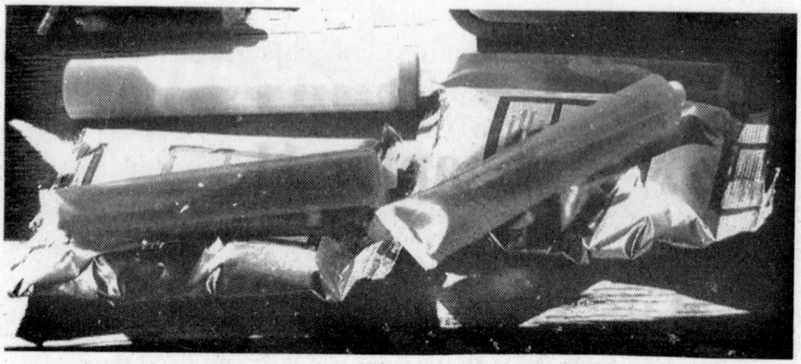

Fig. 34 The Magic Called Kinepak: Components Before Mixing

Only three steps are required in using it. Attach the arming needle to the Liquid Tube. Insert the arming needle into the Solid Component and dispense Liquid. Attach the detonator or the detonating cord. The liquid content percolates through the powder and is allowed to stand for a short period until fully absorbed. Detonation can be achieved using safety fuse with a non-electric detonator; by an electric detonator; or by using a shock tube.

When the mixture is placed in a one-pound foil bag it is known as **Kinepouch**; when placed in a 1/3-, 1/2-, or 1-pound cylindrical plastic tube, it is known as **Kinestik**. In them the tube containing liquid is opened and poured into the bottle (Kinestik Solid) or into the plastic pouch (Kinepouch). The Kinestik or Kinepouch is primed with a detonator and is detonated in the normal way. The Kinestik Solid bottle has a detonator well molded into its wall, to accept a detonator. The Kinepouch Solid pouch must be pierced in order to pour in the liquid component and insert the detonator.

Fig. 35 Using Two Kinestiks For Neutralizing An Anti-Tank Mine

Kinepak is not suitable for most military uses except for neutralizing mines as the quantity is so small. It was designed strictly for civilian use. Kinepak makes for an excellent booster that would reliably initiate most blasting agents, both bulk and cartridge. Using Kinepak saves greatly on the trouble and expenses of transporting and storing high explosives.

(11.2.2) Binex

Binex, invented by **Hansen** and others uses a two component system of an aqueous solution of sodium perchlorate and aluminum powder. When these two components are combined, a liquid explosive is formed that is cap sensitive. There is some doubt whether binex can become a viable replacement for cartridged explosives because of the high cost and the environmental concerns with the sodium perchlorate solution. However, there is a current military application where this product is used to blast fox holes in conjunction with an entrenchment kit for soldiers. It is known that this explosive has detonation velocity that is much lower than Kinepak and other commercial cartridged explosives. In the case of the military application, this is an advantage as lower velocity explosives are generally better for cratering in soil.

(11.2.3) The Brilliant Invention Called FIXOR: An Explosive That Self-Neutralizes After Some Time

Made by **MREL Specialty Explosive Products Limited**, Canada, in 1988 **FIXOR** is a two-component explosive, based on a flammable liquid and a powder. FIXOR is an acronym for **Field-friendly Inexpensive uneXploded Ordnance Remover**. The exact composition has not been made known but it is believed that the liquid is nitromethane and the powder is microbead. It has been designed to replace plastic explosives and block TNT as traditional demining and **UXO (unexploded explosive ordnance)** explosive charges.

Like other binary explosives, FIXOR too has the great advantage that it can be transported around the world by any method, including commercial passenger aircraft. However, what makes FIXOR truly a great invention is its unique feature of **self-neutralization**. FIXOR comes in 500 ml bottles, one containing the red liquid and the other the white powder. Before use, all you have to do is to mix them and get the orange mixture indicating that it a high explosive now. It can be detonated by No. 8 detonator containing a minimum of 0.8 g of secondary charge or a knotted detonating cord. But you must use the mixture within two hours. The components separate after that and then you have to remix them by shaking before detonating. The detonation velocity is 14100 fps and the TNT equivalency is about 80-85% by weight.

The West has always had a morbid fear of terrorists making use of binary explosives. FIXOR takes care of that concern. Leave the prepared explosive mixture for some time and it will become non-explosive. This means that even if someone made a booby trap or an IED out of it, he could not possibly leave the device for more than 2 hours in place and go home as the explosive would become ineffective! Thus it is of no use to those who have to make IEDs and booby traps. It does not, however, preclude its use on aircrafts. It is not known whether the company uses some taggants on the liquid and the powder or not. Even if they do employ taggants, once somebody analyzes the components, he can try to produce an improvised equivalent of FIXOR sans the taggants. Such a binary explosive can easily be carried on board aircrafts and used for explosion in directly controlled or timer devices as the duration could be easily less than 2 hours.

FIXOR is in operational use in many landmine affected countries and has been proven effective at destroying a large variety of anti-personnel and anti-tank landmines as well as unexploded ordnance (UXO). When you want to use FIXOR for UXO disposal, you may require 1 to 4 bottles depending on what exactly you are seeking to destroy. Small anti-personnel mines can be destroyed with one bottle, 81 mm mortar bombs and anti-tank mines require two bottles each, and 155 mm HE artillery shells would require four bottles each.

Fig. 36 One Bottle Of FIXOR Being Used To Neutralize An AP Mine

Fig. 37 Two Bottles of FIXOR Being Used To Neutralize An AT Mine

As such it is a wonderful invention. But the problem is that it cannot substitute those military explosives which can be stored for long.

(11.2.4) HELIX

HELIX (acronym for **High Energy Liquid eXplosive**) is a binary explosive consisting of nitromethane and aluminum powder. It can be used as a shaped charge also. This gives a detonation velocity of 20,600 fps. The mixing time is less than one minute. The container is made of PVC plastic and contains no ferrous parts. The liner is made of powdered or stamped metal, primarily copper. The shaped charge will penetrate up to 4 in. of mild steel.

To activate HELIX, the liquid component is poured into the black HDPE (high density polyethylene) bottle containing the activator powder until the bottle is half full or the powder floats to the top. Once the cap is secure, the bottle is shaken vigorously until the powder is completely distributed throughout the liquid. The cap is removed and more liquid is poured in until the bottle is full up to the neck. The cap is replaced and the bottle is shaken briefly, making HELIX active. Once active, the liquid is poured into the shaped-charge container for use against the targets.

HELIX has found good use in neutralizing mines. After placing the explosive-filled container next to the landmine, HELIX is initiated using any standard blasting cap placed in the specially designed holder in the top of the container or by a detonating cord inserted into the blasting cap holder.

Fig. 38 HELIX Being Used To Neutralize An Anti-Tank Mine

(11.2.5) Other Binary Explosives

There are many other possible candidates for use as binary explosives. However, most of them may not be viable for consideration as commercial products on account of the toxicity of the components and/or detonation products; the question of stability of the components before and after mixing; shelf life of the ingredients and the prepared mixture; cost; and the ease or difficulty of mixing. Since everything needs to be evaluated against some standard, it will have to be seen whether a new mixture offers any substantial benefit over ammonium nitrate/nitromethane (**AN-NM**) systems (e.g., Kinepak) which have become the de facto standard. Other factors that need to be considered are detonation velocity, density, gas production, etc. Although other systems besides AN-NM exist, there has not been a commercially viable product available as a substitute for conventional small diameter cartridge explosives.

(11.2.6) Relative Merits And Demerits Of Binary Explosives

They have the following advantages over conventional explosives:

- The components are not explosives before mixing;
- The components do not have to be transported as explosives; and
- The components do not have to be stored as explosives and therefore do not require expensive storage "magazines".

However, there are a few disadvantages:

- Mixing can be time-consuming;
- Shelf life of the ammonium nitrate powder can be short depending on ambient conditions, particularly temperature and humidity; and
- They can cost 2 to 3 times more than conventional explosives.

❑❑❑

Chapter 12

LIQUID AND FOAM EXPLOSIVES: AMAZING, IST'T IT?

(12.1) EVER EXPECTED EXPLOSIVES TO BE IN LIQUID OR FOAM FORM?

The creation of liquid and foam explosives is a most wonderful invention. They defy the popular perception of explosives and yet have extremely attractive qualities. They can be used to plant explosives where you do not get enough space to place a solid charge, such as in crevices, cramped spaces, and small holes. They come in extremely handy in special and covert operations and in destruction of military mines. You have only to tax your imagination a bit and you will realize that there are innumerous ways you can use liquid and foam explosives to great advantage.

(12.2) LIQUID EXPLOSIVES BASED ON NITROMETHANE

Nitromethane was first synthesized by **Kolbe** in 1872. It is so insensitive that it was not until 1938 that its detonation property was revealed by **McKittrick**. Pure nitromethane will not usually detonate unless it is subjected to extreme shock and/or confinement at elevated temperatures. Pure nitromethane is actually a powerful explosive. However, without the addition of some additives or modifiers, it is so insensitive that it is classified as a 'flammable liquid' for transportation purposes.

Once this detonation property was discovered, research was initiated to find sensitizers to increase its ease of detonation. World War II research produced **sensitizers**, primarily amines, which made nitromethane detonatable with a blasting cap. In 1945, **Ericksen** and **Rowen** listed over one dozen **nitromethane-amine mixtures** along with their explosive ability. Now several explosive compositions comprising nitromethane and a sensitizer for the nitromethane are well known. These compositions are formed by combining nitromethane with a sensitizing chemical compound. Various chemical compounds serve as effective sensitizers for nitromethane: for example, liquid explosive compositions containing nitromethane sensitized with amines or polyamines such as diethylamine, triethylamine, ethanolamine, ethylenediamine and morpholine. The so-called **PLX**

(Picatinny Liquid Explosive) developed by the Picatinny Arsenal during the WWII for clearing of minefields was nothing but 95% nitromethane and 5% ethylenediamine.

Nitromethane can be sensitized by **non-chemical means** also. Some of the methods include **air-entrapping structures**, such as **resin or glass micro-balloons** and **open-celled polymeric foam.**

(12.2.1) The Mysterious Explosive Called Astrolite

Astrolite is the trade name of a family of explosives, invented by **Gerald Hurst** in the 1960s during his employment with the Atlas Powder Company. The Astrolite family consists of two compounds, **Astrolite G** and **Astrolite A**. Both are two-part liquid-state high explosive mixtures, comprising ammonium nitrate oxidizer and hydrazine rocket fuel. They still find some use in commercial and civil blasting applications, but have mostly been superseded by cheaper and safer compounds, largely due to the expense and exceptionally poisonous nature of the hydrazine component. We, however, considered it necessary to mention it because of the hype associated with Astrolite among the uninitiated.

Astrolite explosives are formed when ammonium nitrate is mixed with anhydrous hydrazine. Extensive solvolysis occurs with the liberation of large amounts of ammonia gas and a new compound (hydrazonium nitrate) is formed and remains in solution. This produces a clear liquid explosive. It has a detonation velocity of approximately 28,200 fps. It has been widely (and incorrectly) referred to as the 'world's most powerful non-nuclear explosive', a misconception caused largely by hype among the uninformed, who cite its high detonation velocity in comparison to nitroglycerine or TNT, while failing to notice its low density. Conventional high explosives like RDX etc. have a far higher brisance and material density than Astrolite G, making them far more powerful in actual practice.

Astrolite G, the most common type of Astrolite, is a mixture of ammonium nitrate and hydrazine at a ratio of 2:1 by weight.

Astrolite A (or A-1-5), a secondary (and less common) type of Astrolite, is synthesized by the addition of finely powdered aluminum to the Astrolite G mixture. Though it has a lower detonation velocity (approximately 25,000 fps) than Astrolite G, the addition of the aluminium increases both its density and brisance, moderately increasing its overall effectiveness. The aluminum powder does not get dissolved in the solution but remains suspended.

A very notable characteristic of Astrolite is its remarkable degree of persistency for a liquid explosive compound. Due to its low volatility, it can be dispersed in an area, be absorbed by the soil, and still retain its full explosive characteristics for a period of approximately 4 days in spite of rain!

It appears that Astrolite could find use for Special Forces operations rather than standard military operations. One could create an invisible minefield by dispersing some Astrolite in the ground and set it off remotely or otherwise. At least for anti-personnel applications, it would be a frightening prospect. It will literally make the ground under your feet explode!

(12.3) MULTI-COMPONENT LIQUID EXPLOSIVES: VERY CLEVER DEVICES

As we saw above, most of the efforts to make a usable nitromethane based liquid explosive have centered on adding dangerous amine compounds (such as ethylenediamine) and/or incorporating entrapped air bubbles by some means, such as micro-balloons (resin, glass, or open-celled polymeric foam material etc.). These air bubbles, while having the desired result of sensitization, have the undesired result of decreasing the density of the material. Further, since these air bubble means are non-energetic, the per unit volume energy is also decreased. Not only that, since the balloons will float to the surface of pure nitromethane, a thickening (gelling) agent such as fumed silica must be added to prevent this.

The search for a better liquid explosive therefore continued. One composition consists of nitromethane, a nitethane sensitizer and an energetic compound. The energetic compound is selected from the group consisting of dihyro-nitroisoxazole, dinitropropane, dinitrocyclopropane and dinitroethane. The addition of an energetic compound to a mixture of nitromethane and sensitizer forms an explosive composition that displays a significantly higher detonation pressure than the combination of nitromethane and sensitizer alone.

As noted above, liquid nitromethane based explosive compositions may have to be converted into a semi-solid or gelled state by the addition of a gelling agent. This conversion increases the density and, therefore, the detonation pressure of the semi-solid or gelled compositions. But they defeat the very purpose. If you want a liquid explosive for some purpose of making it flow along some crevice or whatever, you want a liquid explosive; how will a gelled explosive help you? You might as well use a solid explosive then.

One very clever composition that circumvents most of the above shortcomings contains a powder consisting of aluminum preferably having an average particle size of 5 to 50 microns and a surface area of 0.5 to 2 square meters per cubic centimeter containing 0.1 to 5% stearic acid by weight. Coating aluminum with stearic acid gives it a hydrophobic property. This causes air bubbles to cling to the surface of the aluminum particles. As noted before, the incorporation of air bubbles into explosive mixtures increases their sensitivity. The coated aluminum powder is mixed with nitromethane to form a liquid explosive formulation detonable at a wide range of temperatures and diameters with a standard commercial No. 8 blasting cap. The mixture should contain a minimum of 5% of aluminum powder in relation to nitromethane by weight. In practice, they use about 1 to 1.2 ounces of aluminum powder to about 6 ounces of nitromethane, by weight. The aluminum powder acts as a sensitizer to the nitromethane, causing the mixture to become cap-sensitive, the aluminum powder then reacts with the products of the detonation behind the reaction front thereby adding increased energy to the total explosion. It requires no further thickening agents in order to keep the aluminum powder in suspension in the nitromethane.

The resultant explosive is a blasting agent requiring a one pound booster for initiation, and not a cap-sensitive mixture that could be used in small diameter bore holes.

(12.3.1) Liquid Explosive Pouch

The proprietary product called **Liquid Explosive Pouch** consists of nitromethane and diethylenetriamine sensitizer.

Liquid Explosive Pouch is a hand-held, binary-explosive system consisting of flexible plastic pouches and two commercial, non-explosive liquid chemicals: nitromethane and an amine sensitizer, **diethylenetriamine (DETA)**. Nitromethane becomes an explosive only after it has been "sensitized" by the addition of the DETA. Before filling the pouch with nitromethane a metered amount of the sensitizer DETA is injected in to the pouch with a plastic syringe. A dye indicator included in the nitromethane changes to purple when the nitromethane is sensitized; creating a positive visual warning that the liquid mixture is an explosive. The explosive liquid mixture is equivalent to TNT on a weight basis; it has a detonation velocity of about 21,000 fps. Initiation is achieved by taping a No. 8 or equivalent blasting cap to the outside of the pouch next to the binary-liquid fill.

Fig. 39 Liquid Explosive Pouch (Half A Pound) Being Used To Neutralize An Anti-Tank Mine

For most applications a standard one-pound tub-shaped pouch is used. However, pouches can be made in different shapes—some of them can be made to bend around the contours of the ordnance being disposed so as to achieve a '**lens effect**' or they can be used in pairs to achieve the '**ear muff effect**'.

(12.4) FOAM EXPLOSIVES

(12.4.1) Lexfoam

Lexfoam, developed by the Canadian firm **MREL** (Mining Resource Engineering Limited), is a nitromethane-based stock solution and mixture of propellants that is packaged, stored and transported in two aerosol cans as Class 3 flammable liquid. It is not explosive until operators combine the two units and dispense the material on the target. Once dispensed, it forms a foam which is explosive. The white self-supporting foam can adhere to walls, uneven and rounded surfaces where using traditional HE demolition blocks would be very difficult. Lexfoam is resistant to both mild rain and large temperature variations.

Lexfoam stock solution is nearly 90 % NM. The remaining 10 % is believed to be surfactant, thickeners, emulsifiers, and fuzed silica powder. The stock solution is charged with an appropriate amount of a blend of hydrocarbons which is also a flammable liquid, then pressurized with nitrogen gas at 180 psi. The liquefied blend hydrocarbons are gases under normal room temperature and pressure. In Lexfoam stock solution the blend hydrocarbons are present as liquid and therefore are confined at a temperature above their boiling point as a result of the pressure in the Lexfoam container. When the Lexfoam solution is dispensed into the atmosphere, the propellants flash from a liquid to a gas. Lexfoam becomes an explosive only after it has been dispersed in foam form. It has the looks and consistency of shaving foam. The foam is cap-sensitive. The voids in the foam are believed to sensitize the nitromethane. The detonation velocity is 12960 fps.

The lexfoam dispensing system consists of two containers. The stock solution is kept in pressurized containers that vary in size from a ½ liter can to a truck-mounted system. The second is a smaller container with the blend liquid hydrocarbon propellants (propane, n-butane and isobutane). In the field, the blended liquid hydrocarbon propellant would be injected into the Lexfoam stock solution before use.

Lexfoam has found great use in neutralizing mines. One has to spray the foam over the mine and then insert a detonator in the foam. It is able to neutralize mines buried one inch below the surface with a two inch coating over the surface of the ground. One full tank of the backpack system weighing 79 lbs can be used to neutralize between 20 and 40 mines, depending on the conditions. The aerosol can of 2.2 lbs is expected to neutralize up to five mines. The handheld system became necessary because it was found that for many users the backpack system was too heavy. Then the eight foot long lexfoam delivery trap could turn out to be too long to maneuver in a one meter wide clearing lane. Further, nitrogen and propane gas tanks are not easily available in many parts. In addition, plenty of water is needed to clean the unit at the end of the day.

(12.4.2) NMX-foam

The proprietary product NMX-foam (NitroMethane eXplosive foam) is quite similar to lexfoam and is used also in a similar manner.

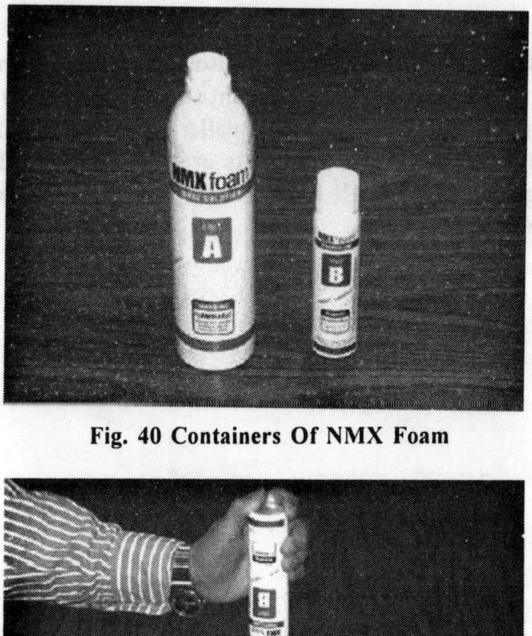

Fig. 40 Containers Of NMX Foam

Fig. 41 That's How The Contents Of The Two Containers Are Mixed

Fig. 42 NMX Foam Being Used To Neutralize An Anti-Tank Mine

It is formulated from nitromethane stock solution (nitromethane, surfactants, and silica powder) and a mixture of hydrocarbons (propane and isobutane), which form the propellant. Thus it is quite similar to lexfoam. The delivery system consists of 505 g (±5 g) of nitromethane stock solution in a disposable aerosol can and a second, smaller container containing approximately 50 g of the propellant. Separately, both components are classified as flammable liquids for transportation and storage purposes. NMX-foam is used in a manner exactly similar to that of lexfoam.

❑❑❑

Chapter 13

MILESTONES IN THE DEVELOPMENT OF EXPLOSIVES

(13.1) INDELIBLE FOOTPRINTS MADE IN THE ROARING 19TH CENTURY

a. 1800. Mercury fulminate was discovered by Edward Howard, an English chemist, who described its detonating properties in a paper before the Royal Society.

b. 1801. Fireworks formulas containing metal salts and ammonium chloride were published by C. F. Ruggieri, an Italian artificer. These salts produced brighter flames because volatile metal chlorides were formed during the combustion reaction.

c. 1807. The principle of using mercury fulminate as a primer was patented by Rev. Alexander J. Forsyth of Aberdeenshire, Scotland.

d. 1823. Chlorates, discovered by C. F. Berthollet in 1786, were first used in fireworks by John Cutbush, an English artificer.

e. 1825. Benzene, creosote, and naphthalene were isolated from coal tar (obtained from bituminous coal) by Rev. Dr. Clayton in England. Most of his products were later nitrated to produce explosives, such as trinitrobenzene and nitronaphthalenes.

f. 1826. Aniline was first prepared by Unverdorben. Later, nitration of aniline resulted in the very powerful explosive called tetranitroaniline.

g. 1831. **Safety fuse** for blasting was invented by William Bickford of Cornwall, England. The **Bickford fuse** consisted of a core of black powder tightly wrapped in textiles (mainly jute yarn). The timing, or burning time, for a given length of fuse was very accurate and consistent. This fuse soon replaced the less reliable fuses which were made of straws or quills filled with black powder, thus greatly reducing the hazard of accidental explosions in mining or construction. Later, the Bickford safety fuse was waterproofed by applying a coat of asphalt covered with textile.

h. 1832. Ammonium perchlorate was first prepared by E. Mitscherlich, a German chemist, who also nitrated benzene and toluene in 1833 or 1834.

i. 1833. **Nitrostarch** was discovered by **Henri Braconnot**, a French chemist, while dissolving starch in concentrated nitric acid to form a flammable nitric acid ester of starch. This is considered to be the **beginning of modern high explosive technology**.

j. 1837-1838. Nitrated paper, soluble in ethyl alcohol, and nitrated cotton fibers, insoluble in ethyl alcohol, were prepared by Prof. Theophile Jules Pelouze, a French chemist. These products were highly combustible.

k. 1841. Ammonium picrate was first prepared by Marchand, and was used in 1869 by Brugere as a propellant.

l. 1843. Picric acid, discovered but not identified by Peter Woulfe in 1771, was rediscovered by A. Laurent who found that some of the salts were explosive.

m. 1845. **Guncotton** was prepared by Christian F. Schoenbein, a German chemist, while at the University of Basel, Switzerland. Upon nitrating cotton with mixed nitric acid and sulfuric acid, he demonstrated that the resulting nitrocellulose was up to four times as powerful as black powder for blasting. The guncotton burned without leaving an ash. Schoenbein also prepared the first nitrosugar, which deflagrated suddenly and violently when heated to a certain point. Schoenbein and R. Boetter of Frankfurt, Germany, who had discovered guncotton independently, obtained patents in 1846 and 1847 under the title: 'Improvement in Preparation of Cotton Wool and Other Substances as Substitutes for Gunpowder'. Serious accidental explosions during early attempts to manufacture guncotton in 1847 delayed its acceptance as an explosive; however, guncotton eventually became accepted universally for use in blasting and for smokeless powder when mixed with nitroglycerin.

n. 1846-1847. Nitroglycerin, an ester of glycerin and nitric acid, was invented by Asconio Sobrero, an Italian chemist. Because nitroglycerin is very sensitive to slight shocks, commercial use was delayed until the invention of dynamite and blasting gelatin. Nitroglycerin was also used later to make smokeless powder.

o. 1847. Mannitol hexanitrate or hexanitromannitol was first prepared by A. Sobrero, but the explosive properties were not examined until 1878 when N. K. Sokoloff of Russia investigated these characteristics.

p. 1847. Cyanuric triazide was first prepared by Auguste A. T. Cahours, a French chemist.

q. 1850. A testing device for measuring explosive pressure in cannons was developed by Col. George Bomford of the US Ordnance Department. Bomford inserted pistol barrels in holes drilled along the gun tube and measured velocities of bullet blowout. Based on these data, the weight of guns could be reduced.

r. 1850. An improved testing device for measuring explosive pressures in cannons was designed by Thomas J. Rodman of US Ordnance Department. Pistons with punch-type ends were installed at intervals on the cannon and were forced by the explosion of the propellant into copper blocks. Measuring the depth of the indentations indicated the relative pressures along the length of the gun tube. Rodman also developed a process of casting guns hollow and cooling them from within. The Rodman process put the metal surrounding the bore of the gun under a permanent compressive strain that greatly increased the gun's resistance to interior pressures.

s. 1853. Nitroglycerin absorbed in magnesia was developed by V. F. Petrushevskii, a Russian officer, for use in mining gold in Siberia. This was a precursor of dynamite.

t. 1857-1858. Sodium nitrate deposits found in Chile were originally converted to saltpeter (potassium nitrate) by reaction with potassium chloride, which was also in plentiful supply. Lammont du Pont, an American industrialist, introduced a new formula for black powder using sodium nitrate in lieu of potassium nitrate. Black powder containing potassium nitrate was called blasting powder A, and black powder containing sodium nitrate was called blasting powder B. Blasting powder B was less expensive and more hygroscopic, which made it more suitable for blasting. Powder formulas were considered more suitable for firearms, safety fuses, and fireworks.

u. 1858. Diazo compounds were discovered by Peter Griess, who also prepared dinol or diazodinitrophenol.

v. 1859. Ethylene oxide, ethylene glycol, and ethylene glycol ether were prepared by Charles A. Wurtz. A century later, ethylene oxide would be used in fuel-air explosives developed by the US Armed Forces. Diethylene glycol was also developed simultaneously in 1859 by Wurtz and A. V. Laurenco.

w. 1860. A **case-less propellant**, consisting of cakes of compressed black powder cakes or grains with perforations, was developed by T. J. Rodman while working at Frankford Arsenal, Philadelphia. This led to the use of grains shaped into hexagonal prisms about 1 inch long and 1 inch across. Perforations and grooves in the grains channel the flame from the explosion. As the grains burn inward from the prism surfaces and outward from the perforations, the burning gases flow continuously, thereby permitting the explosive energy to build up gradually until the maximum power is exerted as the projectile leaves the muzzle of the gun. This acceleration imparts a more efficient thrust to the projectile and extends the range of the gun.

x. 1861. Guanidine was formed by the oxidation of guanine by Strecker. Guanine is a derivative of guano, which was formed from manure, especially that of birds and bats.

y. 1863. Trinitrotoluene (TNT) was prepared by J. Willibrand, German scientist. Although TNT was used for many years in the dye industry, TNT was not used as an explosive until 1904. Thereafter, TNT became one of the most used high explosives.

z. 1862-1866. Nitroglycerin production plants were designed for commercial applications by Alfred Bernard Nobel, a Swedish chemist, based on studies conducted with his father, Emmanuel, in 1859-61. In 1862, they built their first plant at Heleneborg, Sweden. Despite the tragic loss of his brother Emil in an explosion at that plant in 1864, Alfred erected two more plants in 1865 and another in 1866.

aa. 1865. Magnesium was first used for pyrotechnics.

bb. 1865. Nitrocellulose was purified by Sir Frederick A. Abel, an English chemist, who also successfully manufactured granulated nitrocotton. Abel pulped, washed, and compressed the nitrocellulose into blocks, sheets, discs, and cylinders, which were particularly useful for blasting.

cc. 1865. Smokeless powder, used as a sporting propellant, was invented by Schultze, a Prussian officer, through nitrating wood, washing the wood with a dilute solution of sodium carbonate, and drying.

dd. 1866-1867. Dynamite was invented by Nobel by absorbing nitroglycerin in kieselguhr (a diatomaceous earth). Kieselguhr, an inactive ingredient, stabilized the nitroglycerin, and made dynamite a much safer explosive to handle and use.

ee. 1867. The **fulminate blasting cap**, a device consisting of mercury fulminate in a copper tube used to detonate explosives, was invented by **Nobel**. This cap was crimped to one end of a safety fuse (e.g. Bickford fuse), and then inserted into the dynamite cartridge.

ff. 1867. Ammonium nitrate, pulverized and mixed with sawdust or charcoal, was proposed as an absorbant for nitroglycerin by J. H. Norribin and C. J. Ohlsson, Swedish inventors.

gg. 1868. H. **Julius Smith** makes the first **electric blasting cap**. On the other hand, dry compressed guncotton was detonated with a mercury fulminate detonator by E. A. Brown, an English chemist, who also discovered that wet guncotton could be exploded by the initiation of a small quantity of dry guncotton. This important discovery, on the principle of a booster explosive, led to the use of large blocks of wet guncotton in naval mines with comparative safety.

hh. 1869. Dynamites with an active base were patented by A. Nobel. To create an active base dynamite, Nobel mixed nitroglycerin with combustibles (sawdust, charcoal, rosin, and starch), and oxidizers (sodium nitrate or potassium nitrate). This led to a more efficient explosive than guhr dynamite.

ii. 1870. Ohlsson and Norribin's patent for ammonium nitrate dynamite was purchased by A. Nobel, who subsequently began production of ammonium nitrate dynamite.

jj. 1871. Sprengel explosives, which consisted of mixing various combinations of oxidizing agents (chlorates, nitrates, and nitric acid) with combustibles (nitronaphthalene, benzene, and nitrobenzene), were patented by Hermann Sprengel, a German chemist. Oxidizing agents and liquid fuel were mixed on the site before using.

kk. 1875. **Perry Gardiner** and **Julius Smith** make the first detonator that had separate primary and secondary explosives. The **modern detonator** is very similar to the Gardiner and Smith caps. On the other hand, the 'Explosives Act' was passed by the British Government after a disastrous explosion at Birmingham killed 53 persons. The **Explosives Act** established Inspectors of Explosives, who were given the power to inspect all magazines and factories to ensure that operations were carried out in a reasonably safe manner. As a result, the number of deaths in explosives factories was greatly reduced. The provisions of this Act have been adopted by many other countries.

ll. 1875. Gelatinous dynamites and blasting gelatins were patented by A. Nobel.

mm. 1877. Nitroguanidine was first prepared by Jousselin, but was not used in propellants until 1900.

nn. 1879. Tetryl, a booster explosive, was first prepared by Michler and Meyer of Germany.

oo. 1879. Ammonium nitrate gelatins, known as ammongelatins, were patented by A. Nobel.

pp. 1882. Brown or cocoa gun powder was made by the Germans using partially burned charcoal from rye straw. The grains were made in single, perforated hexagonal or octagonal prisms. **Brown powder** was of higher quality, but was more sensitive to friction than black powder. The United States used brown powder during the Spanish-American War.

qq. 1883. Tetryl was synthesized from picryl chloride and potassium methylnitramine by van Romburgh, a German chemist.

rr. 1883-1884. Colloided rifle powder was prepared by Max von Duttenhofer by nitrating brown charcoal before mixing the ingredients. This brown powder was adopted by the German Army in 1884.

ss. 1884. Smokeless powder, for military rifles, was invented by Paul Vieille of France. This powder was a completely colloided single-base nitrocellulose propellant. The French Army adopted Vieille's Poudre B in 1866.

tt. 1884-1885. High-speed photographic cameras were used for ballistic studies by Eugene Mach and P. Salcher of Austria.

uu. 1885. Two flash-less powders, ammonpulver and amidpulver, were patented and made by Gaens of Hamburg from ammonium nitrate, potassium nitrate, and charcoal.

vv. 1885. The Hell Gate blast in New York Harbor used 34,000 kg (75,000 lbs) of dynamite, and 11,000 kg (24,000 lbs) of potassium chlorate soaked in nitrobenzene (rack-a-rock) to remove Flood Rock, a menace to navigation. This was the largest amount of explosives used in a single blast until then.

ww. 1885-1888. **Eugene Turpin**, a French scientist, patented the process of **melt-pouring picric acid** into artillery shells as a high explosive filler. Sprengel had discovered that picric acid could be initiated by a powerful detonator, but had not exploited this knowledge. The French Government adopted a high explosive shell designed by Turpin using picric acid filler. They designated the picric acid filler as **melinite**. In 1888, the British adopted picric acid, which they called lyddite as a high explosive filler.

Fig. 43 Literally A Forest Of Shells: In Spite Of The Problems Associated With Picric Acid, It Was Used In Millions Of Artillery Shells Throughout The World In The Late 19th Century And Till The WWI

xx. 1888. The shaped charge effect was accidentally discovered by Charles E. Munroe, an American scientist, at the Naval Academy. Munroe discovered that a concave-shaped explosive charge pierces a steel plate.

yy. 1888. Cyclotrimethylene trinitrosamine was discovered by Griess & Harrow and by Mayer, yet was not used until the Germans, who called it R-salz, used the composition in World War II.

zz. 1888. Ballistite, the first double-base smokeless propellant, was prepared by Alfred Nobel by replacing camphor with nitroglycerin. In 1889, Nobel prepared ballistite from collodion cotton and nitroglycerin using a solvent-less method.

aaa. 1888-1889. **Cordite**, the second successful double-base smokeless propellant, was prepared by Dr. **W. Kellner** in F. Abel's English laboratory. Sir James Dewar, British chemist and physicist, and Abel patented cordite for the British Government, which adopted cordite as a military propellant named Cordite Mark I or CSP (cordite smokeless powder).

bbb. 1890. Tetranitrocarbazole, used in igniter and pyrotechnic compositions, was first prepared by Karl Graebe, a German chemist. The current method of preparation was developed in 1912 by the Leopold Cassella Company in Frankfurt, Germany, a division of the Interessen Gemeinschaft (I. G.) Farben corporation.

ccc. 1889-1890. The development of explosive devices was greatly enhanced by the work of **Theodor Curtius**, German Nobel laureate, who discovered hydrazine in 1889 and hydrogen azide in 1890. He subsequently suggested to the German military the use of **metal azides for primer** application.

ddd. 1892. Tetranitrooxanilide was prepared by A. G. Perkin, an English scientist, and used as a component in black powder type explosives and pyrotechnic compositions.

eee. 1894. Butanetriol trinitrate, used a half century later as an explosive plasticizer for nitrocellulose, was first prepared by Wagner and Ginsberg.

fff. 1895. Liquid oxygen explosive (LOX), carbon black packed in porous bags and dipped in liquid oxygen, was introduced by Karl P. G. von Linde of Germany. LOX was widely used by Germany during World War I, and in the United States Midwest for strip-mining in the 1950's.

ggg. 1896. A layered smokeless propellant, which was prepared by glueing and compressing together several layers of different components, was patented by A. B. Bofors, Nobelkrut of Sweden. The outer layers were slow-burning, while the burning speed increased progressively towards the inner layers.

hhh. 1897-1898. **Cheddite**, an explosive for blasting consisting of potassium chlorate or perchlorate mixed with a nitro compound and castor oil, was invented in England by E. Street. Cheddite derives its name from Chedde, France, where it was made.

iii. 1899. Potassium dinitrobenzfuroxan, used in priming compositions, was first prepared by P. Von Drost.

jjj. 1899. Cyclonite (RDX) was first prepared by Henning of Germany. Henning used RDX, which he called Hexogen, for medical purposes. The first to recognize the value of RDX as an explosive was E. Von Herz of Germany. The first to prepare RDX in quantity was G. C. Hale of Picatinny Arsenal, NJ, in 1925. However, it was not until World War II that the most important work in manufacturing processes and applications of RDX was done.

kkk. 1899. **Aluminized explosives**, containing aluminum to increase the performance of explosives, were first proposed by **R. Escales** of Germany and were patented by **G. Roth**. Ammonal, their first explosive, consisted of ammonium nitrate, charcoal, and aluminum.

(13.2) THE FURIOUS 20TH CENTURY KICKS OFF WITH GIANT STRIDES MADE IN THE PURSUIT OF THE FIRST WORLD WAR

With two World Wars in the first half itself, the 20th century was nothing but conducive for great developments in the field of explosives. In this section we will review the developments made in the pursuit of the World War I. About **2.5 million tons of high explosives** were used by the contending powers during the War, resulting in an estimated 10 million casualties.

This period saw the use of cordite smokeless powder and repeating rifles by the British during the Boer War (1899-1902) in South Africa; large-scale naval battles and full-scale use of machine guns in the Russo-Japanese War (1904-1905); and the introduction of submarine warfare, armored tanks, aerial bombing of civilians by dirigibles, and strafing and bombing by aircraft during World War 1 (1914-1918). World War I saw concentrated machine gun fire, artillery barrages, firing at long ranges from railroad guns and from large guns on battleships, firing torpedoes, laying mines, using depth charges in anti-submarine warfare, and signaling and illuminating with flares. Soldiers were issued steel helmets and gas masks for individual protection.

a. 1900. Cordite smokeless powder, used by the British in the Boer War, caused excessive corrosion in gun barrels. As a result, the formula for cordite was changed to reduce the amount of nitroglycerin from 58% to 30% and to increase the amount of nitrocellulose to 65% with the amount of mineral jelly remaining at 5%.

Fig. 44 Cordite Was The Main Propellant Of The Guns Of the First World War

b. 1900. Picric acid (British lyddite) shells were used in the Boer War but did not detonate completely. This was attributed to faulty construction of the detonators.

c. 1900. The **Contact Process for manufacturing concentrated sulfuric acid** and oleum was discovered by Badische Anilin-und Sodafabrik of Germany. This process was less expensive than the chamber method previously used. As such acid was required for mixed nitric-sulfuric acid to nitrate aromatic hydrocarbons to polysubstitution; trinitrobenzene and trinitrotoluene (TNT) could be inexpensively manufactured.

d. 1900. Chromatography, a physical method of separation in which the components are partitioned between two phases, was invented simultaneously and independently by American geologist D. T. Day and Russian botanist M. S. Tsvet. The qualitative method of separation of ingredients by adsorption was known for centuries.

e. 1900. Regular-delay electric blasting caps for rotational shooting (0.5 to 12 seconds) developed.

f. 1901-1904. A trinitrotoluene (TNT) manufacturing process consisting of three phases was developed by C. Haussermann of Germany. This method, which proved to be more economical than direct nitration of toluene to TNT, was used in the United States until World War II. The German industry adopted this process in 1902, when they adopted TNT as the filler for high explosive shells. They also used TNT for demolition purposes, replacing picric acid.

g. 1902. **Detonating cord**, developed in France, was improved, leading to adoption in the United States by the Ensign-Bickford Company in 1912.

h. 1902. **Antifreeze** for dynamite was researched extensively in many countries following a disastrous explosion during defrosting dynamite in Greisnau, Germany. The problem of producing a satisfactory antifreeze remained unsolved until **S. Nauckhoff** of Sweden published his work in 1905, listing requirements for a satisfactory antifreeze, and compounds which met those requirements.

i. 1903. A muzzle-flash suppressor using vaseline with 1-2% sodium bicarbonate was proposed by Rottweil Fabrik of Germany. This was replaced in 1905 by soap or rosin.

j. 1904. The Obermueller Stability Test, which consisted of heating 1-2 grams of nitrocellulose in a small tube under vacuum at 135-140° and measuring the pressure of evolved gas by vacuum manometer, was introduced.

k. 1904-1905. Methyl picrate (trinitroanisole) was patented for use in smokeless propellants. The powder consisted of an equal amount of methylpicrate and pyrocellulose.

l. 1905. Nitrostarch was produced in a stable form. Nitrostarch is similar to nitrocellulose but is lower in strength. Nitrostarch does not cause headaches from skin contact as do TNT, nitroglycerin, dynamite, and many other explosives.

m. 1906. Various substituted ureas, called centralites and akardites, were developed at the Central Laboratory for Scientific and Technical Research at Neubabelsberg, Germany, for chemical stabilizer application in smokeless powders.

n. 1906. **Lead azide** was proposed by **Col. A. A. Solonina** of Russia for use in compound **detonators** in lieu of mercury fulminate.

o. 1906. Hexanitroazobenzene was prepared by E. Grandmougin and H. Lehmann of France, and recommended by Tenny L. Davis for use in boosters.

p. 1907. Diphenylamine, was approved as a stabilizer for French naval cannon propellants and other large caliber guns. This stabilizer replaced amyl alcohol after the disastrous explosion of the battleship Iena in 1907.

q. 1907. Mixed glycol esters, such as dinitroacetin with dinitroformin, were patented as dynamite antifreeze by Vezio Vender of Italy.

r. 1908. The **detonating cord (cordeau detonant)** containing TNT instead of black powder was patented in France. This detonating cord had a detonation velocity of 16,000 feet per second.

s. 1909. **Ammonium picrate (Dunnite** or **Explosive D)** was standardized in the United States as a bursting charge for armor-piercing (AP) shells. These AP projectiles could be fired through 12 inches of armor plate, and could be detonated on the far side by an insensitive primer.

t. 1909. Low-nitrated aromatic hydrocarbons, because of their stability and insensitivity to shock and to ignition, began to be used in coal mining explosives compositions.

u. 1909. **Solvent-less powder/propellant** was discovered by **Brunswig** and **Thieme** of Germany, although many others contributed to the development of such propellants.

v. 1910. Preparation of tetranitroaniline, a powerful explosive, was patented by Flurscheim. This explosive was used during World War II by the Germans and Japanese.

w. 1910. Tetrazene was first prepared by K. A. Hofmann in Germany. More extensive studies of the compound were conducted in 1931 by Rinkenbach and Burton.

x. 1910-1913. **Laminated solvent-less propellants** were prepared by Dr. **C. Claessen**, a German scientist. Making propellants into large solvent-less grains was less time consuming than making solvent propellants.

y. 1912. TNT was adopted as the standard bursting charge in high explosive shells for the field artillery of the US Army.

z. 1912. Nitroisobutylglycerol trinitrate or trimethylol nitromethane trinitrate, used as an explosive gelatinizer for nitrocellulose, was first prepared by Hofwimmer.

aa. 1912-1913. Tetranitroaniline (TNA) was patented by Dr. B. J. Flurscheim, an English scientist.

bb. 1914. Lead styphnate (trinitroresorcinol), an initiating explosive, was first prepared by E. von Herz of Germany. Col. A. A. Solonina of Russia was the first to propose using lead styphnate in compound detonators.

cc. World War I. The high-explosive technology of the contending powers greatly influenced the conduct of World War I. At the outbreak of war in the summer of 1914, the German General Staff had planned to fight a high-explosives war of great intensity. Their plan was to mount a sudden and overwhelming attack using vast stores of artillery ammunition accumulated over the preceding forty years. After the battle of the Marne in 1914, the Germans realized that explosives production must be expanded; and they turned to their dye industry for this expansion. When the war began, Germany's chemical industry was producing 85% of the world's dyes. The Germans were able to rapidly convert their plants to production of synthetic ammonia, nitric acid, and sulfuric acid, which were required for the manufacture of explosives and chemical warfare agents. When Germany's access to Chilean saltpeter was cut off by the Allied blockade, the key to the **German war production of explosives** was the **Haber process for producing ammonia** from atmospheric nitrogen. By expanding their nitrogen plants at Oppau and Merseburg, Germany was able to become independent of foreign countries for the supply of ammonia and nitric acid. Because these chemicals were so essential for the manufacture of explosives and fertilizer for food products, it is doubtful if Germany could have continued the war beyond 1916 without the amazing speed with which the I. G. Farben concern was able to convert their dye plants to explosives plants. For example, at Leverkusen, a TNT plant producing 250 tons per month was put into operation in six weeks! The major powers began accelerating the search for substitutes or supplements to TNT. The most important were mixtures of high explosives, such as TNT, TNX, and TNN, with ammonium nitrate and aluminum. By 1914, the French had already begun replacing picric acid (melanite) with TNT. In 1915, the British developed amatol (TNT plus ammonium nitrate) to extend the supply of TNT. In 1917, Austrian ammonal T (TNT, aluminum, and ammonium nitrate) was developed by R. Forg to be used in underwater explosives, such as torpedoes and depth charges. Lead azide was introduced as an initiator, and tetryl was used as a booster explosive for artillery shell charges.

dd. World War I. **High-explosive aerial bombs** were first dropped from German Zeppelins during the 31 May 1915 raid on London.

ee. World War I. **Incendiary bombs** were also first used during the German Zeppelin raid over London on 31 May 1915, when one dirigible dropped 90 incendiary bombs. The early incendiaries used by the Germans in their air raids over London and Paris were made from a mixture containing manganese dioxide and magnesium. Later, the French used sulfur-bound thermite, a mixture of iron oxide and powdered aluminum, in an incendiary drop bomb.

ff. World War I. **Flamethrowers** were first introduced by the Germans on the French front on 25 June 1915. Their weapon was a cylindrical steel vessel about 2 feet high and 15 inches in diameter, fitted with carrying straps. Internally, the cylinder was divided into a compression chamber and an air reservoir. A short hose carried the fuel to the nozzle where it was ignited. The propellant was nitrogen. The fiery spurt of burning oil, the roar of the flame, and the billowing clouds of black smoke had a terrifying effect on troops in trenches. Using this prototype of the modern portable flamethrower, the Germans drove back a British rifle brigade from trenches near Ypres, Belgium, in 1915. The Americans used a heavy viscous tar and a more fluid, flammable liquid such as crude benzene for their flame fuel mixture. A hydrogen pilot lamp at the flamethrower nozzle was used to ignite the flame fuel. The World War I flamethrowers had a range of about 30 meters. Although effective for assaulting fortifications, the flamethrower was also very hazardous for the operator because of its short range.

gg. World War I. **Smoke munitions** were used both on land and sea by the belligerent powers in World War I. In July 1915, the British were the first to use smoke pots, which were filled with pitch, tallow, black powder, and potassium nitrate. The first large-scale smoke operation occurred on 20 September 1915, when the Canadians fired several thousand smoke shells from trench mortars during the attack against Messines Ridge. Although very successful against the Germans, the British and French tanks of World War I were capable of traveling only four miles per hour, which made them very vulnerable to artillery fire. The British tank attack at Cambrai on 20 November 1917 was supported by intense artillery smoke shelling, which enabled 350 tanks to break through the German lines. On 15 June 1918, the Austrians used smoke to screen their crossing of the Piave River, from Vidor into San Giovanni, from Italian machine gun concentrations. The Allies generally used white phosphorus (WP) to fill their smoke shells; whereas the Germans, because of a shortage of phosphorus, used a mixture of sulfur trioxide in a chlorosulfonic acid solution. The Germans used this type of smoke (FS) to successfully screen their warships during the battle of Jutland in 1916. Later, the United States adopted FS smoke for use in spotting rounds, for liquid chemical agent shots, and in airplane smoke tanks. FS was reclassified obsolete in November 1975 because WP was considered better for marking targets and FS droplets from aerial spray destroyed the nylon fabric in parachutes. Titanium tetrachloride (FM) was also used as a filler for spotting rounds during both world wars, but was removed from the US inventory because of its high cost. In 1918, tanks carried containers filled with chemicals, which when injected into the vehicle exhaust system formed a cloud of screening smoke. Sulfuric acid, chlorosulfonic acid, oleum, and crude oil were used as the smoke screening agent in these early smoke generators. WP hand and rifle grenades were used by the infantry for screening movements. The main disadvantage was the danger of the user being exposed to falling particles of burning white phosphorus, which is also an incendiary. WP grenades were reclassified obsolete in November 1975 by the US Army.

hh. World War I. **Incendiary artillery shells** and rifle bullets were introduced in 1915 to set fire to buildings, supply dumps, and enemy observation balloons. The Germans used sodium in their artillery shells. They also used thermite or an aluminum-thermite mixture in their other artillery rounds. The Allies used thermite most frequently. Thermite incendiary grenades were issued to raiding parties for hasty destruction of captured enemy weapons that could not be retrieved.

ii. World War I. **Pyrotechnics for battlefield illumination** were used extensively during World War I. The British alone used 10 million position light flares per month. The art of making illuminants had been neglected after the reign of Louis XIV until the twentieth century. In 1913, the German Krupp Works developed an illuminating projectile, which they called a searchlight shell and which later became known as the star shell. The first star shell contained six small tubes, called stars, loaded with an illuminating composition. The mixture used was similar to that in the white Bengal lights for fireworks displays; but ribbons of magnesium were substituted for antimony powder. A folded parachute was placed in the bottom of each tube. The projectile was fired from a cannon or mortar. When the projectile reached a height of 300 meters, a mechanical time fuze set off a small black powder charge releasing the tubes from the shell and igniting the pyrotechnic mixture. Springs were used to open the parachute instantly, and the tube with its burning star floated slowly to earth. The illuminating material burned very brightly for 45 seconds to several minutes depending upon the model. The French constructed a similar projectile containing eight cylinders, which was fired from their 155 millimeter howitzer. However, they preferred using burning-type fuses to set off the powder charge in their star shells. Later, the Germans replaced the star tubes with a six-sided prism, which was more compact and could be adapted for use with rifles, special pistols, and large caliber carbines. By using plates to retard the fall of the star, more space was made available in the star for storing the illuminating mixture; and a light of greater intensity or longer duration was obtained. The Italians used a hand-held, heavy pasteboard tube to discharge a sheet-iron star shell tube, which expelled its illuminating parachute flare. This flare burned for about 40 seconds. Many different illuminating devices were developed and used by both sides, but perhaps the simplest and most useful for trench warfare was the illuminating grenade. Thus every soldier on night patrol had his own device for lighting his sector, and exposing the advancing enemy to aimed gunfire.

jj. World War I. During World War I, **pyrotechnics for signaling** included colored smoke cartridges, flares, streamers or trails, star signals, and night signals. Four general designs of fireworks devices were in constant use by the French and the British armies. These devices were Very-type signal pistols, rifle lights, rockets, and ground or position signals (e.g. candles). (The same type of devices was used for illumination.) The signal-pistol cartridges used by the French were fitted with delay fuzes, which effected the ignition when the projectile reached its maximum height. Some realization of the development in pyrotechnic construction due to the war may be had by considering the latest production of the French 35- millimeter signal cartridge, which included 14 varieties or combinations of signals. Upon entering the war, the United States adopted many of the French signals for Army use. This was mainly because the American forces were to operate in the French sector, which necessitated an identical system of signals. Generally, the fireworks and devices employed by the Germans were similar to those of the French, but there was an important difference in the method of construction for the signal-pistol cartridge. The German cartridge was ignited just after emerging from the barrel of the pistol. Rockets mounted on sticks for stabilization during flight were employed by the Allies. The Germans followed Congreve's construction in the use of metal, but modified the design to eliminate the use of a stick. In the German rocket, the openings for escaping gas were in the base. Before the outbreak of the war, the art of pyrotechnics had reached such a point that little difficulty was encountered in securing adequate mixtures for the various colors required in signaling. Different colored flares had been used for many years commercially. The railroad companies regularly employed pyrotechnic devices, such as red railroad fusee, while blue light pyrotechnics were used for signaling at sea. The colors used by the Allies included red, green, yellow, blue, black, various numbers of white stars, and changing colors. The Italians used a tri-color parachute flare, which emitted red, white, and green colors. The Italians also developed a hissing flare, which, when suspended from a parachute, burned with a hissing noise audible over an area of 3,000 meters (3,284 yards) in diameter. This was found particularly effective as a gas alarm. Some of the ingredients used in the WW I pyrotechnics were aluminum, potassium nitrate, sulfur, antimony sulfide, antimony metal, dextrine, meal powder, barium chlorate, potassium chlorate, fine charcoal, barium nitrate, red gum, paris green, shellac, sodium oxalate, magnesium carbonate, strontium carbonate, calomel, copper oxychloride, calcium carbonate, sawdust and grease mixture, sugar, lamp black, ultramarine blue, paranitraniline red, lactose, auramine, chrysoidin, synthetic indigo, phosphorus, paraffin, and arsenic.

(13.3) WITH THE SECOND WORLD WAR LOOMING LARGE, DEVELOPMENTS CONTINUE AT A FEVERISH PITCH

With the impetus gained by the First World War, research into explosives continued at a feverish pitch as the world prepared for another round of hostilities. The victors had to secure their gains; the vanquished had to restore their lost pride! The World War II was going to be a huge consumer of explosives. The Americans alone used over **six million tons of explosives in air-dropped munitions,** that is, more than twice the amount of explosives used by all the powers put together during the First World War!

During these two decades, RDX, PETN, EDNA, DEGN, lead styphnate, and dextrinated lead azide were developed as military explosives. The development in the United States of processes for producing toluene from petroleum removed limitations on the availability of TNT, and permitted development of the powerful and castable explosives like Composition B and pentolite. Flash-less propellants were developed in the United States, and low erosion DEGN propellants were developed in Germany and Italy. Diazodinitrophenol was developed as an initiator in the United States, and tetrazene was developed in Germany as a new explosive ingredient of priming compositions. World War II (1939-1945) saw the development of rocket propellants based on nitrocellulose-nitroglycerin or nitrocellulose-DEGN mixtures, and the use of nitroguanidine in artillery propellants. New special-purpose binary explosives, such as tetrytol and picratol, were developed for use in demolitions, chemical bombs, and semi-armor-piercing bombs. A number of plastic explosives used for demolition work were developed in Great Britain and the United States, the most important being the C-3 composition based on RDX. The discovery and great value of the blast effect of explosives led to the development of tritonal, torpex, and minol, which contain powdered aluminum and have powerful blast effects. Incendiaries were found to be even more effective than high-blast explosives in the destruction of cities. By using improved fire bombs filled with thickened or gelatinized incendiary oil or gasoline mixtures, and large clusters of incendiary bomblets filled with thermate (an improvement of the WW I thermite), the allied air forces were able to literally set fire to cities like Dresden and Tokyo. Flamethrowers and smoke munitions were used extensively by both sides. Finally, the extensive use of armor during the war led to the development of anti-tank weapons, armor-piercing shells, and shaped charge munitions. The application of the shaped-charge principle resulted in the use of special explosives of the pentolite type.

a. 1918-1939. Considerable research was devoted to the problem of preserving propellants.

b. 1918-1939. Development of the art of rocketry and rocket propellants misses some golden opportunities. Just before the Armistice in 1918, Dr. Robert H. Goddard, Professor of Physics at Clark University, demonstrated a shoulder-fired "recoilless gun" or "rocket gun" at Aberdeen Proving Ground, MD. The results of Dr. Goddard's demonstration were summarized by Ordnance Department witnesses as proving the validity of his concept of modern rocketry. However, the lack of suitable powder and the need for further work on the electrical firing mechanism along with the Armistice led the Ordnance Department to shelve the project. Goddard died without receiving any acclaim for this pioneering work, although comparison of his rocket gun with the bazooka adopted 24 years later shows how closely the 1918 model resembles the later weapon. During the 1920's, metriol trinitrate or trimethylolethane trinitrate, used as an ingredient in rocket propellants in World War II, was prepared in Italy and Germany. In 1931, the US Army created a one-man rocket unit by assigning Captain Leslie A. Skinner to study the possible use of rockets. Skinner was handicapped by limited funds for research, and by the indifference of superiors. During the mid-thirties, the British began more intensive research on anti-aircraft rockets in anticipation of massive bomber air raids upon their cities. In the meantime, the Germans made great strides in the development of rockets and missiles and the attendant propellants. This was especially due to the work of Wernher von Braun in the late 1930's.

c. 1920. Cardox, a device for breaking coal in gaseous mines by the pressure produced on heating liquefied carbon dioxide, was developed. Similar devices were developed in the United States in 1930, and were improved upon in 1955 (Hydrox) and 1959 (Chemecol).

d. 1927. Diethyleneglycol dinitrate was thoroughly examined by W. H. Rinkenbach, and the current method of preparation was patented in 1928 by A. Hough.

e. 1930-1931. Dextrinated lead azide was adopted for use by the US Army, and began replacing mercuric fulminate in priming compositions.

f. 1931. Dinitrotoluene (DNT), obtained earlier as an impurity in crude TNT, was used in some explosive mixtures.

g. 1935. From the 1920's onward, scientists at the Picatinny Arsenal had been trying to find a compound that would have the high brisance of RDX without the sensitivity to friction and impact. Research on this problem, principally by Dr. George C. Hale, chief chemist, led to the discovery of ethylenedinitramine (**EDNA or haleite**) the first entirely American high explosive. More powerful than TNT, EDNA was slightly less powerful then RDX but was also less sensitive. EDNA's stability gave it an important advantage in considerations of manufacturing, loading, storage, transportation, and field use. Designated haleite, in honor of Dr. Hale, this new explosive could be press-loaded into small shells without a desensitizing agent, and its derivative, ednatol, a mixture containing 42 percent TNT, could be melt-loaded into large shells as easily as amatol. These advantages were offset in pre-war days by the high manufacturing cost of one of its intermediates, ethylene urea. Delay in solving manufacturing problems prevented haleite from getting into combat in World War II.

h. 1936. **Primacord**, the **modern detonating cord** based on a French patent, was developed by the Ensign-Bickford Company of USA. Primacord consisted of PETN covered with textiles, waterproofing material, and plastics. This detonating cord had a velocity of 21,000 fps, and has been used extensively by the armed forces for demolition work.

i. 1937. Baratol, a mixture of barium nitrate and TNT, was developed by the British and used extensively in WW II.

j. World War II. **High explosive bombs** are unleashed in all their fury.

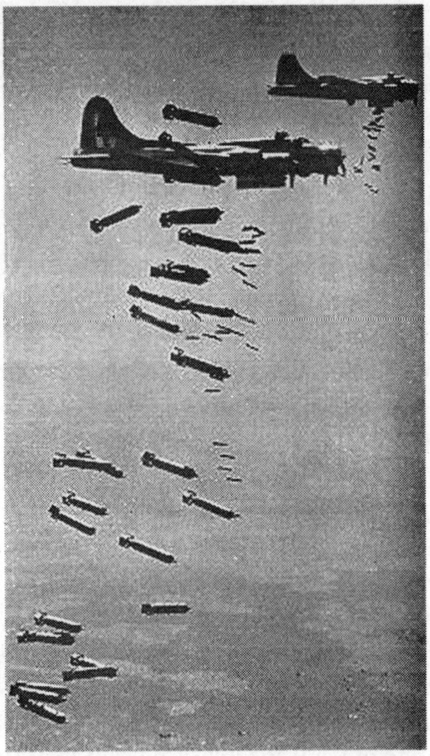

Fig. 45 B-17 Flying Fortresses Unleash Their Load Of Bombs: The USAF Dropped Over 6 Million Tons of Bombs In The Second World War

k. World War II. **Reverse Nitration of TNT**. During 1941 and 1942 the inadequate supply of TNT for high explosive bombs and shells was a major problem for Ordnance. Because the shortage had been foreseen, plans were made to use a substitute explosive called amatol (a mixture of TNT and ammonium nitrate) for shell or bomb loading until new TNT plants came into full production. Even so, a critical shortage of TNT developed despite these plans. But the shortage suddenly disappeared when a new process appeared on the scene almost by accident. During an inspection of a small Canadian TNT plant at Beloeil near Montreal in 1941, John P. Harris of Ordnance discovered that the plant was 'doing things backward' by putting toluene into the acid instead of putting acid into the toluene. Despite some resistance by US TNT producers, the new process was tried at the partly built Keystone Plant at Meadville, PA. The result was a tripling of TNT output. Lines designed to turn out 16 tons a day produced more than 50 tons a day. The need for TNT substitutes vanished, and the cost per unit was cut in half.

l. World War II. **Toluene from Petroleum**. Development of a new means of producing toluene, the basic material from which TNT is made, was another highly significant technological advance of World War II. The importance of this chemical stems from the fact that nearly half of every unit of TNT (trinitrotoluene) must come from toluene. In World War I, toluene was derived from coal as a by-product of coke ovens, and some was extracted from natural gas. Yet the supply in 1917-18 was so limited that the Assistant Secretary of War, Benedict Crowell, later called this shortage "the greatest and most pressing of all the problems in regard to existing raw materials." In contrast, during World War II high-explosives production was never seriously hampered by lack of toluene. As early as 1927, Standard Oil of New Jersey had obtained patent rights from a German firm to use a process for producing toluene from petroleum. In 1939, a contract with Standard Oil was placed by Picatinny Arsenal to prepare toluene by this process at Baytown Ordnance Works, which was erected next to the Humble (a Standard Oil subsidiary) refinery in Baytown, TX. By October 1942 this plant was producing toluene at the rate of 246 million liters (65 million gallons) per year, which when compared with a total toluene production of less than 34 million liters (9 million gallons) in the US during 1918 shows us the value of this process.

m. World War II. **Artillery Propellants**. The problems of obtaining artillery propellants that were **both flash-less and smokeless** continued to plague the U.S. Army.

n. World War II. **Wood Pulp and Cotton Linters**. Before the war, the standard practice for making smokeless powder called for the treatment of bleached cotton linters with a mixture of nitric and sulfuric acid. In the summer of 1941, it became apparent that even with a good cotton linters crop and capacity operation of bleacheries, the supply of linters for powder would fall short of requirements. Ordnance chemists, therefore, turned to the use of a special wood pulp that was available in quantity, was suitable for most kinds of powder, and was cheaper than cotton linters. The Hercules Powder Company had made smokeless powder from wood pulp for a number of years, and Ordnance found the product completely acceptable. Use of wood pulp as a supplement of, but not a replacement for, cotton linters was approved promptly except for rifle powder and certain other uses. Soon most plants had one or two lines for nitrating cotton, an equal number of lines for nitrating wood pulp, and several 'swing lines', which were adaptable for use of either material. From January 1942 to the end of the War, Ordnance plants used roughly equal quantities of cotton linters and wood pulp. There was never a concurrent shortage of both materials, although there were times when the supply of one ran low or was expected to run low. At such times, Ordnance drew upon its inventories while converting the swing lines to meet the situation. The use of wood pulp doubled the existing supply of cellulose for powder, and eliminated a serious potential bottleneck in ammunition production.

o. World War II. Research is undertaken into **rocket propellants**.

p. World War II. **New explosives** are tried **for artillery projectiles**. Considerable power combined with sensitivity, which made for easy loading, stability, and safety in handling and transport, made TNT and amatol the preferred high explosive fillings for most high-explosive artillery shells at the outset of World War II. This was largely due to their availability in large quantities. As the war progressed and ammunition became more complex in design and more specialized in function, demand arose for improved explosives. This demand could not be met to any extent because the explosives developed between wars did not get into large-scale production in time. Nevertheless, throughout the war the US Army Ordnance Department sought ways of using more powerful new explosives, and ways to adapt old ones to special purposes. There were at hand several explosives of higher shattering effect, or brisance, than TNT. The most important were, first, cyclotrimethylene trinitramine, which the Americans called cyclonite and the British called RDX; second, pentaerythritol tetranitrate or PETN; and third, EDNA (later called haleite). RDX and PETN were too sensitive to be used in the pure state in a shell. Therefore, to form usable compositions it was necessary to combine them with oils or waxes or with other explosives. The British had managed to desensitize RDX by adding nine percent beeswax to form Composition A for press-loading into shells; with 39.5 percent TNT and one percent beeswax to form Composition B, chiefly for bomb loading; and with 11.7 percent of a plasticizing oil to form Composition C, for demolition work. These formulas were provided by the British to the US in 1940, and development work was undertaken by the DuPont Company under contract of the Ordnance Department. The most sensitive of all high explosives was PETN, which was even more readily detonated than RDX. PETN was desensitized by mixing with TNT to form a composition named pentolite, which was been extensively used in detonators, bazooka rockets, rifle grenades, boosting devices, and in the shaped charges of antitank shells. Because of production difficulties, haleite, the most promising of the new high explosives, never saw much use in World War II except in the form of ednatol in mixtures with other explosives.

q. World War II. New explosives are tried for **armor-piercing and anti-tank projectiles**. Much of the work of improving high explosive compositions was directed toward finding the most efficient filling for anti-tank shells. For armor-piercing projectiles, relatively insensitive ammonium picrate, "Explosive D," had long been preferred. As it was not likely to detonate on impact, the shell could penetrate the armor plate before exploding. But experience with heavy German tanks in North Africa showed that something more was needed in the way of power and fragment action coupled with greater incendiary effect within the tank. Chemists at Picatinny Arsenal accordingly tried several expedients. In armor-piercing shell, addition of a small amount of powdered aluminum to cyclotol (a mixture of RDX and TNT), to ednatol, or to TNT produced more brisance than Explosive D and increased sensitivity to impact. In high-explosive antitank shell fillings, conversely, the difficulty was the exact opposite. The high sensitivity of pentolite made it liable to detonation on target impact so that the problem was to desensitize it to the proper degree. Several possibilities considered were the addition of wax to the pentolite, the reduction of the PETN content, and the substitution of ednatol or Composition B. None was entirely satisfactory. The search for an explosive composition of the greatest possible power and brisance took a new turn after analysis of foreign explosives at Picatinny Arsenal during 1943. Hitherto, research had been concentrated on binary explosive compositions such as pentolite, Composition B, or ednatol. The examination of a Soviet 76 mm high explosive armor-piercing round suggested the possibility of employing ternary mixtures. Tests revealed that castable ternary explosive mixtures, such as RDX-Tetryl-TNT and Haleite-PETN-RDX, offered promise not only for armor-piercing projectiles but as fragmentation ammunition for weapons designed to produce blast, and for demolition charges. Further study showed that haleite ternaries were unstable. The best combination seemed to be a mixture of PETN-RDX- TNT designated PTX-2 (Picatinny Ternary Explosive). More brisant than any of the binary compositions, PTX-2 was more stable than 50/50 pentolite and less sensitive to impact. Preliminary firings at Picatinny indicated that PTX-2 would be particularly adaptable to shaped-charged ammunition; but PTX-2 was still in the testing stage at V-J Day.

r. World War II. Research continues into the field of **shaped charges**.

v. World War II. **Flame and Incendiary Munitions**. In the intervening years from World War I to World War II, the development of flame weapons was studied by most nations except the United States. During the Ethiopian War in 1935 and 1936, the Italians employed the first armored vehicle-mounted flamethrowers. The Germans employed a flame tank during the Spanish Civil War and in 1939 used portable flamethrowers in Poland. These instances proved to the United States that the development of flame weapons could no longer be denied. In 1940, the US Army took steps toward developing a flamethrower and a flamethrower tank. These flamethrowers served an important role in the war in the Pacific, particularly in dislodging the determined Japanese soldiers from caves and fortifications. On Okinawa alone, flamethrowers are credited with the capture or death of 20,000 Japanese soldiers who were holed up in the caves while less than 100 American flamethrower operators were killed in the operations. The development of napalm ("nap" for the napthenic acids and "palm" for the coconut fatty acids which made up the first type of this thickner) by the United States in 1941 made possible the greater use of gasoline for use in flame weapons. This gelled mixture made it possible for aircraft to deliver fire bombs over hard-to-get pinpointed areas and for flamethrowers to project a flame rod over a greater distance than was previously obtainable. By August 1944 the United States AAF Board had come to the conclusion that "where there is vulnerability to fire, the damage by fire is greater than by demolition," a conclusion, to be sure, that observers had reached during the London blitz of 1940 but which had not been acted upon. The decision of the Board in 1944 was to increase the incendiary bomb program to the highest priority. The United States Army CWS provided the Army Air Force with about **48 million incendiary bombs,** which were effectively used, especially on the large cities of Japan. Most of these bombs were clustered bomblets containing thermate (TH3 and TH4), a new incendiary mixture, which replaced the original thermite mixture. When ignited, thermate burns intensely at 2200°C. Molten iron from burning thermate will melt metal, burn through steel plate, and ignite flammable materials. Oil and metal incendiary mixtures, such as PT1 and PTV, were also used as fillings in large thin-cased bombs. PT1 is a complex mixture based on "goop," a paste comprised of magnesium dust, magnesium oxide, and carbon, with a sufficient amount of petroleum distillate and asphalt to form the paste. PTV is an improved oil and metal incendiary mixture of polybutadiene, gasoline, magnesium, sodium nitrate, and p-aminophenol. However, napalm proved to be one of the most effective aerial incendiary agents in the war. On the evening of 9 March 1945, more than 300 Superfortress (B29 bombers) swarmed over Tokyo dropping about 2,000 tons of incendiaries, mostly clusters of M69 six-pound bomblets. The M69 bomblet was a tube which contained a black powder propellant charge that ignited and projected the napalm filling from the tube. Aerial photos indicated 16 square miles had been burned out. Tokyo police records, examined after the war, showed that more than 250,000 buildings were destroyed (about one fourth of Tokyo) in this one raid. More than 100,000 tons of incendiaries were dropped on the cities of Japan during World War II. Most of these were M69 bomb clusters. For destroying materiel to prevent enemy use, the CWS developed a thermate grenade, which was standardized as the AN-M14 TH3 incendiary grenade for use by the Army and Navy. Harvard University also developed a napalm fire starter (MI), for use by troops to ignite fires in wet climates. The armament of the American heavy bombers was usually the .50 caliber machine gun. During World War II, enemy fighter aircraft did not have armor plating capable of deflecting a .50 caliber bullet, although, bullet penetration of the plane's armor did not necessarily knock out aircraft. The enemy's use of self-sealing fuel tanks necessitated the development of effective incendiary ammunition. Most urgently needed was an effective .50 caliber incendiary round. The first acceptable design was the work of the Remington Arms Company whose staff already had considerable experience in working on Swiss patents for incendiary ammunition. The Remington development was based upon the British .303 B Mark VI Z, and was adopted in September 1941. The bullet was a flatbase type with lead base closure and steel body, and was charged with 35 grains of a incendiary mixture consisting of 50 magnesium alloy and 50 percent barium nitrate. An improved design was developed shortly thereafter by Frankford Arsenal and was standardized for United States use.

(13.4) THINGS NOVEL AND EXOTIC RESULT FROM POST-WORLD WAR II RESEARCH AND DEVELOPMENT

The nuclear bombs that ended World War II did not end requirements for conventional explosives and ammunition. As was shown by the Korean War (1950-1953) and the Vietnam War (1964-1973), the need for the United States to develop even more effective munitions and more powerful explosives or energetic materials became more apparent than ever. In fact explosives were consumed on an unprecedented scale during the war in Indochina. The Americans threw in as much as **15.3 million tons of explosives** in the small areas of **Indochina**, that is, more than twice the amount they had used in the Second World War in all the theaters of war!

In spite of intensive effort, chemists at the US Army's laboratories and arsenals had failed during World War II to develop a new explosive composition for shell loading that was both satisfactory and readily available in quantity. The obstacles were often disheartening. The characteristics of an explosive might be considerably affected by impurities that existed in the raw material or which were admitted during manufacture. For example, the instability of PETN was probably due to impurities in the raw material, pentaerythritol. Other variables that had to be taken into account were the different methods of testing compositions and the differences in interpreting results. Assuming a composition had been found that promised to combine greater brisance with less sensitivity, there was still the question of whether it could be economically manufactured, safely handled, and made unchanging in character in temperatures ranging from arctic cold to tropical heat. In the search for explosives with special properties, much work had been done in the field of aluminized explosives, but, although aluminized TNT (tritonal) was used in bombs, much remained to be done.

a. 1945. PLX (Picatinny Liquid Explosive), consisting of 95 percent nitromethane and five percent ethylenedinitramine, was developed at Picatinny Arsenal by L. H. Eriksen and J. W. Rowan for use in mine clearing. The Germans had used similar liquid explosives during World War II under the name Myrol.

b. 1945. Tripentaerythritol octanitrate was patented by J. A. Wyler of Trojan Powder Company, PA, for use as a high explosive and possible gelatinizer for nitrocellulose.

c. 1946-1949. **Composition C-4**, known as Harrisite, was developed by **K. G. Ottoson** at Picatinny Arsenal as a replacement for Composition C1, C-2, and C-3. Composition C-1, C-2, and C-3 were the United States standard plastic RDX demolition explosives, and were originally derived from Britain's Composition C, which was developed during World War II.

d. 1946-1949. **Benite**, a black powder substitute, was developed at Picatinny Arsenal for use in igniter compositions of artillery primers, or in base igniter bags for separate-loading ammunition. Benite is in the form of extruded strands consisting of ingredients of black powder in a matrix of nitrocellulose.

e. 1946-1949. Composition D-2, which is not an explosive but serves as an emulsifier and desensitizer of explosives like HBX-1, was developed. Composition D-2 consists of 84 percent wax, two percent lecithin, and 14 percent nitrocellulose.

f. 1950-1953. When World War II ended, the United States had a tremendous inventory of munitions on hand, but unfortunately it was not a balanced stock. There were enormous quantities of some types of munitions, and only small amounts of others. The Korean War demonstrated the point that there is a need for better industrial mobilization planning, and for closer contact with industry in the development and maintenance of a viable production base.

g. 1950-1973. During the Korean conflict, both **portable and mechanized flamethrowers** were used and found to be very effective. Flame mines, using incendiary bursters to set off 55-gallon drums of napalm, were used in a defensive role. The large batches of napalm required for these operations were mixed in truck-mounted service units. M1 fuel thickening compound was used to gel the gasoline for flamethrowers and landmines. An undesirable feature of M1 thickener was that it absorbed moisture rapidly from the air, which degraded the gelling properties so that the gel became unstable. Army chemists at Edgewood Arsenal, MD, began work on a non-hygroscopic thickener, which was standardized later as M4 flame fuel thickening compound. The M4 flame fuel thickening compound is a di-acid aluminum soap of iso-octanoic acids derived from iso-octyl alcohol or iso-octyl aldehyde, which are obtained from the oxidization of petroleum. Because of the large World War II surplus inventory of M1 thickener, M4 thickener was not produced in quantity until the Vietnam War. The Air Force used tremendous quantities of fire bombs. These bombs were essentially large fuel tanks filled with napalm, which were ignited by pyrophoric igniters (WP or sodium) upon impact. Continuous process mechanical mixers were developed for use at Air Force bases to mix and fill the fire bombs on the spot. M1 thickener could be mixed only in batches, so a new thickener, M2, was developed and fielded. M2 flame fuel thickening compound is an intimate mixture of 95 percent M1 thickener and five percent de-volatilized silica aerogel or other approved anti-agglomerant. M2 thickener was an improvement of the M1 thickener for use in fire bombs. This is because of the free-flowing and fast-setting characteristics, which allow its use in continuous mechanical mixers. M2 thickener was not suitable for use in flamethrowers, which require an aged gel obtained in the batch process. Napalm B, a special fire bomb fuel formulation, was developed by the Air Force to replace M2 thickener. Napalm B currently consists of 46 percent polystyrene, 21 percent benzene, and 33 percent gasoline by weight. Napalm B withstands high velocity impact dissemination more effectively than M2 thickened gasoline gels, resulting in less fuel burned en route and more fuel delivered on target.

h. 1950. **MOX (metal oxidizer explosives)** were developed by National Northern Technical Division of the National Fireworks Ordnance Corporation, West Hanover, Mass., for use mostly in small caliber antiaircraft shells.

i. 1952. Military medium velocity dynamite was developed by W. R. Baldwin, Jr. at Hercules Powder Company Laboratory.

j. 1952. **PB-RDX (plastic-bonded RDX)** was developed by the Los Alamos Scientific Laboratory of the University of California for use as a mechanical strength explosive. PB-RDX consisted of 90 percent RDX, 8.5 percent polystyrene, and 1.5 percent dioctylphthalate.

k. 1957. Military low velocity dynamite was developed at Picatinny Arsenal by H. W. Voigt.

l. 1958. HTA-3 consisting of 49 percent RDX, 29 percent TNT, and 22 percent aluminum was prepared at Picatinny Arsenal.

m. 1958. Octols 70/30 and 75/25, mixtures of HMX and TNT, were developed at Northern Corporation as fillers for bombs and shells. Octols are standard high explosive fillers, which have been adopted for Army use.

n. 1958. Slurry explosives were developed by M. A. Cook and H. E. Farnham by adding water to ammonium nitrate to form slurries of the oxidizer salt, ammonium nitrate and sodium nitrate, and a solid fuel sensitizer. Slurry explosives provide three to six times the detonation force of ammonium nitrate-fuel oil (ANFO) explosives.

o. 1960. **Detaclad process** or **explosive cladding**, which is also called **explosive bonding**, was developed by the DuPont Company.

p. 1960. Detacord, detaflex flexible cord explosive, and Detasheet flexible sheet explosives were developed by the DuPont Company.

q. 1964-1973. The Vietnam War, as it dragged on, spurred research in many fields and even new munitions were developed. For example, although flamethrowers were used extensively, a stand-off capability to project flame was needed to reduce the vulnerability of the operator to enemy fire. From this requirement, a 66 mm multi-shot flame rocket launcher (M202 and M202A1) was developed. The resulting M74 incendiary TPA (thickened pyrophoric agent) incendiary 66 mm rocket was fielded and is now a standard item. The rocket is filled with TPA, which consists of triethyl aluminum (TEA) thickened with polyisobutylene (PIB). Other significant developments were a broad range of various 40 mm cartridges and signaling devices, improved mines, and pyrotechnics for battlefield illumination.

r. 1973. The problem of managing a rapidly increasing number of conventional ammunition items in the Army inventory was addressed at the conclusion of the war in Vietnam. Hundreds of obsolescent and non-essential items of ammunition and explosives were eliminated. Among these were FS smoke and M1 thickener.

s. 1973-1976. To protect tanks, the United States Army devoted considerable attention to the problem of rapidly screening tanks from enemy antitank rounds and missiles. One of the first items adopted for Army use was a British smoke grenade launcher, which fires salvos of red phosphorous (RP) UK L8A1 **smoke screening grenades** from armored vehicles. After launch, these grenades detonate in the air to form an immediate, dense white smoke cloud that obscures the tank from enemy view. Smoke grenade launchers have been developed for use in many of the armored vehicles in use or development today, including the M1 tank.

□□□

Chapter 14

PROPELLANTS: TO SHOVE HARD BUT NOT TO SHATTER

(14.1) THE DESIRE TO MAKE A SMOKELESS POWDER

For centuries black powder remained the propellant of both small arms and cannons. And yet, smoke from the black powder was a major problem. It was a dead giveaway of your position whether you happened to be an artilleryman or an infantryman. Gun smoke is due to particulate, usually inorganic, material emerging from the gun behind the projectile. Primer and igniter combustion products, barrel scrapings, and solid combustion products from the propellants are the major constituents of gun smoke. Then the smoke left acidic residues also which led to corrosion of the gun barrels. Another pain was keeping the black powder dry. In the end, people desired something more powerful too. Hence people wanted a powder that would be smokeless. The first solution to these problems came in the form of nitrocellulose. Pure nitrocellulose was used for some time but it was rather volatile and eroded barrels quickly. Since then propellants have come a long way. Let us undertake this marvelous journey once again.

(14.2) SINGLE-BASE PROPELLANTS

The first smokeless powder called **Poudre B** (Poudre Blanche—white powder) was invented by a French chemist Paul Vieille in 1886. It was made of two forms of **nitrocellulose** (NC) (collodion and guncotton) softened with alcohol and ether and kneaded together. The Poudre B is an example of what came to be known as **single-base propellants (SB propellants)**.

In 1833, Braconnot had nitrated starch, and in 1838, Pelouze, continuing the experiments of Braconnot, also nitrated paper, cotton and various other materials but did not realize that he had prepared nitrocellulose. Work on the nitration of cellulose to produce nitrocellulose (also known as guncotton) was undertaken by different workers, notably **Schonbein** at Basel and **Bottger** at Frankfurt-am-Main during 1845-47. With the announcement by Schonbein in 1846, and in the same year by Bottger that nitrocellulose had been prepared, the names of these two men soon became associated with the discovery and utilization of nitrocellulose.

Many catastrophic accidents occurred during the preparation of nitrocellulose, and manufacturing plants were destroyed in France, England and Austria. The accidents were attributed to insufficient purification of the product. Sir Frederick Abel discovered in 1865 that the amount of acid retained in the fibers of nitrocellulose was responsible for the lack of stability. Thus one of the most important problems in nitrocellulose technology became the purification process which is responsible for increasing the stability of the product as far as possible. The formation of sulfuric acid esters was assumed to be responsible for the low stability of the nitrocellulose. Abel showed that by a process of pulping, boiling and washing the stability of nitrocellulose could be greatly improved. Boiling removes the acid occlusions or breaks the esters down. Still, nitrocellulose was not used in military and commercial explosives until 1868 when Abel's assistant, E. A. Brown discovered that dry, compressed, highly-nitrated nitrocellulose could be detonated using a mercury fulminate detonator, and **wet, compressed nitrocellulose** could be exploded by a small quantity of dry nitrocellulose **(the principle of a booster)**. Thus, large blocks of wet nitrocellulose could be used with comparative safety. Nitrocellulose is insoluble in water, which allows its preparation, stabilization, and transport by quenching, using the appropriate quantity of water.

The energetic performances of NC depend on the nitrogen content. The higher the nitrogen content, the higher the heat of explosion. More nitrogen is present in the molecule and more oxygen can be used to oxidize the reducing agents (carbon and hydrogen). This leads to a better oxygen balance in the molecule. For military purposes, the propellants with the highest nitrogen content are used because the highest performances are required. Therefore two main products are employed:

- Grade A called **pyrocellulose**: nitrogen content between 12.3 and 12.8%
- Grade B called **guncotton**: nitrogen content higher than 13.35%.

The main ingredient of SB propellants as they exist today is basically nitrocellulose (or more correctly, various nitrates of cellulose), which has been made colloidal by the action of a suitable solvent. The rest of the composition consists of plasticizers. They exist, for example, in an inert form like dibutyl phthalate, dibutyl sebacate , dibutyl adipate, and camphor or as energetic ones like the isomeric mixture of dinitrotoluene and 2,6-dinitrotoluene . The role of the plasticizers is to act as surface moderants and at the same time as coolants. The principle of a surface moderant is to retard the initial burning rate, the initial gas generation rate, and the initial flame temperature. Typical concentrations of plasticizers are between 1 and 10%.

(14.2.1) Manufacturing Single-Base Propellants

Wet nitrocellulose from the manufacturing process is dehydrated after the moisture content has been reduced to approximately 28 percent by wringing. Dehydration is accomplished by pressing the nitrocellulose at low pressure so as to squeeze out some water, adding 95 percent ethanol, and pressing at about 3,500 psi. The nitrocellulose is transferred to a water-cooled mixing machine of the dough-mixer type and, while this is in operation, ether equal to approximately two-thirds of the weight of dry nitrocellulose is added. Any plasticizing agents and stabilizers to be included in the composition are dissolved in or mixed with the ether prior to addition to the nitrocellulose. After addition of the ether is complete, materials such as potassium nitrate are added. Mixing of the ingredients is continued for about one hour. This effect is further increased and uniformity of the mixture is improved by subjecting it to a pressure of about 3,500 psi in a macaroni press. Here the material is squeezed through a series of screens and perforated plates and emerges in a form resembling that of macaroni. This is pressed again to form a block of well colloided material. This is placed in a graining press and extruded through a carefully designed die by the application of pressure. The material emerges as a cord with one or more cylindrical perforations. The cord is cut into pieces of predetermined length.

(14.3) DOUBLE-BASE PROPELLANTS

Single-base propellants were followed by **Ballistite** made by none other than Alfred Nobel. It consisted of nitrocellulose, nitroglycerin and camphor. Since there are two high explosives, it had greater power than nitrocellulose alone. This was called a **double-base (DB) propellant**. The idea of mixing camphor was that it reacted with any acidic products of the chemical breakdown of the two explosives and thus theoretically increased the stability. But it tended to evaporate over time, leaving a potentially unstable mixture.

DB propellants in general contain nitroglycerin or other nitroglycol compounds and nitrocellulose. Nitroglycerin or other nitroglycol acts as plasticizer that is energetic as well. Typically, double base propellants contain 50.60% nitrocellulose and 30.49% nitroglycerin. The immediate advantage of the DB propellants over SB propellants was that due to the solvent-less production technology for the DB propellants, bigger size propellant grains with large wall thickness (web) and bigger blocks of propellant with different and complicated geometries became possible.

The formulations, however, continued to suffer from the problem of instability in storage. The British resolved this problem in 1889. Frederick Abel and James Dewar patented what became famous as **Cordite**. This was made of nitrocellulose, nitroglycerin and Vaseline. Using acetone as a solvent, it was extruded in the form of cords and hence the name Cordite. The early composition was somewhat corrosive and they changed it a little to result in **Cordite MD** (for modified). Shortage of acetone led to the development of what is known as **Cordite SC** (solvent-less cordite). In its various forms, Cordite remained the main propellant of the British forces until the 1930s.

(14.3.1) Manufacturing Double-Base Propellants

Double-base propellants are manufactured by two methods. The **solvent process** is similar to that used for single-base powders except that a mixture of ethanol and acetone is used as the solvent and the solvent recovery procedure is omitted because of the hazard involved in recovering solvents containing nitroglycerin. The **solvent-less process** is used when the nitroglycerin and any other colloiding agents constitute approximately 40 percent of the composition. In this process the wet nitrocellulose is blended with the nitroglycerin in a tank filled with water. Ethyl centralite is mixed in and the bulk of the excess water is removed by centrifuging. The resulting paste is put in cotton bags and subjected to heated air currents to reduce the moisture content. The remaining constituents are then blended with the partly dried paste. Repeated rolling between heated steel rollers removes the rest of the water and completes colloiding of the nitrocellulose. The thickness of the sheet formed is controlled carefully and varies with use. If the sheet is to be cut into flakes for use in small arms or mortars, the thickness is between 0.003 and 0.0125 inch. Sheets to be extruded in the form of large grains for use in rockets may be as thick as 0.125 inch.

(14.4) TRIPLE-BASE PROPELLANTS

Cordite was fine otherwise and served for a long time but it had a fundamental problem. The burning temperature was quite high and that caused erosion of the barrel of the gun. Then there was the problem of muzzle flash also. The problems of obtaining artillery propellants that were **both flash-less and smokeless** continued to plague the armies of the world.

Several attempts were made to get around this problem. People tried using diethylene glycol dinitrate instead of nitroglycerin in DB propellants. Diethylene glycol dinitrate is much less sensitive to shock than NG and its heat of explosion amounts to only 4480 J/g as against 6100 J/g for NG. The people tried triethylene glycol dinitrate also. It delivers an even lower heat of explosion (3140 J/g). Still, they remained only expedient solutions.

The problem was eventually solved by the US Navy. Early in the World War II, the US Navy seized upon nitroguanidine as the one feasible answer to their problems. For the first time, American ships in the Pacific were preparing to fight in small harbors where maneuvering was all but impossible. Flash at night betrayed the vessel's position, and smoke by day made second rounds inaccurate. Nitroguanidine was added to Cordite to produce a propellant called **Cordite N** which had a lower burning temperature and thus reduced barrel erosion. Muzzle flash was also reduced as nitroguanidine produces large amounts of oxygen when heated. DB propellants with picrite or nitroguanidine added to the formulation are considered as a separate class and are called **Triple Base (TB) propellants**. When used by the Germans in World War II in antiaircraft guns, a nitroguanidine propellant increased the barrel life from 1,700 firings to about 15,000 firings.

Fig. 46 From The Smallest Pistols To Great Guns Such as This, Propellants Drove Them All

(14.4.1) What Is Nitroguanidine?

Nitroguanidine is also known as picrite or guanylnitramine. It may be classified as one of the least sensitive of military explosives. At least two crystalline forms exist for nitroguanidine; alpha and beta. The alpha form is obtained by the reaction of guanidine and nitric acid to form the nitrate. Several methods for the preparation of nitroguanidine are known. The earliest method was by the direct nitration of guanidine thiocyanate with mixed acids. However, this method of production also produced sulfur compound impurities which attacked nitrocellulose. This lowered the stability of propellant compositions to an unacceptable degree, thus precluding early use of the compound as an ingredient in nitrocellulose based propellants. A more pure form of nitroguanidine that does not contain the sulfur compound impurities can be prepared in one of the several known ways. In one method equimolecular quantities of urea and ammonium nitrate are fused. The product is then re-crystallized from boiling water.

(14.4.2) Manufacturing Triple-Base Propellants

The manufacturing process used for the nitroguanidine triple-base propellants in the United States has been uniformly solvent extrusion. The amount of solvent used is quite low so the propellant is very soft during extrusion. The soft strands may require partial drying before cutting in order not to deform the cross section at the cut. Removal of solvent from the triple-base propellant is rapid, possibly due to diffusion of solvent within the grain along the crystal-plastic interfaces. In order to make a good quality grain, lower drying temperature gradients are used in order to avoid steep solvent gradients resulting in distortion and cracking.

(14.4.3) Going Beyond Nitroguanidine

Whatever the advantages of nitroguanidine, neither it nor any other composition was ideal for all purposes. Even in conventional artillery and small arms ammunition, where ballisticians understood propellant behavior better than in rockets and recoilless rifle ammunition, compromises were inescapable. The primary requisite for one weapon or one particular use tended to be different from every other weapon or use. In addition to these problems, World War II introduced the problem of climate extremes at which firing had to take place when Allied troops were fighting in arid deserts, damp jungle heat, or in the sub-zero winter weather of northern Europe and of the Aleutian Islands. Therefore, a series of propellants were needed to cover widely varying contingencies. Since basic research as well as prolonged applied research was necessary, many problems remained unsolved till the end of the War. But the field was explored more thoroughly than ever before in the United States, and lines of investigation were clarified for post-war development. In processing propellants, industry and the Ordnance Department made considerable advances during the course of the war. One new method developed by the Hercules Powder Company for washing nitrocellulose in a continuous filter instead of in large tubs by the old "settle and decant" system, washed the nitrocellulose more thoroughly and thus improved the stability of nitrocellulose. The DuPont Company found that using preheated alcohol during the winter to dehydrate nitrocellulose reduced the hydration time cycle, improved the yield, and made for a more uniform product. This in turn produced a much better powder. On the other hand, the Americans found at the end of the War that Germany had developed several processes more effective than those of the United States. The most novel German method was one of casting propellant grains by adding a paste of moist nitrocellulose and diethyleneglycol dinitrate (DEGN) to molten TNT and pouring the mixture into steel molds to cool. Grains as large as 1,000 millimeters were cast this way. After the war, complete sets of the German equipment deemed most useful and novel went to Picatinny Arsenal for study.

Nitroguanidine gives off obnoxious ammoniacal fumes, but a good substitute is hard to come by. Adding potassium sulfate to propellants for antiaircraft fire, where flash-lessness was all important, helped alleviate the problem to some extent.

(14.5) HOW DIFFERENT TYPES OF PROPELLANTS ARE USED FOR DIFFERENT JOBS

In nearly all calibers of handgun, rifle, machine gun, aircraft- and antiaircraft gun, cannons and howitzers, SB propellant grains or sticks of any geometric shape can be found. In bigger caliber cannons and howitzers, SB propellant grains or sticks can be part of the whole charge. Single-base propellants are also used in case-less ammunition for small caliber guns. NC with low nitrogen content is still used in combustible cartridge cases and in combustible containers for modular charges for big caliber guns.

The following geometric forms of powder grains are manufactured:

- perforated long tubes
- perforated tubes
- cut short multi-perforated tubes
- flakes
- strips
- ball powder
- cubes
- rods
- cut short rings

Ball powders can be found in ammunition for pistols, rifles and machine guns; ammunition for rifles can also contain cylindrical grains or short tubes. Flake powders are restricted to ammunition for antitank guns and mortars. For aircraft and antiaircraft guns, grains in the geometry of short tubes or hollow cylinders are used. In modern aircraft and antiaircraft guns the use of the more progressive burning 7-perforated grains is normal. Tank guns use the even faster burning 19-perforated grains. Long stripe- and cross-stick propellants are found in ammunition for cannons and howitzers.

Finer-grained powders are used for portable firearms; tubular powder is mostly employed for guns; powders in the form of flakes and short tubes are employed for mortars, howitzers, and other high-angle firearms. Finer-grained powders can be improved in their ballistic behavior by W Surface Treatment. Phlegmatizers are infiltrated in the outer layer of the powder grains; the burning rate in the weapon chamber begins slowly and turns progressive.

Fig. 47 Propellants Are Made In Different Shapes And Sizes For Different Purposes

Large caliber cannons and howitzers use DB (or Triple Base) propellant in the shape of tubes or slotted tubes. Flake powders are found in ammunition for pistols and grenade launchers. Amongst rockets and missiles, the anti-tank missile Dragon has rolled up DB propellant sheets in its numerous small motors. The ejection of the missile from the launcher is done by a charge of short tubes of DB propellant. The launch motor of the anti-tank missile TOW is constructed of big size tubes. The launch motor of the famous Russian anti-tank rocket RPG-7 consists of DB sticks and the flight motor consists of a hollow DB cylinder. To produce bigger DB rocket motors, casting technology is used. An example of this propellant type is the anti-aircraft missile Rapier with its launch and flight phase propellant cast into the same motor.

(14.6) THE EFFECT OF PROPELLANT GRAIN GEOMETRY ON PROPELLANT PERFORMANCE

The shape of the grain determines how much surface area is exposed as the burning proceeds. For example, as a cord burns, the surface area of the exposed propellant decreases. This is called a **degressive grain**. As a single perforated grain burns, the exposed surface area remains nearly constant because the outer surface area decreases as the inner surface area increases. This is called a **neutral grain**. In a **progressive grain**, for example the multi-perforated and rosette shaped grains, the surface area increases during burning. The burning time of a propelling charge that contains a propellant with a high burning rate and strongly degressive grain geometry could be equal to the burning time of a charge that contains a propellant with a low burning rate and strongly progressive grain geometry.

European nations have favored the use of single perforation, strip, and cord propellants. The United States prefers single perforation and multi-perforation propellants. The single perforation grains may be slotted or un-slotted. The slotted grain has the desirable characteristic of venting gas during combustion. In a multi-perforation or rosette grain that consists of a propelling compound with a specific linear burning rate, the burn time can be varied by varying the web thickness. The web thickness is the average dimension of the outer web and inner web. In general, the finer the web the shorter is the burning time of the propellant charge.

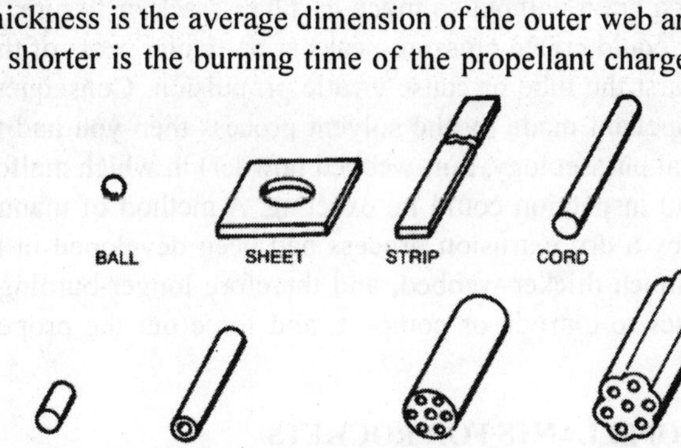

Fig. 48 Different Geometries Of Propellant Grains

The grain dimensions also influence the burning time. The finer the web the higher the pressure produced. Cannon is designed to operate with a maximum pressure. This pressure must be taken into account when designing the grain web. The same projectile velocity can be attained with a smaller amount of propellant with fine web and with a larger amount of propellant with a coarse web. However, the pressure with the former is considerably greater than the latter. In general, the heavier the projectile or the greater the prescribed velocity, the larger the powder grain and web used.

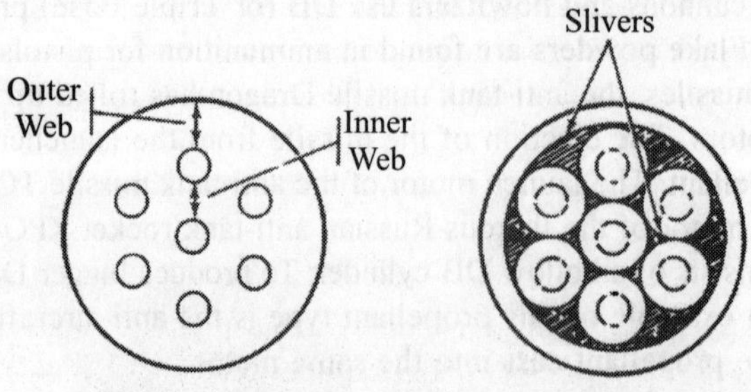

A - Unburned Grain B - Burning Grain

Fig. 49 The Effect Of Web Thickness On Propellant Burning Characteristics

(14.7) THE QUEST OF SUITABLE ROCKET PROPELLANTS IN THE EARLY YEARS

Difficult though the US program was for improving propellants for conventional weapons, the question of developing suitable rocket propulsion was still harder to answer. From the very beginning everyone concerned with rocket research agreed that single-base powders would lack the necessary energy and that double-base powders, that is nitrocellulose and nitroglycerin combined, must be used. Double-base powders, made by the solvent process, had been manufactured in the United States for a number of years for use in intermediate and large caliber artillery. The method used a solvent of acetone and alcohol to make a colloid, from which grains or flakes could then be formed. About 1939, the Hercules Powder Company found a way of producing a solvent-less double-base propellant, which was plasticized by heat and pressure and then rolled out into sheets for mortar increments. But neither was adaptable for rockets where a solid stick or grain, not sheet propellant, was needed, and where the essential quality of uniform burning precluded using a grain without so much as a hair crack or fissure. Even a tiny crack in a burning grain of powder could create pressure peaks at particular spots of the casing of the rocket motor and thus either burst the tube or cause erratic propulsion. Consequently, if you had to get a safe usable rocket propellant made by the solvent process then you had to use grains of small cross-section (in technical phraseology, thin-webbed powder) in which malformations occurring in drying would be few and inspection could be exacting. A method of manufacturing solvent-less double-base propellant by a dry extrusion process had been developed in the late 1930's by the British. This produced much thicker-webbed, and therefore longer-burning, grains, but required enormously heavy presses to extrude or compact, and force out the propellant into the desired shape.

(14.8) COMPOSITE PROPELLANTS FOR ROCKETS

The size, dimension, and geometry of propellant grains based on SB and DB propellant is limited. For long-range rockets or missiles other types of propellants were needed. This eventually led to the development of the composite propellants during the WWII. **Composite propellants** are based on an oxidizing solid, commonly a perchlorate, together with an organic binder, which acts as fuel and gives also adequate mechanical strength to the resulting propellant. The two main oxidizing compounds used in composite propellant manufacture are ammonium perchlorate (AP) and ammonium nitrate (AN). Ammonium perchlorate is more frequently used. The excellent mechanical properties allow propellant grains to be manufactured in dimensions larger than are possible with SB or DB propellants. This and the high-energy content of composite propellants make them suitable for use in medium and large size solid rocket motors. Ammonium perchlorate has the added advantage that it gives a suitably low burning rate. Because rocket propellant must burn for a relatively long time, the burning rate has to be low and a high heat of explosion is not important.

In the beginning, the binders-cum-fuel used were tar or rubber. Today, a curable polymeric binder is loaded with an oxygen-rich, crystalline solid (mostly perchlorates), and a metal (mainly aluminum). After mixing, the highly viscous mass is cast into big and complicated molds. After curing either a **case-bonded** or **free-standing solid rocket motor** is obtained. A case-bonded motor means that the case and the cast-cured motor are one unit, whereas a free-standing motor is a propellant block which is later applied into a case.

Later, with the evolution of polymer technology, more modern polymeric fuels were introduced like polymethacrylates or polybutadienes. The polymers used in rocket propellant technology are quite similar to those used in high explosives technology (PBX). As the polymer is only partly responsible for the reducing part of the redox reaction in the formulation, some metals are added in the composition. The most popular ones are aluminum, zirconium (high density), beryllium (very energetic but toxic), boron, and magnesium.

During the burning process of ammonium perchlorate a lot of hydrochloric acid is produced. With the moisture in the air, this creates an intense white smoke. Now you understand why there is such a tremendous amount of smoke when the Space Shuttle is launched. The weight of the composite propellant of a single booster motor for the Space Shuttle is 480-499 tons! The immense clouds of smoke look pretty impressive when non-military rockets like the Space Shuttle and the Ariane are fired. Smoke is, however, a liability for military rockets, particularly relatively short range tactical rockets like anti-aircraft or anti-missile missiles. The reason is that the trajectory and the starting point of the missile can easily be observed. To reduce the amount of smoke you must add substances like magnesium-aluminum alloys or sodium nitrate that will suppress the hydrochloric acid.

Fig. 50 The Immense Amount Of Smoke Produced By The Solid Rocket Boosters Of The Space Shuttle—They Have 480-499 Tons Of Composite Propellant And Provide 2.8 Million Pounds Of Thrust

Ammonium nitrate on the other hand, does not create the problem of hydrochloric acid production, but its main drawbacks are the different crystal modifications. A crystal modification change in a rocket motor is fatal. It will influence the physical properties and cracks in the propellant could occur thus enlarging the burning surface. Under operational conditions the propellant may produce more gases than the nozzle can manage and the rocket or missile can finally explode. The use of AN for rocket motor technology also needs the addition of a crystal structure stabilizer, e.g. potassium nitrate.

The only use of composite propellant in guns is restricted to the so-called Base Bleed projectiles. For extended long range shooting, mostly with 155 mm caliber howitzers, a composite propellant grain is burned off during the flight of the shell. The emitting gases of the slow-burning Base Bleed grain do not deliver thrust, but limit the drag at the end of the projectile. This reduction of drag results in a significantly longer reach of the projectile.

(14.8.1) Manufacturing Composite Propellants

A composite propellant is manufactured by a simple mixing operation, cured, and molded into grains or poured directly into the rocket case and cured by heating in controlled ovens. Rockets may have to operate from temperatures as low as -40 degrees to +60 degrees Celsius. The high solid content requires great resin strength and good elastic properties to withstand low temperatures. Precise control of the oxidizer particle size is needed to control ballistic properties. Still, solid propellant technology has been perfected to such an extent that rockets up to 21.6 feet in diameter and weighing over a hundred thousand pounds have been fired successfully.

(14.9) THE PROBLEM OF PRESERVING PROPELLANTS

Ammunition research in the first post-World War I years was inspired not only by new requirements, but also by the problems encountered in **preserving propellants**, stored after the Armistice. Research proceeded in three areas: determining the stability of smokeless powder, studying the effects of long term storage on stability, and creating more efficient methods of drying. In 1926 the Picatinny Arsenal investigated the feasibility of the vapor method of drying, which reduced drying time from months or weeks to days. But a more permanent solution of some phases of the powder storage problem would have been to develop new non-hygroscopic powders, which because of their chemical composition would not absorb enough moisture to affect their ballistics or chemical stability even when stored in a damp atmosphere. If, at the same time, flash-less and smokeless qualities could be incorporated, the advantages would be even greater. Thus the search for **flash-less non-hygroscopic (FNH) powders** was vigorously pushed. The DuPont Company, under a special agreement with the Ordnance Department, followed one route of investigation, while Picatinny Arsenal followed another. Each achieved considerable success. The peacetime development of complete single-base and double-base non-hygroscopic powders, flash-less in many weapons, was one of the most useful accomplishments of the Ordnance Department before 1940. A sample of British MD cordite containing petroleum jelly or petrolatum as stabilizer was found to be of apparently unchanged stability after 30 years of temperature-climate storage.

(14.10) STABILIZATION OF PROPELLANTS

Nitrocellulose has poor stability. The degradation of nitrocellulose proceeds by two chemical reactions. In the first degradation reaction, nitrocellulose loses nitrogen oxides. The loss of nitrogen oxides is an intrinsic tendency of the nitrocellulose molecule. The reaction rate varies with the temperature, but even at low temperatures the loss of nitrogen oxides occurs. In the second degradation reaction, which occurs initially as a result of the first, the nitrogen oxides that are formed attack the nitrocellulose molecule. The reaction between the nitrocellulose and nitrogen oxides produces more nitrogen oxides. The increased number of nitrogen oxides and increased temperature cause the reaction rate to accelerate. After a few years the reaction rate in samples of sufficient size is so fast that the propellant self-heats above the flame point and may self-ignite. Otherwise there may be unacceptable variations in ballistic and mechanical properties of the propellant. The most effective means of controlling the degradation reaction is to include a **chemical stabilizer** in the propellant composition. A stabilizer has a greater affinity for the nitrogen oxides than nitrocellulose, and so absorbs them before they can attack the nitrocellulose molecule. This reduces the net rate of propellant decomposition by taking care of the second mechanism of degradation. The first degradation reaction remains unaffected by the stabilizer. However, the stabilizer is consumed as it absorbs nitrogen oxides. Therefore the increase in time the propellant will remain stable is directly proportional to the amount of stabilizer added. Some propellants can be stabilized in this manner for 30 or 40 years.

The origin of stabilizers goes back to the invention of Poudre B by Paul Vieille in 1886, when he first recommended the use of amyl alcohol, but abandoned it after 1905 for diphenylamine, considered a better stabilizer. Presently, three stabilizers are of interest for military applications; diphenylamine (DPA), ethyl centralite (EC), and the akardites (AK). DPA is used in single-base propellants but is incompatible with nitroglycerin and so cannot be used in double-base and triple-base compositions. EC is used for double and triple-base propellants which use nitroglycerin as the gelatinizing agent for the nitrocellulose. Akardites are used in propellants that contain DEGN rather than nitroglycerin.

Since stabilization is a simple chemical process, a study of the different nitro derivatives of diphenylamine gives precise information about the ageing state of a propellant. Some of the degradation products are: nitrosophenylamine, 2-nitro-diphenylamine, and 4-nitro-diphenylamine. It was also found that if the quantity of diphenylamine introduced into DB propellants reaches or exceeds 3.4%, an incompatibility problem occurs and thus poses a limitation on the amount of stabilizer that you can use. However, it was also found that one of the degradation products 2-nitrodiphenylamine still possesses stabilizing properties and is compatible with nitroglycerin (NG). Hence it can be used in some double base propellants.

The discovery of more recent stabilizers which are completely compatible with NG and other nitroglycols, allows new DB and triple base propellants to be produced which are better stabilized and may be stored for a longer period.This also leads to a prolongation of the shelf life of ammunition in a fully safe manner. Some of the modern stabilizers used include: diethyl diphenylurea centralite, dimethyl diphenylurea centralite, diphenyl ethyl methylurea centralite, nitro methylaniline, diphenylurea acardite, diphenyl methylurea acardite, and diphenyl ethylurea acardite,

(14.11) BURNING ACCELERATORS OR REGULATORS

As the name implies, burning accelerators or regulators modify the rate of burning or the combustion velocity. The use of inorganic and metallic salts may enhance the combustion velocity of propellants. For example potassium cryolite, an aluminum fluoride compound, is able to accelerate the burning rate of propellant. Copper chromite, which is used as catalyst for the degradation of perchlorate, also accelerates the rate of burning of propellants. The mechanism of this process is not yet known.

For composite propellants, ferrocene and ferrocene derivatives, are known to be efficient burning rate accelerators. It seems that the ferrocene group activates the decomposition of perchloric acid in the first step of the burning reaction. The commercial compound Butacene from the French company SME Propulsion is often used in composite propellants. Catocene is a ferrocene derivative often used in the contemporary propellants. Liquid catalysts have a negative influence on the rheological properties of the propellants. To avoid this problem, a new generation of catalysts, based on carborane molecules has been introduced. An example is n-hexylcarborane.

The introduction of carbon black in the propellant formulation, obtained by calcinations of hydrocarbons, regulates the burning rate of propellant, leading to a more efficient thermal exchange.

(14.12) MUZZLE-FLASH REGULATORS

Two components contribute to muzzle flash: the primary flash and the secondary flash. The primary flash is caused by incandescent hot gases emerging from the muzzle or heated to incandescence by the shock following the projectile out of the gun. The secondary flash is caused by the burning of combustible muzzle gas when mixed with air and heated to ignition by the shock wave. The secondary muzzle-flash is due to a post-combustion of gases, carbon monoxide, and hydrogen, resulting from the combustion in the chamber. The incorporation of compounds able to liberate a sizeable quantity of inert gas during the combustion will reduce this post-combustion. As we saw, this is what nitroguanidine does. Various chemical additives, especially alkali metal salts, are very effective and widely used to suppress secondary flash. The salts may be incorporated into the compositions, or they may be loaded as separate components in the charge. Potassium sulfate, for example, is a very good anti-catalyst for the post-combustion phenomenon. The disadvantage of potassium salts is, however, their high water solubility.

(14.13) SOME NOVEL PROPELLANTS

Two very fast burning materials, **BTATz** and **DHT**, have been successfully synthesized and are under evaluation as enhanced performance gun and rocket propellant ingredients. Metastable Intermolecular Composites (MIC) developed under this program were the first successful examples of nano-structured energetic materials with significantly enhanced performance. They demonstrated that tailored, ultra-fine reactant particles could dramatically increase the energy release rate of thermite-like materials and provide twice the total energy of high explosives. The first application of this technology is for lead-free percussion primers for small arms ammunition, and this program is now in engineering development. The current focus is on the optimization of this material for other weapons applications via better diagnostic and measurement methods.

(14.14) BI-MODULAR CHARGE SYSTEM

When the artillery is small, say 105 mm or less, the projectile and the propellant are contained in a single unit. This becomes a round--the familiar brass casing. When you are using

155 mm or bigger guns, and if you still want to contain the propellant charge and the projectile in one container, the round will become too big and too heavy for handling. Hence you separate the propellant charge. The propellant charge, for quite some time was contained in cylindrical bags of very fine silk—the silk burnt completely leaving no residue or embers.

The next development was to dispense with the bag. They developed propellants which could be cast in the form of a cylinder and which was strong enough to withstand handling. This is very much in use even now. Then it was desired that if you could have the cast charge in modules you could use them in increments as desired--for example you could use the same elevation of the gun for firing at different ranges simply by changing the amount of charge used as that will change the velocity of the projectile. In other words, using the modular charge makes the charge more versatile. It also increases the rate of firing substantially.

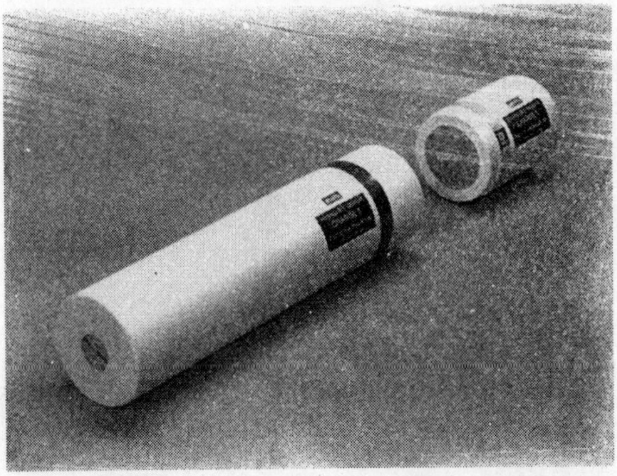

Fig. 51 Non-Modular Charge System For 155 MM Artillery

Fig. 52 Bi-Modular Charge System For 155 MM Artillery

Bi-modular charge system simply means that there are two types of propellants and you can use different combinations of the two. For example the US Army's modular charge system uses two modules of M231 charge for low zone firing and four modules of M232 charge for high zone firing.

❑❑❑

Chapter 15

Initiating Devices: Safety Fuse, Detonators and Detonating Cord

(15.1) SAFETY FUSE: BURN SLOW, BURN STEADY

Do not belittle this simple but important invention. Before the safety fuse was invented, lighting a charge and getting away to safety before it exploded in you own face was a serious problem and numerous accidents used to take place. An open powder trail used to burn very fast and if you used a wick or something of that sort instead for carrying the flame, it was not reliable under field conditions. The safety fuse, invented by **William Bickford** of England in 1831 accomplished two things. In the first place, it provided for reliable ignition under field conditions. More importantly, the safety fuse enabled one to get not only sufficient but controllable delay in igniting the charge so that you could get away to safety—and hence the name of 'safety fuse'! For historical reasons it is also called **cannon fuse**. That made the field use of gunpowder infinitely easier and also paved the way for many innovative uses of gunpowder.

Bickford was inspired by watching a friend making rope! In its original form it was indeed a hemp cord consisting of gunpowder and a water-resistant cover. Many refinements were made over the years. These days the safety fuse consists of a core of gunpowder enclosed in paper tape or strands of wound jute or cotton yarn, and then covered with one more layer of jute or cotton yarn. It is then coated with bitumen/asphalt and a coloring composition. It may also be covered with suitable waterproofing material such as PVC (poly vinyl chloride) or LDP (low density plastic), etc.

When you light it, the powder burns along the length of the fuse and thus carries the flame forward. The burning rate varies between 30 to 45 seconds per foot. Ever wondered why the safety fuse does not burn as fast as an open powder trail? After all that was the beauty of the invention! This is what happens in a safety fuse. The rate of burning of gunpowder depends on the ambient pressure on it. For an open powder trail, it is only the atmospheric pressure and hence it burns fast. Take it where the atmospheric pressure is low and it will burn faster. In a safety fuse, as the black powder core burns it produces gases, which must escape. The covering resists this escape. But the heat produced by the combustion melts some of the bitumen and gutta-percha and thus produces some side ventilation through the textile layers. The eventual pressure of gases on the powder is that resulting from the equilibrium between the generation of gases and the loss of gases. It is this pressure, being higher than atmospheric pressure which makes the fuse burn slower than an open powder trail.

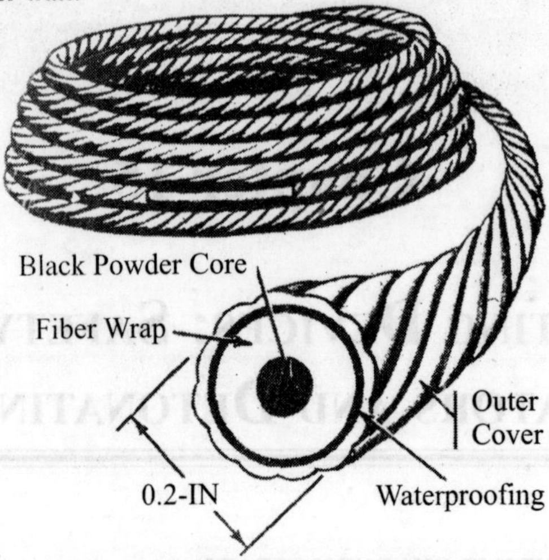

Fig. 53 Construction Of Safety Fuse

Modern safety fuses can burn underwater also. For military purposes fuses are now lit by a **pull ring type fuse igniter**. This device is for igniting timed blasting fuse in all weather conditions, even underwater, if properly waterproofed. Pulling the pull ring releases the striker assembly, allowing the firing pin to initiate the primer, igniting the fuse. When the fuse igniter is not available, the fuse can still be lit by a matchstick or two. For this the fuse is split slightly near the end and the match stick is inserted in it so that the match head only projects from one side. The match head is lit in the usual manner. It is the initial conflagration of the match head that does the trick. Merely holding the flame of a match against the fuse may not ignite it dependably as then the match releases heat rather slowly.

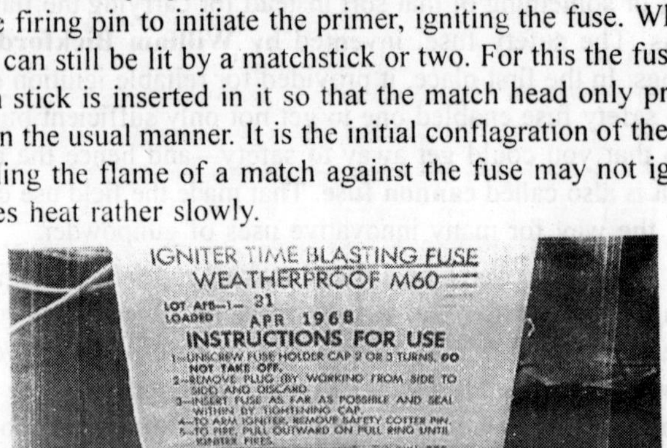

Fig. 54 Pull Ring Type Fuse Igniters

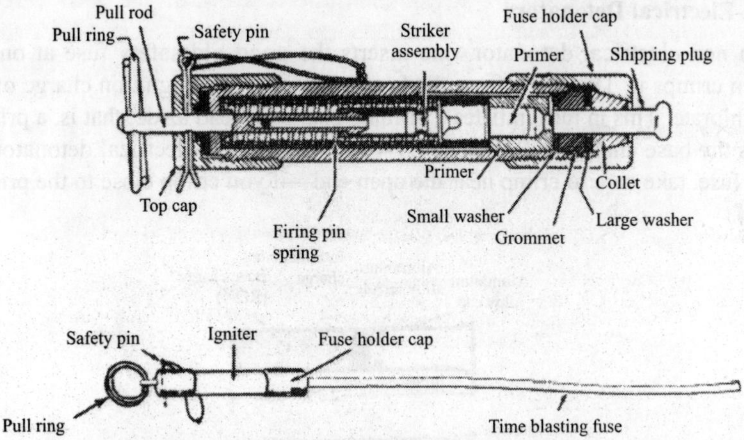

Fig. 55 Internal Structure Of A Fuse Igniter

Devices used for starting burning or deflagration are called squibs. A **squib** is a small metal tube which is filed with say, gunpowder. The powder is ignited by a filament such as that of a flashlight bulb. The squib provides a jet of flame to ignite the main charge. The squib gave rise to the idiom 'damp squib' as a damp squib won't set off a deflagration.

(15.2) DETONATORS ARE FOREVER

The safety fuse carried a flame. Thus it could be used for setting off charges made of low explosives only which would deflagrate. High explosives, as we have seen, are made rather insensitive and need a critical amount of thermo-mechanical shock to set them off. The safety fuse by itself alone did not work for high explosives. The problem was solved by the invention of the detonator or the blasting cap. The detonator or the **blasting cap** is a device that provides this critical shock to set off high explosives reliably. Without it the use of high explosives would have been a very cumbersome process. The blasting cap is therefore an invention as great as the high explosives themselves. Detonators use the principle of **explosive train**. That is, a small amount of primary explosive is used to initiate a small amount of high explosive which is sensitive enough to be initiated by that small amount of primary explosive. It is the explosion of this high explosive which provides the critical shock to set off the main charge.

A detonator or blasting cap is a small thin (.01") metal tube about .25 inch in diameter and 1.5 to 2.5 inches long containing about just a gram of explosives in all. The No. 8 blasting cap has typically, for example, 800 mg of PETN base charge pressed to 4000 psi and 300 mg of lead azide primary charge. Detonators can be non-electric or electric depending on how the primary explosive is initiated. They are very cheap and consumed by the thousands of millions throughout the world. From the oldest to the newest high explosives, from the simplest explosive device to the most advanced nuclear weapons, it is the detonator which makes each one of them work! You cannot even think of using high explosives without the detonator. The human race would remain eternally grateful to the genius of **Alfred Nobel** in inventing (1867) the **first fulminate-based non-electric blasting cap**, to **Julius Smith** for inventing (1868) the **first fulminate-based electric blasting cap**, and to **Perry Gardiner** and **Julius Smith** for inventing (1875) the **first truly modern detonator** in which both primary and secondary explosives were used in the form of an explosive train. Thank you, gentlemen—you were simply awesome! More than 132 years have passed but your genius lives on!

(15.2.1) Non-Electrical Detonators

In a non-electrical detonator, one inserts the good old safety fuse at one end which is open and then crimps it. The flame from the fuse ignites a flash or ignition charge of lead styphnate and barium chlorate. This in turn initiates a priming charge of lead azide, that is, a primary explosive. That initiates the base charge of RDX or PETN etc. With a non-electrical detonator, when you are crimping the fuse, take care to crimp near the open end—if you crimp close to the primary explosive, it may go off.

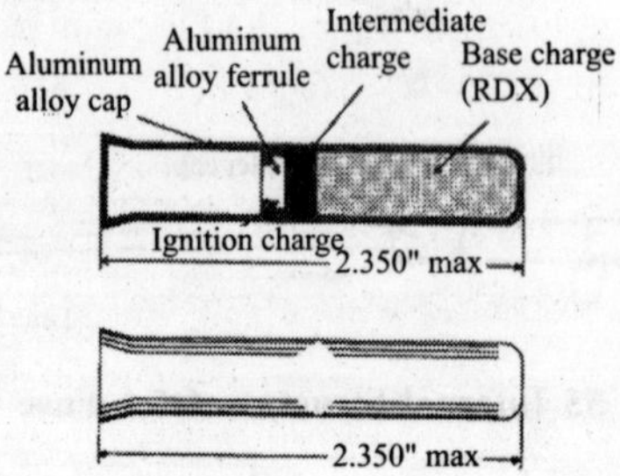

Fig. 56 Internal Structure Of A Non-Electrical Detonator

(15.2.2) Electrical Detonators

In an electrical detonator there is a filament or bridge wire which, quite like the filament of the bulb of a flashlight, heats up when current flows through it. This heat ignites the flash charge. After that the construction and functioning of the detonator is essentially the same as that in a non-electrical detonator. There is no open end in an electrical detonator. The open end is sealed with a plastic or rubber plug. An insulating material like sulfur insulates the wires joining the bridge wire. These wires protrude out of the detonator and are usually made of a length of 12 to 15 feet. To obviate the possibility of the circuit picking up static electricity, the ends of the wires are kept short-circuited by a small metal tab called a 'shunt'. This shunt must be removed before using the detonator. Enclosing the fuse head in an electrostatic protective sleeve also decreases the risk of unintentional initiation due to static discharge.

One can have a pyrotechnic delay element also in the train between the bridge wire and the primer to give a precise, pre-determined delay. The final delay time of an electric detonator is the sum of the delays in the delay element, the intermediate composition, used for fine tuning of the accurate delay time, and the I-element (initiation element).

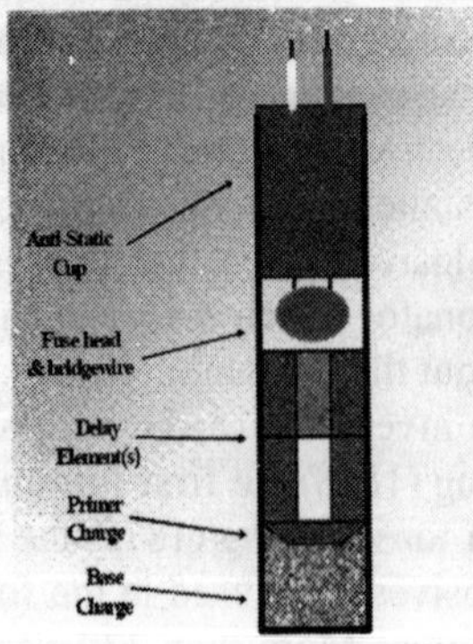

Fig. 57 Internal Structure Of An Electrical Detonator

The delay times generally vary in steps between 25 ms (0.025 seconds) and 5000 ms (5 seconds). Electric detonators are manufactured with two types of delay, millisecond delay (MS) and half-second delay (HS). In the MS-detonators the delay time increases in increments of 25 ms between each period number while the increment is 500 ms (half a second) in the HS-detonators. Electric HS-detonators are designed for use in underground operations as longer delay times are needed in tunnel rounds to give enough time for the rock to break and be thrown from the tunnel face. HS-detonators are prohibited for use in surface operations as the delay times are too long and can cause flyrock. Electric MS-detonators are used in surface operations and mostly in smaller operations with low bench heights, i.e. bench and trench blasting. The extended MS-series detonators are used in underground operations and bench blasting with large burdens (8 - 10 m).

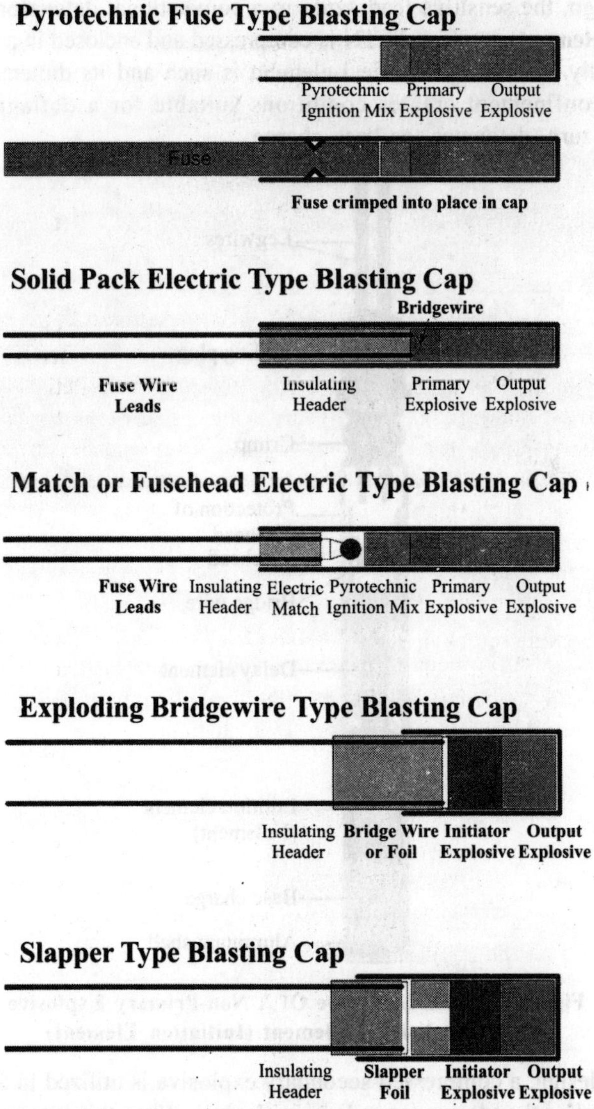

Fig. 58 An Overview Of Various Types Of Non-Electrical And Electrical Detonators

(15.2.3) NON-PRIMARY EXPLOSIVE DETONATORS (NPEDs)

Detonators contain primary explosives and hence require very careful handling. If they are crushed or banged about or exposed for long to heat and sunlight, they may go off. The presence of primary explosives means that the detonator becomes potentially a risky thing. Many innovators have solved it by developing detonators that do not have primary explosives.

They are known as **NPED** (Non Primary Explosive Detonator). This means they do not contain any primary explosive (e.g. lead azide). NPEDs are considerably less sensitive to impact and rough handling than detonators that contain primary explosive. How did they do it? There are several solutions to this problem.

In one design, the sensitive lead azide in a conventional detonator is replaced by an **I-element (initiation element)** in which PETN is compressed and enclosed in a steel tube. The PETN is initiated electrically. The design of the I-element is such and its dimensions and strength so balanced that the confinement creates conditions suitable for a deflagration to turn into a detonation. This, in turn, detonates the base charge.

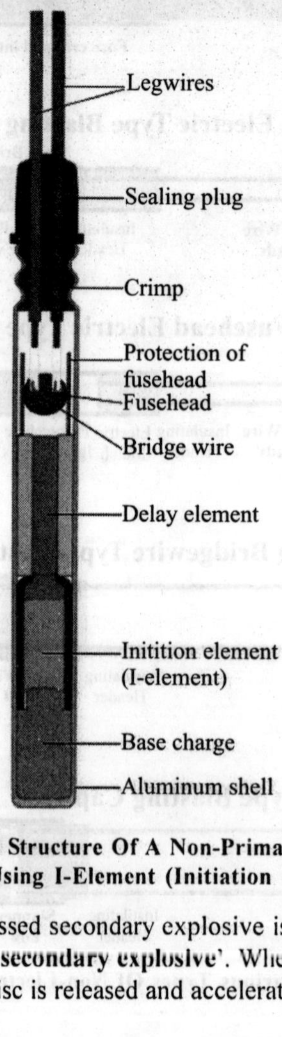

Fig. 59 Internal Structure Of A Non-Primary Explosive Detonator Using I-Element (Initiation Element)

In another design, a compressed secondary explosive is utilized in the first stage with an **impactor disc**. It is called the **'donor secondary explosive'**. When this 'donor secondary explosive' explodes, a portion of the impactor disc is released and accelerated when the said donor secondary explosive is initiated by a bridge wire. The impactor disc strikes an **'acceptor secondary explosive'** with sufficient velocity to produce detonation of the 'acceptor secondary explosive'.

NPEDs of the **exploding wire** or **exploding foil type** are able to produce a shock of sufficient strength to directly induce detonation in secondary explosives when exposed to high momentary electric currents.

Yet another device uses a laser pulse that passes through a fiber optic bundle and a focusing bead to impinge upon a charge of a secondary explosive which is set into low order detonation. This, in turn, initiates a low order detonation into a second secondary charge. This second charge is, however, loaded in a gradient of increasing density. Hence the velocity of the reaction increases very rapidly and a high order detonation is obtained.

One device uses an electrical initiator wherein polymeric solid sulfur nitride $(SN)_x$ is utilized as the sole explosive initiating means thanks to its ability to act as an explosive as well as an electrical conductor.

Each of these designs has its limitations. Devices in which you progress from deflagration to detonation, these conditions normally include heavy confinement of rather large amounts of the explosive, which adds to cost, size and weight when compared to conventional primary explosive detonators. Exploding wire type detonators are normally not suitable in civilian applications since expensive and elaborate blasting machines are required for the high currents needed. Further, they are not compatible with ordinary pyrotechnical delay devices. The forces involved in impactor disc type devices necessitate that they should be large and mechanically complicated. They are not very reliable either.

Broadly speaking, the successful commercialization of non-primary explosive detonators has been restricted due to at least three reasons. The first is the requirement for complex design or heavy confinement, which adds to both material and manufacturing cost when regular production equipments cannot be used. Secondly, while it is possible to obtain some function with various non-primary detonator designs, it is very difficult to reach the very high initiation reliability of primary explosive detonators. Reliability is extremely important. For example, in mining it becomes a very dangerous task to deal with un-detonated borehole charges. Finally, conventional detonators are so cheap and easy to use that they more than offset the little risk involved in their handling. NPEDs have shown that it is possible to design detonators without using primary explosives but life is so much simpler with the old design!

(15.2.4) Plastic Detonators

One of the greatest advances in the field of explosives that is sure to gladden the hearts of bombers across the world is the invention of the plastic detonator. Hitherto the problem was that even if you could manage some explosive with extremely low vapor pressure making it virtually undetectable by the usual explosive detectors, the detonator had to have metal in it. Hence if the explosive escaped detection, the detonator gave away the bomb's existence. It is not that only the non-state actors were keen on finding a solution to the problem. Even the militaries of the world were deeply interested in developing landmines that would escape detection by the traditional minesweepers. Technically the issue was exactly the same—how to get rid of the metal in the detonator. The plastic detonator solved all these problems! The Yugoslavian **PMA-1 AP mine** had earlier a plastic body but a conventional metal detonator. They quickly realized the mistake and replaced it with a plastic detonator. Many nations still produce plastic bodied mines with metal detonators—they reduce the possibility of detection but do not eliminate it.

shown actual size (approx.)

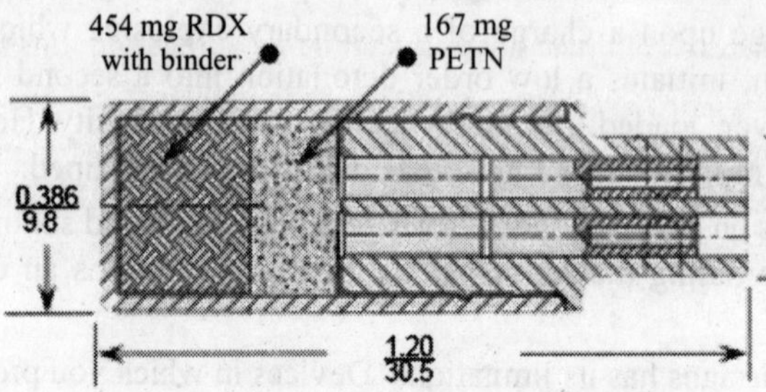

Fig. 60 External And Internal Structures Of A Plastic Detonator

The only thing metallic in the plastic detonator is the brass contacts for the wires and, of course, the exploding bridge wire of gold—but the quantity of metal is extremely small and can easily be masked by other metallic objects present in the vicinity.

The **RP-503 EBW** (Exploding Bridge Wire) **plastic detonator**, for example, has strength equivalent to a No. 8 detonator. It contains an initial pressed charge of 167 mg PETN and 454 mg of RDX plus binder as the main charge. At 0.386 inches diameter, it is thicker than most metallic detonators though the length is approximately the same at 1.20 inches. This means that if a charge has a factory-made detonator well made for an old electric detonator then it would not fit into that. Yet another problem with the plastic detonator is that it requires a threshold burst current of 180 amps and a threshold voltage of 500 volts.

To counter plastic detonators, companies like SILab (Seals and Identification Techniques Laboratory) have patented a detector based on an array of electrostatic antennas that can identify hidden plastics objects. The principle of its functioning is that an electrostatic detector detects the variations of the electrostatic field when passing over a plastic object. A prototype of an array of static antennas has been setup in such a way that the shape of hidden plastic object can also be visualized. Metal, stones, wood etc. do not affect the electrostatic signal; therefore the system detects just plastic objects.

(15.2.5) Electronic Detonators

In mining, the controlled sequence of detonations in various boreholes is one of the very critical parameters having a direct effect on overall blast performance in many ways. The measure of the potential effectiveness (that is, the available energy from the explosives has to both break and displace the rock mass) is directly proportional to the effective burden that energy must overcome. This relationship is a crucial element in basic blast design. Any variation in hole detonation timing results in that hole being fired prior to or after its nominal firing time. This results in burden-to-energy relationships that have adverse impacts on the performance of a blast. The results of these impacts include: poor rock fragmentation; large amounts of oversize; high ground vibration levels; high air blast levels; flyrock incidents; increased need for secondary blasting; and increased excavation and crushing costs.

This means that for highly efficient blasting you must have very high accuracy detonators so that you may ensure proper timings. Conventional electrical detonators are generally satisfactory for ordinary works but you require something better when you need highly accurate timings. The standard pyrotechnic delay elements do after all involve burning of chemicals and the burning rates may differ, albeit slightly, depending on various factors like their compaction, and exposure to elements like moisture and heat in storage, etc.

Electronic detonators eliminate the pyrotechnic fuse train and thus improve timing accuracy and safety. In the electronic detonators, typically an integrated circuit and a capacitor system internal to each detonator separate the leg wires from the base charge. The integrated circuit chip is used to control delay timing which uses electrical energy stored in one or more capacitors to provide power for the timing clock and initiation energy. Delay is therefore achieved electronically and not pyrotechnically.

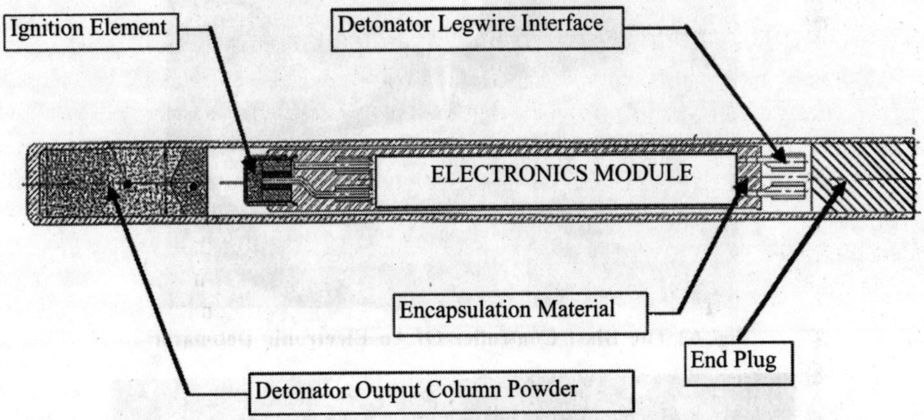

Fig. 61 Internal Structure Of An Electronic Detonator

The advantages of electronic detonators include precise timing, reduced vibrations, a reduced sensitivity to stray electrical currents and radio frequencies, and a great reduction in misfires through more precise circuit testing. Electronic detonators also make it possible to test the functionality of the circuit before the blasting. Since they cannot be initiated by conventional blasting units, this reduces the chances of their being misused by those who have conventional blasting units only. They are also relatively safer from extraneous electrical or electromagnetic influences. They are, however, costlier and require more skilled personnel to use.

Each electronic detonator has an assigned traceable ID number held on the internal chip and printed on a tag which is attached to the leg wire connector. A specially designed blast controller unique to each manufactured system transmits a selectable digital signal to each wired electronic detonator. The signal is identified by each electronic detonator and the detonation firing sequence is accurately assigned. The manufacturer's control unit will show any incomplete circuits during hookup prior to initiation of the explosive round. The wired round won't fire until all detonators in the circuit are properly accounted for with respect to the current blasting plan layout.

Fig. 62 The Blast Controller Of An Electronic Detonator

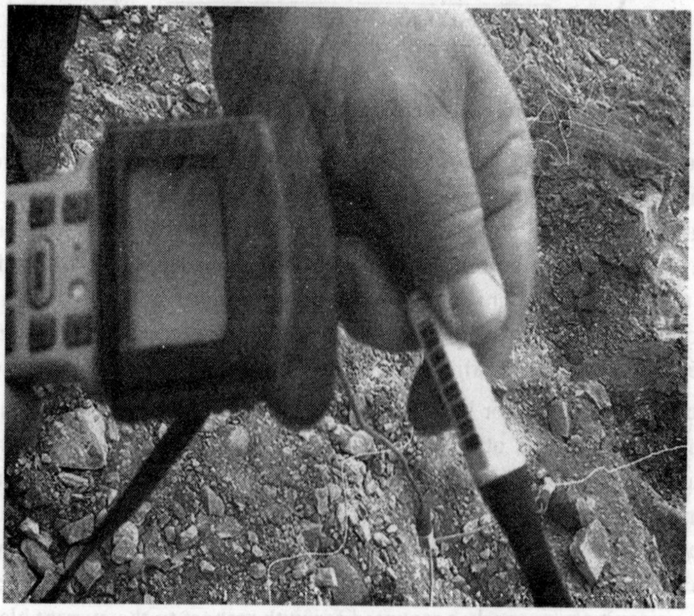

Fig. 63 The Bar Code Reader Of An Electronic Detonator

The HotShot system, for example, is programmed to communicate the digital signal received by one detonator to the next, eliminating the need to program each detonator individually. The software provided with the system not only programs the detonators, but provides a new level of safety by highlighting design and connector errors as well as detecting and warning the blaster about detonator and harness faults.

(15.3) DETONATING CORD: WHAT CLEVER GENIUS!

The detonating cord made a quantum jump over the safety fuse. The safety fuse was meant for deliberately introducing delays in initiation. However, if you wanted instantaneous initiation of a charge and that too from a safe distance, there was nothing you could use. This seemingly insurmountable problem was solved by the invention of the detonating cord. The detonating cord is a great invention; in fact one of the cleverest inventions in the history of explosives. The first rudimentary detonating cord was developed in France in 1902. The genius was a man called **Cordeau**. His invention consisted of a tin tube filled with picric acid. The detonating cord was materially different from the safety fuse. The safety fuse carried a flame forward at a slow speed; the detonating cord carried a detonating explosion forward at a great speed! The detonating cord actually explodes along its length rather than burn and thus transmits an explosion along its length instead of a flame. You can in fact think of it as a long flexible detonating cap! You initiate it at one end and you can use the explosion carried to the other end to detonate a secondary high explosive!

Then in 1908 another patent was taken in France which consisted of a lead tube, about the same diameter as safety fuse, filled with a core of TNT. It was made by filling a large tube with molten TNT that was allowed to solidify. The tube was then passed through successively smaller rolls until it reached the specified diameter. In France the product was called **cordeau détonant,** elsewhere shortened to cordeau. The velocity of propagation of explosion in it was about 16,000 feet per second. The Ensign-Bickford Company of USA adopted cordeau in 1912. In 1936 the **Ensign-Bickford Company** developed the **modern detonating cord** which they called **Primacord**, based on French patents and constituting a core of PETN covered with various combinations of textiles, waterproofing materials, and plastics. That made it a perfectly flexible device and thus added enormously to its operational value. The velocity of propagation of explosion in it was in excess of 21,000 feet per second.

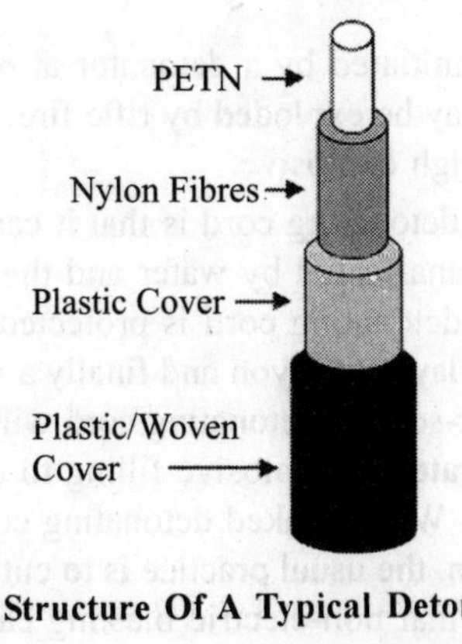

Fig. 64 Structure Of A Typical Detonating Cord

The modern detonating cord is a thin flexible tube containing a PETN core. You can have lighter or heavier cords depending on your intended purpose—whether you require the cord to have some explosive effect of its own or not. For example, high-yield detonating cord has been used by divers to remove old dock pilings and other underwater obstructions. Creating a slipknot from detonating cord yields a field improvised device that can be quickly employed to cut a locked doorknob off a door—this has been found to be of immense use in urban combat and in special operations. The possibilities are limited only by one's imagination. The amount of compressed PETN used in it varies from 4 grains per foot to 400 grain per foot in nearly 100 brands and types. The 50 grains to 250 grains per foot types are, however, the most prevalent. That comes to about 10 grams to 50 grams per meter.

The cores of detonating cords are now made of other explosive materials also. To identify the type of explosive that is used, look at the cross section.

- PETN core - white in color
- RDX core - dyed pink
- HMX core - dyed pink in color

Velocities of propagation of explosion vary from 22,000 fps to 27,000 fps.

The most striking and important use of detonating cords is in reliably and inexpensively chaining together multiple explosive charges in demolitions, mining and drilling, etc. Suppose you have to connect several charges of high explosives and you want them fired simultaneously or at time intervals differing only slightly from each other but in a certain order. What do you do? You cannot use the safety fuse. The safety fuse with black powder in the core burns with a speed of just 30 to 45 seconds per foot and it may vary for different rolls under different atmospheric and climactic conditions. Obviously this will not be of any use when you have to fire several charges that may be spaced hundreds of feet apart (such as in a demolition) simultaneously. There you use detonating cord. Because its core contains a high explosive like PETN, the detonating cord transports the explosion extremely fast and for all practical purposes, almost instantaneously. The detonation velocity is great enough to use it for synchronizing multiple charges to make them detonate almost simultaneously even if the charges are placed at fairly long distances from the point of initiation. The propagation of the explosive shock wave in the detonating cord takes place at a reliable rate of say 23,000 fps or higher. In fact the precisely known rate of propagation also enables us to use it as a timing mechanism. This is particularly useful in controlled demolitions or building implosions where structural elements may have to be destroyed in a specific order to control the collapse of the building.

Detonating cord is usually initiated by a detonator at one end. As such it is insensitive to friction and ordinary shock, but may be exploded by rifle fire. It also detonates sympathetically with the detonation of an adjacent high explosive.

One great advantage of the detonating cord is that it can be very reliably used to set off underwater charges. PETN itself is unaffected by water and there is water-proof coating also in the cord. The explosive core of the detonating cord is protected first by a seamless cotton tube, then a layer of asphalt followed by a layer of rayon and finally a coating of polyethylene to render it waterproof. However, even a water-soaked detonating cord will detonate if there is a dry end to allow initiation. Moisture can penetrate the explosive filling to a maximum distance of 6 inches from any cut or break in the coating. Water-soaked detonating cord will detonate if there is a dry end to allow initiation. For this reason, the usual practice is to cut off and discard the first 6 inches of any new or used detonating cord that non-electric blasting caps are crimped to.

(15.4) THE ART OF USING ELECTRICAL DETONATORS

(15.4.1) The Importance Of The Strength Of The Detonator

The strength of a detonator depends on the explosives used in it. Several strengths are commercially made. The No. 6 detonator will detonate most dynamites but will not be able to detonate military explosives. The No. 8 detonator will detonate pressed TNT used in military demolition blocks but not cast TNT or plastic explosives. The Special Military detonator will detonate all explosives. It is nearly twice as powerful as the No. 8 which in turn, is nearly twice as powerful as the No. 6. Intelligent bombers, who do not get the Special Military detonator, apply a little mind and use two No. 8 detonators side by side. Smart, isn't it?

(15.4.2) Detonators: Identification And Accessories

Electric detonators are color-coded by different colors on the leg wires. The factor common to all groups is that one of the leg wires of MS-(millisecond) detonators is green and one of the leg wires of the HS-(half second) detonators is red. One of the leg wires of the so-called instantaneous detonators is white.

A firing cable is used to connect the electric round to the blasting machine. Resistance in the firing cable should be as low as possible as high firing cable resistance will decrease the capacity of the blasting machine.

A connecting wire is used for extending detonator leads down blast holes, away from faces, or otherwise connecting lead wires to the firing line. The connecting wire is a single, well-insulated, wire with low resistance. If the resistance is too high it may affect the capacity of the blasting machine, reducing the numbers of detonators that may be connected in the circuit. The connecting wire should be used only once. Used connecting wire may cause misfires.

(15.4.3) Correct Placement Of The Detonator

When using detonators with high explosives it is necessary that the detonator is placed centrally within the explosive mass. With plastic explosives the insertion should be about one inch and there should be about half an inch of peripheral mass along all the sides of the tube. It is common for militaries to provide a **'cap well'** in their cast or hard pressed explosive charges in which the detonator may be inserted so as not to leave scope of any mistake by some blockhead. It is also necessary to ensure that the detonator has a tight fit with the explosive. You may squeeze the explosive if necessary. If the fit is not tight, there may be a misfire or a low order detonation.

(15.4.4) Ensuring Correct Current Supply

For a successful simultaneous initiation of a large number of detonators, sufficient electrical power must be delivered to all detonators within a few milliseconds. The time required to heat up the bridge wire until the flash charge deflagrates or the primary charge explodes is a function of the voltage and the current. It is important that all detonators in a firing circuit are initiated instantaneously. If one detonator in the round fires before any other round, the circuit will break and one or several detonators will not get the firing impulse and misfires will occur. Hence the so-called instantaneous detonators or those with very short delay times should not be connected in series.

The currents required for firing various types of detonators are as follows:

- **Class 1/Group 1 detonators**: The no fire current is 0.25 A (ampere) or less. A minimum of 0.6 A firing current is required to initiate one Group 1 detonator and a minimum of 1.0 A is required to reliably initiate a series of Group 1 detonators.

- **Class 2/Group 1A detonators**: The no fire current is 0.45 A or less. A minimum of 1.0 A firing current is required to fire one Group 1A detonator and a minimum of 1.5 A is required to initiate a series of Group 1A detonators.
- **Class 3/Group 2 detonators**: The no fire current is 1.2 A or less. Group 2 detonators are detonators with a high degree of safety. Even though they are much safer in the presence of electrical hazards than Group 1 detonators, safety precautions must still be taken close to strong radio transmitters, radar, power lines over 70 kV and during thunderstorms. A minimum of 2.2 A is required to fire this detonator and a minimum of 3.5 A is required to fire a series of them. The resistance of such detonators is independent of leg wire length and is 3.6 ohms + 0.3 ohms at 20 degrees Celsius.
- **Class 4/Group 3 detonators**: No fire current is 4.0 A or less. This is a detonator with a very high degree of safety against electric hazards. A minimum of 6.0 A is required to fire one Group 3 detonator and a minimum of 25 A is required to fire a series of Group 3 detonators.

(15.4.5) RF Hazards With Electrical Detonators

If the electric detonator wires are located in a strong RF (radio frequency) field (say for example, near a transmitter that is radiating RF power), the usually insulated but electromagnetically unshielded leg wires or circuit wires will act as antenna similar to that on a radio or TV set. That happens irrespective of whether the circuit wires are connected to a blasting machine or not, or whether they are shunted (short circuited ends) or not shunted (open ends). This antenna will absorb RF energy from the transmitter RF field and the electric current transmitted to the detonator wires will flow into the detonator. Depending on the strength of the RF field and the antenna configuration formed by the detonator wires and its orientation, sufficient voltage may be induced in the wires to fire the electric detonator. It must be kept in mind that this problem of induced voltage arises only when the detonators are taken out for use. In their original containers, the leg wires of the detonators are folded or coiled so as to provide effective protection against current induction.

The possibility of premature explosions of electric detonators due to RF energy is remote. Each year throughout the continental United States approximately 100 million such detonators are used with few mishaps. However, there have been authenticated cases in which detonators were prematurely initiated by RF transmission to the detonator wires. Subsequent investigations revealed that the instances would not have occurred if proper **safe distances from the RF sources** had been maintained.

Radio frequency transmitters commonly encountered include citizens band (CB) radios, cellular telephones, AM and FM radios, radar, and television. Commercial AM broadcast transmitters (0.535 to 1.605 MHz) are potentially the most hazardous RF energy source. They combine high power and low frequency so that there is little loss of induced RF energy in the detonator lead wires. Frequency-modulated FM and TV transmitters are not likely to create hazardous situations. Although their power is extremely high and the antennas are horizontally polarized, their high frequency currents are rapidly attenuated in the detonator or leg wires. This RF source employs antennas on very high towers, which have the additional effect of reducing the electromagnetic field at ground level because of the distance. The hazards of RF energy from microwave relays are small because they operate at a very high frequency, and have a restricted radiation pattern. Mobile radios and cellular telephones that transmit RF energy must be rated as a high potential hazard because, although their power is low, they can be brought directly into a blasting area. Similarly, mobile service base stations are also potentially hazardous sources. Radar installations radiate high power levels through the use of high gain antennas. Radar installations pose a hazard if blasting is done within the radar beam range.

The detonators can pick up RF energy in three ways; the **dipole pick-up circuit**, the **long wire pick-up circuit** and the **loop circuit**. The dipole circuit presents the most hazardous conditions when:

- the circuit wiring and/or electric detonator leg wires are elevated several feet off the ground;
- the length of this wiring is equal to one-half the wave length of the radio wave; or
- the electric detonator is located at a point where the RF current in the circuit wiring is at a maximum

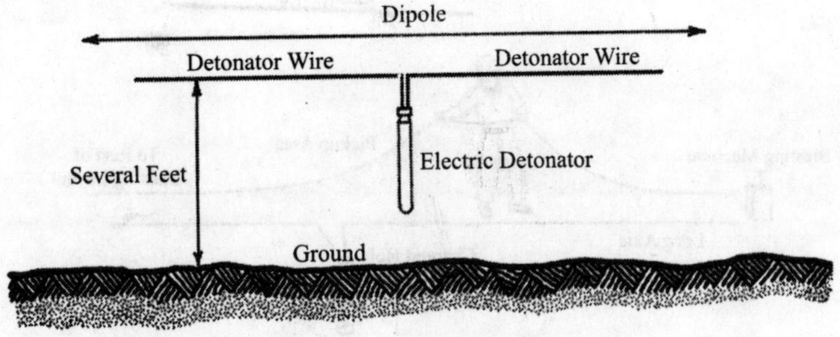

Fig. 65 A Dipole Pick-Up Circuit

The long wire circuit condition occurs when the electric detonator is at one end of the wiring that:

- is elevated in the air;
- has a length equivalent to one-quarter of the wavelength of the radio wave; or
- is grounded to the earth through the electric detonator

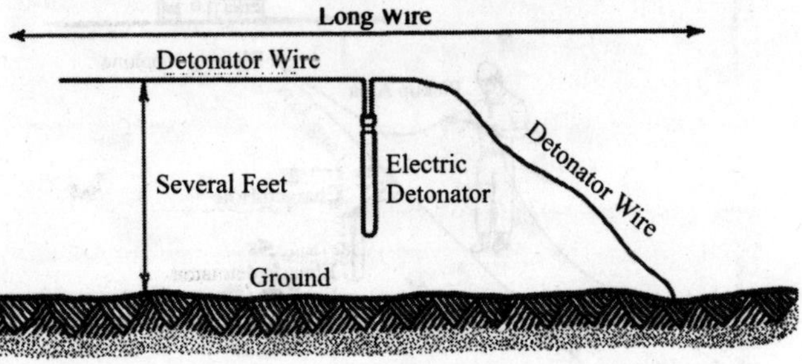

Fig. 66 A Long Wire Pick-Up Circuit

Both the circuits require that the lead line or detonator wires be suspended above the ground. Both of the circuits (antennas) achieve their maximum current pickup when they are parallel to a horizontal transmitting antenna (FM, TV, or amateur radio) or are pointed toward a vertical antenna (AM, mobile, etc.).

Another sensitive RF pickup circuit, and one commonly encountered in blasting operations, is the **loop pick-up circuit**. This circuit is sensitive to the magnetic portion of the electromagnetic wave. The loop circuit receives the maximum pickup when its long axis is placed in the direction of the transmitting antenna. In general, the loop areas can be reduced by picking up both lead wires as in a duplex wire circuit and making all wire splices as close to the ground as possible.

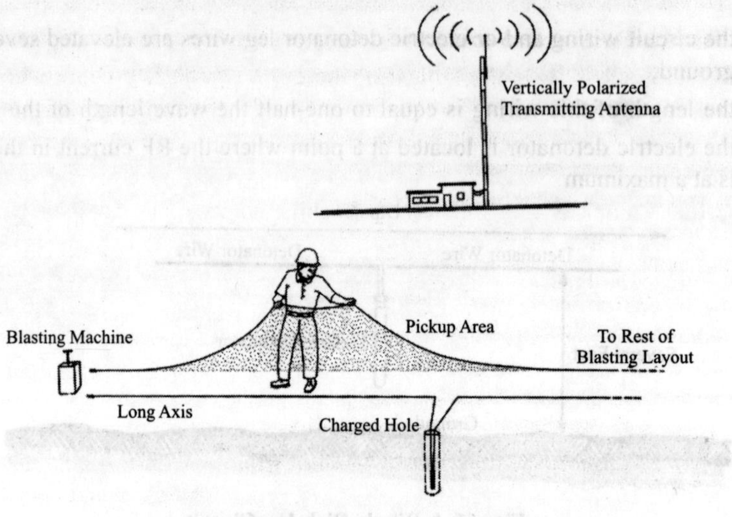

Fig. 67 A Loop Pick-Up Circuit: Preferred case

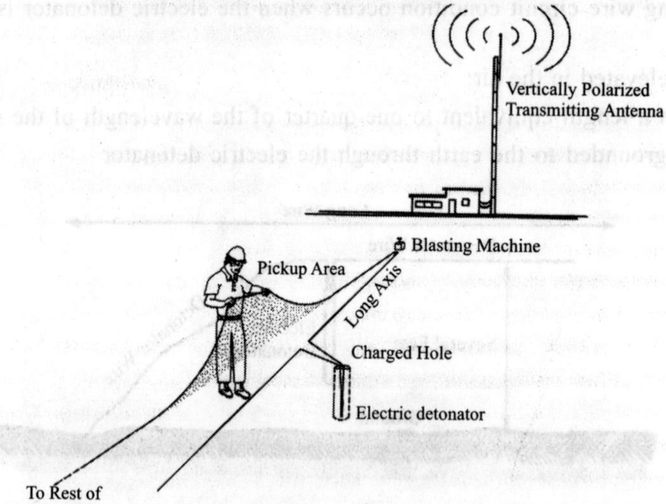

Fig. 68 A Loop Pick-Up Circuit: Less Desirable Case

Calculations have been done for finding out the **safe distances** when you have to operate in the presence of RF sources. A rough idea can be had from the following. For a commercial AM broadcast in the frequency range 0.535 to 1.605 MHz, the safe distance for a 4000 watt transmitter is 800 feet—for transmitters of higher power, it will obviously be higher. For transmitters in the frequency range up to 50 MHz the safe distance will be 800 feet even for a 100 watt transmitter. For

FM transmission the safe distance of 800 feet would be for a 1000 watt transmitter. For VHF TV transmission the safe distance for a 1000 watt transmitter would be 600 feet. For UHF TV transmission 600 feet applies to a transmitter of 10,000 watts. For VHF transmitters in the Citizens Band (2-meter) the safe distance for a 1000 watt transmitter would be 310 feet. For cellular phones the safe distance is reduced to 180 feet for a 1000 watt transmitter. For 5 watt VHF sets in the 2-meter Citizen Band, the safe distance is 20 feet and for cellular phones it is 10 feet.

(15.4.6) Blasting Machines

Modern **capacitor discharge blasting machines** generally come with a two-handed operation so that there is no unintended initiation. Typically the capacitors have a capacitance of 200 microfarad. They require a **charging time** of 15 seconds from a dc source of 12 or 14 volts. They fire at 540 volts. Some idea of their capability can be had from the following.

The **CB 20 VA or CB 400 NT blasting machine**, for example, can initiate up to 170 Group 1 detonators in one series and 400 in 4 parallel series. This assumes the firing current to be minimum 1 A, the firing impulse to be less than 5 mJoules/ohm, the resistance of each detonator to be 2.5 ohm and the firing cable resistance to be 5 ohm. It may be kept in mind that the resistance of Group 1 detonators depends on the leg wire length. The same machine will blast 70 Group 1A detonators in one series and 120 in 3 parallel series. In this instance the firing current is assumed to be 1.5 A, the firing impulse to be less than 16 mJoules/ohm, the resistance of each detonator to be 3.5 ohm and the firing cable resistance to be 5 ohm. With Group 2 detonators the number comes down to 20 assuming the firing current to be less than 3.5 A, the firing impulse to be less than 140 mJoules/ohm, the resistance of each detonator to be 3.5 ohm and the firing cable resistance to be 5 ohm. It may be noted that the resistance of Group 2 detonators is independent of the length of the leg wire. For Group 3 detonators the number would come down to 3 assuming the firing current to be less than 25 A, the firing impulse to be less than 2500 mJoules/ohm, the resistance of each detonator to be 0.6 ohm and the firing cable resistance to be 5 ohm. Group 3 detonators are connected only in a single series circuit. It may also be kept in mind that the resistance of Group 3 detonators depends on the length of the leg wire.

Fig. 69 A Hand-Cranked Capacitor Discharge Blasting Machine (Model CI 160A): Representative Of Its Class

Blasting machines like the **CI 160A** are charged by a **hand-cranked generator**. The maximum voltage (1950 V) is reached after approximately 20 seconds. The capacitance of the capacitor used is 105 microfarad. The CI 160 VA can initiate up to 160 Group 2 (VA) detonators connected in 4 parallel series and a firing cable resistance not exceeding 2.5 ohm. The charging level is shown on a meter which is divided in 4 zones (I - IV). The required charge level for a given size of the round is indicated on a plate on the blasting machine. Thus the firing current may be adapted to the number of detonators in the round, and the machine's maximum firing current is used only when necessary. The CI 160A can initiate 145 Group 1 detonators in one series at charge level I. At charge level III it can initiate 500 Group 1 detonators in two parallel series. For Group 1A detonators the numbers come down to 60 and 300 respectively. For Group 2 detonators the figures are 10 and 45 respectively. When using charge level IV, it can initiate up to 50 Group 3 detonators.

(15.5) SHOCK TUBES OVERCOME MANY SHORTCOMINGS OF ELECTRIC DETONATORS AND DETONATING CORDS

Electric detonators, as we saw, are sensitive to extraneous sources of electricity (static or current) and electromagnetic waves. Detonating cords are free from this hazard but the initiation is noisy, there is a large amount of cord movement and there is always the risk of the shock wave getting reflected if there are, by any chance, sharp angles, kinks, or even loops all of which could direct the explosion back toward the oncoming line of detonation. Shock tubes overcome this problem. The shock tube offers all the advantages of electric detonators and detonating cords but none of their disadvantages. The only drawback of the shock tube is that the system is costly and that circuit testing is not possible.

The shock tube is a totally non-electric system. It transmits a shock wave to the detonator. The shock wave results from the explosion of a coating of reactive powders in the tube. The reactive coating usually consists of HMX and aluminum powder. Some others use PETN for the reactive coating.

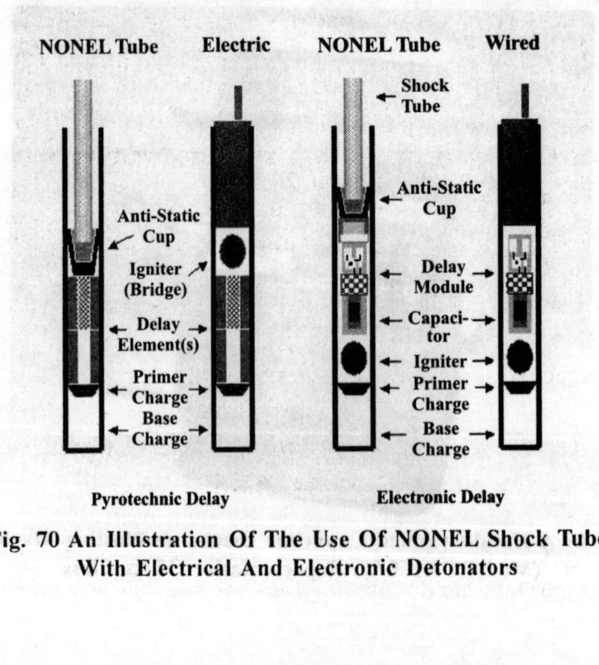

Fig. 70 An Illustration Of The Use Of NONEL Shock Tube With Electrical And Electronic Detonators

The shock tube is a small diameter (0.12 inch) plastic laminate tube made of polyethylene or Surylyn and coated with a very thin layer of reactive material at one pound per 100,000 feet of tube. The **NONEL shock tube** of Dyno Nobel, for example, is a small diameter, three-layer plastic tube coated on the innermost wall with a reactive explosive compound. When initiated, the shock tube transmits a low energy signal or propagation wave from one point to another. This shock wave is similar to a dust explosion and is able to propagate through most sharp bends, knots and kinks in the tube. The detonation is sustained by such a small quantity of reactive material that the outer surface of the tube remains intact during and after the propagation of the shock wave through it. The shock wave in a shock tube travels, in comparison to the detonating cord, at a much lower speed of 6500-6900 fps. The shock tube can be initiated by small shot shell-type primers, detonating cord or detonators.

(15.5.1) DigiDet

DigiDet of Dyno Nobel is the world's only electronic initiation system that is initiated by shock tube. It is easy to deploy, flexible and combines the accuracy of an electronic detonator with the simplicity of a standard shock tube system. This novel technology consists of a length of shock tube crimped inside an electronic detonator. When the shock tube is initiated, it delivers a signal to the DigiDet electronic detonator, which then activates an electronic timer inside the detonator. This gives the DigiDet the simplicity of a shock tube system while benefiting from the accuracy of electronic timing. DigiDet detonators are factory programmed with a range of delay times up to 1000 milliseconds.

(15.6) MODERN DEMOLITION INITIATOR (MDI)

The Modern Demolition Initiator (MDI), fielded first in 1996, is a completely self-contained suite of initiating components used to activate all standard military demolitions and explosives in the US Army. The suite is slated to eventually replace all electric and non-electric firing systems for conventional forces while maintaining compatibility with existing Army systems. The MDI enables the soldiers to pre-prime the charges and carry them to the demolition site. With MDI, soldiers can successfully complete demolition missions in a safe, quick and easy manner. Being completely non-electrical it is immune to RF hazards and to electromagnetic pulse (EMP).

The MDI consists of non-electric blasting cap assemblies with an integral time delay initiator, a time fuse or shock tube, and a 'J' hook that can be attached to a detonating cord. Timing can be set to fire immediately or up to a 20-minute delay. The MDI's blasting cap assemblies consist of five high-strength blasting cap assemblies and two high strength booster assemblies. These assemblies can be used to prime standard military explosives or initiate the shock tube or detonating cord of other MDI components. MDI also contains an igniter that can activate either time fuse or shock tube. The M151 and M152 booster assemblies introduced recently consist of a booster, which contains only insensitive secondary explosives and a length of low-powered detonating cord (5 grains/foot as against 50 grains/foot for high strength) with a high-strength booster in lengths of 10 feet and 30 feet respectively. As such, the low-powered detonating cord will not cut itself or other detonating cords if crossed. The boosters at the end of the M151 and M152 assemblies look like blasting caps, but they are not blasting caps because they contain no primary explosives.

Although the boosters have the same strength as No. 12 commercial blasting caps, they can be treated like detonating cord because they contain only secondary explosives. The booster end piece (the part that looks like a cap) is added to the low-strength detonating cord to fit the system into standard military cap wells. Since the booster assemblies contain no sensitive initiating element, they can be used to safely initiate underground charges also.

❑❑❑

Chapter 16

How Explosions Do What They Do: The Mechanics of Damage by Explosions

(16.1) THE GENERATION OF A SHOCK WAVE

When a high explosive detonates, it is converted almost instantly into a gas of very high pressure and temperature. Typically you might get up to 200 kilo bars of pressure (1 bar equals 1 atmosphere) and temperature in the range of 3000 to 5000 degrees Celsius for a high-explosive weapon. Under the pressure of the gases thus generated, the weapon case expands and eventually breaks into fragments. The excessive pressure generated by the decomposition of the explosive compresses the air surrounding the casing. It is compressed and a shock (blast) wave is transmitted into it. Shock waves are fundamentally different from sound waves. The shock front moves outward with an initial velocity much greater than that of sound but, after a short distance when the pressure drops, the velocity also decreases rapidly. Of the total energy available, as much as half may be used to expand the bomb casing and the remainder is used to compress the surrounding environment.

The shock wave generated by the explosion is a **compression wave**, in which the pressure rises from atmospheric pressure to peak overpressure in a fraction of a microsecond. This layer of compressed air is bounded by an extremely sharp front called the **shock front** in which the pressure rises abruptly. It is followed by a much slower (hundredths of a second) decline to atmospheric pressure. This portion is known as the **positive phase** of the shock wave. The pressure continues to decline to sub-atmospheric pressure; much like a pendulum continuing its swing up even after it has passed the lower-most point, and then returns to normal. This portion is called the **negative or suction phase**. The durations of these two phases are referred to as the **positive** and **negative durations**. While the duration of the positive pressure portion of the shock wave is about one-fourth that of the portion at less than atmospheric pressure, the amplitude of the pressure at the shock front is much greater than the amplitude of the decrease in pressure below atmospheric. The duration of the positive phase varies with the explosive charge and distance from the point of origin. To illustrate by real example, in one experiment, the duration of the positive phase was found to be of the order of 0.006 second at a distance of 51 feet from the bomb.

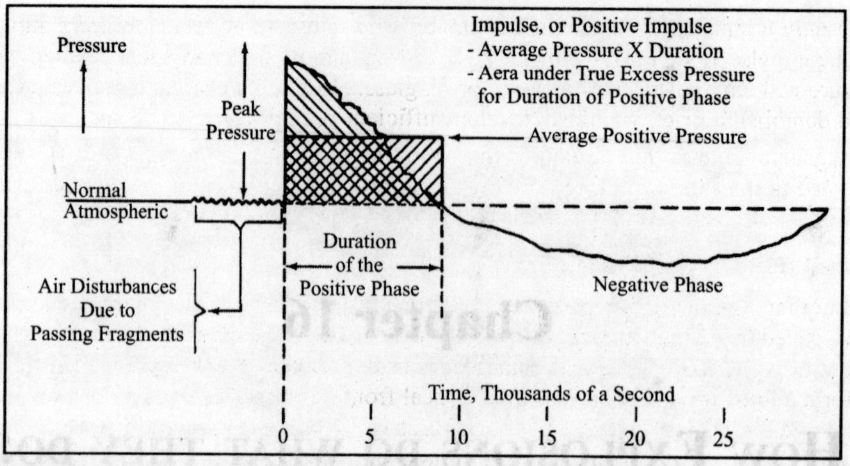

Fig. 71 The Variation Of Pressure With Time For The Shock Wave Of A Typical Explosion

(16.1.1) The Effects Of The Shock Wave

When the pressure becomes less than atmospheric, the wind reverses in direction and blows backward toward the point of detonation. Any light object when struck by the shock front, is rapidly carried away from the point of detonation and then is blown back almost to the original position when the wind reverses. The result of this positive/negative pressure variation is a push-pull effect upon the target. A wall, for example, struck by such a shock wave is first pushed forward by the short, sharp, hammer-like blow of the positive pressure phase and then pulled backward by the much longer negative phase. Whether the wall falls forward or backward depends upon circumstances.

The effect of the blast depends on two factors, **peak pressure** and **impulse.** Peak pressure is the pressure increase at the shock front or the highest pressure in the shock wave minus atmospheric pressure. The peak pressure represents a measure of the maximum force exerted against a surface by a blast wave, since force is equal to the product of pressure and area. Impulse, on the other hand, is mathematically equal to the area under the time-pressure curve for the duration of the positive phase. The area under the pressure-time curve during the positive phase represents the **positive impulse** and that during the negative phase, the **negative impulse**. Those without a background in mathematics may understand it this way. Since the pressure is changing you may take the average pressure in place of the pressure. Thus impulse becomes approximately half the peak pressure multiplied by the duration of the positive phase.

For a fixed-weight explosive, the peak pressure and positive impulse decrease with distance from the explosion. This is due to the attenuation of the blast wave. The rate of attenuation is proportional to the rate of expansion of the volume of gases behind the blast wave. In other words the blast pressure is inversely proportional to the cube of the distance from the blast center. As the shock front moves outward, the peak pressure decreases and the duration of the positive pressure phase increases. The rates of these changes vary with the magnitude of the peak pressure, but the net effect is a decrease in impulse. With increase in distance from the bomb, impulse decreases approximately as the reciprocal of the radius.

In general, two criteria must be satisfied in order to demolish any structure. Both peak pressure and impulse must exceed certain minimum values which depend upon the type of structure. Window glass, for example, requires a moderately high peak pressure but only a low impulse value since the positive pressure need not last long to cause fracture. On the other hand, a brick wall withstands only a small peak pressure, but the pressure must be of relatively long duration and therefore the impulse value must be high. The blast from almost any charge that develops sufficient peak pressure will have sufficient impulse to break glass. Most bomb charges that develop sufficient impulse to demolish a brick wall also develop sufficient peak pressure to do so.

A general rule is that structures that are strong and light in weight, with respect to the area presented to the shock front, require high peak pressure but no great impulse for demolition. Heavy but relatively weak structures require considerable impulse but not such a high peak pressure.

(16.1.2) The Effect Of Mach Reflections

Another aspect of overpressure occurring in air bursts is the phenomenon of **Mach reflections**, called the '**Mach Effect**'. When a bomb is detonated at some distance above the ground, the reflected wave catches up to and combines with the original shock wave, called the incident wave, to form a third wave that has a nearly vertical front at ground level. This third wave is called a '**Mach Wave**' or '**Mach Stem**', and the point at which the three waves intersect is called the '**triple point**'. What exactly happens is as follows. You see, blast waves are reflected from solid surfaces but not in the same manner as sound and light waves are reflected. When the incident and reflected waves intersect on the ground, they do not make equal angles with the ground as do sound waves. At small angles of incidence (the angle between the perpendicular to the blast front and the perpendicular to the reflecting surface), the angle of reflection is smaller than the angle of incidence, and the pressure behind the reflected wave is greater than behind the incident wave. Such reflection of a blast wave from a bomb that has detonated above the ground gives rise to an increased pressure wave at ground level. When the angle of incidence is greater than an extreme angle (40 degrees for strong shocks to a value approaching 90 degrees for very weak shocks) the incident and reflected waves do not intersect on the ground. This is known as **irregular** or **Mach reflection**. In comparison, reflection that occurs when the angle of incidence occurs at the extreme angle is known as the extreme regular reflection, whereas, reflection occurring where the angle of incidence is smaller than the extreme is known as regular reflection. When the incident angle exceeds the extreme angle, the reflected and incident waves intersect at some point above the ground. The lower portion of the reflected and the incident waves fuse into one wave which is known as the **Mach stem**, the point of intersection being known as the **triple point**.

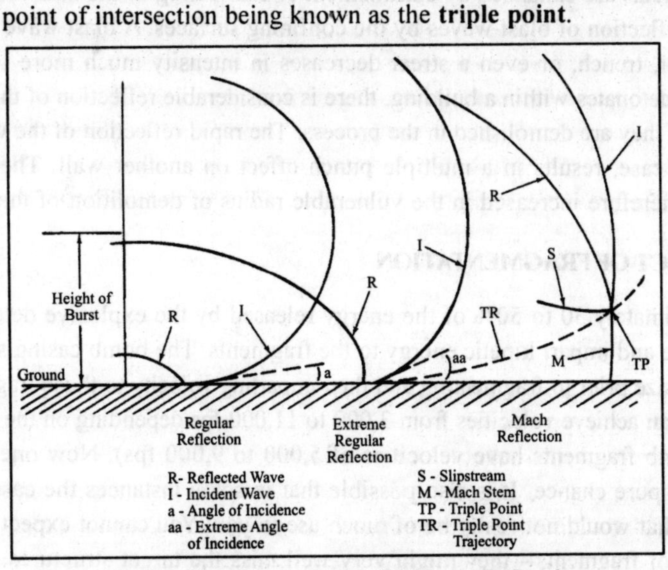

Fig. 72 The Phenomenon Of Mach Reflection

The resulting existence of three waves, termed the Mach reflection, causes a density discontinuity and a region of high pressure termed the Mach region is formed. The pressure behind the front of the Mach region is approximately twice that of the incident wave. The top of this high-pressure region, the triple point, travels away from the reflecting surface. As pressure and impulse appear to have their maximum values just above and below the triple point, respectively, the region of maximum blast effect is approximately that of the triple point. The region of maximum blast effect is determined by the height above the ground at which the blast occurs. At distances that are large compared to the height of the burst, the incident and reflected waves have fused and proceed outward as a single shock. After the point of complete fusion of the waves, the shock wave is the same as one produced by a double charge detonated on the ground. As the distance from point of the blast increases, the Mach stem becomes normal to the ground.

The Mach Stem grows in height as it spreads laterally, and as the Mach Stem grows, the triple point rises, describing a curve through the air. In the Mach Stem the incident wave is reinforced by the reflected wave, and both the peak pressure and impulse are at a maximum that is considerably higher than the peak pressure and impulse of the original shock wave at the same distance from the point of explosion.

Using the phenomenon of Mach reflections, it is possible to increase the radius of effectiveness of a bomb considerably. By detonating a warhead at the proper height above the ground, the maximum radius at which a given pressure or impulse is exerted can be increased, in some cases by almost 50%, over that for the same bomb detonated at ground level. The area of effectiveness, or damage volume, may thereby be increased by as much as 100%. Now you understand why almost all nuclear warheads are designed to explode at a certain height in the atmosphere and not on the ground.

(16.1.3) The Effect Of Confinement Or Channeling Of Blast Wave

Blast effects are enhanced by confinement or channeling of the blast wave. This happens because of the reflection of blast waves by the confining surfaces. A blast wave traveling through a tunnel, corridor, trench, or even a street decreases in intensity much more slowly than in the open. If a bomb detonates within a building, there is considerable reflection of the blast wave from the walls, even if they are demolished in the process. The rapid reflection of the wave from various walls, in such a case, results in a multiple punch effect on another wall. The overall effect of confinement is therefore increased in the vulnerable radius of demolition of the bomb charge.

(16.2) THE EFFECT OF FRAGMENTATION

Approximately 30 to 50% of the energy released by the explosive detonation is used to fragment the case and impart kinetic energy to the fragments. The bomb casing swells to up to 1.5 times its normal size prior to fragmenting and then imparting velocity to those fragments. Fragments of a bomb case can achieve velocities from 3,000 to 11,000 fps depending on the type of bomb (for example GP bomb fragments have velocities of 5,000 to 9,000 fps). Now one should not leave fragmentation to pure chance. It is also possible that in some instances the case would split into just two parts. That would not really be of much use to you. You cannot expect much damage by a small number of fragments—they might very well miss the target structures. Then you do not want the case to be blown to smithereens either. If the fragments are too small in size, they would once again be not of much use. Very small pieces carry little energy individually and cannot inflict much damage. The problem is akin to selecting the right load for a 12 bore shotgun in combat. If you select a load with a very large number of very small pellets such as the No. 1 (nearly 100 pellets of 0.16 inch each), the pellets are so small that they don't achieve any significant penetration in human skin, not to speak of killing them. On the other hand if you decide upon the slug, it can fell even a lion but the chances are that you won't be able to score a hit at 50 yards! Hence you require a load which has got pellets of a respectable size so that they may achieve sufficient penetration and, at the same time, there are enough of them so that the probability of scoring a hit is reasonably high. That's how Special Forces of the world generally select the SSG load for combat which has got 12 pellets or so of 0.31 inches each. The same holds true for a bomb also. If you are making a weapon, you should know the number and size of fragments for you to be able to predict the performance of the bomb. How do you do it? Simple! Suppose you make a score on the casing. This would make the casing weak along the line of the score, much as you have scores on drug tablets so that you may take half the tablet should that be necessary. The case would break along the score. Then you can make several scores also and the case would break along those scores. Several grenades use a winding of wire in their casings. The wires have nicks made in them at set intervals. This means that the wire would break along those nicks and you will have fragments of a known size. Making scored casings or using pre-cut fragments in the bomb enables one to get fragments in predictable number and sizes, otherwise fragmentation remains a statistically random phenomenon.

The fragments are propelled at high velocity, and after a short distance they overtake and pass through the shock wave. The rate at which the velocity of the shock front accompanying the blast decreases is generally much greater than the decrease in velocity of fragments, which occurs due to air friction. Therefore, the advance of the shock front lags behind that of the fragments. The **radius of effective fragment damage**, although target dependent, thus exceeds considerably the **radius of effective blast damage** in an air burst. As we saw, the blast effects are attenuated in inverse proportion to the cube of the distance. The fragmentation effects, on the other hand are attenuated in inverse proportion to the distance or the square of the distance, depending upon the specific design of the payload. This should help you understand why good bomb makers of general purpose bombs would not like their bombs to depend entirely on blast effect and place great emphasis on fragmentation effect. Herein lies the principle advantage of a fragmentation payload: it can afford a greater miss distance and still remain effective because its attenuation is less. It is only for special purpose bombs that fragmentation effects are sought to be minimized at the cost of blast effects.

One more thing. The initial velocity of the fragment depends on the amount of the charge in relation to the weight of the casing or more precisely, the charge-to-metal ratio. However, the way a fragment will lose its velocity in air due to drag depends on its own mass and cross-sectional area. A lighter fragment tends to lose velocity more rapidly. Assuming that the initial fragment velocity is 6000 fps, a five-grain (.324 grams) fragment would lose half its initial velocity after traveling just 37 feet, whereas a 500-grain (32.4 grams) fragment would travel 175 feet before losing half its velocity. Now you understand why it is necessary to score the casing in such a way so as to yield fragments of an **optimal size**!

The fragments of a warhead travel outward in a nearly perpendicular direction to the surface of its casing. Life is simple for a spherical warhead. But in real life there are no spherical warheads. For a cylindrical warhead there is a 7- to 10-degree lead angle. The tail and nose spray are frequently referred to separately as the 'forty-five degree cone', which is an area of less dense fragmentation. If this payload were to be detonated in flight, the dense side spray would have a slight forward thrust with an increased velocity equal to missile flight velocity. The angle of the side spray would be defined as the beam width of this fragmenting payload. **Fragment beam width** is defined as the angle covered by a useful density of fragments. Beam width in general depends on the shape of the warhead and the placement of the detonator(s) in the explosive charge.

The latest air target warheads are designed to emit a narrow beam of high-velocity fragments. This type of warhead, called an **Annular Blast Fragmentation Warhead (ABF Warhead)**, has a fragmentation pattern that propagates out in the form of a ring with tremendous destructive potential. A newer type of fragmentation warhead is the **Selectively Aimable Warhead (SAW Warhead)**. This **'smart' warhead** is designed to aim its fragment density at the target. This is accomplished by the fuzing system telling the warhead where the target is located and causing it to detonate so as to maximize the energy density on the target.

(16.3) THE EFFECT OF EXPLOSIONS ON BUILDINGS

Consider an explosion in the vicinity of a building. The air-blast shock wave is the primary damage mechanism in an explosion. What happens when a shock wave strikes the surface of a building? The building may give way immediately if the pressure exceeds its strength. Or if the building does not give way immediately, the shock wave is reflected.

(16.3.1) Direct Air-Blast Effects

Damage due to the air-blast shock wave may be divided into **direct air-blast effects** and **progressive collapse**. In terms of the timing of events, the building is engulfed by the shockwave and direct air-blast damage occurs within tens to hundreds of milliseconds from the time of detonation. If progressive collapse is initiated, it typically occurs within seconds. Direct air-blast effects are damage caused by the high-intensity pressures of the air-blast close in to the explosion and may induce the localized failure of exterior walls, windows, floor systems, columns, and girders. The first step in predicting blast effects on a building is to predict blast loads on the structure. Because the blast pressure pulse depends on several factors like the stand-off distance, the angle of incidence, and the reflected pressure over the exterior of the building, predictions of the blast load can be very complex. Experts use sophisticated methods such as Computational Fluid Dynamics (CFD) computer programs to predict blast loads. A complete analysis of structures subjected to the effects of an explosion is very complex and requires an understanding of structural engineering, dynamics, strengths of materials, and specialized training in explosive effects.

The pressures the shock wave exerts on building surfaces may be several orders of magnitude greater than the loads for which the building is designed. The shock wave also acts in directions that the building may not have been designed for, such as upward on the floor system. In terms of sequence of response, the air-blast first impinges on the weakest point in the vicinity of the device closest to the explosion, typically the exterior envelope of the building. The explosion pushes on the exterior walls at the lower stories and may cause wall failure and window breakage. As the shock wave continues to expand, it enters the structure, pushing both upward and downward on the floors. In chronological order, the blast wave first breaks windows and exterior walls are blown in; the columns may get damaged. This is followed by the blast wave forcing the floors upwards. As the blast wave surrounds the structure, it exerts downward pressure on the roof and inward pressure on all sides.

Floor failure is common in large-scale vehicle-delivered explosive attacks, because floor slabs typically have a large surface area for the pressure to act on and a comparably small thickness. Glass is often the weakest part of a building, breaking at low pressures compared to other components such as the floors, walls, or columns. In the case of large explosions, glass breakage can be found even miles away!

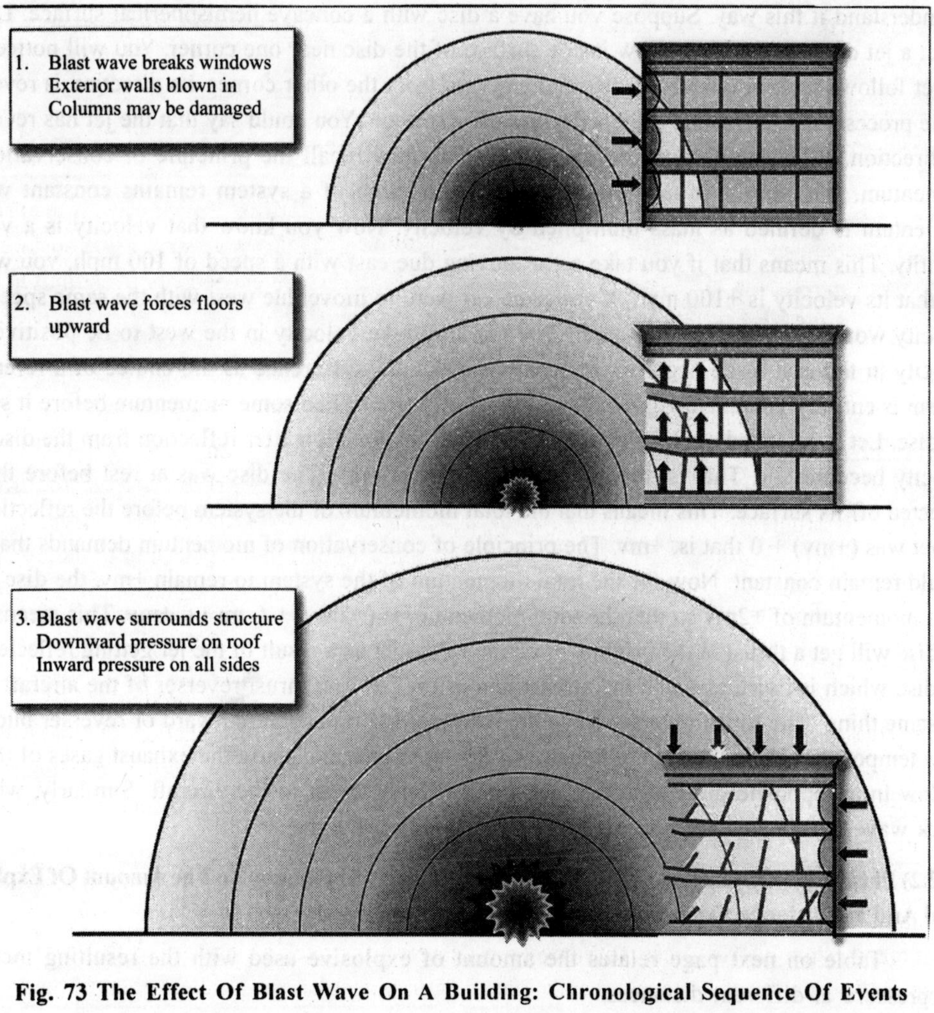

Fig. 73 The Effect Of Blast Wave On A Building: Chronological Sequence Of Events

As we have seen, late in the explosive event, the shock wave becomes negative, creating suction. Behind the shock wave, where a vacuum has been created, air rushes in, creating a powerful wind or drag pressure on all surfaces of the building. This wind picks up and carries flying debris in the vicinity of the detonation. In an external explosion, a portion of the energy is also imparted to the ground, creating a crater and generating a ground shock wave analogous to a high-intensity, but very short-duration earthquake.

What happens when the shock wave is reflected? A very interesting physical phenomenon takes place. The surface gets twice the shock! How it happens?

We will explain it with the help of an example. Ever wondered how does a jet airliner stop after landing? It cannot apply brakes to its wheels. The aircraft is far too heavy and its speed far too high to be stopped by brakes. The wheels would skid! Well, it makes use of something called 'thrust reverser'. What is a thrust reverser? It is a mechanism by which the thrust of the jet engine is reversed in direction—that amounts to applying brakes on the aircraft. How does it work? Let us understand it this way. Suppose you have a disc with a concave hemispherical surface. Let us direct a jet of water on the hollow inside surface of the disc near one corner. You will notice that the jet follows the contour of the disc and emerges from the other corner. Its direction is reversed in the process. It is quite like light reflecting off a surface. You could say that the jet has reversed its direction. What happens to the disc? Well, you may recall the principle of conservation of momentum. Put simply, it says that the total momentum of a system remains constant where momentum is defined as mass multiplied by velocity. Now you know that velocity is a vector quantity. This means that if you take a car moving due east with a speed of 100 mph, you would say that its velocity is +100 mph. If the same car were to move due west with the same speed, its velocity would be taken as –100 mph! You can also take velocity in the west to be positive and velocity in the east to be negative. It does not make any difference as the choice of a reference system is entirely yours. Now get back to the jet of water. It had some momentum before it struck the disc. Let it be called +mv. When the jet reverses its direction after reflection from the disc, the velocity becomes –v. That is, the momentum becomes –mv. The disc was at rest before the jet reflected off its surface. This means that the total momentum of the system before the reflection of the jet was (+mv) + 0 that is, +mv. The principle of conservation of momentum demands that this should remain constant. Now for the total momentum of the system to remain +mv, the disc must get a momentum of +2mv so that the total momentum is (+2mv) + (–mv) = +mv. This means that the disc will get a thrust in the original direction of the jet as a result of the jet getting reflected off the disc which is twice as much the momentum of the jet. The thrust reverser of the aircraft does the same thing. The thrust reverser has either reverser blades angled outward or reverser buckets. They temporarily close the regular exhaust of the jet engine and cause the exhaust gases of the jet to flow in an opposite direction. This imparts a reverse thrust to the aircraft. Similarly, when a shock wave is reflected off a surface, the surface gets twice the shock!

(16.3.2) Illustration Of The Extent Of Damage On Buildings In Relation To The Amount Of Explosive Used And Its Distance From The Target Building

Table on next page relates the amount of explosive used with the resulting incident overpressure at different distances.

Charge weight (lbs of TNT)	Incident overpressure (psi)	At approx. distance (feet)
100	0.5	325
100	1.0	175
100	2.0	100
100	10.0	40
1000	0.5	775
1000	1.0	480
1000	2.0	250
1000	10.0	70
10,000	0.5	1650
10,000	1.0	1000
10,000	2.0	560
10,000	10.0	165

The overpressures may be used to estimate the extent of damage caused:

Incident overpressure (psi)	Extent of damage
	Typical window glass breakage
0.5-1.1	Minor damage to some buildings
1.1-1.8	Panels of sheet metal buckled
1.8-2.9	Failure of concrete block walls
Over 5.0	Collapse of wooden framed buildings
4-7	Serious damage to steel framed buildings
6-9	Severe damage to RCC structures
10-12	Probable total destruction of most buildings

This means that a typical truck bomb carrying about 1000 lbs of TNT equivalent, if detonated even over 250 feet away from a concrete building would be able to cause failure of the concrete block walls!

(16.3.3) Estimation Of Human Casualties

Human casualties as a result of explosions outside buildings depend on several factors. The high pressure of the air-blast that enters through broken windows can cause eardrum damage and lung collapse. As the air-blast damages the building components in its path, secondary missiles are generated that cause impact injuries. Airborne glass fragments typically cause penetration or laceration-type injuries. Larger fragments may cause non-penetrating, or blunt trauma, injuries. Finally, the air-blast pressures can cause occupants to be bodily thrown against objects or to fall. Lacerations due to high-velocity flying glass fragments have been responsible for a significant portion of the injuries received in explosion incidents. In the bombing of the Murrah Federal Building in Oklahoma City, for instance, 40 percent of the survivors in the building had suffered injuries from glass. Even for those inside nearby buildings, about 25 to 30 per cent survivors had injuries from glass.

(16.4) DESIGNING BUILDINGS TO RESIST EXPLOSIONS

You cannot design a building to resist all possible explosive threats for the simple reason that it would not be economical! How much you would like to protect a building must depend on the threat you perceive for the building. It is a different thing that you may reduce some of the threats by taking suitable precautionary measures. For example, if you secure a large perimeter, you reduce the threat from car bombs. If you institute screening procedures at entrance, you reduce the threat from bombs carried in briefcases. None of these would, however, protect your building from an aerial bomb attack!

(16.4.1) Façade Protection

The building's exterior is its first real defense against the effects of a bomb. How the façade responds to this loading will significantly affect the behavior of the structure. If you want to harden the façade fully, it may require a dramatic change to the exterior appearance of the structure such as smaller window sizes and more rugged attachments. Most people would not accept it. It serves little purpose to make changes which would protect only partly. Hence most experts would concentrate on improving the post-damage behavior of the façade.

To improve the post-damage behavior of the glazing system, one could use laminated glass for new construction or apply anti-shatter film to existing glazing. While these features do little to improve the strength of the glass, they attempt to hold the shards of glass together and protect the occupants better from hazardous debris. Laminated glass possesses the best post-damage behavior, may be used with a wide variety of glazing materials and thickness, and provides the highest degree of safety to occupants. The effectiveness of Mylar films, on the other hand, depends on the method of application and the thickness of the film. Then if you are using laminated glass or anti-shatter film, you must ensure that the window frames are much stronger than the glass, otherwise the glazing itself will dislodge from the housing intact and cause serious damage or injury. Then the mullions (vertical bars between panes of glass) may be designed to span from floor to floor or tie into wall panels. The mullion support would absorb a portion of the blast energy and improve the performance of the glazing.

(16.4.2) Curtain Wall Protection

A curtain wall is a non-load-bearing exterior enclosure that is supported by a building's structural steel or concrete frame and holds either glass, metal, stone, or pre-cast concrete panels. Lightweight and composed of relatively slender extruded aluminum members, curtain-wall façades are considerably more flexible than conventional, hardened punched window systems. The design of curtain-wall systems to withstand the effects of explosive loading depends on the performance of the various elements that comprise the system. In addition to hardening the individual members that comprise the curtain-wall system, the attachments to the floor slabs or spandrel beams require special attention.

(16.4.3) Energy-Absorbing Catch Systems

An alternative approach to blast protection takes the concept of a flexible curtain-wall system one step further by making full use of the flexibility and capacity of all the window materials to absorb and dissipate large amounts of blast energy while preventing debris from entering the occupied space. Energy-absorbing catch systems or the so-called **Cable Protected Window Systems**

(CPWS) work in such a way that as the glass is damaged it bears against a cable catch system, which in turn deforms the window frames.

(16.4.4) Floor Slab Reinforcements

A reinforced-concrete, flat-plate structural slab is an economical system that provides for maximum use of vertical space, particularly for buildings in areas with height restriction. However, when subjected to a blast load, punching shear and softening of the moment-resisting capacity of the slabs reduce the lateral-load-resisting capacity of the system. Once the moment-resisting capacity of the slabs at the columns is lost, the ability of the slab to transfer forces to the shear walls is diminished and the structure is severely weakened. In addition to the failure of the floor slab, the loss of contact between the slab and the columns may increase the unsupported column lengths, which may lead to the buckling of those columns. Furthermore, the lateral load resisting system—which consists of the shear walls, the columns, and the slab diaphragms that transfer the lateral loads—may be weakened to such an extent that the whole building may become laterally unstable.

The slab should be designed to prevent a punching shear failure that may in turn develop into a progressive collapse. Increased protection against blasts can be obtained by paying more attention to the conventional flat-plate design and detailing of exterior bays and lower floors, which are most susceptible to a bomb exploding outside the building, and the design of the spandrel beams, which tie the structure together and enhance the response of the slab edge. Drop panels and column capitols may be used to shorten the effective slab length and improve the punching shear resistance. If vertical clearance is a problem, shear-heads embedded in the slab will improve the shear resistance and improve the ability of the slab to transfer moments to the columns. The inclusion of beams to span over critical sections of the slab will greatly enhance the ability of the framing system to transfer lateral loads to the shear walls. The slab-column interface may be designed to contain closed-hoop stirrup reinforcement properly anchored around flexural bars within a prescribed distance from the column face. Bottom reinforcement may be provided continuous through the column. This reinforcement serves to prevent brittle failure at the connection and provides an alternate mechanism for developing shear transfer once the concrete has punched through.

Furthermore, the blast pressures that enter the structure through shattered windows and failed curtain walls will load the underside and subsequently the top surfaces of the floor slabs along the height of the building. Both the delay in the sequence of loading and the difference in magnitude of loading will determine the net pressures acting on the slabs. Consequently, there will be a brief time in which each floor will receive a net upward loading. This upward load requires that the slab be reinforced to resist loads opposing the effects of gravity.

(16.4.5) Column Reinforcements

Due to direct blast pressures, the columns of a typical building, which are designed primarily to resist gravity loads with no special detailing for ductility demands, may experience severe bending deformations in addition to the axial loads that the columns support. To enhance protection, the columns must be designed to be sufficiently ductile to sustain the combined effects of axial load and lateral displacement. In conditions that cause uplift—the net upward load on the slab—the column's tension will experience a brief tensile force. Conventional reinforced concrete columns not designed to resist the combined effects of bending may be prone to damage under these conditions. The lower-floor columns must therefore be designed with adequate ductility and strength to resist the effects of direct lateral loading from the blast pressure and impact of explosive debris. Reinforced concrete columns may be designed to resist the effects of an explosion by providing adequate longitudinal reinforcement, staggering the bar splices, and providing closely spaced ties at plastic hinge locations. Steel columns may be sized to withstand the lateral loads and column splices may be detailed to develop the plastic moments of the section. Existing concrete columns may be encased in a steel jacket or wrapped with a composite fiber to confine the concrete core and increase the shear capacity. On the other hand, existing steel columns may be encased in concrete to add mass and prevent a premature buckling of the thin flanges.

(16.4.6) Preventing Progressive Collapse

Progressive collapse occurs when an initiating localized failure causes adjoining members to be overloaded and fail, resulting in an extent of damage that is disproportionate to the originating region of localized failure. In addition to façade debris hazards, building occupants may be vulnerable to heavier debris resulting from structural damage. A protective design may avoid structural systems that either facilitate or are vulnerable to a progression of collapse resulting from the loss of a primary vertical load-bearing member.

New facilities may be designed to accept the loss of an exterior column for one or possibly two floors above grade without precipitating further collapse. In these cases, the design requirements are intended to be threat-independent to protect against an explosion of indeterminate size that might damage a single column, which results in adequate alternate load paths in the structure should damage occur due to an unspecified abnormal loading. But the upgrading of existing structures to prevent localized damage from developing into a progressive collapse may not be easily accomplished through the alternate path method. This is because the loss of support at a column line would increase the spans of all beams directly above the zone of damage and require different patterns of reinforcement and different types of connection details than those typically detailed for conventional structural design.

Alternatively, columns may be sized, reinforced, or protected to prevent critical damage as a result of the design threat charge weight that may be located in close proximity to them. The vulnerable concrete columns may be jacketed with steel plate or wrapped with composite materials and the vulnerable steel columns may be encased in concrete to protect the cross sections and add mass. For upgrading existing structures, these approaches are better for preventing progressive collapse than supplementing the capacity of the connecting beams and girders, or upgrading them using the alternate path method.

Transfer girders and the columns supporting transfer girders are particularly vulnerable to blast loading. Damage to the girder may leave several lines of columns, which terminate at the girder from above, totally unsupported. Transfer girders typically concentrate the load-bearing system into fewer number of structural elements, which goes against the concept of redundancy or alternate paths desired when coping with a blast. Transfer girders, therefore, create critical sections the loss of which may result in a progressive collapse. So if a transfer girder is required and is vulnerable to an explosive loading, then the girder should be designed to be continuous over several supports. There should be substantial structure framing into the transfer girder to create a two-way redundancy, and thereby an alternate load path in the event of a failure. The column connections, which support the transfer girders, should be designed as Type 2 connections to provide sustained strength despite inelastic deformations.

(16.4.7) Overall Lateral Resistance

The conventional lateral loads—wind and seismic zone 1 forces—to which most buildings are designed are minimal. These minimal lateral load requirements may be resisted by a combination of shear walls, braced frames, and moment-resisting frame action. At each floor level, the slab diaphragms transfer the lateral loads to the lateral-load resisting system. Blasts present a very different scene. The ability of structures to resist a highly impulsive blast loading depends in great measure on the structural detailing of the slabs, joists, and columns that provides for the ductility of the load-resisting system. The structure has to be able to deform inelastically under extreme overload (i.e., dissipate large amounts of energy) prior to failure.

In addition to providing ductile behavior, there needs to be a well-distributed lateral-load resisting mechanism in the horizontal floor plan. The use of several shear walls distributed throughout the building will improve the overall seismic as well as the blast behavior of the building. If adding more shear walls is not architecturally feasible, a combined lateral-load resisting mechanism can also be used. A central shear wall and a perimeter moment-resisting frame will provide for a balanced solution. The perimeter moment-resisting frame will require strengthening the spandrel beams and the connections to the outside columns. This will also result in better protection of the outside columns.

(16.4.8) Using Alternative Construction Materials To Resist Explosive Threats

A variety of materials, not traditionally used in building construction, may provide alternatives to conventional blast hardening solutions. Among these alternatives there are **Shock Attenuating Chemically Bonded Ceramics (SA/CBC),** and composite systems comprised of carbon, aramid and polyethylene fibers and resin. SA/CBC materials are well-developed systems currently in use for the prevention of sympathetic detonation of explosives in munitions storage depots as well as in the seismic retrofit of reinforced concrete columns in highway bridges in California (carbon fiber wrapping). In the latter application, carbon fiber wrappings were found to have advantages over conventional steel jacketing of columns due to problems with weld seams and corrosion. **Spray-on elasto-polymers** have been demonstrated to protect un-reinforced masonry walls by providing a ductile membrane that enables these brittle elements to sustain large deformations without fragmenting and throwing hazardous debris.

In retrofit scenarios where conventional structural treatments may be too heavy or too labor intensive, composite materials may be attractive alternatives because of their lightweight and high tensile strength.

(16.4.9) Seismic Protection Vs. Blast Protection

Many people think that blast damage would be reduced if 'seismic-like' construction standards were adhered to. It is true that seismic building design details enhance the ductility of structures and thereby increase their capacity to sustain plastic hinges and withstand large rotations. However, the required extent of confinement and ductility, and the location of the stress concentrations which form as a result of blast loading will not be the same as for structures subjected to a seismic event. Furthermore, lateral loads resulting from strong ground motions are proportional to the mass, which is distributed throughout the building. Conversely, blast design relies, to some extent, on the inertial resistance of massive structural elements. Finally, seismic resistance is distributed globally throughout the structure whereas blast hardening must provide protection against localized explosive loads. Therefore, it does not follow that a structure adhering to the standard building codes' recommended provisions for seismic design or designed to withstand a strong ground motion would also be sufficient to resist the prescribed blast loading or prevent subsequent progressive collapse.

(16.5) THE EFFECT OF UNDERGROUND EXPLOSIONS

An explosive charge, if detonated while buried, exerts pressure on the surrounding earth and causes movement effects over a distance known as radius of rupture. If the depth of burial is less than the radius of rupture, the explosion products blow through the surface of the ground and form a roughly circular depression known as a crater. An explosion on the surface of the ground makes a shallow crater which may be of a diameter greater than that produced when the explosive is buried. This is due to the scouring action of gases projected downward from the explosive charge. Important factors in crater formation are the type of earth cratered and the type of explosive used. Explosives of a moderate charge in swampy ground make a relatively huge crater, while in a rock formation only a small crater will be formed. An explosive of low detonation velocity that produces a large volume of gases may form a larger crater than TNT if the charge is well below the surface and the ground is not too hard or rocky. If the charge is on the surface, a low-rate explosion makes a smaller crater than TNT because of the weaker scouring action of the lower velocity gases. If an explosive is buried at a depth exceeding the radius of rupture, the compression effect downward and horizontally and the lifting with subsequent subsidence cause a depression on the surface known as a camouflet.

(16.6) THE EFFECT OF UNDERWATER EXPLOSIONS

The mechanism of an underwater blast presents some interesting phenomena associated with a medium denser than air. An underwater explosion creates a cavity filled with high-pressure gas, which pushes the water out radially against the opposing external hydrostatic pressure. This creates a bubble. In addition, the heat causes a certain amount of water to vaporize, adding to the volume of the bubble. This action immediately begins to force the water in contact with the blast front in an outward direction. As it happens with the shock wave in air experiencing a negative impulse, the inertia of the water causes the bubble to overshoot the point at which its internal pressure is equal to the external pressure of the water. The bubble then becomes rarefied, and its radial motion is brought to rest. The external pressure now compresses the rarefied bubble and the compression takes it beyond the equilibrium point. Like a pendulum this sets the bubble into an oscillation. The result is a pulsating bubble of gas slowly rising to the surface, with each expansion of the bubble creating shock wave. Approximately 90% of the bubble's energy is dissipated after the first expansion and contraction. This phenomenon explains how an underwater explosion appears to be followed by other explosions. The time interval of the energy being returned to the bubble (the period of pulsations) varies with the intensity of the initial explosion.

The shock wave is similar in general form to that in air, although if differs in detail. Just as in air, there is a sharp rise in overpressure at the shock front. However, in water, the peak overpressure does not fall off as rapidly with distance as it does in air. Hence, the peak values in water are much higher than those at the same distance from an equal explosion in air. The velocity of sound in water is nearly one mile per second, almost five times as great as in air. Consequently, the duration of the shock wave developed is shorter than in air. The close proximity of the upper and lower boundaries between which the shock wave is forced to travel (water surface and ocean floor) causes complex shock-wave patterns to occur as a result of reflection and rarefaction.

Also, in addition to the initial shock wave that results from the initial gas bubble expansion, subsequent shock waves are produced by bubble pulsation. The pulsating shock wave is of lower magnitude and of longer duration than the initial shock wave.

fig. 74

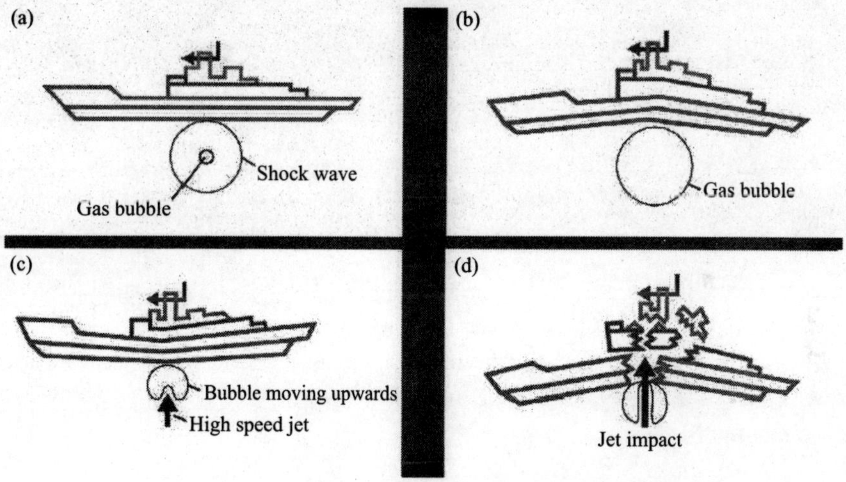

Fig. 74 The Effect Of An Underwater Explosion Beneath A Ship: Chronological Sequence Of Events

The potentially devastating effect of an underwater explosion on a ship is shown in the figure above. At the moment of explosion, a shock wave propagates through the water (in a few milliseconds). This shock wave will have a damaging impact on the ship body [Part (a)].This figure also shows the creation of a bubble at the epicenter of the explosion. The high pressure gas bubble expands very rapidly. At the same time, the ship will bend upwards and its hull is weakened [Part (b)].After reaching its maximum volume, the bubble collapses eventually due to the effect of inertia. The pressure will now be lower than the surrounding reference pressure, causing the ship body to bend downwards [Part (c)].During this collapse phase the bubble is attracted to the ship body. In the final phase of the collapse, a high-speed jet is formed. This high-speed jet can easily cut through the ship body. This mechanism has been identified as being a second cause of damage (and probably even the most devastating one) that appears during an underwater explosion [Part (d)].After the bubble has reached its second minimum a similar cycle can begin. However, as the bubble has lost much of its energy, the following oscillations will not have as great destructive power in general.

Another interesting phenomenon of an underwater blast is surface cutoff. At the surface, the shock wave moving through the water meets a much less dense medium--air. As a result, a reflected wave is sent back into the water, but this is a rarefaction or suction wave. At a point below the surface, the combination of the reflected suction wave with the direct incident wave produces a sharp decrease in the water shock pressure. This is surface cutoff. This negative pressure phase is of short duration and can result in decrease in the extent of damage sustained by the target. The time interval between the arrival of the direct shock wave at a particular location (or target) in the water and that of the cutoff, signaling the arrival of the reflected wave, depends upon the depth of burst, the depth of the target, and the distance from the burst point to the target. It can generally be said that a **depth charge** should be detonated at or below the target and that a target is less vulnerable near the surface.

❑❑❑

Chapter 17

Oh, The So Fascinating Job of Demolitions!

(17.1) WHAT MAKES CONTROLLED DEMOLITION A SCIENCE?

Military demolition or controlled demolition is not a wanton activity. Given a large quantity of explosives, any fool can bring down any structure. There is nothing great about it. Controlled demolitions are fascinating because it is a science in itself. It amazes one to no end as to how massive structures can be brought down with a small quantity of explosives if they are applied properly. It entails as much knowledge of explosives as of structures.

Most people have gross misconceptions about the charges required to be used in demolitions. It has become an almost arcane matter and the wrong notions have been strengthened further by depictions both in the print, the TV and the films where American soldiers are shown bringing down two-storied structures with demolition charges no bigger than a hand grenade and with magnets affixed to them so that all they require is bringing them close to some iron object! This chapter will demystify the business of demolitions.

There is one more reason one should learn about demolitions. It lets you know what the vulnerable points in a structure are. For example, once you know what the vulnerable points in a bridge are, you may protect precisely those points when defending a bridge!

(17.2) THE BASIC PRINCIPLES OF DEMOLITION

(17.2.1) Destructive Effects Of Explosions On Solid Objects In Immediate Vicinity

A high-explosive charge detonated in direct contact with a solid object causes the following destructive effects.

- **Deformation**: The charge's shock wave deforms the surface of the object directly under the charge. When the charge is placed on a concrete surface, it causes a compressive shock wave that crumbles the concrete in the immediate vicinity of the charge, forming a crater. When placed on a steel surface, the charge causes an indentation or depression about the size of the contact area of the charge.

- **Spall**: Suppose the charge is exploded close to a target which has got a free surface on the other side. Now even if the charge's shock wave is not able to rupture the target, what happens is that because of the vast difference in density between the target and the air, the charge's compressive shock wave reflects as a tensile shock wave from the free surface of the target on the other side. This subjects the target to alternate compression and expansion longitudinally. The result is that small fragments of the target are chipped away from the free surface. This is called spalling and the phenomenon is used in the HESH (High Explosive Squash Head) anti-tank warheads. The effect is observed on concrete also. The HESH warheads don't bother to penetrate the armor of the tank. Rather they seek to decapitate the crew inside by the spall fragments flying inside the tank from the inner surface of the armor. To defeat it modern tank armor is made of layered composites which is a poor conductor of shock waves and also uses 'spall liners' made of Kevlar on the inner surface to trap any spall that could be formed.
- **Radial Cracks**: If the charge is large enough, the expanding gases can create a pressure load on the object that will cause cracking and therefore displace the material. This effect is known as radial cracking. When placed on concrete walls, the charge may crack the surface into a large number of chunks and project them away from the center of the explosion. When placed on steel plates, the charge may bend the steel away from the center of the explosion.

(17.2.2) Types Of Charges To Be Used In Various Situations

- **Internal Charges**: Internal charges are used when you can make a hole into the target. Inserting the charge inside the body of the target ensures that the explosive energy is utilized in the best manner. However, if you leave the hole open, some of the explosive energy would be dissipated in the air through that hole. Hence this hole must be sealed. In the field this is accomplished by confining the charge with tightly packed sand, wet clay, or other some other material which would not give way easily. This is called **tamping** or **stemming** the charge and is vital.
- **External Charges**: External charges are placed on the surface of the target when you can't make a hole into it. Still you must ensure that as much explosive force is directed on to the surface of the target as possible. If you leave the charge open, it would dissipate its energy in 360 degrees and a very small part of it would actually be working on the target. Hence you must cover the target so that it reflects the shock wave back on the target. For this purpose in the field you may cover the charge with sandbags or similar heavy things. To be most effective, make the thickness of the tamping material at least equal to the breaching radius. Small breaching charges on horizontal surfaces are tamped with several inches of wet clay or mud.

(17.2.3) The Significance of Charge Dimensions

The force of an explosion depends on the quantity and power of the explosive. The destructive effect depends on the direction in which the explosive force is directed. To transmit the greatest shock, the charge must have an optimal relationship of contact area and thickness to target volume and density. This is vital. Understand it this way. Suppose you spread a calculated charge too thinly. This means that you have not provided enough space for the shock wave to reach full velocity before striking the target. Further, in improperly configured explosives (for example, too thin or wrong strength), the shock wave tends to travel in a parallel rather than a perpendicular direction to the surface. As a result, the volume of the target will be too high for the resulting shockwave to do any significant damage. On the other hand, a thick charge with too small a contact area will transmit a shock wave over too small a target area, with much lateral loss of energy.

(17.2.4) The Significance of Charge Placement

The destructive effect of an explosive charge also depends on the location of the charge in relation to the target size, shape, and configuration. Any significant air or water gap between the target and explosive will lessen the force of the shock wave. We have already discussed the importance of tamping. If the target has an awkward shape, it becomes difficult for charges in the form of blocks or sticks to be placed in close juxtaposition to the target. That's where plastic explosives or explosives in the form of sheets come in handy because they can be cut and/or molded in the desired shape.

(17.2.5) The Significance Of The Direction Of Initiation

It is not only the placement of charge with respect to the target which is important; the location of the detonator and the direction of initiation are also important. As you will see in the diagram, the shock waves travel longitudinally for an elongated target if the initiator is also placed longitudinally. On the other hand, if the detonator is placed transverse, the shock wave will also travel transverse. This means that for the same charge placement, the charge will not be able to cut the target if the detonator was to be placed parallel to the target.

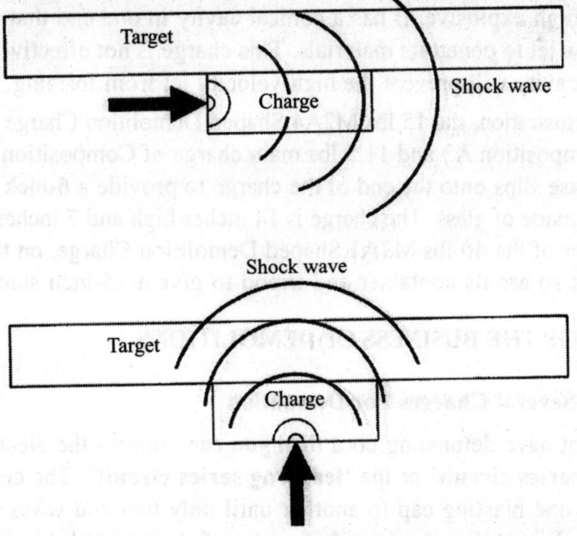

Fig. 75 The Effect Of The Location Of The Detonator And The Direction Of Initiation

(17.2.6) The Destructive Power Of Demolition Charges: Some Quantitative Examples

To illustrate the science of demolitions and to clear any mistaken notions which the readers might be harboring, we would give a concrete idea of what exactly demolition charges can accomplish. The charges mentioned are standard demolition charges of the US Army. The 15 lbs shaped charge, for example, can penetrate 12 inches of armor; the 40 lbs charge can penetrate 20 inches. The diameters of the holes produced in the two cases are 1.5 inches and 2.5 inches respectively. When used against reinforced concrete the 15 lbs charge can breach a wall 30 inches thick whereas the 40 lbs charge will breach a wall 60 inches thick. The diameters of the holes are 2.75 inches and 3.50 inches respectively. When the 15 lbs charge is used against soil or graveled road with a stand-off distance of 30 inches, it can create a hole of 7 feet depth with a diameter of 7 inches. The 40 lbs charge with a stand-off distance of 48 inches will create a hole of 9 feet depth. The hole will have a diameter of 7 inches in graveled road and 14.5 inches in soil.

(17.2.7) Explosives Suited For Demolitions

Most high explosives can be used for demolitions. However, given the universal fame of TNT, calculations by scientists are made for TNT. If you are using other high explosives, the relative effectiveness factor has to be applied for getting the amount of the other explosive. Explosives with high detonation velocities are useful for cutting and breaching. It is for such works that they score over explosives with lower detonation velocities. In view of its excellent **cratering effects** and low cost, ammonium nitrate is a component of most **cratering** and **ditching charges**, however. But 30 lbs of ANFO would require 10 lbs of TNT based booster charge if the charge has to be detonated properly and to result in a militarily useful ditch. PETN is excellent for underwater demolitions because it is almost insoluble in water.

(17.2.8) Shaped Demolition Charges

A shaped demolition charge's primary use is for boring holes in earth, metal, masonry, concrete, and paved and unpaved roads. Its effectiveness depends largely on its shape, composition, and placement. The shaped demolition charge used in military operations is a cylindrical block of high explosive. It has a conical cavity in one end that directs the cone-lining material into a narrow jet to penetrate materials. This charge is not effective underwater, since any water in the conical cavity will prevent the high-velocity jet from forming.

By way of illustration, the 15 lbs M2A4 Shaped Demolition Charge contains a 0.11 lbs (50 gram) booster of Composition A3 and 11.5 lbs main charge of Composition B in a fiber container. A cylindrical fiber base slips onto the end of the charge to provide a 6-inch standoff distance. The cavity liner is a cone made of glass. The charge is 14 inches high and 7 inches in diameter, including the standoff. The liner of the 40 lbs M3A1 Shaped Demolition Charge, on the other hand, is made of metal instead, and so are its container and tripod to give a 15-inch standoff distance.

(17.3) THE BASICS OF THE BUSINESS OF DEMOLITIONS

(17.3.1) Connecting Several Charges For Demolition

If you do not have detonating cord then you can connect the electrical detonators in the so-called **'common series circuit'** or the **'leapfrog series circuit'**. The common series circuit is made by connecting one blasting cap to another until only two end wires are free. The leapfrog series circuit is made by starting at one end of a row of charges and priming alternate charges to the opposite end and then priming the remaining charges on the return leg of the series. This method eliminates the necessity for a long return lead from the far end of the line of charges.

(17.3.2) Techniques Of Proper Priming Of Demolition Charges

A 6-inch length of detonating cord equals the power output of a blasting cap. However, detonating cord will not detonate explosives as reliably as a blasting cap because its power is not as concentrated. Therefore, one should always use several turns or a knot of detonating cord for priming charges. Detonating cord cannot be used for priming shaped charges; they require detonators of appropriate strength.

For priming plastic explosives one should use the **Uli knot**, the **Double Overhand knot** or the **Triple Roll knot**. A notch must be made in the explosive large enough to insert the knot formed. The knot must be placed in the notch and then it should be covered with the explosive which had been removed for the notch. Everything can then be secured with tape. There should be at least half an inch of explosive on all sides of the knot and the notch must be cut keeping this in mind.

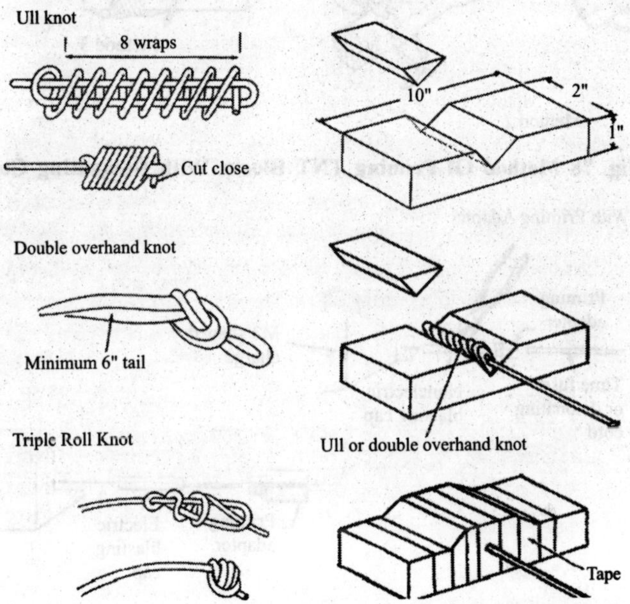

Fig. 76 Method Of Priming Plastic Explosives With Detonating Cord

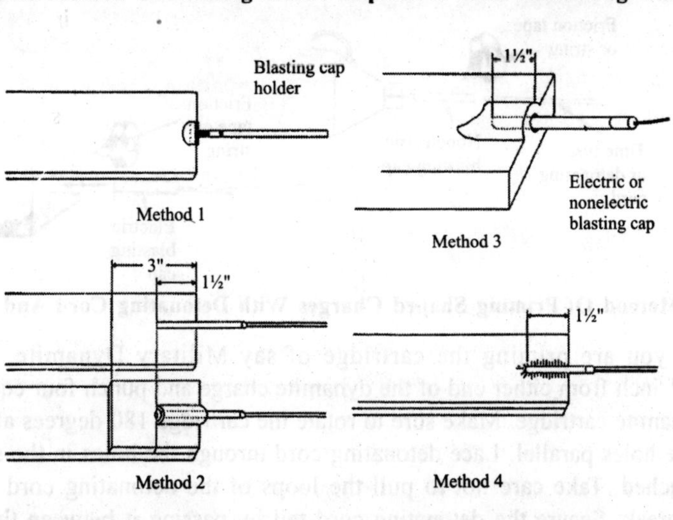

Fig. 77 Method Of Priming Sheet Explosives With Detonators

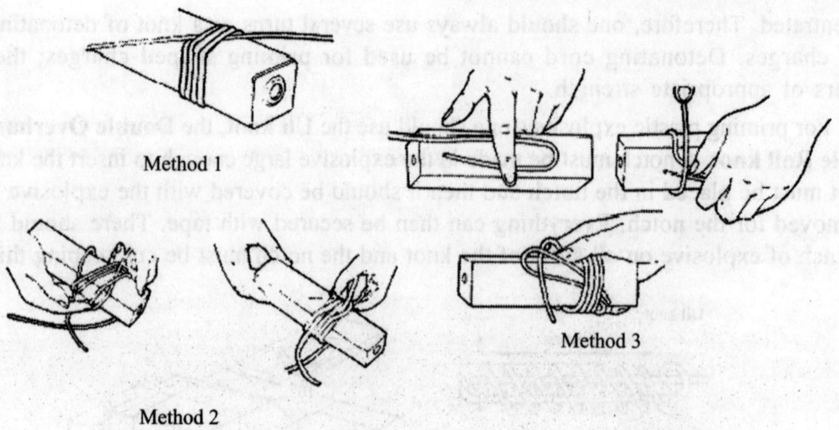

Fig. 78 Method Of Priming TNT Blocks With Detonating Cord

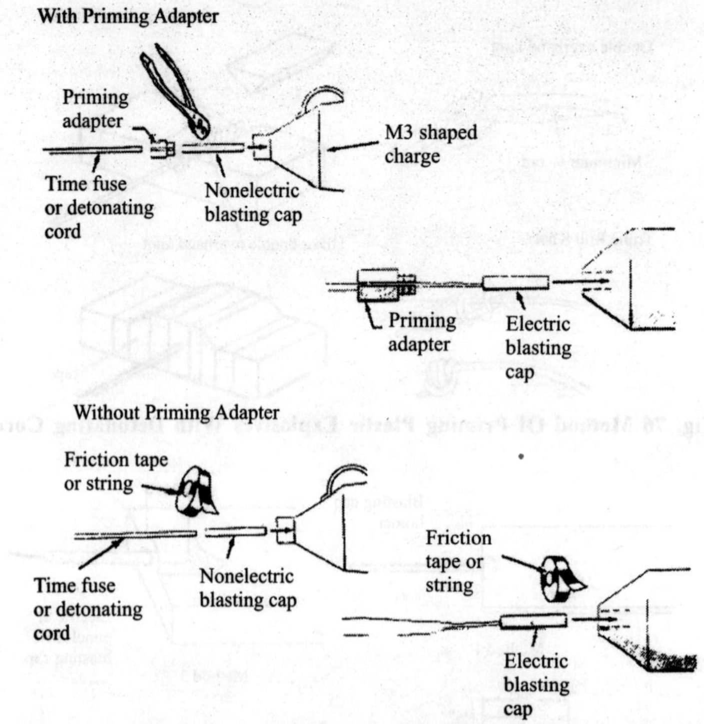

Fig. 79 Method Of Priming Shaped Charges With Detonating Cord And Detonators

When you are priming the cartridge of say Military Dynamite, you should start approximately 1 inch from either end of the dynamite charge and punch four equally spaced holes through the dynamite cartridge. Make sure to rotate the cartridge 180 degrees after punching each hole to keep the holes parallel. Lace detonating cord through the holes in the same direction the holes were punched. Take care not to pull the loops of the detonating cord too tightly or the dynamite will break. Secure the detonating cord tail by passing it between the detonating cord lace and the dynamite charge.

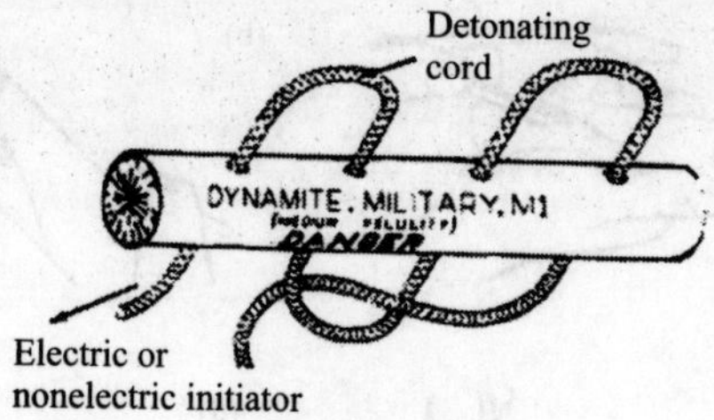

Fig. 80 Method Of Priming Dynamite With Detonating Cord

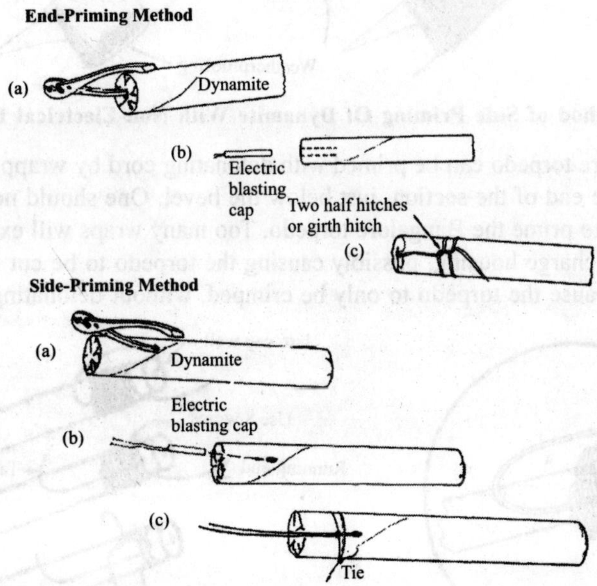

Fig. 81 Method Of End Priming Of Dynamite With Electrical Detonator

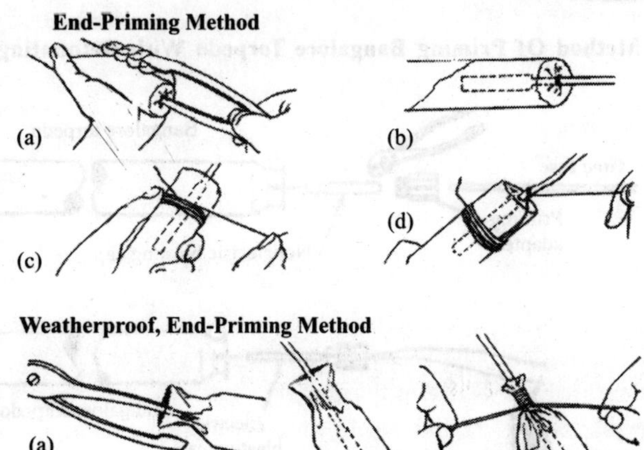

Fig. 82 Method of End Priming Of Dynamite With Non-Electrical Detonator

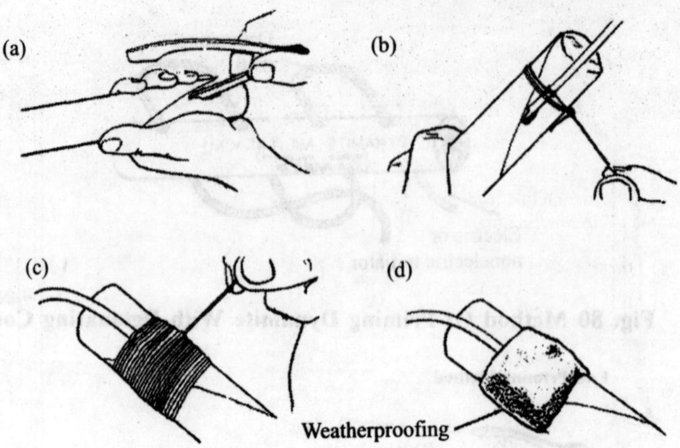

Fig. 83 Method of Side Priming Of Dynamite With Non-Electrical Detonator

The Bangalore torpedo can be primed with detonating cord by wrapping detonating cord eight times around the end of the section, just below the bevel. One should not use more than or less than eight wraps to prime the Bangalore torpedo. Too many wraps will extend the detonating cord past the booster charge housing, possibly causing the torpedo to be cut without detonating. Too few wraps may cause the torpedo to only be crimped, without detonating.

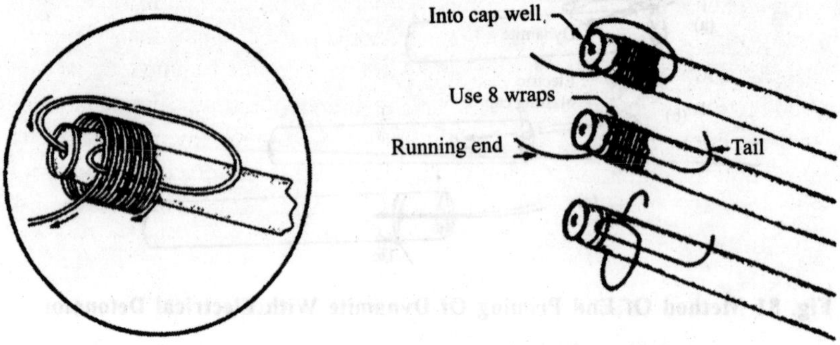

Fig. 84 Method Of Priming Bangalore Torpedo With Detonating Cord

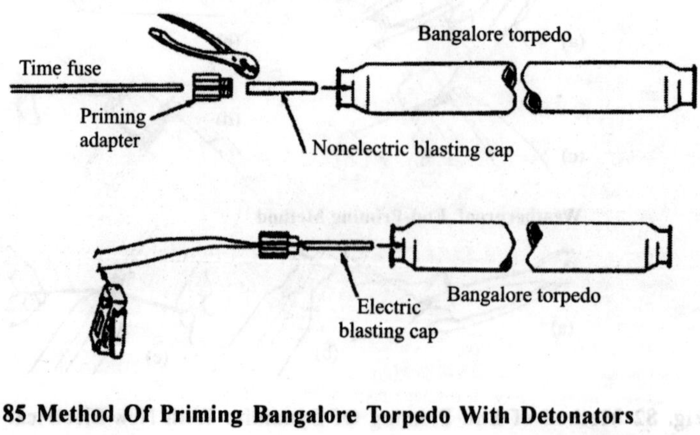

Fig. 85 Method Of Priming Bangalore Torpedo With Detonators

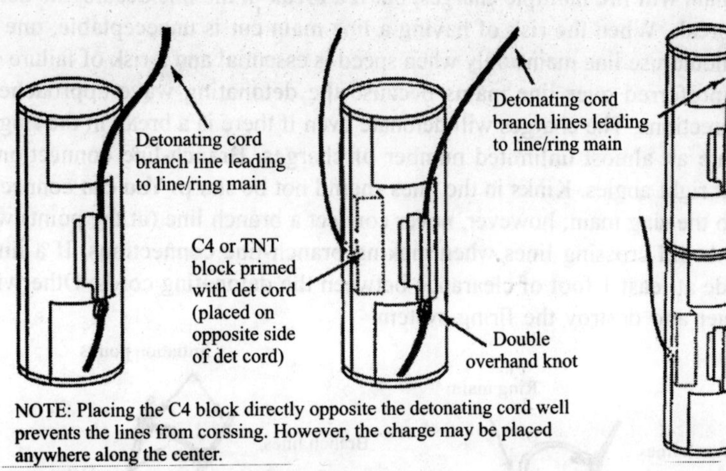

Fig. 86 Method Of Priming Ammonium Nitrate Based Cratering Charge

(17.3.3) Various Types Of Firing Circuits

(a) Single-Initiated, Single-Fired, Single-Primed

In this circuit each charge is singly primed with a branch line. The **branch line** is tied to the **line main or ring main**. If possible, branch lines should not be longer than 12 feet from the charge to the ring or line main. A longer branch line is too susceptible to damage that may isolate the charge. The connections of branch lines and ring or line mains should intersect at right (90-degree) angles. If these connections are not at right angles, the branch line may be blown off the line main without complete detonation. (Tying to the ring main is preferred but construction of a ring main may not be possible because of the amount of detonating cord. The ring main decreases the chances of a misfire should a break or cut occur anywhere within the ring main.)

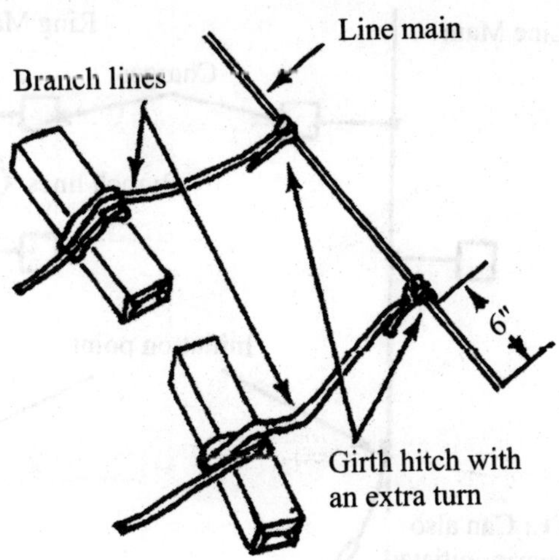

Fig. 87 Line Main With Branch Lines

A line main will fire multiple charges, but if a break in the line occurs, the detonating wave will stop at the break. When the risk of having a line main cut is unacceptable, one should use a ring main. One should use line mains only when speed is essential and a risk of failure is acceptable. Ring mains are preferred over line mains because the detonating wave approaches the branch lines from two directions. The charges will detonate even if there is a break in the ring main. A ring main will detonate an almost unlimited number of charges. Branch-line connections at the ring main should be at right angles. Kinks in the lines should not be sharp. You can connect any number of branch lines to the ring main; however, never connect a branch line (at the point) where the ring main is spliced. Avoid crossing lines when making branch-line connections. If a line crossing is necessary, provide at least 1 foot of clearance between the detonating cords. Otherwise, the cords will cut each other and destroy the firing system.

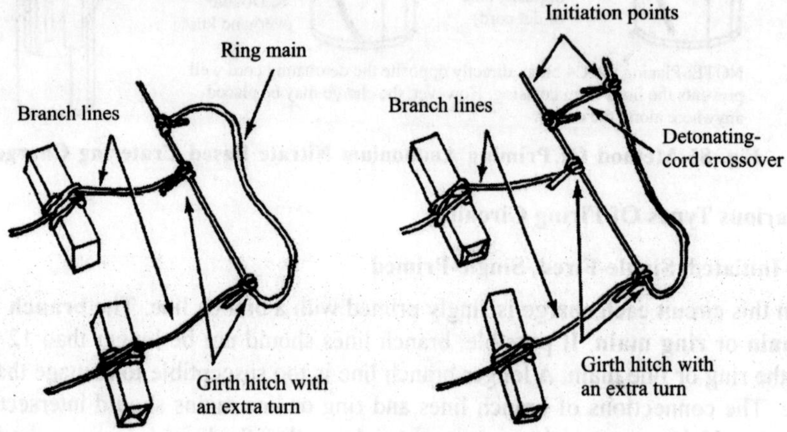

Fig. 88 Ring Main

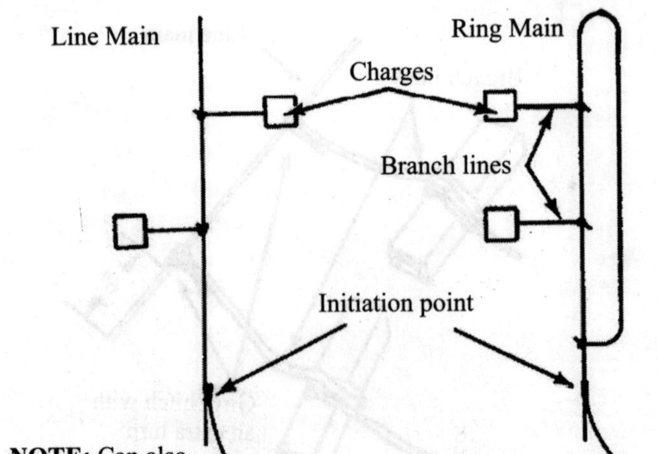

Fig. 89 Single-Initiated, Single-Fired, Single-Primed Circuit

(b) Dual-Initiated, Single-Fired, Single-Primed

When using a combination initiation system, the electric initiation system is always the primary means of initiation. When using dual, non-electric initiation systems, the shorter time fuse is the primary initiation system.

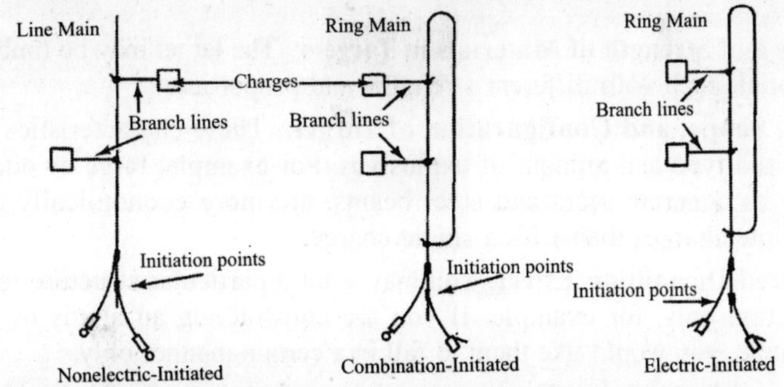

Fig. 90 Dual-Initiated, Single-Fired, Single-Primed Circuit

(c) Dual-Installed, Dual-Fired, Dual-Primed

In this circuit, each charge is dual-primed with two branch lines. One branch line is tied to one firing system, and the other branch line is tied to an independent firing system. Line mains or ring mains may be used; however, they should not be mixed. To help prevent misfires, one should use detonating-cord crossovers. Crossovers are used to tie both firing systems together at the ends. The initiation systems are taped in—the primary initiation system goes to one firing system, the secondary goes to the other.

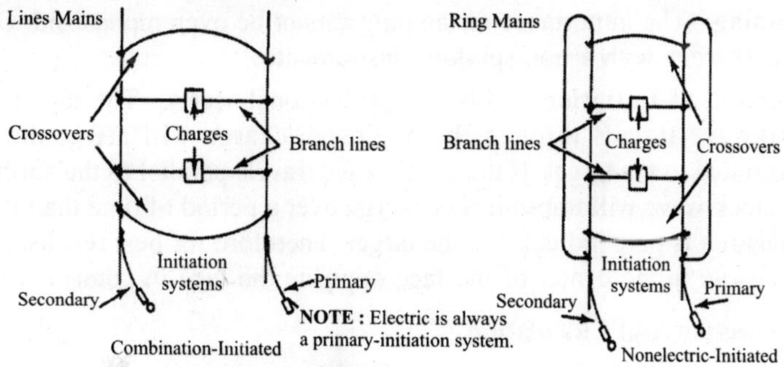

Fig. 91 Dual-Installed, Dual-Fired, Dual-Primed Circuit

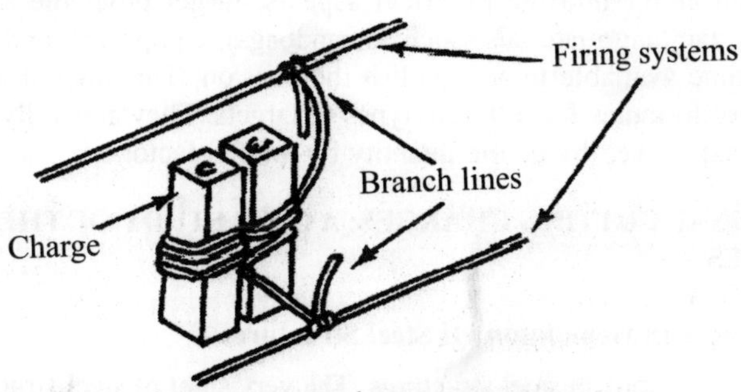

Fig. 92 A Dual Primed Charge

(17.4) CALCULATING THE RIGHT QUANTITY OF CHARGE FOR A PARTICULAR JOB

(17.4.1) Factors Which Determine The Amount Of Charge Required

The amount of explosives required for any demolition project depends on the following critical factors:

- **Type and Strength of Materials in Targets**: The target may be timber, steel, or other material, each with different strengths and properties.
- **Size, Shape, and Configuration of Target**: These characteristics all influence the required type and amount of explosives. For example, large or odd-shaped targets, such as concrete piers and steel beams, are more economically demolished with multiple charges than with a single charge.
- **Desired Demolition Effect**: You may want a particular structure to fall in a certain direction only, for example. If you are constructing an abatis by felling trees for example, you would like them to fall in a certain manner only.
- **Type of Explosive**: The characteristics of each type of explosive determine its application for demolition purposes.
- **Size and Placement of Charge**: When you are using external charges in a hurry and you do not have time for calculating its size and placement you can use a flat, square charge with a thickness-to-width ratio of 1:3. In general, charges of less than 5 pounds should be at least 1 inch thick. Charges from 5 to 40 pounds should be 2 inches thick. Charges of 40 pounds or more should be 4 inches thick. Fasten charges to the target using wire, adhesive compound, tape, or string. Prop charges against targets with wooden or metal frames made of scrap or other available materials or place the charges in boreholes.
- **Tamping**: The importance of tamping cannot be overemphasized. One must always tamp charges with a non-sparking instrument.
- **Direction of Initiation**: This is a professional secret. The direction in which the shockwave travels through the explosive charge will affect the rate of energy transmitted to the target. If the shock wave travels parallel to the surface of the target, the shock wave will transmit less energy over a period of time than if the direction of detonation is perpendicular to the target. Therefore for best results, one must initiate the charge in the center of the face opposite the face in contact with the target.

(17.4.2) Charge Selection And Calculation

Selection of explosives for successful demolition operations is a balance between the critical factors listed above and the practical aspects: target type; the amount and types of explosives available, tamping materials (such as sandbags), equipment, and personnel available; and the amount of time available to accomplish the mission. The amount of charge required is calculated by standard formulas for different types of targets. They generally refer to TNT. If you are using any other explosive, divide the quantity by the RE factor.

(17.5) CONVENTIONAL CUTTING CHARGES: A CASE STUDY OF THE DEMOLITION OF STEEL STRUCTURES

(17.5.1) Critical Factors In Demolition Of Steel Structures

The target is to demolish steel structures. The very sight of steel structures literally oozing strength presents a formidable challenge to the layman but not for the demolition expert! How would you go about it? Several factors must be taken into consideration before you attach your bundle of explosives to them.

- Firstly, the configuration of the steel in the structure determines the type and amount of charge necessary for successful demolition. I-beams, wide-flange beams, channels, angle sections, structural tees, and steel plates used in building or bridge construction would require different treatments.
- The second important factor is the composition of the steel. You would have to know whether the structure uses high-carbon steel, alloy steel, an easy customer like cast iron or a difficult customer like Nickel-molybdenum steel. Nickel-molybdenum steel, for example, cannot be cut easily by conventional steel-cutting charges. The jet from a shaped charge will penetrate it, but cutting it would require multiple charges or linear-shaped charges. Nickel-molybdenum steel shafts can be cut with a diamond charge which we have discussed in the next section. However, the saddle charge will not cut nickel-molybdenum shafts. Unless you know such things you would be wasting your time and the explosives on them. Therefore, you will have to use some method other than explosives to cut nickel-molybdenum steel, such as thermite or acetylene or electrical cutting tools.
- Thirdly, in steel-cutting charges, the type of the explosive to be used is important. High detonation velocity explosives like C4 are ideal for the job. Most importantly, plastic explosives also enable you to easily cut and shape them to fit tightly into the grooves and angles of the target.
- Fourthly, even with an explosive of choice, correct placement is of vital importance. You must ensure that the charge is continuous over the complete line of the proposed cut and there is close contact between the charge and the target.
- Fifthly, the charge dimensions must be correct. The width of the charge's cross section, for example, must be between one and three times its thickness. Do not use charges more than 6 inches thick because you can achieve better results by increasing the width rather than the thickness.
- Sixthly, you must ensure proper initiation in the entire charge. Do not leave anything to theory. In practice, for example, long charges must be primed every 4 to 5 feet. If butting C4 packages end to end along the line of the cut, you must prime every fourth package. Then you must ensure that the direction of initiation is perpendicular to the target.

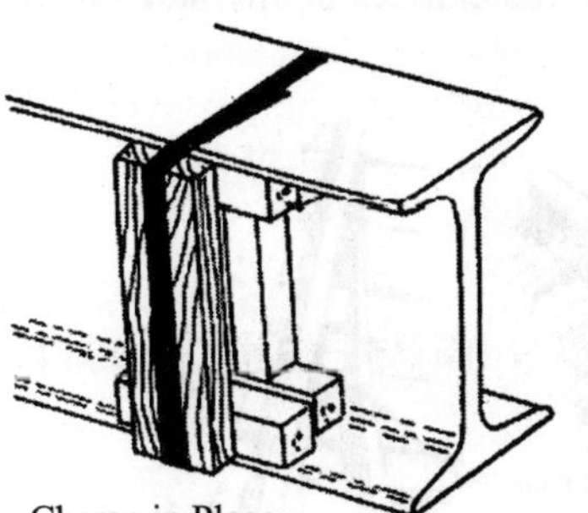

Charge in Place on One Side

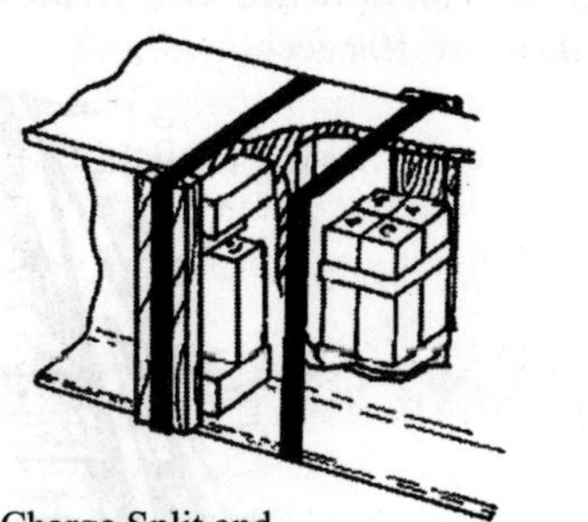

Charge Split and Placed on Two Sides

Fig. 93 Placement Of Charges For Steel Members

Finally, the size of the charge is dictated by the target steel's type and size and the type of charge selected. Remember that each steel configuration requires a unique charge size.

(17.5.2) Amount Of Explosives Required For Cutting I-Beams, Built-Up Girders, Steel Plates, Columns, Steel Bars, Rods, Chains, Cables, And Railroad Rails

For quick action in the field scientists have worked out formulas to guide the demolition expert. The following formula, for example, will give you the size of charge necessary for **cutting I-beams, built-up girders, steel plates, columns**, and other structural steel sections. Keep in mind, however, that when calculating cutting charges for steel beams, the area for the top flange, web, and bottom flange must be calculated separately. Built-up beams also have rivet heads and angles or welds joining the flanges to the web. You must add the thickness of one rivet head and the angle iron to the flange thickness when determining the thickness of a built-up beam's flange. Use the thinnest point of the web as the web thickness, ignoring rivet-head and angle-iron thickness. The lattice of the lattice-girder webs must be cut diagonally by placing a charge on each lattice along the line of the cut.

$P = 0.375A$

Where P is the amount of TNT required (in pounds) and A is the cross-sectional area of the steel member (in square inches).

For high carbon steel the formula is

$P = D^2$ where D is the diameter or thickness of the section to be cut in inches.

Now suppose you have to **cut steel bars, rods, chains, or cables** (up to 2 inches). The size of these materials makes proper charge placement difficult. If the explosive is long enough to bridge both sides of the link or is large enough to fit snugly between the two links, use one charge. If the explosive is not large enough to bridge both sides, use two charges. For materials up to 1 inch in diameter or thickness, use 1 pound of explosive. For materials between 1 and 2 inches in diameter or thickness, use 2 pounds of explosive. For steel bars, rods, chains, and cables (over 2 inches) where the target diameter or thickness is 2 inches or greater, you can use the equation $P = 0.375A$. When the thickness or diameter is 3 inches or greater, you must place half of the charge on each side of, the target and stagger the placement to produce the maximum shearing effect.

For **cutting railroad rails** the height of the railroad rail is the critical dimension for determining the amount of explosive required. For rails 5 inches or more in height, crossovers, and switches, use 1 pound of C4 or TNT. For rails less than 5 inches high, use 1/2 pound of C4 or TNT. **Railroad frogs** require 2 pounds of C4 or TNT. Place the charges at vulnerable points, such as frogs, curves, switches, and crossovers, if possible. Place the charges at alternate rail splices for a distance of 500 feet.

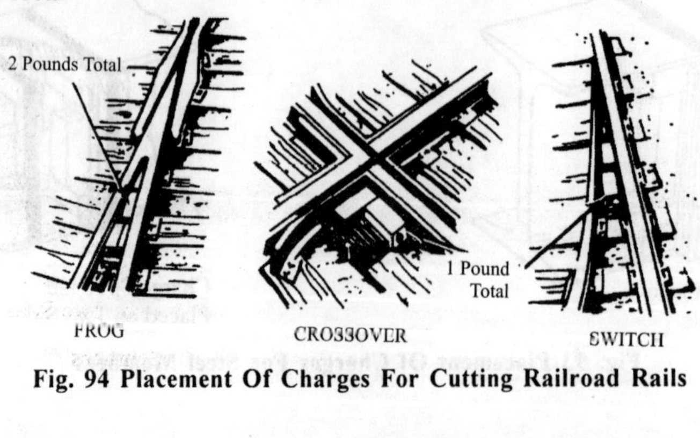

Fig. 94 Placement Of Charges For Cutting Railroad Rails

Fascinating, isn't it?

(17.6) SPECIAL CUTTING CHARGES: A CASE STUDY OF THE DEMOLITION OF STEEL STRUCTURES

You must use the special cutting charges (ribbon, saddle, and diamond charges) instead of conventional cutting charges whenever and wherever time and circumstances permit their use. These charges require extra time to prepare, since they require exact and careful target measurement to achieve optimal effect. The major advantage of special cutting charges is that they use considerably less explosive than conventional charges. For such charges **sheet explosive** is more suitable than C4, since sheet explosive is more flexible and requires less cutting. C4 requires considerable cutting, shaping, and molding, which may reduce its density and, therefore, its effectiveness. Sheet charges being thin require blasting caps crimped to a detonating-cord branch line for initiation. A detonating-cord knot will work but is difficult to place and can ruin the advantage of the special charge shape.

(17.6.1) Ribbon Charge

Ribbon charges are used to **cut flat, steel targets** up to 3 inches thick. A standard ribbon charge is made as follows:

a. Charge Thickness: The charge thickness must equal one half the target's thickness; however, it should never be less than 1/2 inch.

b. Charge Width: The charge width must be three times the charge thickness.

c. Charge Length: The charge length equals the length of the desired cut.

The ribbon charge must be detonated from the center or from either end. The detonating-cord branch lines must be the same length and must connect in a junction box.

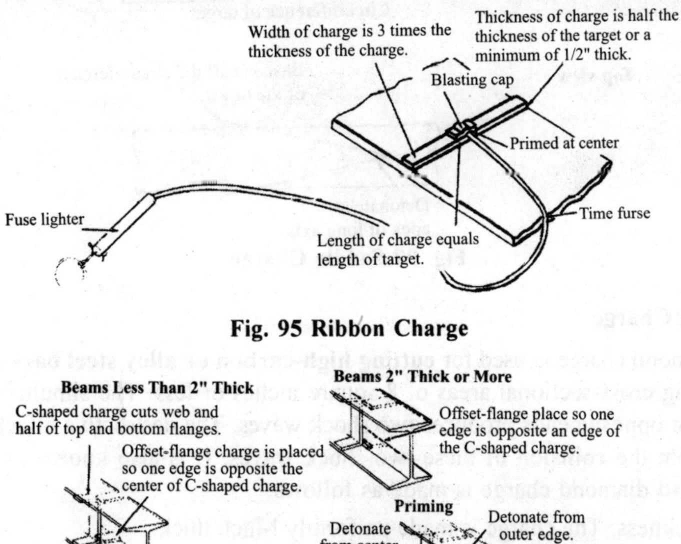

Fig. 95 Ribbon Charge

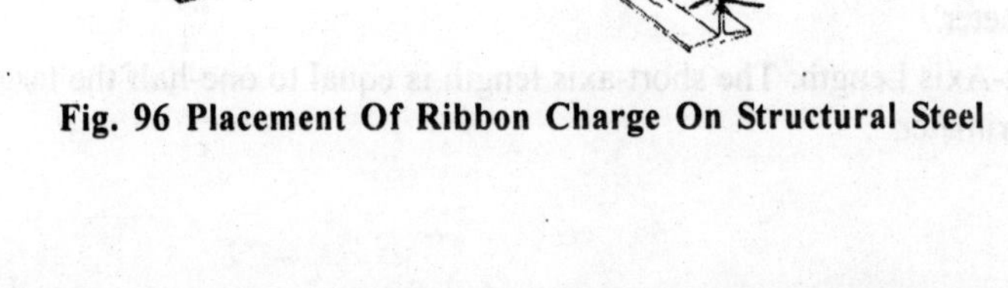

Fig. 96 Placement Of Ribbon Charge On Structural Steel

(17.6.2) Saddle Charge

This steel-cutting method uses the destructive effect of the cross fracture formed in the steel by the base of the saddle charge (end opposite the point of initiation). The saddle charge is used on **mild steel bars; whether round, square, or rectangular** in shape, up to 8 square inches or 8 inches in diameter. A standard saddle charge is made as follows:

a. Thickness: The charge is made uniformly l-inch thick.
b. Base Width: The base width is equal to one-half the target circumference or perimeter.
c. Long-Axis Length: The long-axis length is equal to the target circumference or perimeter.
d. Priming: The saddle charge is detonated by placing a blasting cap at the apex of the long axis.
e. Placement: The long axis of the saddle charge should be parallel with the long axis of the target. You must cut the charge to the correct shape and dimensions and then place it around the target. You must also ensure that the charge maintains close contact with the target by taping the charge to the target.

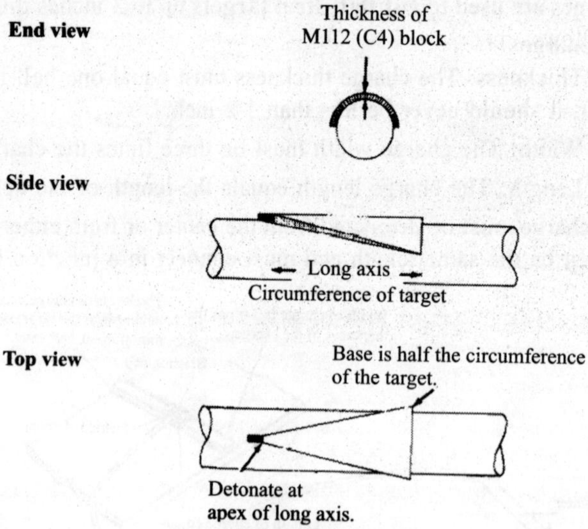

Fig. 97 Saddle Charge

(17.6.3) Diamond Charge

The diamond charge is used for **cutting high-carbon or alloy steel bars** up to 8 inches in diameter or having cross-sectional areas of 8 square inches or less. The simultaneous detonation of the charge from opposite ends produces two shock waves. The destructive effect of the diamond charge comes from the collision of these two shock waves. It is also known as the stress-wave method. A standard diamond charge is made as follows:

a. Thickness: The charge is made uniformly l-inch thick.
b. Long-Axis Length: The long-axis length is equal to the target circumference or perimeter.
c. Short-Axis Length: The short-axis length is equal to one-half the target circumference or perimeter.

d. Priming: The diamond charge is primed with two detonating cord branch lines using one of the following methods, namely, detonating cord knots; two electric blasting caps in a series circuit; or two non-electric blasting caps. When using detonating cord knots or non-electric blasting caps, the branch lines must be the same length.

e. Placement: You must place the explosive completely around the target so that the ends of the long axes touch. You may have to slightly increase the charge dimensions to do this. To ensure adequate contact with the target, tape the charge to the target.

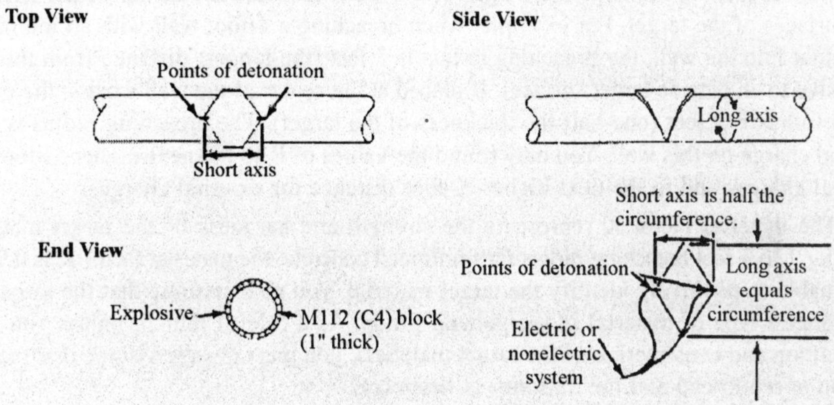

Fig. 98 Diamond Charge

(17.7) BREACHING CHARGES: A CASE STUDY OF THE DEMOLITION OF CONCRETE STRUCTURES

(17.7.1) Critical Factors In Demolition Of Concrete, Masonry, Rock Or Similar Structures

If steel structures frighten you with their strength, concrete structures tend to overawe the layman with their sheer mass and huge size and seem to scorn any attempt of destruction. That does not, however, dishearten the demolition expert in the least. Breaching charges are used to destroy massive concrete structures like bridge piers, bridge abutments, and permanent field fortifications. Critical factors that must be taken into consideration for attacking concrete structures include the size, shape, placement, and tamping or confinement of the breaching charges. Of these factors, the size and confinement of the explosive are most important. If you can find the time to drill holes and place internal charges, it would make the job much easier. But that may be often impractical in the field. Breaching charges do not really intend to blow up concrete structures. That would require a huge amount of explosives and is really not necessary. The intent of breaching charges is to produce and transmit sufficient energy to the target to make a crater and create spalling. Remember that breaching charges placed against reinforced concrete will not be able to cut the steel reinforcing bars therein. You will have to remove or cut the steel reinforcement with a steel-cutting charge after the concrete is breached.

(17.7.2) Amount Of Explosives Required For Breaching Concrete Structures

Once again scientists have developed formulas to enable the demolition experts to carry out their job quickly in the field. We have:

$$P = R^3 KC$$

Where P is the amount of TNT required (in pounds); R is the breaching radius (in feet); K is the so-called material factor, which reflects the strength, hardness, and mass of the material to be demolished; and C is the tamping factor, which depends on the location and tamping of the charge.

The breaching radius R for external charges is equal to the thickness of the target being breached. For internal charges placed in the center of the target's mass, the breaching radius is reduced to half the thickness of the target. If the charge is placed at less than half the mass thickness, the breaching radius is the longer of the distances from the center of the charge to the outside surfaces of the target. For example, when breaching a 4-foot wall with an internal charge placed 1 foot into the wall, the breaching radius is 3 feet (the longest distance from the center of the explosive to an outside target surface). If placed at the center of the wall's mass, the explosive's breaching radius is 2 feet (one-half the thickness of the target). The breaching radius is 4 feet for an external charge on this wall. You may round the values of R to the next-higher ¼-foot distance for internal charges and to the next-higher ½-foot distance for external charges.

The material factor K represents the strength and hardness of the target material. For example, for 1 to 3 feet breaching radius for reinforced concrete, the material factor K is 0.96. When you are unable to positively identify the target material, you must assume that the target consists of the strongest type of material in the general group. As a rule of thumb, unless you know the exact condition and construction of the target materials, you must always assume that the concrete before you is reinforced and the masonry is first-class.

The tamping factor C depends on where you have placed the charge and on the materials you have used for tamping. You must not consider a charge tamped with a solid material (such as sand or earth) as fully tamped unless you cover the charge to a depth equal to or greater than the breaching radius. For example, C is equal to 1 if the charge is placed in the center of the mass. If the charge is placed outside and tamped with earth by a thickness equal to the breaching radius, C is equal to 2. If the charge is placed on the ground next to the target and is un-tamped, C is equal to 3.6. C, however, becomes 1.8 if the un-tamped charge is elevated above the ground.

The number of charges required for demolishing slabs, piers or walls is calculated by the formula:

$$N = W/2R$$

Where N is the number of charges; W is the width of the pier, slab, or wall (in feet); and R is the breaching radius (in feet). If N comes to less than 1.25, you must use one charge; if N comes to 1.25 but less than 2.5, you should use two charges; if N is equal to or greater than 2.5, you must round it off to the nearest whole number and use that many charges. The first charge must be placed at a distance of R from one side of the target. The remaining charges are spaced at a distance of 2R apart.

(17.7.3) Placement And Configuration Of Breaching Charges

Piers and walls offer limited locations for placing explosives. Unless a demolition chamber is available, you must place the charge (or charges) against one face of the target. Placing a charge above ground level is more effective than placing one directly on the ground. When the demolition requires several charges to destroy a pier, slab, or wall and you plan to use elevated charges, distribute the charges equally, no less than one breaching radius high from the base of the target. Doing this takes maximum advantage of the shock wave. If possible, place breaching charges so that there is a free reflection surface on the opposite side of the target. As you would recall, this free reflection surface allows spalling to take place. If time and situation permit, you must tamp all charges thoroughly with sandbags. The tamped area must be equal to or greater than the breaching radius. For piers, slabs, or walls partially submerged in water, you must place charges equal to or greater than the breaching radius below the water line, if possible.

As far as the configuration of breaching charges is concerned, you must place the explosive charge in the shape of a flat square for maximum effectiveness. The charge width should be approximately three times the charge thickness. The thickness of the charge would, of course, depend on the amount of explosive required.

(17.8) ILLUSTRATIVE EXAMPLES OF SOME DEMOLITION CHARGES IN REAL LIFE SITUATIONS

It would be highly illustrative as well as illuminative to have a look at some real life examples. That will help you visualize the situation in a better manner. Just keep in mind that the figures given are theoretical values. In practice C4 comes in packages of 1.25 lbs (volume 20 cubic inches) and demolition experts consider it a safe practice to use a little more than the theoretical value by calculating the number of packages required (either by the theoretical weight or volume of the charge required) and then rounding off the figure, i.e., the number of packages required to the next higher number.

An 'I' shaped steel beam with top and bottom flanges five inches wide and half an inch thick and the web being 11 inches long and 3/8 inches thick would require 2.9 lbs of C4 to cut it.

A 14 inch wide and 2 inch thick steel plate would require 2.6 lbs of C4 in a ribbon charge for cutting.

A seven inch diameter steel bar would require a saddle charge of 7.6 lbs of C4.

A nine inch diameter high-carbon steel bar would require 12.5 lbs of C4 in a diamond charge.

For breaching a five feet thick and 30 feet wide reinforced concrete pier, you would require three charges of 90.67 lbs each of C4 provided they are un-tamped and placed at a height of five feet.

To counterforce a concrete cube three feet thick, you would require a total charge of 4.5 lbs of C4.

(17.9) USE OF COUNTERFORCE CHARGES AGAINST CONCRETE COLUMNS

This method, comprising two charges placed on the opposite sides of the target, derives its effectivity from the simultaneous detonation of two charges placed directly opposite each other and as near the center of the target as possible. The target is therefore subjected to forces from opposite sides. The method, however, suffers from the limitation that it requires accessibility to both sides of the target so that the charges fit flush against their respective target sides. This special breaching technique is effective against **rectangular masonry or concrete columns** 4 feet thick or less. It is not effective against walls, piers, or long obstacles. Thc obstacle also must have at least three free faces or be freestanding. If constructed of plastic explosives (C4) and properly placed and detonated, counterforce charges produce excellent results with a relatively small amount of explosive.

The thickness or the diameter of the target determines the amount of plastic explosive required. The amount of plastic explosive equals 1½ times the thickness of the target in feet (that is, 1½ pounds of explosive per feet). You must round off fractional measurements to the next higher half foot before multiplying. For example, a concrete target measuring 3 feet 9 inches thick requires 6 pounds of plastic explosive (1.5 lb/foot x 4 feet).

Placement of the counterforce charges is simple. Split the charge in half. Place the two halves directly opposite each other on the target. You must prime both charges on the face farthest from the target. You can join the ends of the detonating-cord branch lines in a junction box. The length of the branch lines from both charges must be equal to ensure simultaneous detonation.

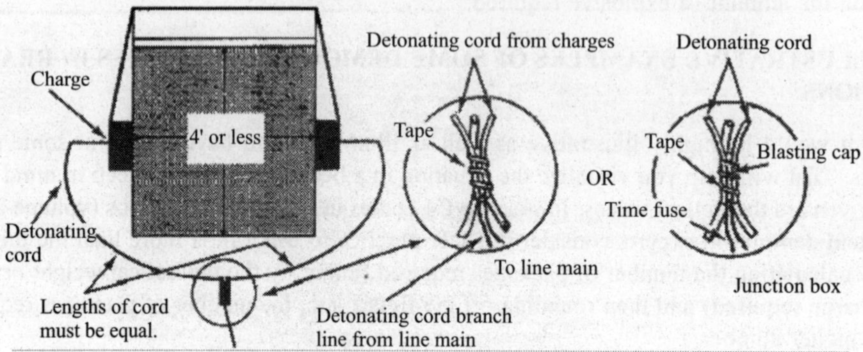

Fig. 99 Counterforce Charge

(17.10) CRATERING AND DITCHING CHARGES

All military explosives can create **anti-tank craters**. As mentioned earlier, a 40 lbs ammonium nitrate based cratering charge is a good option for making craters. To be effective obstacles, road craters must be too wide for track vehicles to span and too deep and steep-sided for any vehicle to pass through. It must be kept in mind, however, that blasted road craters will, however, not be able to stop modern main battle tanks indefinitely. Road craters are effective anti-tank obstacles if a tank requires three or more passes to traverse the crater, thereby providing enough time for anti-tank weapons to stop the tank. A tank, making repeated attempts to traverse the crater, will pull soil loose from the slopes of the crater, filling the bottom and reducing both the crater's depth and angle of slope. Road craters should be large enough to tie into natural or constructed obstacles at each end. You must cut road craters across the desired gap at a 45-degree angle from the direction of approach. This angled cut will increase the tank's tendency to slip sideways and ride off its track. To achieve sufficient obstacle depth, you must make craters in multiple or single rows, enhancing some other obstacle, such as a bridge demolition. When creating more than one row of craters, you should space them far enough apart so that a single armored vehicle launch bridge (AVLB) will not be able to span them.

For making a crater about 25-30 feet wide and 7-8 feet deep with side slopes of 30-37 degrees, you will require four charges of 80 lbs each planted 7 feet deep and two charges of 40 lbs each planted at a distance of 5 feet from the other. You will have to place two 80 lbs charges in the middle flanked by 40 lbs charges and two 80 lbs charges at the ends.

When you encounter hard-surfaced pavements, you must first dig a hole for the cratering charge. This can be done by exploding tamped charges on the pavement surface. It will be necessary to use a l-pound charge of explosive for each 2 inches of pavement thickness. Tamp the charges twice as deep as the pavement thickness. Shaped charges also are effective for breaching hard-surfaced pavements. A shaped charge will readily blast a small-diameter borehole through the pavement. Blasting the boreholes with shaped charges would speed up the job of cratering by first eliminating the need to breach the pavement with explosive charges and then digging the hole for the cratering charge.

(17.10.1) Safe Distances To Be Observed While Demolishing Structures

The following will give some idea of the safe distances for exposed personnel in the open from various types of missiles/flying objects created by bare charges on the ground regardless of the type or condition of the soil.

Charge weight 27 lbs (or less)	985 feet
50 lbs	1211 feet
80 lbs	1415 feet
100 lbs	1526 feet
200 lbs	1920 feet
300 lbs	2199 feet
400 lbs	2418 feet
500 lbs	2625 feet

For charges heavier than 500 lbs the safe distance is calculated by the formula

Safe distance (in feet) = 300 x (charge weight in lbs) $^{1/3}$

When charges are placed on targets instead of the ground the safe distance increases considerably. For example, the minimum safe distance when a charge is placed on concrete structures would be 1641 feet. It becomes 3281 feet when the targets are predominantly metallic.

(17.11) DEMOLITION OF BRIDGES

Demolition of bridges is one of the most fascinating parts of the entire business of demolitions. It really gives a strange feeling to see such massive and strong structures that have taken months to be erected crumble under a few well-placed charges.

The complete demolition of a bridge would require the destruction of all the components (spans, piers, and abutments). Complete demolition may be justified when the terrain forces the enemy to reconstruct a bridge on the same site. However, complete destruction is not normally required to meet tactical objectives. All you need is to damage the bridge to such an extent that the enemy is not able to use it in the intended manner. It does not matter that most of its parts remain intact and standing. The tactical purpose is served by making gaps large enough to make repair uneconomical and to force the enemy to construct other bridges on other sites. The minimum gap required must exceed the enemy's assault bridging capability by 5 meters. For planning purposes, armies across the world use 25 meters as the minimum gap size, but 35 meters is better. The gap may be less than 25 meters if enemy forces must depend on the demolished bridge components to bear their assault bridging and there is insufficient bearing capacity in the remains to carry the loads. You must therefore select the method of attack that achieves the tactical goal, with the minimum quantity of explosives, effort and time.

(17.11.1) Different Categories Of Bridges

The first step in any efficient bridge demolition job is to categorize the bridge correctly. The term categorization has been adopted to avoid confusion with classification, which is concerned with the load-carrying capacity of bridges. The correct categorization of bridges, coupled with an elementary knowledge of bridge design, allows you to select a suitable attack method. All bridges fit into one of three categories:

a. **Simply Supported**: In simply supported bridges, the ends of each span rest on the supports; there are no intermediate supports.

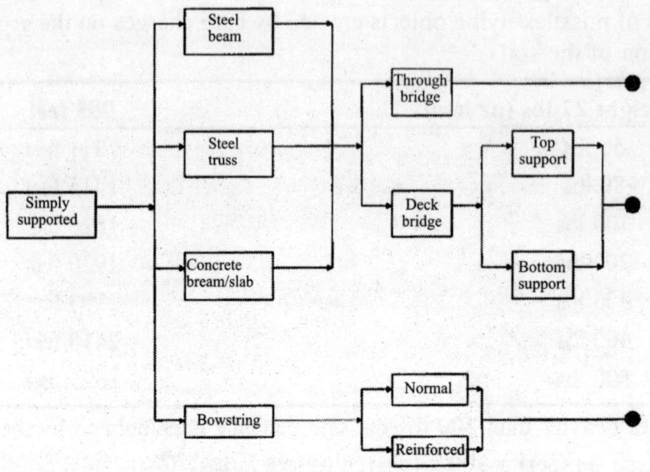

Fig. 100 Various Categories Of Simply Supported Bridges

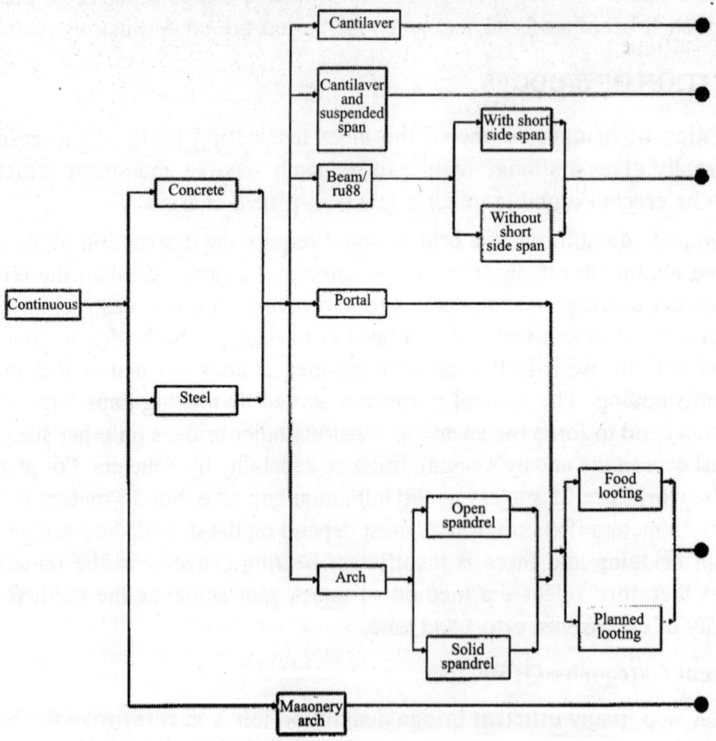

Fig. 101 Various Categories Of Continuous Bridges

b. **Miscellaneous**: Miscellaneous bridges form a small proportion of bridge structures. The theoretical principles governing these bridges determine the appropriate methods of attack. Examples of bridges in this category are suspension, lift, and cable-stayed bridges.

c. **Continuous**: If a bridge does not fit in the miscellaneous category and is not simply supported, categorize it as a continuous bridge. Hence, continuous has a wider meaning than multi-span, continuous-beam bridges, as is normally implied.

(17.11.2) Stages Of Destruction

When designing the demolition of a bridge, the first priority is to create a gap. Accomplishing this may require one or two attacks. Further actions that improve the obstacle may follow, if the situation permits. There are two minimum conditions for successful bridge demolition:

- You must design a proper **collapse mechanism.**
- You must ensure the attacked span will be free to move far enough, under its own weight, to create the desired obstacle.

Under normal conditions, a bridge is a stable structure. In demolition of bridges, the goal is to destroy the appropriate parts of a bridge so that it becomes unstable and collapses under its own weight. In other words, you form a collapse mechanism. This may involve either cutting completely through all structural members or creating points of weakness in certain parts of the bridge. At times, making bridges unstable by attacking their piers rather than their superstructures is easier, but it is still possible for bridges not to collapse, even if they lose the support provided by one or more of their piers. To avoid this type of demolition failure, you must place the charges on the structural members of the superstructure, immediately above the piers being attacked.

There are three basic types of collapse mechanisms: **see-saw collapse mechanism, beam collapse mechanism,** and **member-without-collapse mechanism.**

Bridge demolitions may **fail** due to the following reasons:

The formation of cantilevers is a typical example of a no-collapse mechanism being formed. The likelihood of this occurring is high when attacking continuous bridges. Or, jamming may take place. The span, once moved by the collapse mechanism, jams before moving far enough to create the desired obstacle. The most likely causes of jamming are the formation of a three-pin arch or a cranked beam. When attacking bridge spans, always consider the possibility of jamming during bottom and top attacks.

(17.11.3) Basic Methods Of Attacking Bridges

(a) Bottom Attack: In the bottom attack, the hinge forms at the top. As the span falls, the cut ends at the bottom move outward. The span may form a three-pin arch and fail to fall completely if the distance the cut ends must move is greater than the total end clearance between the span ends and the pier or abutment faces. If a three-pin arch situation is likely, one should not attempt a bottom attack.

(b) Top Attack: In a top attack, the hinge forms at the bottom. As the span falls, the cut ends at the top move inward. Some bridges may jam along the faces of the cut before the ends of the span have fallen off the abutments, forming a cranked beam. One must ensure the length of span removed at the top is sufficient to prevent the formation of a cranked beam.

(17.11.4) The Importance Of Concrete-Stripping

It is important to strip the concrete on the bridge. The purpose of concrete-stripping is to:

- Remove all concrete from above the main reinforcing steel;
- Remove all concrete from below the main reinforcing steel (spalling);
- Damage the main reinforcing steel to some extent; and
- Destroy the minor reinforcing steel near the surface under the charge.

Concrete-stripping charges are bulk, surface-placed charges designed for removing concrete from reinforced-concrete beams and slabs and exposing the steel reinforcement. Although these charges cause some damage to the reinforcing steel, you will not be able to predict the extent of this damage. These charges are effective against reinforced-concrete beams and slabs up to 2 meters thick.

The quantity of charge required for concrete-stripping depends primarily on the beam or slab plus roadway depth. The formula is:

$$P = 3.3\,(3.3h + 0.5)^3$$

Where P is the required charge size, in pounds of TNT per meter of bridge width; and h is the beam or slab plus roadway depth, in meters (the minimum being 0.3 meters and maximum 2 meters for this formula to be valid). For beam-and-slab bridge spans (T-beam and I-beam bridges), you must determine the charge sizes for the beams and slab separately. If tamping with two filled sandbags per pound of explosive is used, the charge requirement is reduced by one third.

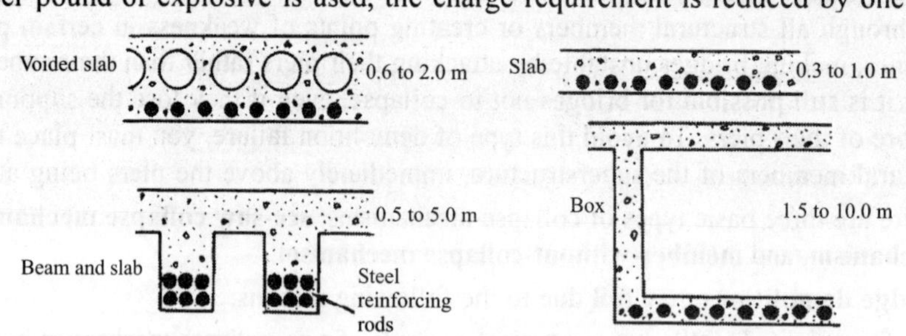

Fig. 102 Mid-Span Cross-Sectional View Of Typical Concrete Bridges

The charge will produce a ditch across the width of the bridge. Let that be denoted by W_d. Now there is a critical length denoted by L_c. For all simply supported concrete bridges, the stripping of all concrete over a specified L_c will cause collapse. If L_c is equal to or less than W_d, use one row of charges as specified by P. If L_c is greater than W_d, but less than twice W_d, increase the size of charge by 10 percent. If L_c is twice W_d, double the charge and place them in two lines, side by side. Place charges in a continuous line across the full width of the bridge at the point of attack. The shape of the end cross section of the charge should be such that the width is between one and three times the height.

(17.11.5) Attacking Intermediate Supports And Abutments

To demolish abutments, place charges in the fill behind the abutment. This method uses less explosive than external breaching charges and also conceals the charges from the enemy. The disadvantage is the difficulty in placing the charges. Abutments less than five feet thick require placing a line of 40-pound cratering charges, on 5-foot centers, in boreholes 5 feet deep, located 5 feet behind the face of the abutment (triple-nickel-forty method). Abutments over five feet thick require breaching charges in contact with the back of the abutment. Abutments over 20 feet high require placing a row of breaching charges at the base of the abutment on the gap side, in addition to the charges mentioned above.

As far as intermediate supports are concerned, concrete and masonry piers need to be demolished with internal or external charges. Placing internal charges has the advantage that you can manage the affair with much less explosive but the preparation time is higher and demands drilling equipment. A 2-inch-diameter borehole holds approximately 2 pounds of explosive per foot of depth. The steel reinforcing bars, however, make drilling in heavily reinforced concrete impractical. External charges must be placed at the base of the pier or higher, and do not space the charges by more than twice the breaching radius. Stagger the charges to leave a jagged surface to hinder future use. Thoroughly tamp all external charges with earth and sandbags, if time, size, shape, and location of the target permit.

(17.12) ATTACKING SIMPLY SUPPORTED BRIDGES

There are four main **subcategories of simply supported bridges: steel beam, steel truss, concrete beam and slab**, and **bowstring**. The first three types can be either of the thorough bridge or the deck bridge design. Then for the deck bridge design you can have either top support type or the bottom support type.

Steel-beam bridges may be constructed of normal steel-beam, plate-girder, or box-girder spans. Steel truss bridges have diagonal members in the trusses. In a concrete beam and slab bridge, also known as a girder bridge, there is a horizontal concrete slab supported at each end. For the demolition expert, the point to be noted is that at mid-span, the majority of steel reinforcing rods or tendons are located in the bottom portion of the superstructure.

Simply supported, steel-truss through bridge

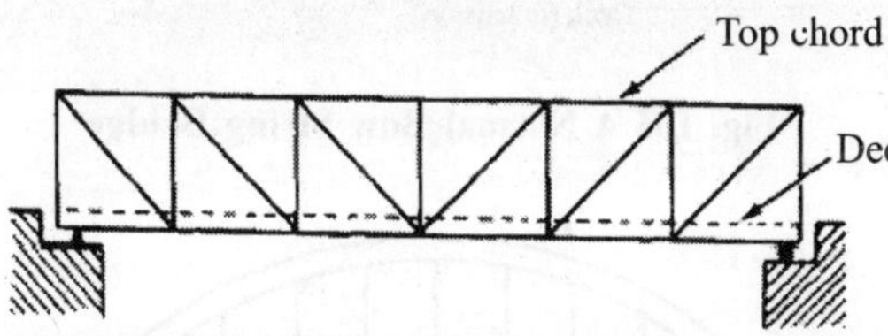

Simply supported, steel-truss deck bridge (bottom support)

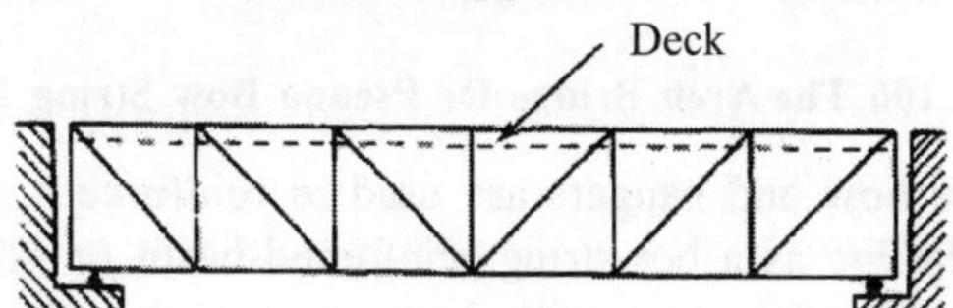

Simply supported, steel-truss deck bridge (top support)

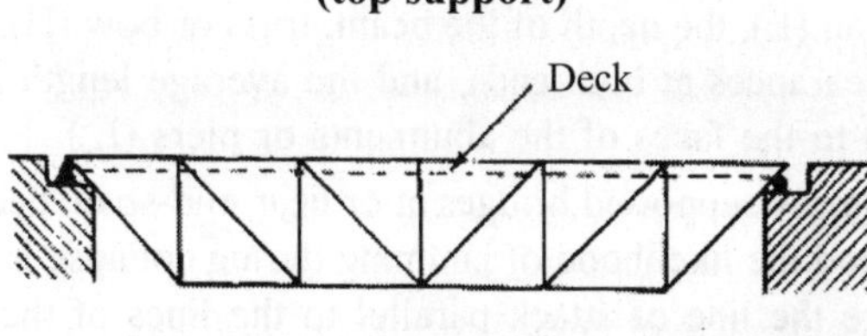

Fig. 103 Side Elevation Of Steel Truss Bridges

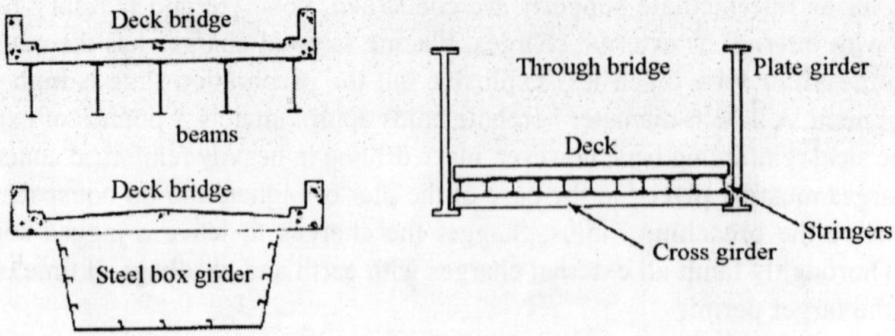

Fig. 104 Typical Cross-Sections Of Steel Beam Bridges

In a bowstring bridge the bow is in compression. The bow may be a steel beam, box girder, concrete beam, or steel truss. The bow's depth (thickness) is larger than or equal to the depth of the deck support members. The deck acts as a tie and resists the outward force applied by the bow. The deck is designed as a weak beam supported by the hangers. There is no diagonal bracing between the hangers.

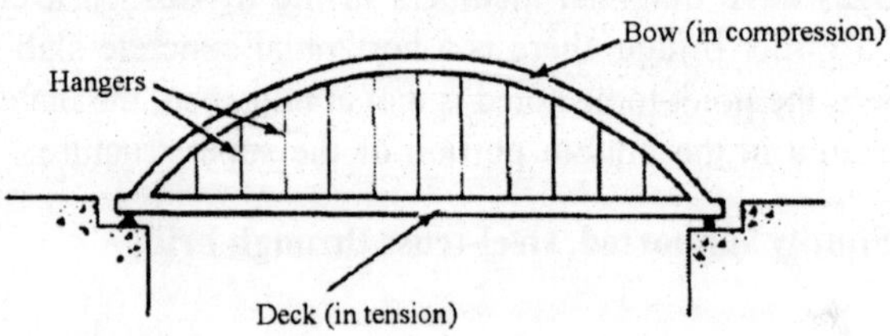

Fig. 105 A Normal Bow String Bridge

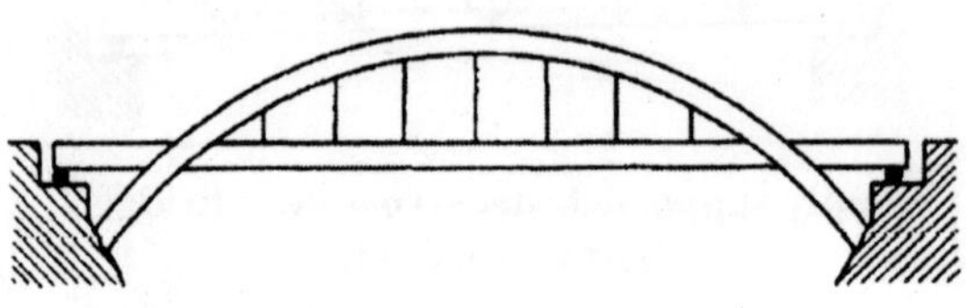

Fig. 106 The Arch Bridge Or Pseudo Bow String Bridge

Occasionally the bow and hangers are used to reinforce a steel-beam or-truss bridge. Categorize this type of bridge as a bowstring reinforced-beam or -truss bridge. In this type of bridge, the depth (thickness) of the bow will always be less than the depth of the deck support members.

Before attacking the bridge, you must reconnoiter and have the parameters of the bridge, such as the length of the span (L), the depth of the beam, truss or bow (H), the total end clearance (E), i.e., the sum of end clearances at both ends, and the average length of the bearing supports from the ends of the spans to the faces of the abutments or piers (L_s).

One must attack simply supported bridges at or near mid-span, because bending moments are maximum at mid-span and the likelihood of jamming during collapse is reduced if the bridge is attacked at mid-span. Make the line of attack parallel to the lines of the abutments. Doing this reduces the risk that the two parts of the span will slew in opposite directions and jam.

The three recommended ways of attacking simply supported spans are bottom, top, and angled attacks.

Use the **bottom attack** whenever possible, as it leaves the roadway open and enables you to use the bridge, even when the demolitions are at a ready-to-fire state. Reinforced and pre-stressed (tension) beams are very vulnerable to bottom attack, as the steel cables and reinforcing bars run along the bottom portion of the beam and are thus covered by less concrete. The major disadvantages of the bottom attack are the increased amount of time and effort necessary for placing and inspecting the charges. Because it is generally impracticable to place sufficient explosive below a reinforced or pre-stressed slab to guarantee a cut deeper than 0.15 meters, use the top or angled attacks for these types of bridges.

For **top attack**, L_c must be removed from the top of the bridge to prevent jamming. This action, by itself, should cause collapse. There is no requirement to cut steel reinforcing rods.

For **angled attacks**, cut all members (span, hand-rails, service pipes, and so forth) of the bridge. Make the angle of attack approximately 70 degrees to the horizontal to prevent jamming. The location of the charge should be between the mid-span point and a point L/3 from the end. Although an angled attack is effective on any type of bridge, it is essential when the bridge must be kept open to traffic, or when there is ample time to prepare demolitions.

(17.13) ATTACKING CONTINUOUS BRIDGES

There are six main **subcategories of continuous bridges: cantilever, cantilever and suspended span, beam or truss, portal, arch, and masonry arch.**

For the fist five types, you can have them in concrete or in steel. They will call for different methods of attack. If a continuous bridge is of composite construction (for example, steel beams supporting a reinforced-concrete deck), the material that comprises the main longitudinal load-bearing members, will determine the method of attack.

A cantilever bridge has a mid-span shear joint. Cantilever and suspended span bridges are cantilever bridges with a suspended span. Beam or truss bridges can be of two types. They could be bridges with spans of similar lengths and those with short side spans. A short side span is one that is less than three quarters of the length of the next adjacent span. Portal bridges could have two types of footings; fixed or pinned. Arch bridges could have an open or solid spandrel and fixed or pinned footings. All of them would call for different methods of attack.

Before attacking a continuous bridge, you must have the following parameters: length of the span L; and the rise H, i.e., the height from the springing or bottom of the support leg to the deck or top of the arch, whichever is greater.

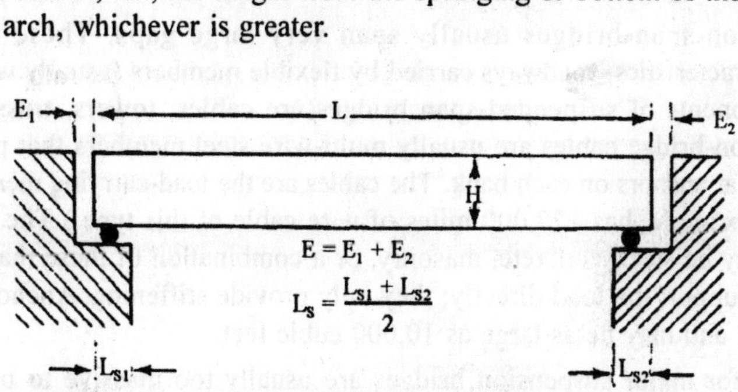

Fig. 107 Measurements For Simply Supported Spans

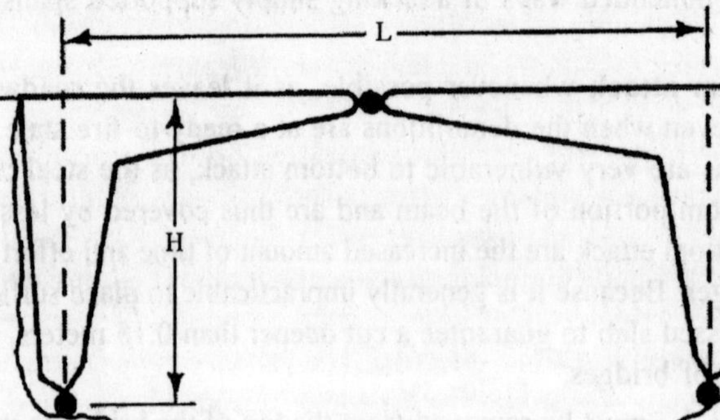

Fig. 108 Measurements For Continuous Bridges

No common point-of-attack rule exists for all categories of continuous bridges, but the line-of-attack rule applies to all continuous bridges. That is, the line of attack must be parallel to the lines of the abutments, and twisting must not occur during the demolition. When attacking continuous-span steel bridges, you must use the **see-saw** or **unsupported-member collapse mechanism**. Both mechanisms produce complete cuts through the span. Provided you can properly place charges, you may be able to demolish these bridges with a single-stage attack. However, on particularly deep superstructures (concrete decks on steel beams), charges designed to sever the deck may not cut through all of the reinforcing steel. Therefore, during reconnaissance, always plan for the possibility of a two-stage attack on deep, composite superstructures. Make angle cuts at about 70 degrees to the horizontal to prevent jamming during collapse.

Continuous concrete bridges are the most difficult to demolish and hence are poor choices for demolitions. Even when construction drawings are available and there is ample time for preparation, single-stage attacks are rarely successful. Consider using a bottom attack for this type of bridge.

For arch bridges and portal bridges with pinned footings, collapse can be guaranteed only by removing a specified minimum span length.

(17.14) ATTACKING MISCELLANEOUS BRIDGES

(17.14.1) Suspension-Span Bridges

Suspension-span bridges usually span very large gaps. These bridges have two distinguishing characteristics: roadways carried by flexible members (usually wire cable) and long spans. The components of suspended-span bridges are cables, towers, trusses or girders, and anchors. Suspension-bridge cables are usually multi-wire steel members that pass over the tower tops and terminate at anchors on each bank. The cables are the load-carrying members. (The Golden Gate Bridge, for example, has 127,000 miles of wire cable of this type.) The towers support the cables. Towers may be steel, concrete, masonry, or a combination of these materials. The trusses or girders do not support the load directly; they only provide stiffening. Anchors hold the ends of the cables in place and may be as large as 10,000 cubic feet.

Anchors for major suspension bridges are usually too massive to be demolished. The cables are usually too thick to be effectively cut with explosives. The most economical demolition method is to drop the approach span or a roadway section by cutting the suspenders of the main or load-bearing cables. The enemy's repair and tactical bridging capabilities determine the length of the target section. When reinforced-concrete towers are present, it may be feasible to breach the concrete and cut the steel of the towers.

The two vulnerable points on minor suspension bridges are towers and cables. Towers can be destroyed by placing tower charges slightly above the level of the roadway. You will have to cut a section out of each side of each tower. Place the charges so that they force the ends of the cut sections to move in opposite directions, twisting the tower. Doing this will prevent the end of a single cut from remaining intact. Demolition chambers, provided in some of the newer bridges, make blasting easier, quicker, and more effective. The cables can be destroyed by placing charges as close as possible to anchor points, such as the top of towers. Cables are difficult to cut because of the air space between the individual wires in the cable. Ensure the charge extends no more than one half the cable's circumference. These charges are usually bulky, exposed, and difficult to place. Shaped charges, on the other hand, are very effective for cable cutting.

(17.14.2) Movable Bridges

(a) Swing-Span Bridges: A swing span is a continuous span capable of rotating on a central pier. The arms of a swing-span bridge may not be of equal length. If the arms are not of equal length, weights are added to balance them. Rollers that run on a circular track on top of the central pier carry the span's weight. The swing span is independent from any other span in the bridge. Identify a swing-span bridge by its wide, central pier. This central pier is much wider than the one under a continuous-span bridge that accommodates the rollers and turning mechanism.

In principle, swing-span bridges are continuous bridges and hence a method applicable for demolishing them can be applied to the latter also. For partial demolition, open the swing span and damage the turning mechanism.

(b) Drawbridges Or Bascule Bridges: These bridges usually have two leaves that fold upward, but some bascule bridges may have only one leaf. The movable leaves in bascule bridges appear in three general forms: counterweights below the road level (most modern), counterweights above the road level (older type), and no counterweights (lifted by cable or rope; oldest type; usually timber).

The cantilever arms can be demolished with an attack method appropriate for simply supported bridges. For partial demolition, open the bridge and jam or destroy the lifting mechanism.

(c) Vertical Lift Bridges: These bridges have simply supported, movable spans that can be raised vertically in a horizontal position. The span is supported on cables that pass over rollers and connect to large, movable counterweights.

The movable span can be demolished with an attack method appropriate for simply supported bridges. Another method is to raise the bridge and cut the lift cables on one end of the movable span. The movable span will either wedge between the supporting towers or fall free and severely damage the other tower.

(17.14.3) Floating Or Pontoon Bridges

Floating bridges consist of a continuous metal or wood roadway supported by floats or pontoons.

Pneumatic floats are airtight compartments of rubberized fabric inflated with air. For a hasty attack on these bridges, you can cut the anchor cables and bridle lines with axes and the steel cables with explosives. Contrary to popular wisdom, puncturing the floats by automatic fire is not really a very wise idea because using weapons to destroy the floats requires a considerable volume of fire because each float has a large number of watertight compartments. A better method is to make a clean cut through the float, using detonating cord stretched snugly across the surface of the pontoon compartments. One strand of cord is enough to cut most fabrics, but two strands may be necessary for heavier materials. Also, place one turn of a branch-line cord around each inflation valve. This will prevent the raft from being re-inflated if it is repaired.

Rigid pontoons are made of various materials: wood, plastic, or metal. To destroy these bridges, place a ½-pound charge on the upstream end of each pontoon at water level. Detonate all charges simultaneously. If the current is rapid, cut the anchor cables so that the bridge will be carried downstream. Another method is to cut the bridge into rafts. Place ½-pound charges at each end of each pontoon and detonate them simultaneously. To destroy metal tread ways on floating bridges, place cutting charges at every other joint in the tread way and it will damage the bridge beyond use.

(17.14.4) Bailey Bridges

To destroy these bridges, you can place l-pound charges between the channels of the upper and lower chords. Use ½-pound charges for cutting diagonals and l-pound charges for cutting sway bracing. The idea is to cut the bridge in several sections by attacking the panels on each side, including the sway bracing. The angle of attack should be 10 degrees to the horizontal to prevent jamming. In double-storey or triple-storey bridges, increase the charges on the chords at the story-junction line. For further destruction, place charges on the transoms and stringers.

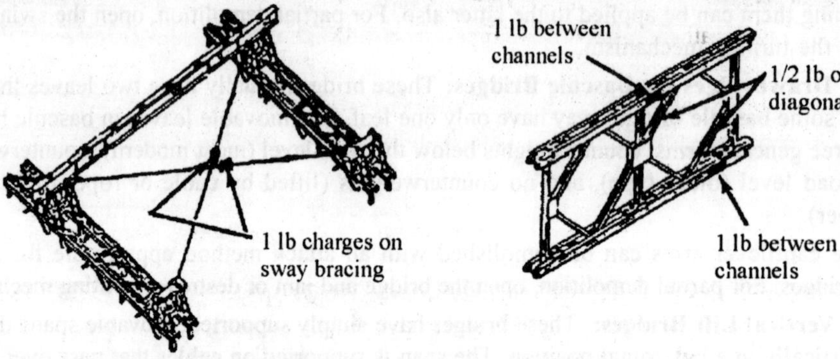

Fig. 109 The Technique Of Attacking A Bailey Bridge

(17.15) A REAL LIFE CASE STUDY OF DEMOLISHING A BRIDGE

It would be illustrative to cite an example of the charges required for attacking real bridges.

Consider a simply-supported reinforced concrete beam-and-slab span. Let the span be 30 m long and let the beam and slab plus the roadway depth be 1.2 m. The beam width is 0.35 m. The slab width is 0.3 m. There are five beams.

You would require 76.47 lbs of C4 per beam and 7.34 lbs of C4 per slab. Five charges each would be required for the beam and the slabs. In all you would require 340 packages of C4. They will create a ditch of width 2.7 m which is substantially greater than the L_c of 1.8 m for this bridge.

(17.16) CONTROLLED DEMOLITIONS AT THEIR VERY BEST: THE MESMERIZING SPECTACLE OF BUILDING IMPLOSIONS

(17.16.1) Why And When Do We Need The Implosions Of Buildings?

There is nothing more fascinating, spectacular and amazing in the entire science of explosives than implosions of buildings. This is the science of demolition at its very best. Huge multi-storied buildings are brought down in such a beautiful manner that buildings situated just a few feet away remain absolutely unscathed! And it is a real constructive work. There might be some huge but old building at some place and you could very well construct a modern building there with a lot more space, amenities and utility. What do you do with the old building? It would obviously be uneconomical to take it down brick by brick. You can't bomb it either. There may be other buildings in the vicinity and you are not allowed to even touch them. What do you do then? Implode the old building!

Demolition of buildings by explosives is not at all difficult if all that you are looking for is bringing down the structure somehow. You can use brute force in using a large amount of explosive or fire the big guns into a building. In counter-insurgency operations in the Third World where specialized weapons for attacking built-up structures and bunkers that restrict their damage to a small area are not available, troops routinely resort to firing rounds after rounds of the 84 mm Carl Gustaf recoilless gun into a building to bring it down. Each round blows a hole into the wall and when the holes become numerous, the wall crumbles. But that's a waste of explosives. If you are cleverer and if you have some knowledge of structures, you could use a smaller quantity of explosive at critical structural points and tamp your explosives to direct their explosive force suitably. You could attack load bearing points directly or demolish some other part in such a manner that the resulting asymmetry in the loads makes the load on that point exceed its stipulated limit when it would give way. It is rather difficult to predict how and where the broken parts of the building, however, thus felled would fall.

Obviously you cannot do this if you have to bring down a building for some constructive purpose. You may have to demolish a building while keeping buildings adjacent to it very much intact and unscathed! In such a situation you cannot afford the luxury of a boom and watching bricks fly all around! When a building is surrounded by other buildings, it may be necessary to **'implode'** the building, that is, make it **collapse down into its footprint**. Strictly speaking, an **implosion** is an event where something collapses inward, because the external atmospheric pressure is greater than the internal pressure. You may recall a childhood experiment. Take a little water and boil it in a tin can of rather thin skin. Close the can tight while the water is still boiling and let it cool down. You would find that as it cools down, the can collapses. Why? Because the steam inside it had driven out the air from it and when the steam cools down to water, the space turns into a vacuum. Now the pressure of the atmosphere outside the can becomes much greater than the pressure inside and if the walls of the can are not strong, the can collapses. A building implosion isn't truly an implosion—atmospheric pressure doesn't pull or push the structure inward; rather gravity makes it collapse. What is really done is a **controlled demolition** so that the debris falls in a limited area only and does not fly around, but the word implosion has somehow become popular. A building implosion is actually one of the most precisely planned, delicately-balanced engineering feats you'll ever see!

(17.16.2) The Fundamental Principles Of The Technology Of Building Implosions

The basic idea of explosive demolition is rather simple: If you remove the support structure of a building at a certain point, the section of the building above that point will fall down on the part of the building below that point. If this upper section is heavy enough, it will collide with the lower part with sufficient force to cause significant damage. The explosives are just the trigger for the demolition. It's really gravity that brings the building down.

In order to demolish a building safely, demolition experts must map out each element of the implosion ahead of time. The first step is to examine **architectural blueprints** of the building, if they can be located, to determine how the building is put together. Next, they must examine the building thoroughly from inside as well as outside and study the support structure on each floor. Once they have gathered all the raw data they need, the blasters work out a plan of attack. Drawing from past experiences with similar buildings, they decide what explosives to use, where to position them in the building and how to **time their detonations**. In some cases, the blasters may have to develop 3-D computer models of the structure so they can test out their plan ahead of time in a virtual world. You can see that it is not an easy job. The demolition expert has to understand the structure of the building as good as any structural engineer and then also how the building would behave under demolition charges at various critical points. Thus his total knowledge has to be higher. Give the job to just an architect or a structural engineer and he would not be able to pull it off!

The main challenge in bringing a building down is controlling the way it falls. If there is space in the vicinity, it would be possible to tumble the building over on to one side, into a parking lot or other open area. This sort of blast is the easiest to execute. Tipping a building over is something like felling a tree. To topple the building to the north, the blasters detonate explosives on the north side of the building first, in the same way you would chop into a tree from the north side if you wanted it to fall in that direction. Blasters may also secure steel cables to support columns in the building, so that they are pulled a certain way as they crumble.

Quite often, though, a building is surrounded by structures that must be preserved. In this case, the blasters proceed with a true controlled demolition, demolishing the building so that it collapses straight down into its own **footprint** (the total area at the base of the building). This feat requires such skill that only a handful of demolition companies in the world will even attempt it. And when they do it it's nothing but pure magic!

Blasters approach each project a little differently, but the basic idea is to think of the building as a collection of separate towers. The blasters set the explosives so that each 'tower' falls toward the center of the building, in roughly the same way that they would set the explosives to topple a single structure to the side. When the explosives are detonated in the right order, the toppling towers crash against each other and all of the rubble collects at the center of the building. Another option is to detonate the columns at the center of the building before the other columns so that the building's sides fall inward.

(17.16.3) The Principles Applied: How To Implode A 20-Storey Building?

For any given building, there are several ways in which demolition experts might bring it down. Generally speaking, blasters will demolish the major support columns on the lower floors first and then a few upper stories. In a 20-storey building, for example, the blasters might blow the columns on the first and second floor, as well as the 12th and 15th floors. In most cases, blowing the support structures on the lower floors is sufficient for collapsing the building, but loading columns on upper floors helps break the building material into smaller pieces as it falls. This makes for easier clean-up following the blast. Smart, isn't it?

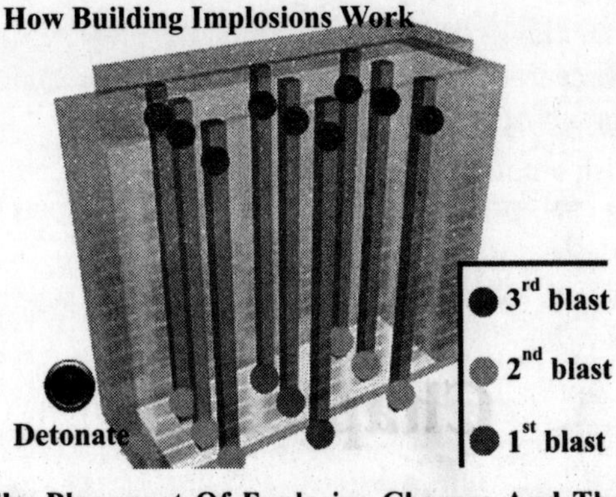

Fig. 110 The Placement Of Explosive Charges And The Sequence In Which They Are Triggered When Imploding A Multi-Storey Building

The first step in preparing to demolish a building is to take out non-load-bearing walls within the building. This makes for a cleaner break at each floor: If these walls were left intact, they would stiffen the building, hindering its collapse. Destruction crews may also weaken the supporting columns with sledge hammers or steel-cutters, so that they give way more easily. Next, blasters start **loading the columns with explosives.** The amount of explosive and the type of explosive required has to be calculated in detail keeping in mind the strength and composition of the column and also the properties of the explosive at hand. Demolition experts determine the quantity of explosive material to be used based largely on their own experience and the information provided by the architects and engineers who originally built the building or from a study of the blueprints if the building is rather old. But most of the time, they won't rely on this data alone. To make sure they don't overload or under-load the support structure, the blasters do perform a test blast on a few of the columns, which they wrap in a shield for safety. The blasters try out varying degrees of explosive material, and based on the effectiveness of each explosion, they determine the minimum explosive charge needed to demolish the columns. This is important. Why? By using only the necessary amount of explosive material, the blasters minimize flying debris, and thus reduce the likelihood of damaging nearby structures.

To ensure proper tamping, it is necessary to bore holes into concrete columns, fill them with the required amount of explosives and seal the hole again. As we have seen, steel support structures are best cut using shaped charges instead of trying to blast them. Additionally, blasters may explode a non-shaped charge on one side of the column to push it over in a particular direction.

To control the explosion sequence, suitable delays have to be employed.

To further reduce flying debris, blasters may wrap chain-link fencing and geotextile fabric around each column. (Geotextiles are special fabrics made for use in 'geological' situations. They are rot-proof and permeable to water. Geotextiles are made of synthetic fibers manufactured in a woven or loose non-woven manner to form blanket-like products.) The fence keeps the large chunks of concrete from flying out, and the fabric catches most of the smaller bits. Blasters may also wrap fabric around the outside of each floor that is rigged with explosives. This acts as an extra net to contain any exploding concrete that tears through the material around each individual column. Structures surrounding the building may also be covered to protect them from flying debris and the pressure of the explosions.

□□□

Chapter 18

How to Use Explosives for Breaching in Tactical Situations like Urban Fighting and Hostage Rescue

(18.1) WHEN DOES EXPLOSIVE BREACHING BECOME NECESSARY?

Entering a building or room in which armed opponents could be hiding is undoubtedly the most difficult task an infantryman could ever encounter in his life. The risks faced in entering a building or a room, particularly when you do not have the liberty of blowing it up, are enormous. The first man who gets in is very likely to be shot up. The biggest problem is: how do you enter a locked room where armed opponents might be waiting to mow you down? It would be of some help if you could lob a fragmentation grenade inside before you enter so that you could inflict some casualties upon the opponents and thus 'soften up' the opposition to some extent. But how do you lob the grenade? You have to have an opening in the room. How do you get that opening? Not every soldier is equipped with the **SIMON rifle grenade** of Rafael, Israel made especially for breaching doors. If it is a wooden door you might consider blowing up the lock with a **12-bore shotgun using SSG load** and then kick in. But it won't help you much if the door is made of stronger material. In fact, one could easily encounter metal doors also that would not be easily taken out by shotgun loads. Remember, if there is even a split-second delay on account of the door not giving way cleanly in one go, the advantage goes to the persons inside. Or you could suddenly find yourself in a building or outside a building from where it is not possible to fire the rifle grenade. All these methods have one more serious limitation that could easily prove fatal to the entire operation. You cannot use them in situations where the armed opponents might be holding some **hostages** in the same room! That rules out lobbing a grenade inside or using a rifle grenade on the door. The hostages themselves might get killed by the fragments of the grenade!

The only solution to this vexatious problem is explosive breaching of the door. The idea is that you should knock off the door in one clean swift blow and then barge in immediately to do the needful. Explosive breaching has the great advantage that it does not create the dangerous fragments that could kill hostages or other innocent persons inside. Explosive breaching is considered to be one of the most advanced skills associated with tactical operations in an urban environment and hostage rescue. In the last 25 years law enforcement and military teams have perfected the art of explosive breaching skills to effectively enter strongholds using a minimum amount of explosives to achieve effective penetration of the target without endangering the lives of possible hostages inside.

The main targets of **breaching charges** are: doors, windows, roofs, walls, metal grills and security bars. Some of the common breaching charges include **frame charges** made of **detonating cords** or **Detasheet, linear charges, water charges** and **window strip charges.**

(18.2) CONSTRUCTING FRAME CHARGES FOR ALL TYPES OF DOORS, WINDOWS AND A VARIETY OF WALLS

Depending on the configuration, the frame charge can be used effectively against solid wood doors, metal doors, windows, bulkheads, and a variety of walls. The frame charge is used to gain entry through a door by creating a hole of entry in the door itself. Frame charges are made of both Detasheet and detonating cords. They are used in the following manner. The charge is placed on the lower half of the door leaving a small band of door below the charge. Placing the charge low on the door ensures the largest opening should the door have horizontal barricade at the mid-point. The charge is placed so that the priming tail is at the bottom.

(18.2.1) Detasheet Frame Charge

The following materials are required to make a Detasheet frame charge: one sheet of cardboard measuring 114 cm x 60 cm with handholds cut into it; two 114 cm x 1.5 cm strips of C2 Detasheet; two 60 cm x 1.5 cm strips of C2 Detasheet; one 15 cm piece of 10.62 g/m cord; and duct tape / double-sided tape.

The following is the sequence of construction:

Step 1	Lay the strips of Detasheet around the edge of the frame. One 15 cm and two 5 cm pieces of 10.62 g/m det cord should be laid in the corner under the last 5 cm of Detasheet so that the firing system can be attached.
Step 2	Cover the explosives with duct tape.
Step 3	Attach double-sided tape or glue packs, to the back of the charge and tab the tape.

(18.2.2) Detonating Cord Frame Charge

The following materials are required to make a detonating cord frame charge: one sheet of cardboard measuring 114 cm x 60 cm, with hand holds cut into it; appropriate rolls of det cord (615 cm of 10.62 g/m per frame); and duct tape / double-sided tape.

The following is the sequence of construction:

Step 1	On the cardboard template, lay the detonating cord around one face of the template near the outer perimeter, and secure with tape.
Step 2	Leave a 15 cm tail of detonating cord exposed at the bottom of the frame to be used for priming.
Step 3	Attach double-sided tape or glue packs, to the back of the charge and tab the tape.

(18.2.3) Tape Charges

Tape charges can be very versatile. The theory behind a tape charge is to attack the weakest part of the door, which are the screws that attach the hinges to the door. It is designed to be used against inward opening wood doors, when the hinges are not exposed to the breacher. The breacher must place the charge on the hinge side of door, to cut the entire door from top to bottom. The tape charge can be placed in almost any manner as long as it cuts the door into 2 pieces. It has been used having been placed vertically left, center, and right side of the door, and having been placed top corner to bottom opposite corner. The best placement is vertical, hinge side of the door, approximately 12 cm from the hinges.

The following materials are required to make the tape charge:

C-2 Detasheet	Det Cord
215 x 5 cm carpet or 3 m tape	215 x 5 cm carpet or 3 cm tape
215 x 5 cm length duct tape	215 x 5 cm length duct tape
215 x 15 cm length of 50 or 25-grain det cord	215 cm length 10.62 g/m det cord
230 cm length 10.62 g/m det cord	226 cm length 10.62 g/m det cord

The following is the sequence of construction:

Step 1	Roll out carpet tape on long flat work surface and remove protective backing from the surface of the tape that is facing up.
Step 2	Place C2 or det cord explosive length-wise down the center of the exposed surface of the carpet tape.
Step 3	Place 15 cm of the 30 cm det cord over the last 15 cm of the C2 explosives at one end.
Step 4	Place the two lengths of the 15 cm lengths of det cord on either side of the det cord attached in step 3. The two 15 cm strips of det cord should be in contact with the C2 explosives for the entire 15 cm length.
Step 5	Place tape over the exposed cut end of the det cord.
Step 6	When constructed with the det cord the larger length will extend beyond the length of the carpet tape to serve as the priming point.
Step 7	Cover det cord with duct tape.
Step 8	Place tape over the exposed cut end of the det cord.

(18.2.4) Priming Charges

The priming system that must be used for virtually all tactical shots should be the so-called "**Red Devil" firing system**. The following materials are required for this: electric or shock tube firing system (for tactical shots, leads will be at least 7.3 m in length); one 25 cm length of 15-grain det cord; one Red Devil det cord junction clip; and electrical tape.

The following is the sequence of construction:

Step 1	Perform appropriate checks for the designated firing system.
Step 2	With the 25 cm length of 15-grain det cord, make a "bite" by folding it in half and taping it together, leaving a loop at one end.

Step 3 Tape the blasting cap to the running end of the det cord loop, ensuring that the entire blasting cap is covered with tape. Fold the lead back onto the blasting cap and tape in place.

Step 4 Tape the firing end of the lead line to the firing machine, ensuring that there is enough lead to connect to the machine at the appropriate time.

Step 5 Wrap the lead lines by "S" folding them at approximately 20 cm and secure by placing a 2.5 x 5 cm PVC tube over the bundle.

(18.3) INTERIOR DOOR CHARGE

It is likely that during an assault, you may encounter a locked door inside the stronghold. You must place the charge between the doorknob and the door jam so that the charge is over the strike of the doorknob. Also, if there is a dead bolt on the door, the charge is placed so that the backing material covers both the strike of the dead bolt and the doorknob.

(18.3.1) Detasheet Strip Charge Construction

The Detasheet strip charge is constructed in the following manner:

- Cut the backing material (Polyethylene or rubber) 30 cm long by 1.9 cm wide.
- Cut a piece of C2 Detasheet 30 cm x 1.9 cm. Lay the datasheet on the backing material.
- Cut 2 pieces of 10.62 g/m detonating cord, one-piece 6 cm, and the second piece 12 cm long. Place these pieces touching and side-by-side on top of the datasheet. The 12 cm piece will hang over the end of the charge 6 cm to be used as a priming tail for the charge.
- Cover the entire charge with tape, leaving the 6 cm tail of detonating cord untapped. Now attach double-sided tape to the backing material on the opposite side from the datasheet. Be sure to make buddy tabs on each end of the double-sided tape.

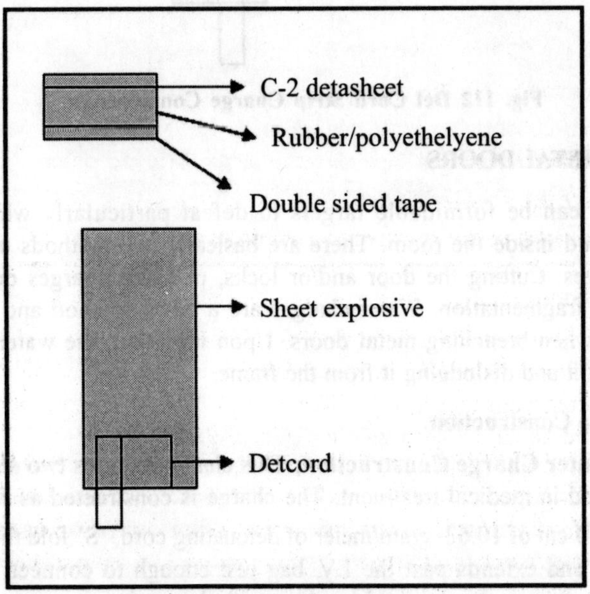

Fig. 111 Interior Door Charge Made Using Detasheet

(18.3.2) Det Cord Strip Charge Construction

The det cord strip charge is constructed in the following manner:

- Cut the backing material the same as for the datasheet charge, 30 cm long by 1.9 cm wide.
- Cut 3 pieces of 10.62 g/m detonating cord: 2 pieces being 30 cm long and the third piece being 35 cm long.
- Lay the detonating cord on the backing material with the longer piece in the center.
- Cover the detonating cord and backing material with tape. Do not cover the 5cm piece of detonating cord hanging over the end of the charge. Attach double-sided tape to the backing material on the opposite side from the datasheet. Be sure to make buddy tabs on each end of the double-sided tape.

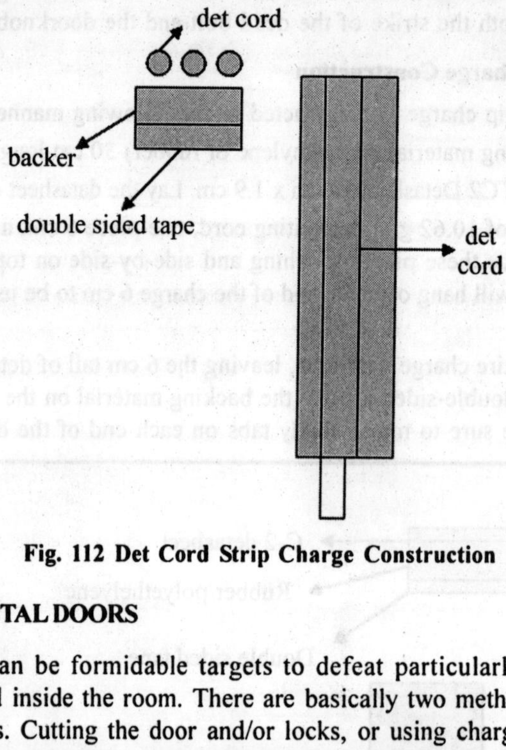

Fig. 112 Det Cord Strip Charge Construction

(18.4) ATTACKING METAL DOORS

Metal doors can be formidable targets to defeat particularly when you don't want fragments flying around inside the room. There are basically two methods used to breach these targets using explosives. Cutting the door and/or locks, or using charges designed to push the door without causing fragmentation. Water charges are a novel method and are often preferred over cutting charges when breaching metal doors. Upon initiation, the water pushes against the metal door, deforming it and dislodging it from the frame.

(18.4.1) Water Charge Construction

(a) Single Water Charge Construction: This charge requires two standard 1000 ml I.V. (intravenous) bags used in medical treatment. The charge is constructed as follows:

- Using 250 cm of 10.62–gram/meter of detonating cord, 'S' fold the detonating cord so that one end extends past the I.V. bag just enough to connect the one outside the I.V. bags. Secure the det cord in place with duct tape.

- **Place the other I.V. bag on top of the detonating cord, sandwiching the detonating cord between the two I.V. bags.**
- **Tape the two bags together tightly with duct tape. Cover the entire charge with tape ensuring that the det cord priming tail extends from the taped charge.**
- **Lay tape, sticky side up, on the construction table; lays a piece of 63 mm plywood (25 cm x 12 cm) on the tape. Place the charge onto the plywood and extend the tape up over the charge. Do not pull the tape tight. Doing so may deform the plywood.**
- **Place double-sided tape on the flat backing material and buddy tab the tape ends.**

(b) Double Water Charge Construction: **The double charge system is used against heavy metal doors. The double water charge system is almost the same as the single water charge but with a second bag connected. It is constructed as follows:**

- **Cut 550 cm of 10.62 g/m detonating cord. 'S' fold 250 cm of the detonating cord and tape it onto one I.V. bag. Leave a 5–6 cm pigtail extending off one end of the I.V. bag, and the remaining 300 cm extending off the other. Place the other I.V. bag on top of the first and complete this charge.**
- **Extend the 300 cm length of detonating cord 50 cm, then S–Fold the remaining 250 cm and tape it onto a third I.V. bag. Place the fourth I.V. bag on top of the third and complete the charge.**
- **Use 63 mm plywood as a backer on each charge and complete in the same manner as a single charge.**
- **The 2–bag charge is primed from the bottom when employed.**

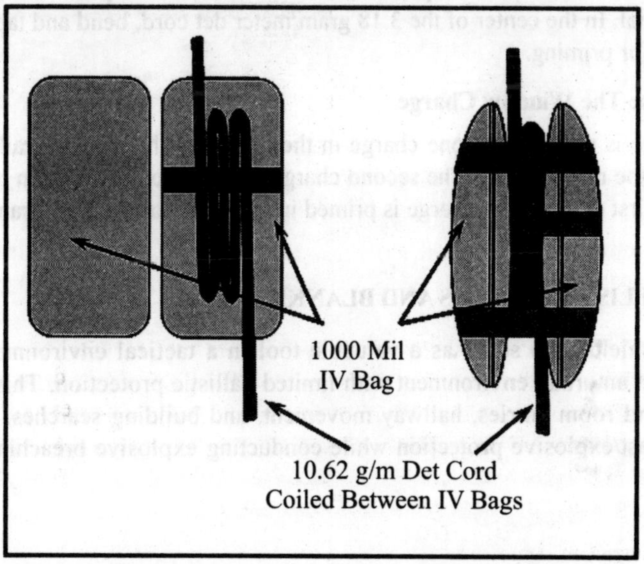

Fig. 113 Double Water Charge For Attacking Metal Doors

(c) How To Use The Water Charge : **The single water charge system is placed centered on the door. When detonated the charge will buckle the door; removing the locking mechanism from the door jam. On an outward opening door hooligan tools may be needed to pry the door open as, though the locking mechanism is no longer engaged, the door may remain closed.**

The double water charge system is placed with the top charge at one-third of the door height from the top of the door, and centered on the door. The second charge is placed directly in line with the first charge, 50 cm below. The charge is primed at the bottom. On an outward opening door hooligan tools may be needed to pry the door open.

(18.5) WINDOW CHARGES

Since entry through the door is a commonly used tactic, it stands to reason that the people inside would also make preparations to defend that, booby trap it or counter it in some other novel way. Using the window for making entry then becomes a good option. The purpose of using the window charge is to remove the glass and the center cross beam assembly of the window frame. This charge may be used against a variety of window types. It can be placed on either side of the window and is designed for use against whatever window you should come up against.

(18.5.1) Window Charge Construction

The window charge is constructed in the following manner:

- Cut two pieces of conveyer belt rubber or polystyrene material as a backer 2 cm x 10 cm.
- For each charge, cut six lengths of 10.62 gram/meter det cord 10 cm long.
- Cut a 71 cm length of 3.18 gram/meter det cord. Attach the detonating cord to the backing material. The 3.18 gram/meter det cord is placed with each end in the center of the 10.62 g/m pieces.
- Attach double sided tape or adhesive to the non-explosive side of the backing material. In the center of the 3.18 gram/meter det cord, bend and tape, forming a 2 cm loop for priming.

(18.5.2) How To Use The Window Charge

The charge is placed with one charge in the corner of the meeting rail. The charge runs horizontally along the meeting rail. The second charge is placed on the bottom rail of the window, directly below the first charge. The charge is primed in the center of the 3.18 gram/meter detonating cord.

(18.6) USE OF BALLISTIC SHIELDS AND BLANKETS

Ballistic shields can serve as a valuable tool in a tactical environment—allowing the operator to move in an urban environment with limited ballistic protection. This environment can include building and room entries, hallway movement, and building searches. Ballistic blankets can be used for blast/explosive protection while conducting explosive breaching.

❑❑❑

Chapter 19

The Esoteric World of Shaped Charges, Explosive Lenses used in Nuclear Weapons and Explosively Formed Projectiles

(19.1) THE SHAPED CHARGE EFFECT

(19.1.1) How The Shaped Charge Effect Was Discovered?

The discovery of what is variously referred to as the **Shaped Charge Effect**, the **Hollow Charge Effect**, the **Cavity Effect**, or the **Munroe Effect**, dates back to the 1880s. Dr. **Charles Munroe**, while working at the Naval Torpedo Station at Newport, Rhode Island, in the 1880s, discovered that if a block of guncotton with letters countersunk into its surface was detonated with its lettered surface against a steel plate, the letters were indented into the surface of the steel. The essential features of this effect were also observed in about 1880 in both Germany and Norway, although no great use was made of it, and it was temporarily forgotten. If letters were raised in relief above the rest of the guncotton then the letters on the plate would also be raised above its surface. The intriguing discoveries should have set the alarm bells ringing but unfortunately it did not happen.

In 1910, **Egon Neumann** of Germany independently made an even greater discovery that a block of TNT containing a conical indentation would cut through a metal plate which would normally only be dented by that quantity of explosive. This came to be known as the **Munroe-Neumann Effect** and Neumann even obtained a British patent for that in 1911. Very curiously, once again no great use could be made of this discovery also and the world had to wait for more than three decades before its immense importance for weapons could be appreciated.

(19.1.2) How Mohaupt Made A Weapon Out Of It

The 'shaped charge' was introduced to warfare as an anti-tank device in World War II after its re-discovery in the late 1930s. In 1935, **Henry Mohaupt**, a chemical engineer (and a machine gunner in the Swiss Army) established a laboratory in Zurich to develop an effective anti-tank weapon that could be used by infantry soldiers. Henry Mohaupt was the inventor of the **lined shaped charge** though it is agreed that **R.W. Wood** of the Physics Department, John Hopkins University had also done pioneering work in the field of metal liners.

Henri Mohaupt designed a rifle grenade to use this principle. He described it as a 'new explosive' and in 1939 the British did evince some interest in it. Surprisingly, great nations of the world haggled for a few dollars when it came to evaluating his path-breaking discovery! The price which Mohaupt quoted for his discovery was considered too high by the British and they lost further interest in his invention. Mohaupt then offered to show the United States particulars of the construction of his device for a fee of $25,000 paid in advance. The Americans too dillydallied until a demonstration by Mohaupt at Aberdeen Proving Ground, MD, was permitted in 1940, doubtlessly because Mohaupt provided 200 grenades for the demonstration. Army and Navy representatives who witnessed the test firings were convinced that this was indeed an important 'new form of weapon'. But the Americans continued to haggle and even cited the 1911 British patent by Egon Neuman to argue that Mohaupt's discovery was not that 'original'. Eventually they forced him to reduce his price!

An adaptation of Mohaupt's design later formed the basis for the **M9A1 anti-tank rifle** grenade, and the legendary 2.6-inch anti-tank '**bazooka**' rocket. Application of the shaped charge principle to artillery naturally proceeded. Before Pearl Harbor, in an atmosphere of great secrecy, work began on a **shaped charge shell**, specifically the '**HEAT**' (high explosive anti-tank) round, for the 76 mm howitzer. The designers, paying careful attention to the length of the ogive, the filler of 50/50 pentolite, and the striking velocity, came up with a round of the same length as the corresponding high-explosive round. The HEAT round, at a muzzle velocity of approximately 305 meters per second, would penetrate 76 millimeters (3 inches) of homogenous armor plate. A similar shell for the 105 millimeter howitzer appeared simultaneously. Standardized in late 1941, HEAT shells were produced in time to take part in the North African tank battles early in 1943.

A very extensive study of these early HEAT shells showed that an increase in plate penetration was directly proportional to an increase, up to approximately three calibers, in 'standoff' distance, that is, the distance from the base of the cone to the target at the moment of detonation. At the same time, the effect of the spin of the projectile on the performance of the shaped charge was also studied.

Shaped charges are now used extensively in civilian applications also, particularly in the oil and gas industry, where they open up the rock around drilled wells. In an oil or gas well, a metallic casing is cemented to the borehole walls to maintain the borehole integrity. Shaped charges are incorporated in a hollow carrier gun or a strip positioned in the casing. The shaped charges are activated to pierce the well casing and the geologic formation at the hydrocarbon producing zone. The hydrocarbons enter the casing through such perforations and are transmitted to the well surface. Shaped charges are used in other industries also to pierce metal, concrete, and other solid materials.

(19.2) HOW MODERN SHAPED CHARGE WARHEADS WORK?

(19.2.1) The Physics Of Shaped Charges

A shaped charge warhead consists basically of a **hollow liner** of metal material, usually copper, aluminum or steel of conical, hemispherical, or other shape, backed on the convex side by high explosive. A container, fuse, and detonating device make up the complete warhead. When this warhead strikes a target, the fuse detonates the charge from the rear. A detonation wave sweeps forward and begins to collapse the metal cone liner at its apex. The collapse of the cone results in the formation and ejection of a continuous high-velocity molten jet of liner material. What happens is as follows.

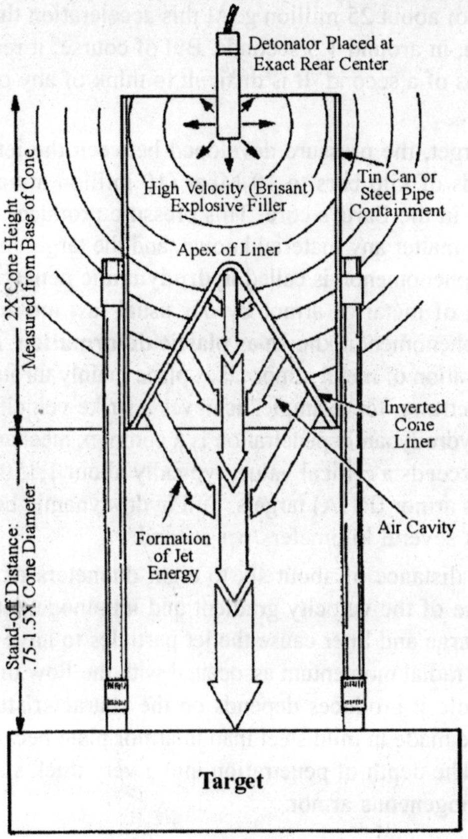

Fig. 114 How A Shaped Charge Works

The liner material is accelerated in a direction roughly perpendicular to the explosive-liner interface. Nearer the apex of the cone, the liner mass-to-charge ratio is lower so the velocity the liner material attains is higher. Also, nearer the apex of the cone, the metal liner does not have to distort as much as further down. This also causes the velocity to be higher toward the apex. When the liner material converges at the center line, or axis, of the charge, the surface material is squeezed out at high velocity. This material forms the jet.

Since the material closest to the apex is at a higher velocity, the portion of the jet which comes from that area is also highest in velocity. Therefore, the jet has a velocity gradient that tends to stretch out or lengthen the jet. The material in the leading tip is moving faster than the material toward the rear. The remaining material, which is the bulk of the liner, forms a heavy slug which follows the jet at a much lower velocity. The stretching occurs at a high strain rate, requiring the cone material to have excellent dynamic ductility at temperatures up to about 450°C and that's why you cannot use just about any metal or material for making the liner of a shaped charge.

The velocity of the tip of the jet is in the vicinity of 8,500 m/s (27,800 fps), while the trailing end of the jet has a velocity in the vicinity of 1,500 m/s (4,920 fps). The slug consists of about 80% of the liner mass. The slug has a velocity of about 600 m/s (1,968 fps). The velocity of the shaped charge jet tip can reach up to 10,000 m/s some 40 microseconds after detonation, giving a cone tip an acceleration of about 25 million g. At this acceleration the tip would reach the speed of light, were this possible, in around 1.5 seconds. But of course, it reaches a terminal velocity in that period of 40 millionths of a second. It is difficult to think of any other terrestrial event as fast as a shaped charge jet tip.

On reaching a target, the pressure developed between the jet tip and the forming crater can be as high as hundreds of kilo bars to 10 Mbar (10 million atmospheres), several times the highest pressure predicted in the Earth's core. This pressure produces stresses far above the yield strength of steel or for that matter any material known, and the target material flows like a fluid out of the path of the jet. This phenomenon is called **hydrodynamic penetration**. This is fundamentally different from penetration of metal or armor by our usual fast moving projectiles. Fast moving projectiles penetrate by a phenomenon known as **plastic deformation**. At strike velocities less than about 1,150 m/s, the penetration of metal armor takes place mainly through the mechanism of plastic deformation. A typical penetrator, for example, achieves a strike velocity of 1,500 m/s to 1,700 m/s depending on the range. Hydrodynamic penetration is a complex mechanism which begins to appear when the strike velocity exceeds a critical value, typically about 1,150 m/s for current penetrators against rolled homogenous armor (RHA) targets. Full hydrodynamic behavior does not occur until the strike velocity reaches several kilometers per second.

After traveling a distance of about six to eight diameters of the original charge, the jet begins to break up because of the velocity gradient and inhomogenieties in the charge and liner. Inhomogenieties in the charge and liner cause the jet particles to have slightly different directions of flight. There is so much radial momentum associated with the flow that the difference in diameter between the jet and the hole it produces depends on the characteristics of the target material. A larger diameter hole will be made in mild steel than in armor plate because the density and hardness of armor plate is greater. The depth of penetration into a very thick slab of mild steel will also be greater than that into homogeneous armor.

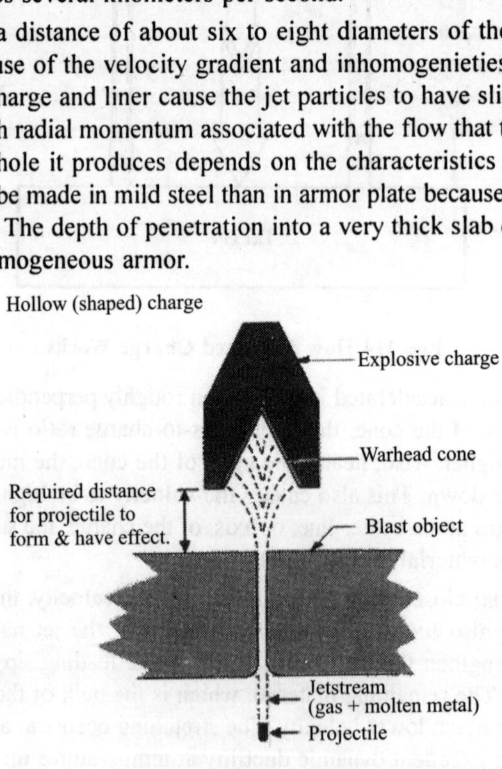

Fig. 115 Penetration By A Shaped Charge

The actual physics of shaped charges is far more complex than this and even today is not completely understood. By modifying the shape and design of the liner, we can control the tip velocity and the mass distribution in the jet to maximize penetration of a target. It has also been found that the shaped charge effect is insignificant if the detonation velocity of the high explosive used is less than 16,400 fps (5000 m/s). These days the explosives commonly used include high detonation velocity plastic explosives like C4, HMX with a plastic binder to result in PBX (plastic/polymer bonded explosive) LX-14, and RDX based compositions as PBX or with wax.

(19.2.2) Factors On Which The Penetration Made By A Shaped Charge Depends

In general, the depth of penetration depends upon five factors: length of jet; density of the target material; hardness of the target material; density of the jet; and precision of the jet (for example, how much it is diverging). The longer the jet, the greater is the depth of penetration. For the jet to be properly formed, the cone has to be at a certain distance from the target. This is known as the **standoff distance**. The standoff distance has an optimal value for a given shaped charge. If the cone is too close, the jet will not get enough distance to be properly formed. If it is too large, the jet will start breaking up even before it has struck the target. As we saw, the jet particulates or breaks up at 6 to 8 cone diameters from the cone base. Particulation is a result of the velocity gradient in the jet, which stretches it out until it breaks up. Jet precision refers to the straightness of the jet. If the jet is formed with some oscillation or wavy motion, then the depth of penetration will be reduced. This is a function of the quality of the liner and the initial detonation location accuracy. The effectiveness cf shaped charge warheads is reduced when they are caused to rotate. That's why spin-stabilized projectiles generally cannot use shaped-charge warheads. Degradation begins at as rotational speeds as low as 600 rpm. The effectiveness of a shaped charge payload is independent of the striking velocity of the warhead. In fact, the velocity of the warhead must be taken into consideration to ensure that detonation of the payload occurs at the instant of optimum standoff distance. The jet can then effectively penetrate the target.

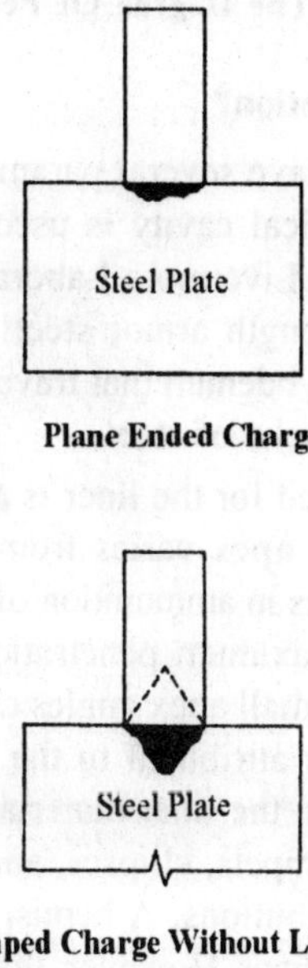

Fig. 116 Illustrating How The Presence Of A Liner Affects The Degree Of Penetration

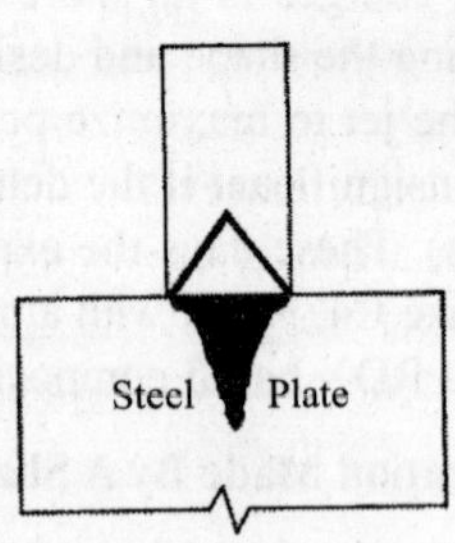

Shaped Charge With Liner

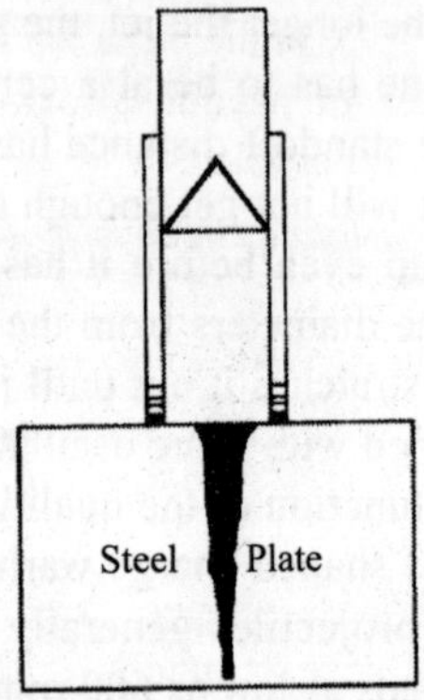

Shaped Charge With Liner and Standoff

Fig. 117 Illustrating How The Availability Of Standoff Distance Affects The Degree Of Penetration

(19.2.3) How To Maximize The Penetration?

To maximize penetration we have several parameters to play with: the shape of the liner cavity; the angle of the cone if a conical cavity is used; the material used in the liner and its thickness, etc. In early 1997, Lawrence Livermore Laboratory successfully tested a shaped charge that penetrated 3.4 meters of high-strength armor steel! The largest diameter precision shaped charge ever built produced a jet of molybdenum that traveled several meters through the air before making its way through successive blocks of steel.

The shape most commonly used for the liner is a cone, with an internal apex angle of 40 to 90 degrees. The **angle of the cone apex** varies from 42 degrees to 60 degrees in American ammunition and 18 degrees to 90 degrees in ammunition of other countries. Theoretically 42 degrees is the optimum angle for producing maximum penetration. Different apex angles yield different distributions of jet mass and velocity. Small apex angles can result in jet bifurcation, or even in the failure of the jet to form at all; this is attributed to the collapse velocity being above a certain threshold, normally slightly higher than the liner material's bulk sound speed. Other widely used shapes include hemispheres, tulips, trumpets, ellipses, and bi-conics; the various shapes yield jets with different velocity and mass distributions. A hemispherical liner, for example, produces jet energies comparable to a cone-shaped liner. However, the velocity and mass distribution in the jet are different. Hemispherical liners turn inside out with almost all of the material projected in the jet. As much as 70 to 80 percent of the material in conical liner stays behind to form the slug. In hemispherical shaped charges, however, jet velocity is only about half of that obtained in a conical shaped charge.

Liners have been made from many **materials**, including glass and various metals. The deepest penetrations are achieved with a dense, ductile metal, and a very common choice has been copper. For some modern anti-armor weapons, molybdenum and pseudo-alloys of tungsten and copper have been adopted. The selection of the material depends on the target to be penetrated; for example, aluminum has been found especially advantageous for concrete targets. For the deepest penetrations, pure metals yield the best results, because they display the greatest ductility, hence postponing the breakup of the stretching jet into particles. In charges for oil-well completion, however, it is essential that a solid slug or 'carrot' not be formed, since it would plug the hole just penetrated and interfere with the influx of oil. In the petroleum industry, therefore, liners are generally fabricated by powder metallurgy, often of pseudo-alloys, which yield jets and slugs that are composed mainly of dispersed fine metal particles.

The **optimum liner thickness** is about three percent of the charge diameter for soft copper. If a different material is used, the thickness should be determined by measuring the weight of soft copper that would be used for the particular charge diameter. This amount of material is then used. Liners made of less dense material will be thicker than the copper counterpart and liners made of more dense material will be thinner. Materials which are commonly used include copper, steel, glass, and aluminum.

When used in projectiles that rotate, shaped charge penetration is considerably reduced. Projectiles that have conical liners may lose as much as 50 percent of their effectiveness due to rotation. In general, penetration is reduced as spin increases from about zero to 200 revolutions per second, after which further increases have little effect. Rotating shaped charges produce holes of greater diameter than static charges, but rotation does not affect the volume of the hole produced. Changing the apex angle of the cone does not eliminate the reduction in penetration caused by rotation of the charge. Hemispherical shaped charges are, on the other hand, less sensitive to rotation than conical shaped charges.

(19.2.4) Linear Shaped Charge

A linear shaped charge (LSC) has a liner with V-shaped profile and varying length. The liner is surrounded with explosive and the explosive then encased within a suitable material that

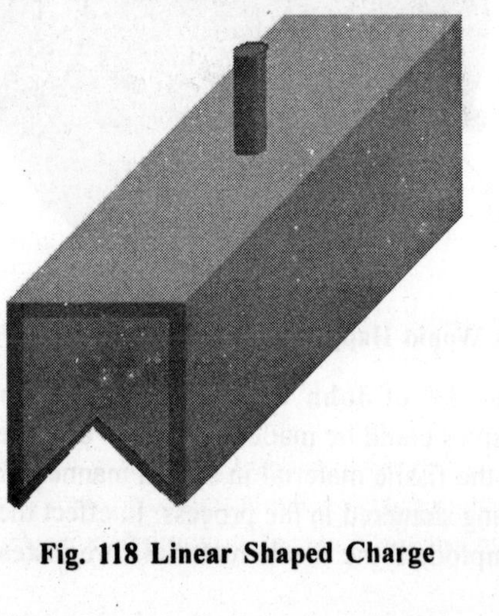

Fig. 118 Linear Shaped Charge

serves to protect explosive and to confine (tamp) it on detonation. The charge is detonated at some point in the explosive above the liner apex. The detonation projects the liner to form a continuous, knife-like (planar) jet. The jet cuts any material in its path, to a depth depending on the size and materials used in the charge. For the cutting of complex geometries, there are also flexible versions of the linear shaped charge, these with a lead or high-density foam sheathing and a ductile/flexible liner material, which also is often lead. LSCs are commonly used in the cutting of rolled steel joists (RSJ) and other structural targets. LSCs are also used to **separate the stages of multi-stage rockets.**

(19.3) IN THE HEART OF EVEN NUCLEAR WEAPONS ARE EXPLOSIVE LENSES

(19.3.1) See How A Genius Invents Explosive Lenses

As the Manhattan Project was belaboring to make the nuclear bomb, it ran into a serious problem. It was found that making a bomb with uranium was easy using a gun type assembly for achieving super-criticality. But there was uranium enough for only one bomb—not enough to even conduct a test besides the bomb to be dropped! Plutonium, on the other hand, did not lend itself to a gun type assembly. They had, however, worked out that a plutonium bomb would work if the plutonium could be imploded, that is, violently compressed from all sides so that it achieved super-criticality. The problem was how to achieve this violent compression so that a dense metal like plutonium could be compressed to over 2.5 times its density—a hitherto unimaginable feat! Earlier people had thought that one could simply surround the plutonium ball with a large quantity of high explosive and that would do the trick. It did not. Tests with other metals showed that such a configuration would result in only shattering the metal ball instead of compressing it uniformly from all sides and without allowing metal to be squeezed out.

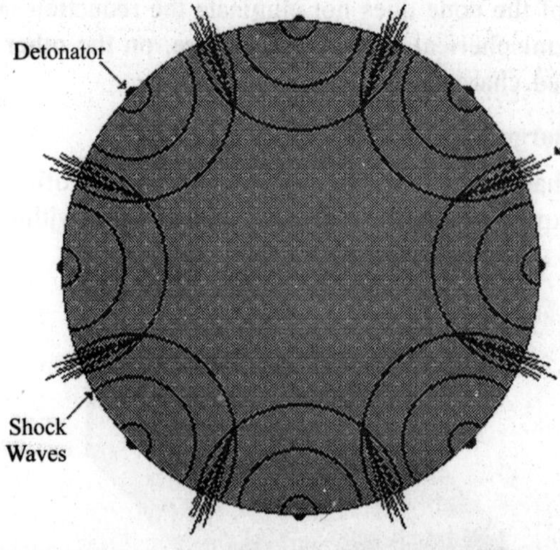

Fig. 119 This Is What Would Happen To The Plutonium If There Is No Explosive Lens

It was the sheer genius of **John von Neumann** that showed that an extremely clever arrangement of high explosives could be made to function as an 'explosive lens' that would focus the explosive energy on to the fissile material in such a manner that it would be compressed to its critical density without getting shattered in the process. In effect the explosive power was harnessed in such a manner that it 'imploded' the fissile material core instead of blowing it up.

Explosive lenses can be visualized in the following manner. Consider two high explosives with different detonation velocities in intimate juxtaposition. Suppose the explosive with the lower detonation velocity is kept on the left side and is detonated first. Now as the shock wave produced by the detonation of the first explosive will strike the explosive with the higher detonation velocity, it will detonate the second one and since the shock wave produced in the second explosive will have a higher velocity, the shock wave produced by the first explosive will be accelerated by it. (Had we detonated the explosive with the higher detonation velocity first, its shock wave would have been decelerated upon encountering the explosive with the lower detonation velocity.). Now if the interface of the two explosives had been so shaped that instead of the ,entire shock front of the first explosive touching the second explosive simultaneously, some part (say the central part) of the wave front of the shock wave from the first explosive makes contact with the second explosive first then that part would be accelerated first, giving it a head start over the wave fronts from the edges. The resulting wave front in the second explosive would then lose the angular symmetry it had in the first explosive and would acquire a pointed character. That is the force of the explosion would be predominantly concentrated towards the center of the second explosive mass instead of wasting its energy in all the directions. In other words, the wave front would be 'shaped' by this arrangement. Different 'shapes' of the resulting wave front can be realized by suitably altering the shapes of the surfaces of the two explosives that are in contact with each other and by keeping the faster or the slower explosive on the first, that is, the initiating side.

This is like treating light to a lens and hence this particular arrangement is known as an 'explosive lens'. You may recall that the lens action of an optical lens becomes possible because light has different velocities in air and in glass. This arrangement of explosives was also called a lens because the detonation wave has different velocities in the two explosives.

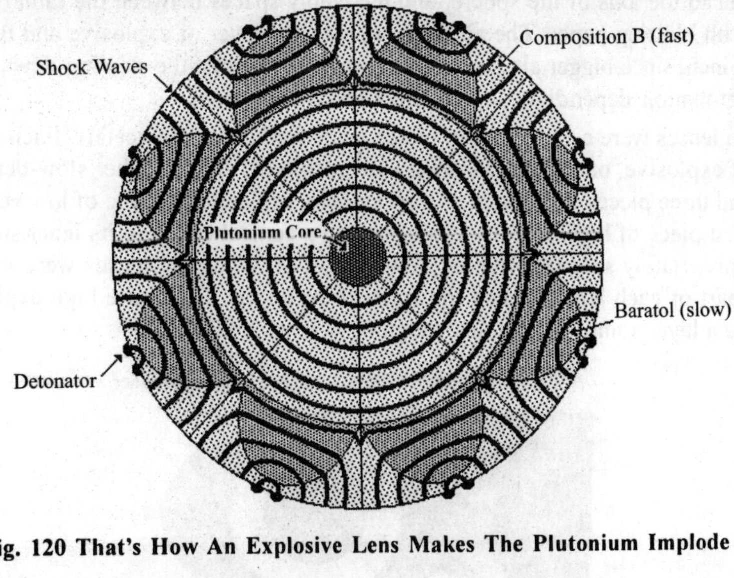

Fig. 120 That's How An Explosive Lens Makes The Plutonium Implode

(19.3.2) And Uses That To Make The Nuclear Bomb!

A complicated arrangement of explosive lenses is used to 'implode' the fissile material (plutonium) to super critical densities in a fission bomb. The whole ingenious scheme was devised by **John von Neumann** when the scientists of the Manhattan Project had been, in the words of the

Nobel laureate Hans Bethe himself, been defeated by the problem. His brilliance made the '**Fat Man**' (the design tested on May 16, 1945 in the Trinity shot and later dropped on Nagasaki on August 9, 1945) possible. The basic principle of compressing the plutonium ball using explosive lenses remains unchanged even now when much more powerful and yet smaller bombs have been designed.

An 'explosive lens', as used in nuclear weapons, is a highly specialized explosive charge. In general, it is a device composed of several explosive charges that are shaped in such a way as to change the shape of the detonation wave passing through it. To convert a spherically expanding wave front into spherically converging by a single boundary between fast and slow explosives, the boundary shape must be hyperbolic; to convert a spherically diverging front into the flat one, it must be parabolic, and so on. Several boundaries must be used to reduce aberrations (deviations from intended shape) of the final wave front, just as we use a combination of optical lenses to produce an aberration-free lens. That's what makes the explosive lenses of a nuclear weapon complex. In a nuclear weapon, an array of explosive lenses is used to change the several approximately spherical diverging detonation waves into a single spherical converging one. The converging wave is then used to collapse the various shells (tamper, reflector, pusher, etc.) and finally compresses the core (pit) of fissionable material to a super-critical state.

In the first implosion bombs, a layer of explosive was formed from 32 specially shaped blocks. The inner center-facing surface of the blocks was spherical with a diameter equal to the outer diameter of the aluminum tamper layer. On the outer surface of the blocks of explosive were grooves shaped to provide for the insertion of 20 hexagon-shaped lenses and 12 pentagon-shaped lenses. A 1/16 inch thick felt pad was placed between the surfaces of the explosive and the lenses perpendicular to the axis of the sphere, and the empty spaces between the radial contact surfaces was filled with blotting paper. The air gaps between the layer of explosive and the lenses did not exceed 1/32 inch, since bigger air gaps could contribute to either the slowing-down or the speeding-up of the detonation depending on the direction of these gaps.

The lenses were cast in special molds made from cellulose acetate. Each lens consisted of two types of explosive, one fast-detonating Composition-B and the other slow-detonating Baratol. Each lens had three pieces: two made of high velocity explosive and one of low velocity explosive. The outermost piece of high velocity explosive had a conical cavity in its inner surface into which fitted an appropriately shaped piece of slow explosive. When the lenses were in place, the fast-detonating part of each lens touched the layer of explosive. The entire high explosive implosion system made a layer some 45 cm thick weighing at least 2500 kg.

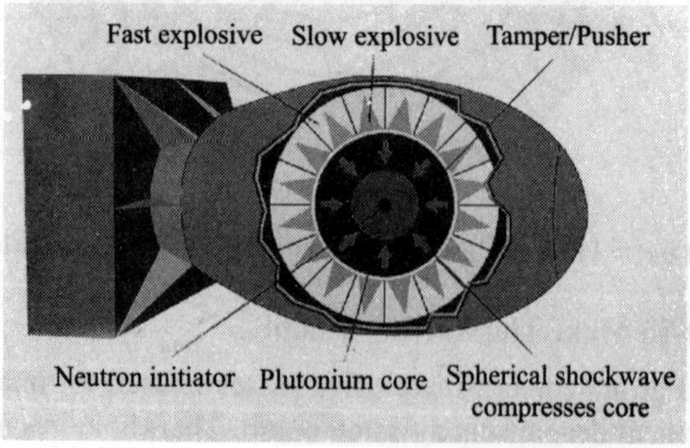

Fig. 121 Schematic Representation Of The Explosive Lenses Used In The Fat Man Type Nuclear Bombs

The successful functioning of this device hinges upon the simultaneous initiation of the wave in each segment, uniformity and precision in the speed of the wave, and correctness and accuracy in the shape of the interface between the two explosives. It is for this purpose that the highly specialized **Exploding Bridge Wire detonator** was invented by **Luis Alvarez** and **Lawrence Johnson.**

It should be plainly obvious that ultimately even nuclear weapons have to depend on high explosives for the 'implosion' of plutonium. Even a hydrogen bomb requires the trigger of a plutonium bomb and it, in turn, is triggered by the explosive lenses. Thus one could say that eventually it is the high explosives which drive even nuclear bombs! Three cheers for high explosives and explosive lenses!

(19.4) HIGH EXPLOSIVES USED IN THE EXPLOSIVE LENSES OF ADVANCED NUCLEAR WEAPONS

We have seen that the explosive lenses of the nuclear bomb would require a fast and a slow explosive. What are the other properties that must be there in the explosives? Obviously the high explosives used in nuclear weapons must lend themselves to very fine machining into intricate shapes. Not only that, they must be dimensionally stable and compositionally uniform. They should be easy to form into parts but resistant to subsequent deformation through temperature, pressure, or mechanical stress. They should be easy to detonate on demand but difficult to explode accidentally. The explosives should also be compatible with all the materials it contacts, and they should retain all their desirable qualities indefinitely.

Much progress has been made since the days of the Manhattan Project. The Los Alamos National Laboratory demonstrated the first use of plastic-bonded explosives in a nuclear explosion in 1956. They are usually machined from a plastic-bonded explosive (PBX) and an inert insert, called a wave-shaper, which is often a dense foam or plastic, though many other materials can be used. This development allowed the shift from precision machined cast explosives to formulations containing high concentrations of high-energy density compounds with reduced sensitivity, more uniformity, and better mechanical characteristics. **Pressed, plastic-bonded explosives** are the key energetic materials in today's enduring stockpile of the USA. Since the 1960s the Livermore Laboratory has also been researching and developing safer high explosives for Livermore-designed weapons.

The plastic coating that binds the explosive granules, typically 5 to 20% of each formulation by weight, is what gives each PBX its distinctive characteristics. Pressing a PBX molding powder converts it into a solid mass, with the polymer binder providing both mechanical rigidity and reduced sensitivity to accidental detonation. The choice of binder affects hardness, safety, and stability.

Too brittle a PBX can sustain damage in normal handling and succumb to extreme temperature swings or thermal shocks; while too soft a PBX may be susceptible to creep and may lack dimensional stability or strength. To achieve safe and stable PBXs, the Laboratory uses two main charge **explosives based on HMX and TATB**. HMX is more energetic than RDX but retains good chemical and thermal stability, important for long-term storage and survival in extreme environments. Sensitivity of any PBX is a complex characteristic strongly affected by HE particle size distribution, visco-elastic properties, binder-to-HE wetting, and storage environment.

Presently the TATB based **LX-17-0** developed by the Livermore laboratory and **PBX-9502** developed by the Los Alamos Laboratory are held in high regards for nuclear weapons as they are truly insensitive high explosives. A new class of insensitive high explosives has been discovered that, unlike previous explosives, are very rich in nitrogen and contain much less carbon and oxygen. The first of these to be synthesized, LAX-112, is less sensitive than TNT and produces a more powerful detonation than TATB. Work is continuing to find an explosive within the new class that is even more insensitive but retains the high performance of LAX-112.

Having perfected their nuclear arsenal the Americans have become concerned about their safety against accidents. Much of the current research into the high explosives meant for nuclear weapons is directed towards that end. For example, a nuclear weapon should not go off even if the missile or bomb is involved in an accident involving very high temperatures or mechanical shock such as those likely to be encountered during crashes. This requires, above other design considerations, insensitive high explosives. They are worried that even if a nuclear yield is totally averted through inherent design features, the explosive's energy release might still disperse radioactive plutonium into the air. Nuclear weapons have long been designed to avoid or drastically reduce such threats. For example, all weapons in the US stockpile are inherently 'one-point' safe; that is; the initiation of the explosive at some random point will not produce a nuclear yield. Weapons have also been tested against the raging inferno of a jet-fuel burn to assure their safe response should, for example, a bomber loaded with nuclear weapons catch on fire. Accident analyses have therefore been extended to address extremely low-probability accidents. Complex, multiple-accident scenarios now being considered include the possibility that after a bomber loaded with nuclear weapons catches on fire, another large plane crashes into it and similar scenarios.

Most accidental insults to a nuclear weapon would not produce shock waves that could initiate the detonation of high explosives. However, exposure to fire along with the impact of a crash might initiate a deflagration which could, in some cases, build up into a full-scale detonation. The deflagration-to-detonation transition is a significant safety consideration in the matter of safety of nuclear weapons against accidents.

(19.5) EXPLOSIVELY FORMED PROJECTILES / PENETRATORS (EFP)

(19.5.1) How Do You Make The EFP And What Are Its Specialties?

Explosively formed projectiles or penetrators (EFP) are used to defeat armored vehicles. They are also known as **self-forging warheads** or **self-forging fragments**. EFP warheads are normally designed to produce a single massive, high velocity penetrator. An EFP has a metal liner in the shape of a shallow dish, backed by a high explosive charge. A typical explosively formed projectile (EFP) consists of a metallic liner, a case, an explosive section, and an initiation train. Very often there is also a retaining ring to position and hold the liner-explosive sub-assembly in place. After detonation, the explosive creates enormous pressures that accelerate the liner while simultaneously reshaping (**dynamic transformation**) it into an **elongated penetrator** or some other desired shape. The force of the blast can mold the plate into any of a number of configurations, depending on how the plate is formed and how the explosive is detonated. The EFP then hits the target at a high speed (typically 1500-2000 m/s), delivering a high mechanical shock to the target. Sophisticated EFP warheads have multiple detonators that can be fired in different arrangements causing different types of wave forms in the explosive, resulting in a long rod penetrator, an aerodynamic slug projectile or multiple high-velocity fragments.

Although both the shaped charge and the EFP have liners and high explosive charges, the phenomena that take place in them are qualitatively different from each other. In an EFP, because of the wide angle cones and other liner shapes such as plates or dishes a jet is not formed. The so-called explosively formed projectile is made by dynamic plastic flow of the metal of the liner.

Obviously the target penetration by the EFP is much less than that of the shaped charge jet, but the diameter of the hole is larger with EFP and there is more spalling. As a rule of thumb, an EFP is generally able to perforate a thickness of armor equal to only about the diameter of its charge, whereas the jet of a typical shaped charge will go through six or more diameters.

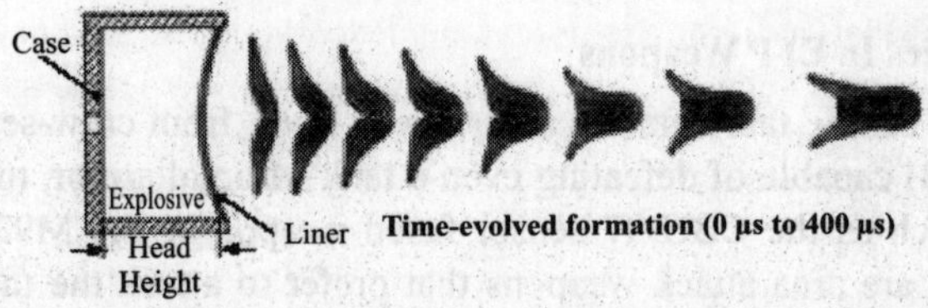

Fig. 122 That's How The Explosively Formed Penetrator Is Formed

What makes the EFP desirable? After all, what we are getting in effect is a mass of metal moving at high speed. Would the same purpose be not served by a gun-fired projectile? No! The reason is that for getting the same velocities for the projectile which you get in the EFP, you will have to use an impractically large gun. Remember that guns use propellants which are low explosives. Here you are using a high explosive. But the high explosive is used in such a clever manner that instead of blowing up the projectile to smithereens, it dynamically transforms the projectile into a mass that flies off at a very high speed and yet retains an aerodynamic shape so that its flight is not erratic and it is capable of hitting a target just like a projectile fired from a gun.

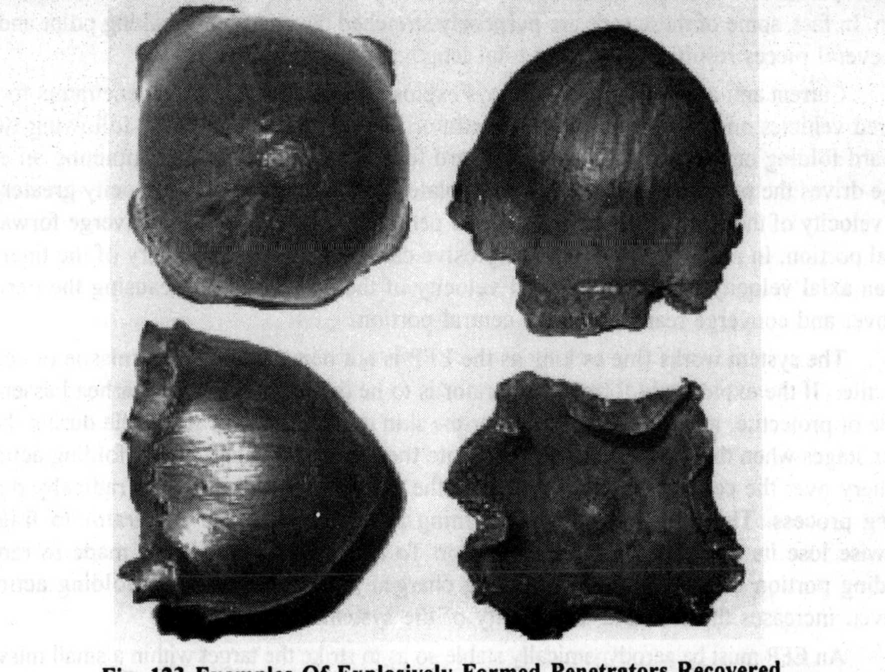

Fig. 123 Examples Of Explosively Formed Penetrators Recovered

There is one more reason for taking interest in the EFP. Although an EFP cannot penetrate as much armor as a conventional shaped charge; the EFP, on account of its larger physical mass, is relatively 'cool' when compared to the plasma jet of a conventional shaped charge, and is therefore less likely to set off any Explosive Reactive Armor (ERA) mounted on the target. You may recall that the ERA was specifically invented to disrupt the jet of a shaped charge and thus reduce its efficiency. The EFP is therefore more difficult to counter. In addition, while vehicles can be modified to resist the effects of EFPs, they cannot typically be modified to defend against both threats as countermeasures which work against EFP are ineffective against shaped charges, and vice versa.

(19.5.2) Recent Advances In EFP Weapons

EFP weapons can be fabricated in a variety of sizes, from crew-served missiles (such as the American TOW 2B) capable of defeating even a tank's frontal armor, to smaller cluster bomb like sub-munitions (such as the CBU-97 sensor-fused weapon or the M93 Hornet sensor-fuzed anti-tank mine), which are area attack weapons that prefer to attack the thinner top armor of an armored vehicle. EFPs have been adopted as warheads in a number of weapon systems, including the BLU-108 air bombs (with the Skeet sub-munition), the M303 Special Operations Forces demolition kit, the M2/M4 Selectable Lightweight Attack Munition (SLAM), and the SADARM sub-munition.

By way of application, EFP warheads are of two types, namely, long-standoff sensor-fuzed sub-munitions and medium standoff, close-over-flight missiles. **Long standoff EFPs** require the formation of a single-piece EFP capable of flying in a stable fashion to the target. This requirement has led to the flared EFP rod and, more recently, to the finned EFP rod designs. For the **medium or short-standoff** applications, a new type of EFP was developed. The need for an aerodynamic shape is not necessary for these applications because of the short distance the EFP must travel, hence, the length of the rod was increased and the flared tail was eliminated from the design. In fact, some of these rods are purposely stretched beyond their breaking point and fracture into several pieces resulting in greater total length.

Current anti-armor ordnance employs explosively formed elongated penetrators for piercing armored vehicles and equipment. Such penetrators are generally one of the following two types: rearward folding or forward folding. In forward folding types a warhead containing an explosive charge drives the periphery of the liner metal plate forward with an axial velocity greater than the axial velocity of the central portion causing the periphery to fold over and converge forward of the central portion. In rearward folding the explosive charge drives the periphery of the liner forward with an axial velocity less than the axial velocity of the central portion causing the periphery to fold over and converge rearward of the central portion.

The system works fine as long as the EFP is not being fired from a missile or some other projectile. If the explosively formed penetrator is to be delivered from the warhead assembly of a missile or projectile, it will have to encounter the skin of the missile or projectile during the critical earlier stages when the liner is being formed into the penetrator shape by the folding action of the periphery over the center. The engagement of the liner with the skin would radically disrupt the folding process. This disruption of the forming process causes the penetrator to fragment or otherwise lose its effectivencss as a penetrator. To avoid this, provision is made to remove the impeding portion of the skin using clearing charges just before the liner folding action. This, however, increases the cost and complexity of the systems.

An EFP must be aerodynamically stable so as to strike the target within a small miss distance and with a small angle of obliquity. In the U.S., extensive work has been focused on forming such EFPs that have canted fins, to induce spin-up. By forming canted fins on an EFP, improvements in aerodynamic stability can be realized.

In order to increase the penetrability, hollow charges differing from conventional hollow charges have also been developed in recent times. These charges can, for instance, comprise an auxiliary body disposed in front of or integrated with the metal cone of the charge so that upon initiation of the charge it generates a slug which follows behind the actual penetration jet and penetrates and enlarges the hole made by the penetration jet. Alternatively, the hollow charge may have a warhead with two complete hollow charges, so-called tandem hollow charges, which after the projectile is fired accompany each other as an integral unit during the greater part of the travel towards the target, only to separate at a predetermined distance from this and to continue towards the target at mutually slightly different velocities along largely the same trajectory and thereafter to hit the target with a sufficient interval of time to enable the charge which reaches the target first to detonate the explosive in any active armor before the second charge reaches the target, so that this latter charge penetration jet is able to work without disturbance and also is assisted by the penetration work already performed by the first charge which has already detonated within the same confined area of the charge.

In order to function in the intended manner each of the two hollow charges in such a tandem hollow charge must have its own ignition system with associated safety device. To separate the two hollow charges, it is also necessary to have a smaller parting charge, e.g. a powder charge, between the two charges in order to impart to each of these its own velocity change.

❑❑❑

PART II
BOMBS:
TO BREAK THE BACK OF NATIONS

Chapter 20

Bombs in a Historical Perspective

(20.1) THE IMPORTANCE OF BOMBS IN WAR

Full-scale wars are won by bombs and not by bullets. It is the ordnance that you unleash on the war-making potential of the enemy that decides the issue. Heroics on the battlefield are fine for medals. They are also quite good material for making films on war and writing novels on war. But heroics don't really break the back of the enemy as a nation. Heroics may result in some enemy soldiers getting killed; tanks getting destroyed; aircrafts getting shot up in the sky or ships getting sunk but have little effect on the capability of the nation to continue the war. You break the back of the enemy only with bombs! Bombs are synonymous with destruction. It is only by bombarding his military-industrial potential that you prevent him from making more tanks and ships. You like management jargon? Well, bullets decide micro-issues; it is the bombs which decide the macro-issues.

Bombing has great effect on the morale of the public at large too. You see, battlefields could be far away from public eye and the nation could well be indulging in propaganda about how it is faring in the war. Most nations are known for distorting the true picture either by denying the media access to the battlefield or, if they are more subtle like the Americans, planting manipulated journalists and presenting a tailored picture. Thus reports regarding battlefields are not reliable in any case. On the other hand, when bombs fall on your territory, even if civilian targets have been spared, there is no mistaken notion about how you are faring in the war. If enemy war planes can penetrate your air defenses and destroy your military related industries, command and control structures, it is for all to see that you are in deep shit. This puts enormous pressure on the government. No government can make a fool of its people even as its cities get relentlessly pounded. Let us put it this way. Whatever arms that one can wield in a battlefield can, at best, defeat armies, but bombs defeat nations. Bombs break the will of a nation to fight. Even if the military value of carpet bombing be limited, the fact remains that it has tremendous impact on the morale of the target population by bringing the war and the might of the enemy right before their eyes.

The Allied bombing of the industrial cities of Germany makes for an excellent case study. In the evening of February 13, 1945 796 Avro Lancasters and 9 De Havilland Mosquitoes of the

RAF were dispatched in two waves separated by three hours and they dropped 1,478 tons of high explosive and 1,182 tons of incendiary bombs by the early hours of February 14 on the city of Dresden. Later during the day on the 14th, B-17 Flying Fortresses of the USAF dropped an additional 771 tons of bombs and followed it up by dropping 466 tons on February 15th. The fire-bombing consisted of by-then standard methods; dropping large amounts of high-explosive to blow off the roofs to expose the timbers within buildings, followed by incendiary devices (fire-sticks) to ignite them and then more high-explosives to hamper the efforts of the fire services. This eventually created a self-sustaining 'fire storm' with temperatures peaking at over 1500 degrees Celsius. Out of the 28,410 houses in the inner city of Dresden, 24,866 were destroyed. An area of 15 square kilometers was totally destroyed. There is a great dispute regarding the number of persons killed, with estimates ranging from 35,000 to 135,000 given that the population of the city was about 642,000 and it had a large number of refugees and wounded soldiers. Bombing helpless civilians was a most despicable act and amounts to a war crime. However, the fact remains that the nation as a whole was utterly demoralized by the complete Allied mastery of the skies and their ability to bomb whatever they wished. We are emphasizing this point because admission of the mere fact that the RAF and the USAF were superior to the Luftwaffe would not have had the same impact on the collective psyche of the nation. People would have thought, okay, so our fighter aircrafts are beaten by their fighter aircrafts, so what—there is much more to war than fighter aircrafts! The difference was made by the bombs. They convinced the people in a way nothing else could do.

Even as late as during the last war on Iraq, according to Iraqi prisoners of war, formations of B-52s dropping general-purpose bombs were one of the most feared aircraft-weapon combinations of the war. Even otherwise the very sight of an unprecedented average of 2555 sorties per day relentlessly for 38 days was absolutely terrifying.

(20.2) BOMBS AND BOMBINGS IN THE TWO WORLD WARS

The aerial bomb was invented during the World War I. The first attacks on ground targets from the air were performed with grenades, artillery shells, or even jars of nitroglycerine, but making finned bombs specifically for the task proved straightforward.

High-explosive aerial bombs were first dropped from German Zeppelins during the 31 May 1915 raid on London. The Germans continued to use these dirigibles, practically unopposed, in their air raids on London and Paris until late in 1916. Then the British and French began to counter these attacks with anti-aircraft gunfire and fighter aircrafts. Soon the Germans began using their large Gotha bombers, which they employed throughout 1917 and 1918. In retaliation, the Allies began producing fleets of larger bombers and various high-explosive bombs.

Incendiary bombs were also first used during the German Zeppelin raid over London on 31 May 1915, when one dirigible dropped 90 incendiary bombs. The early incendiaries used by the Germans in their air raids over London and Paris were made from a mixture containing manganese dioxide and magnesium. Later, the French used sulfur-bound thermite, a mixture of iron oxide and powdered aluminum, in an incendiary drop bomb. The British were the first to develop incendiary bomb clusters, which could release either 144 or 272 bomblets. A De Haviland bomber could dispense up to 860 of these 6.5 oz incendiary bomblets, and a Handley-Page V bomber could dispense up to 16,000. By flying in formation, these bombers could cover a large area with a blanket of burning incendiary materials.

In 1915, Zeppelin raiders over England used thermite bombs to attack British cities. These bombs were conical, consisting of a thermite core surrounded by resin, and wound tight with a layer of rope. High explosive bombs evolved through the war. By the end of the conflict, HE bombs of up to 2200 lbs had been built.

During the course of the World War II the Allied forces dropped about 1.36 million tons of bombs on Germany and slightly more on occupied territories. (In comparison the Pacific theater received only 656,400 tons of bombs with the Japanese home islands getting about 160,800 tons—but you should keep the area of Japan in mind and also the fact that the war in the pacific theater was of a shorter duration.) In Europe, the peak was in April 1945 when 250,000 tons of bombs were dropped in a month alone exclusively on German territory. World War II led to major refinements in bomb design. The different combatants used a wide variety of high explosive, fragmentation, and incendiary munitions. There is little point discussing the whole lot of them as there was nothing great about their design. We will describe only those bomb which were truly great, were far ahead of their times or showed some unusual ingenuity in their design.

(20.3) THE GREATEST NON-NUCLEAR BOMBS OF THE SECOND WORLD WAR

(20.3.1) The Dam Busters

The British developed highly specialized bombs for certain missions. The most famous of these was the **'Upkeep' bouncing bomb**, used for the attacks performed by the elite Royal Air Force (RAF) 617 Squadron on the Moehne, Eder, and Sorpe dams in the night of 16-17 May 1943. The Upkeep weapon was designed by Dr. **Barnes Wallis**, who had been trained as a structural engineer but turned to aircraft and weapons engineering, becoming well known for his geodetic-framed Vickers Wellington bomber. The bombs breached the Moehne and Eder dams. The actual military usefulness of this action is still debated, but the Upkeep weapon worked precisely as designed, even though nothing like it had been done before.

Fig. 124 The Upkeep Dam Buster Bomb Slung Beneath The Wing Of A Lancaster Bomber

Upkeep was really a very special kind of depth charge, and its operation reflected Wallis's knowledge of not only physics but also of massive structures like dams and their weaknesses. It was designed as a 9250 lbs drum, and was carried underneath a modified Lancaster heavy bomber. The bomb was given a 500 rpm backspin before it was released at low altitude over the surface of the water, in front of a dam. Instead of sinking straightaway the bomb would skip repeatedly over the water as flat stones do in ducks-and-drakes, run into the surface of the dam, roll down its face, and explode at a predetermined depth. Therein lay the beauty of the concept! It would have been almost impossible to ensure that a normal bomb dropped from a height would strike the wall of the dam at a certain depth. Striking the wall itself would have been an extremely difficult proposition. Confinement of the explosion to the face of the dam by water pressure multiplied the effect of the blast. The very fact that a mere 5,047 lbs of torpex could breach giant dams was an outstanding engineering achievement! Neither anyone had ever thought of such a feat nor would anyone ever try it again—geniuses are not born everyday!

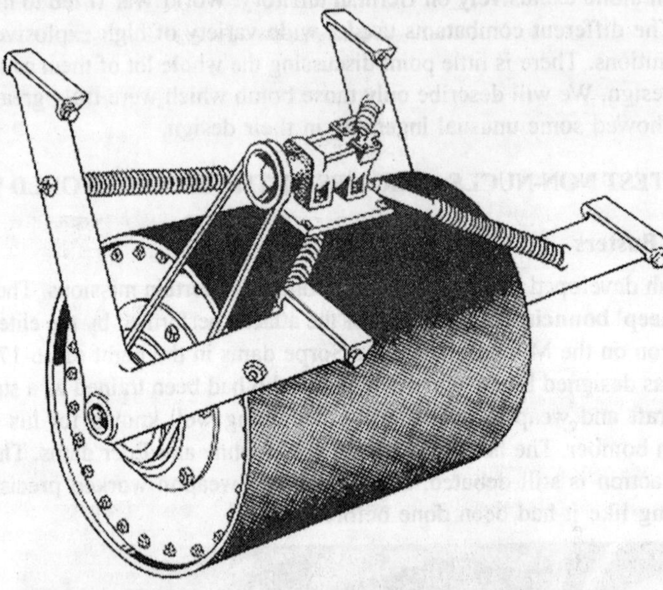

Fig. 125 The Spinning Mechanism Of The Upkeep Bomb

(20.3.2) The Tallboy

Wallis also developed the first serious penetrating bombs. Large high explosive bombs were proving ineffective in attacks on hardened targets such as U-boat pens. Wallis believed that a more effective weapon would be a big, hard-cased, streamlined bomb that could be dropped on such targets from high altitude, with the bomb exceeding Mach 1 before impact to bury itself deeply and then detonate. The first of these bombs was the 12,000 lbs 'Tallboy' bomb, which was first used in combat after the Allied landings in Normandy in June 1944. It was slung beneath modified Lancasters and used for attacks on tunnels, V-weapon sites, and other high priority targets. Precise targeting from high altitude was required, and the Tallboy was only used by highly trained crews of RAF Number 9 and Number 617 Squadrons. Its most important use was in the sinking of the German battleship Tirpitz at anchor in Norway on 12 November 1944.

The bomb was fitted with curved fins to ensure that it would be spin stabilized as it broke through the sound barrier and then plunged into the ground. Structures that could withstand a direct hit by the biggest conventional bomb could be toppled by dropping such a 'Deep Penetration (DP)' bomb next to them. The blast would blow out a huge crater that undermined the target's foundations, and underground installations could be shattered by the shock waves set up by the underground explosion.

Fig. 126 A Tallboy Bomb Falling: The Bombs Broke The Sound Barrier Before They Struck The Targets

(20.3.3) The Grand Slam

Tallboy was followed by a similar but bigger penetration bomb, the 22,000 lbs 'Grand Slam'. This huge weapon was 26 feet 6 inches long. Its hardened casing was cast in a single piece in a sand mold, using a concrete core. Grand Slam was so massive that it required special heavy duty handling even when empty. The Grand Slam was first used in combat in March 1945. It was effective in destroying targets that had resisted other attacks, such as massive viaducts, by blowing away the ground underneath them. In one raid on submarine pens near Bremen, two Grand Slams penetrated 23 feet of reinforced concrete, bringing down the roof. The Grand Slam remains the biggest conventional bomb ever used in action.

Fig. 127 Grand Slam, At 22,000 Lbs The Largest Conventional Bomb Ever Used, Being Loaded Onto A Lancaster Bomber—It Had Penetrated 23 Feet Of Reinforced Concrete—Note The Gigantic Size In Relation To A Man

(20.4) THE NOTORIOUS FIRE-BOMBINGS OF JAPAN BY M-69 INCENDIARY BOMBS

The low level fire-bombing raids began in March 1945, starting with a devastating attack on Tokyo on March 9-10 that killed over 80,000 people (official figure 83,793) and left a million homeless. In that raid 334 B-29 Superfortresses had dropped about 1700 tons of bombs and destroyed an area of about 41 square kilometers. Subsequent raids incinerated Nagoya, Osaka, Kobe, and then worked their way down to the smaller cities. By the time the Americans dropped atomic bombs on Hiroshima and Nagasaki in August, most of Japan's major cities had already been destroyed by the M-69.

The M-69 was a simple, clever weapon. It looked like a length of pipe, and weighed only 6.2 lbs. As handling such a small weapon was inconvenient, and dropping quantities of small bombs from high altitude was wildly inaccurate, it was designed to be incorporated into an **"aimable cluster"**, a type of finned cluster bomb that contained 38 of the M-69 firebombs. The aimable cluster was a bundle of M-69s fitted with a nose shroud and tail assembly. It was dropped from high altitude and then broke apart at about 900 meters (2,000 feet), scattering its M-69s. Each M-69 then ejected a long strip of cloth to stabilize itself, and crashed nose-first into buildings below. On impact, it ignited its payload of napalm, which shot out of the tail of the bomb in a burning jet. The M-69 was small and could not penetrate the roofs of solidly constructed buildings. However, most Japanese buildings were lightly built and extremely vulnerable to fire. A copy of a Japanese residential area was built in the US to test the M-69, and was incinerated in a test bombing.

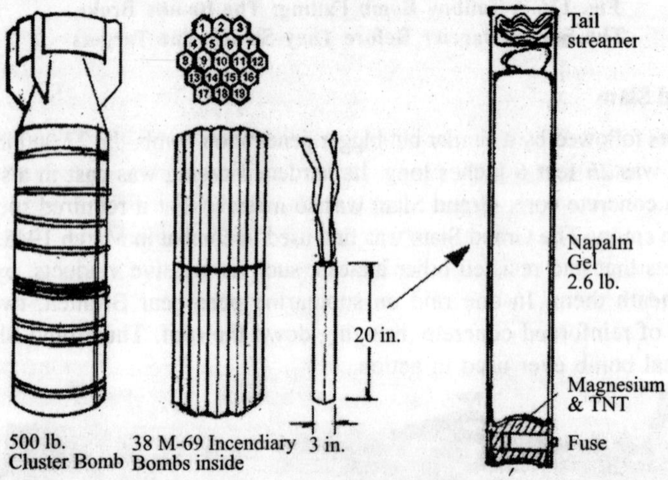

Fig.128 The Notorious M-69 Incendiary Bomb That Had Incinerated Many Cities Of Japan Even Before The Atom Bomb

Chapter 21

THE TECHNOLOGY OF BOMBS: AN OVERVIEW

(21.1) EXPLOSIVES USED IN THE BOMBS OF WORLD WAR II: THE JOURNEY FROM TNT TO COMPOSITION B TO TRITONAL

When World War II began in September 1939, the standard United States filler for high explosive bombs was trinitrotoluene (TNT). On 5 June 1940, the United States released 8,000 tons of TNT to the British under Lend Lease. This came out of a manufacturing surplus of only about 9,000 tons. Most of the War was fought by the United States with bombs standardized in 1941.

The earliest departure from TNT for the fillings was RDX, an explosive known for its great power and brisance but generally considered too sensitive. The British had developed a method of using beeswax to desensitize the RDX, and used this filler with terrible effect in the 2-ton **'blockbuster bombs'** dropped on Berlin in April 1941. During the following summer, Air Marshal Arthur 'Bomber' Harris had pressed for large-scale production of RDX in America. The United States Navy was also interested in the explosive because of its effectiveness underwater, especially in mixture with TNT and aluminum called torpex. But the United States Ordnance Department, while willing to start production for the British and the Navy, held back until May 1943 on the use of RDX in its AAF (Army Air Force) bombs, and then adopted only a less sensitive mixture with TNT, namely, Composition B. This first significant change in bomb-loading came about as a result of AAF insistence that the large fragmentation bombs developed in 1943 would need the greater power of Composition B to burst their thick walls with the greatest effect. Henceforth, Composition B was used in only about 40 percent of the general purpose bombs. The reasons were twofold. First, the short supply caused by the competition between RDX, and high octane gasoline, and synthetic rubber for production facilities and, second, the tendency of Composition B to undergo high-order detonation without fuze action under the shock of impact. RDX was more prone than TNT to prematurely deflagrate (decompose rapidly without detonating) when employed in delayed-action bombs dropped from high altitudes.

In the war-long argument over the relative merits of blast and fragmentation, the pendulum now began to swing back toward blast partly because by 1944 the AAF would have aircraft capable of delivering larger and heavier loads. After 1943, the new aluminized fillings were of far greater interest than either Composition B or ednatol. Until WW II, the use of aluminum in explosives had not been extensive. Tests in 1941 in England had failed to indicate any significant difference between aluminized explosives and amatol or Composition B. In 1943, the discovery that German bombs containing aluminum were extremely effective, spurred research. This led to the development of minol, a mixture of aluminum with amatol, and tritonal, a mixture of aluminum with TNT. For their 2-ton blockbuster bomb, the British favored Minol 2, a mixture of 20% aluminum, 40% TNT, and 40% ammonium nitrate; and they requested that Minol 2 be used in their blockbuster bombs being loaded in the US. The British, by using new methods of blast measurement and interpretation, had learned that Minol 2 produced an area of demolition approximately 30 percent greater than the area obtained with TNT filler. US Army Ordnance technicians had independently arrived at a similarly high opinion of the blast effect contributed by aluminum. They arrived at this conclusion by comparing the performance of one and two-ton blockbuster bombs loaded with minol, TNT, ednatol, and the RDX mixtures. In comparing minol and tritonal, they preferred tritonal, because it contained no ammonium nitrate. When even the slightest degree of moisture was present in the air, aluminum acted on ammonium nitrate and produced 'spewing' (the evolution of hydrogen gas) and explosions. Tritonal was much safer, and the British were won over to it.

The AAF adopted tritonal for large, light-case bombs for jungle warfare, and for all general purpose bombs. Other bombs were filled with explosives suitable to their particular purposes. For fragmentation bombs, Composition B continued to be the preferred filling because it had more brisance than tritonal. The US one-ton semi-armor-piercing bombs developed in early 1944 were loaded with picratol, a mixture of TNT and ammonium picrate, or with Explosive D. Explosive D was the least sensitive to shock and friction, which made it the best explosive to mix with TNT in a bomb that had to withstand severe shock and stress before detonating.

(21.2) MODERN DUMB BOMBS

(21.2.1) Types Of Dumb Bombs

A bomb is nothing but explosive filler enclosed in a casing. Bombs are generally classified according to the ratio of the weight of the explosive material to the total weight. The principal classes are general-purpose (GP), fragmentation, penetration and cluster bombs. Then we have fire bombs like Napalm, Fuel-Air Explosives and other incendiaries also. We will look at these types separately in greater detail. Smart bombs will also be discussed separately at length.

Calling an unguided bomb a dumb bomb is unfair. It is like calling a blonde a dumb blonde. Simply because someone is not smart, you should not call him or her dumb. They have rendered yeoman's service in the past and will continue to do so in the future also. Dumb bombs are cheap and when used by sophisticated strike aircraft, highly effective too. They also come in a bewildering variety of forms, optimized for different missions, and can be used as a basis for smart munitions.

(a) General Purpose (GP) Bombs

Approximately 50 per cent of the General Purpose (GP) bomb's weight is explosive materials. In a 2000 lbs bomb, the explosive filling is typically 965 lbs. The explosives most commonly used are Tritonal, Minol 2 or H6. These bombs usually weigh between 500 and 2,000 pounds and produce a combination of blast and fragmentation effects. The approximately one-half-inch-thick casing creates a fragmentation effect at the moment of detonation, and the 50 per cent explosive filler causes considerable damage from blast effect. The most common GP bombs in the USA are the MK-80 series weapons. General-purpose bombs were the type of ordnance most frequently employed in the Gulf War. GP bombs served as the basic building blocks for many of the other munitions used during the Gulf War.

(b) Fragmentation Bombs

Only ten to twenty per cent of a fragmentation bomb's weight is explosive material; the remainder includes specially scored cases that break into predictably sized pieces. The fragments, which travel at high velocities, are the primary cause of damage.

(c) Penetration Bombs

Penetration bombs have between twenty-five and thirty per cent explosive filler. In a 2000 lbs penetrator you would find about 535 lbs of explosive. The casings are designed to penetrate hardened targets such as bunkers before the explosives detonate. Penetration is achieved by either kinetic energy of the entire projectile or the effects of a shaped-charge.

(d) Cluster Bombs

Cluster bombs are primarily fragmentation weapons. The bomblets of a cluster bomb are so small that they cannot produce any appreciable blast effect—hence they have to depend on fragmentation. Cluster bombs, like GP bombs, can feature mix-and-match components (sub-munitions, fuzes, etc.) to produce the desired effect.

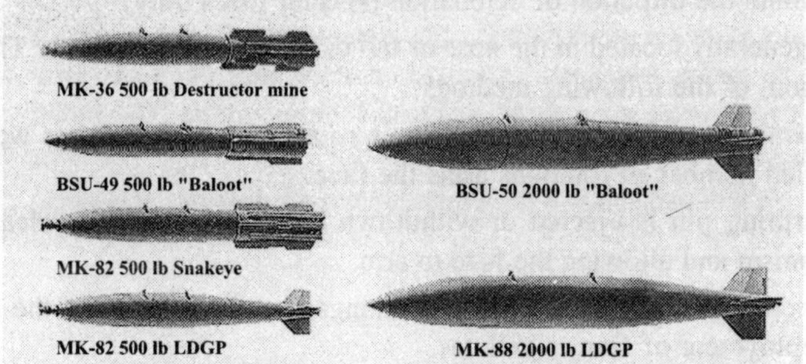

Fig.129 A Comparison Of The Relative Shapes And Sizes Of Various Types Of American Dumb Bombs

(21.3) THE COMPONENTS AND FUNCTIONING OF MODERN BOMBS

Free-fall or gravity bombs have three sections. The bomb body is the casing containing the explosive material. The fuze section can be located in the nose and/or the rear of the bomb and determines the timing of the explosion. The tail section, or the fins, determine how the bomb flies through the air. Desired weapons effects are achieved by selecting a particular combination of bomb body, fuzing, and tail section.

(21.3.1) Bomb Bodies

Bomb bodies vary in size, weight, and thickness of casing. GP bombs have a thinner case and more explosive filler than penetrating bombs, whereas cluster bombs generally come in dispensers that open to release bomblets at pre-determined altitudes. The bomb body casing (except for cluster munitions) houses the explosive filler. Upon detonation, the high-explosive filler ruptures the casing to produce the desired weapons effect.

(21.2.2) Fuzes

Fuzes are extremely critical components. A fuze initiates bomb detonation at a pre-determined time and under the desired circumstances. In the context of weapon system hardware, a fuze and warhead are unique in that they are expected to remain functionally quiescent until a bona fide target is encountered, and then to function as intended in a fraction of a millisecond. Guidance systems may recover from transient malfunctions; target-tracking radars may experience numerous false alarms without significantly compromising their utility; and missile airframes may flex and recover, but the fuze-warhead process is singular and irreversible. The quality required of fuze designs is usually specified by two values: functional reliability, which ranges typically from 0.95 to 0.99 for complex missile fuzes, and to 0.999 for projectile and bomb contact fuzes; and safety reliability, for which a failure rate not greater than 1 in 106 must be proved prior to release of items for service usage.

The weapon fuze system has five basic functions that it can perform:

1. Keep the weapon safe
2. Arm the weapon
3. Recognize or detect the target
4. Initiate the detonation of the warhead
5. Determine the direction of detonation (special fuzes only)

Fuzes are generally located in the nose or tail of the munition, or both. They are armed by one, or a combination, of the following methods:

- The arming vane, a small propeller, is rotated by airflow after weapon release. A specified number of rotations arms the fuze.
- The arming pin is ejected or withdrawn by a spring action releasing the arming mechanism and allowing the fuze to arm.
- The inertia fuze is armed by abrupt changes in the velocity of the bomb caused by the deployment of fins or ballutes.
- The electronic fuze is armed by a time-delay circuit powered by a thermal battery activated by extraction of the arming lanyard upon bomb release.

(21.3.3) Various Types Of Fuzes

(a) Impact Fuze: An **impact fuze** is designed to function on or after impact. Detonation upon impact is selected for targets such as supply dumps when the main destructive energy desired is blast. This sensing device detects the first impact of the weapon with an object. Its output initiates the detonator. Typical sensor mechanisms include displacement of an inertial mass, stressing of a piezoelectric crystal, short-circuiting of a coaxial transmission line, and interruption of an electrical circuit. Point detonating fuzes are of this type. A delay mechanism can also be incorporated that provides a short detonation delay (in microseconds) after weapon contact and allows for weapon penetration of the target to maximize damage effect.

There are also **'all ways fuzes'** that can detonate on impact at any angle. All ways fuzes are generally used on napalm bombs and other un-finned bombs that tumble when dropped. Impact fuzes may also have a short time delay, particularly in penetrating bombs. This allows the bomb to penetrate into the target before detonating. A variation on the simple impact fuze is the **"daisy cutter"** fuze, which resembles a long rod mounted on the nose of an HE or fragmentation bomb. A daisy cutter fuze triggers the bomb a little above the ground before it makes the impact and buries itself in the ground, producing greater destructive effect over the surrounding terrain. The famous BLU-82 Daisy Cutter bomb has a 38-inch fuze probe or extender mounted in its nose.

(b) Delayed Detonation Fuze: For attacking a building, it is desirable that the bomb explodes only after penetrating some floors of the building. That is to say, you require a delayed detonation fuze. A **tail-mounted fuze** is often used on penetrating munitions, since a nose-mounted fuze would be destroyed by the impact.

(c) Time Fuze: In time fuzes, after a predetermined elapsed time, a timing device sends a fire signal to the detonator. The 'sensing' or 'detecting' of the target is pre-determined by the user in that a preset elapsed time is based on calculations of when the target is expected to be within the damage volume. Gun projectiles use this type of fuzing. Bombs can have variable time fuzes that can be set to go off from minutes to hours after striking the ground. Time fuzes may have delays of hours to days. In a time fuze, the delay is normally initiated at bomb release rather than on impact. The timing element is a mechanical or electrical device. Time fuzes are employed to interfere with the work of damage-control teams, or to harass enemy forces trying to pass through a location after an air strike.

(d) Proximity Fuzes: Suppose you have to fire against an aircraft or an airborne missile. Scoring a direct hit on a relatively small moving target like them by manual aiming is hard even if you ignore the effect of wind and other perturbations. Things could improve a bit if you could have shells or missile that would automatically explode after traversing a certain distance or after reaching a certain height. But to set a time- or height-triggered fuze in the shell or the missile one must know the height of the moving target or predict the height of the target at the time you will be able to get a missile or shell in its neighborhood, which is even more difficult.

In tight situations like a surface ship coming under enemy air attack, the ship had very limited options. Generally they resorted to barrage fire. As soon as it could be reasonably expected that their fire might succeed in hitting the enemy planes, a cordon of time-fuzed shells detonating at different altitudes regardless of presence of enemy planes, or their non-existence there, used to be sent up by heavy anti-aircraft fire. This method could hardly be considered very effective, especially since without radar, neither height nor distance could be measured or estimated with great accuracy. In any case, it wasted a great deal of ammunition too.

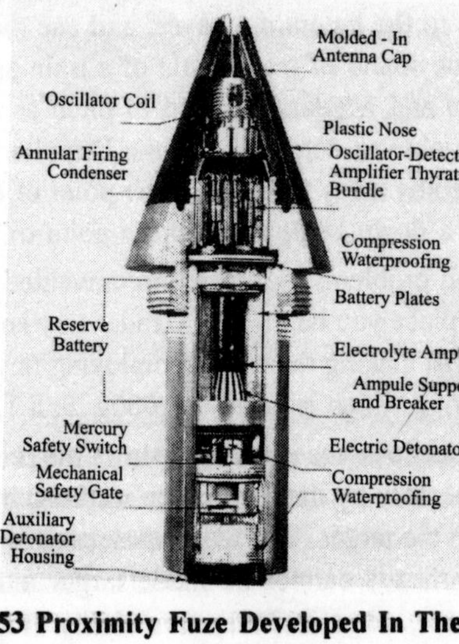

Fig. 130 The MK53 Proximity Fuze Developed In The Second World War

The proximity fuze solves all these problems. This fuze is designed to detonate the warhead automatically when close enough to the target to destroy it. This device does not require target contact, but will sense the physical presence of the target at some distance. It sends a fire signal to the detonator when the target is predicted to be within the damage volume of the warhead. Proximity fuzes are called **target-detecting devices** or **TDDs**. With a proximity fuze, all one has to worry about is getting a shell or missile on a trajectory that, at some time, will pass close by the target. Early proximity fuzes were developed during the Second World War itself. They are also known as **VT (for 'variable time') fuze.**

As to its effectiveness let us place the following on record. VT-fuzed shells made up about two fifths to half of the 127 mm ammunition fired during 1943-1945. For each plane it shot down, it required about 500 rounds, fewer for kamikaze planes, more for conventional attacks, the reason being the simpler trajectory of a kamikaze. In comparison, the non-VT fuzed 127 mm guns had to spend nearly 2,000 rounds per plane, slightly less for regular attacks (probably related to the swift rate of change of the kamikaze's altitude, which the VT fuze needn't care about).

Modern Proximity-fuzed bombs are also used against targets such as troops in trenches, radars, trucks, and other vehicles. A proximity fuze may contain a miniature Doppler radar set that senses height above the ground. When the explosion occurs above the ground, most of the destructive effect is caused by the bomb casing fragments. Hence the proximity fuze is useful for cluster bomb units (CBU) and fuel-air-explosive (FAE) weapons. For use against land targets, the proximity fuze is tuned to detonate the explosive at an optimum height above the ground. Other types of proximity fuzes are based on induced magnetic field, pressure measurement, acoustical impulse, or infrared signal.

We can have three types of proximity fuzes: **active, semi-active** or **passive**.

A **radio proximity fuze** can work in two ways. It can utilize the basic principle of radar, in which the elapsed time between a transmitted and received pulse is a function of range between target and weapon. A range-gate circuit set for a given distance will pass the signal to initiate the warhead when the elapsed time reduces to a predetermined value. Another way is to use the Doppler Effect. In this you send out radio waves that are reflected by the target and compare the frequency of the outgoing waves to the incoming waves, and use that to determine its proximity to a target. You may recall that the sound of the whistle of a train appears shriller or of a higher pitch as the train approaches you and appears lowered in pitch as the train recedes away from you. This means that you will be able to determine by the Doppler shift in frequency whether a target is coming towards you or going away from you. The point of detonation could occur when the Doppler shifts from an up to a down Doppler or at the point of closest approach.

Just imagine the technical problems that were circumvented in the WWII to make radar-based proximity fuze. In the first place you had to make radar tiny enough to be carried in a shell. The main problem in the design was making the radar (employing tiny glass vacuum tubes) rugged enough to survive the shock of firing which is nearly 20,000g in a 127 mm L/38 gun.

In air-target weapon applications, the main function of the proximity fuze is to compensate for terminal errors in weapon trajectory by detonating the warhead at a point calculated to inflict the maximum possible damage on the target. You will appreciate the importance of the proximity fuze better if you recall that warheads cannot be made larger and heavier without seriously compromising missile speed, altitude, range, and maneuverability. If the fragments created by your warhead can damage sensitive parts of the target such as the cockpit, the engine compressor stages, and the flight control system, your purpose will be served.

(e) Magnetic Fuzes: Magnetic fuzes look for magnetic disturbances produced by the target and are used in anti-tank mines and naval mines. **Magnetic anomaly detection** (MAD), measure changes in the earth's magnetic field or the presence of a source of magnetic flux. You can catalog magnetic signatures and make the fuze target specific. Another type employed with mines is an inductor mechanism, frequently called a search cell. This device employs the principle that a magnetic field induces current flow in a conductor as the field changes relative to the conductor. The small voltage induced in the search cell is amplified and then caused to energize the firing circuit, which in turn initiates the detonator. The extreme simplicity of this device makes it highly reliable and difficult to counter.

Magnetostatic fuzing is also used for subsurface targets. The magnetic field disturbance fuze for subsurface targets is also actuated by a change in the surrounding magnetic field. Any change in the magnitude of the magnetic field activates the fuze. A magnetic-type fuze provides the possibility of damaging a target without a direct hit. This is important, as the damage potential is greater when the explosion of the warhead takes place several feet below the hull, rather than at the side or near the surface of the water. The most advanced methods of fuze initiation operated by a ship's magnetic field employ an electromagnetic detecting system. Such a system operates on what can be called the 'generator principle'. An electric generator consists essentially of a coil of wire rotated in a magnetic field to produce a voltage. Similarly a small voltage is developed across a coil of wire (the search coil) when it comes in contact with a moving or changing magnetic field. However, a complex problem can occur, due to the fact that the movement of the torpedo itself through the water creates its own change in the field gradient and may initiate the fuze at an earlier time than intended. This led to the development of the gradiometer, a device attached to the torpedo, which has two search coils approximately one foot apart and connected in opposing series. As the now magnetically balanced torpedo moves in earth's magnetic field, equal and opposite voltages are induced in the coils, and no net voltage results. In the vicinity of a steel vessel, the situation is different. One of the two coils is slightly closer to the vessel than the other, and a slightly different voltage will therefore be induced in it. This difference is small, but when properly amplified, it causes the detonator to explode the warhead.

(f) Acoustic Fuzes: Acoustic fuzes look for acoustic disturbances produced by machinery noises, propellers, movement of hulls, or vibrations in hulls. You can catalog acoustic signatures also like magnetic signatures. Acoustic fuze mechanisms are used in torpedoes as well as mines. The basic acoustic mine employs a hydrophone as a detector to sense the presence of a target. A typical hydrophone functions much the same as the human ear. A diaphragm (corresponding to an eardrum) vibrates from the impact of underwater sound waves. These vibrations are transmitted through an oil medium to a piezoelectric crystal, which produces corresponding voltage. The voltage is amplified to initiate the firing mechanism. The selectivity of the firing mechanism is so critical that only the pulses of the required characteristics are sent to the detonator. Selectivity is necessary because of the varied sounds that are received by the mine. To distinguish among these many sounds, acoustic firing mechanisms must possess a very selective type of hearing. For example, a 2,000-ton tramp steamer may have one large propeller turning over rather slowly, whereas a 2,000-ton destroyer may have two propellers rotating over much faster. An acoustic firing mechanism can distinguish between the steamer and the destroyer, and fire on the selected target. When the firing mechanism detects a sound that has the required characteristics (including intensity and rate of change of intensity), the mechanism initiates the firing circuitry and fires the detonator.

Acoustic fuze mechanisms are used in torpedoes as well as mines. There are two operating modes of acoustic torpedoes, the active and passive types. The passive type has a homing device that guides the torpedo in the direction of the strongest target noise. The active type employs a sonar set that emits a series of sonic pings that are reflected back to a receiver in the torpedo. The principle is similar to that of a radar set. As the torpedo approaches the target, less time is required for a signal to travel to the target and return. At a predetermined critical distance, initiation of the firing circuit begins.

(g) Seismic Fuzes: The seismic fuze sensor is essentially an acoustic fuze, but receives its threshold signal in a lower bandwidth through weapon-case vibration. These sensors can be made extremely sensitive and may provide for both land-based and in-water application. They offer advantages over the pure acoustic mine fuze in that they may be set over a wider range of selectivity and may incorporate some countermeasure features.

(h) Pressure Fuzes: Pressure or hydrostatic pressure fuzes are used in a naval context. Oceans swells and surface waves produce pressure variations of considerable magnitude. Moving ships displace water at a finite rate. This continuous water flow is measurable at considerable distances from the ship as pressure variations that normally exist in the water. Various pressure-measuring mechanisms can be used in fuzes to detect such variations. The pressure differential becomes more pronounced when the ship is moving through confined waters, but is still appreciable in the open sea, even at a considerable depth. This pressure variation, called the 'pressure signature' of a ship, is a function of ship speed and displacement and the water depth. Therefore, to avoid premature firing because of wave action, pressure-firing mechanisms are designed so as to remain unaffected by rapid pressure fluctuations.

Pressure sensors are commonly associated with bottom mines and are extremely difficult to counter through normal influence-mine countermeasure techniques. Pressure-firing mechanisms are seldom used alone, but are generally combined with other influence firing devices. They are also employed in depth bombs used for underwater demolition work.

A bomb may be fitted with **multiple fuzes**. For example, a bomb fuzed to act as a mine may also have a time fuze so that it will eventually self-destruct, either to harass an enemy or to prevent it from being a hazard to advancing friendly forces. Systems involving a combination of influences are available in most mine firing devices. Combinations of magnetic, pressure, and acoustic/seismic systems are used to compensate for the disadvantages of one system with the advantages of another. Mine counter-measure effectiveness can be greatly reduced through use of combination fuzing.

(i) Smart Fuzes: Smart fuzes possess advanced capabilities. Smart fuzes with digital processors and sensors are available that may do things like counting the number of ceilings penetrated in an underground installation and then triggering at a given floor. The so-called **Hard Target Smart Fuze (HTSF)** is an example of this type of fuze. The HTSF is an active decision-making accelerometer-based fuzing system capable of counting layers and voids (floors), as well as calculating distance traveled. When the weapon reaches the pre-determined floor it tells the bomb to explode. It operates in one of three modes: hard-layer detection, void detection, and path-length integration. It also has an adjustable backup time delay that is set in 1-ms increments with a maximum delay of 250 ms. The HTSF uses a void sensing technique to count layers within a structure to initiate fuze function, a depth of burial mode that causes the fuze to function a preset distance after it senses a hard layer, and an integral time delay backup.

Fig. 131 Various Types Of Fuzes

(21.3.4) Safing And Arming (S&A) Of Fuzes

The S&A device is a fuze component that isolates the detonator from the warhead booster charge until the weapon has been launched in the intended mode and has got to a safe distance from the launch vehicle. Then, at that point in the trajectory, the S&A device removes a barrier from the explosive train, allowing the detonator thereafter to initiate the warhead. Some S&A devices function by measuring elapsed time from launch, others ascertain distance traveled from the launch point by sensing and doubly integrating the acceleration experienced by the weapon. Some devices sense and integrate air speed or projectile rotation (centrifugal force), while still others may sense several stimuli characteristic of the weapon trajectory, such as rocket motor gas pressure and acceleration. Most missiles incorporate an S&A device that utilizes acceleration-sensing. This device senses the acceleration from the rocket motor. To maximize the safety reliability of a fuze, the S&A device must ensure that the forces it senses will be unique to the weapon when deployed, and cannot be intentionally or accidentally duplicated during ground handling or pre-launch operations. The rocket motor boost may only induce a 10g force, while the weapon dropped from a sling may easily experience 500g shock forces. Therefore, another feature must be used to prevent arming should the weapon be dropped. This salient feature is that acceleration-sensing systems must experience acceleration for a protracted period of time, such as for at least one second or at least 10 seconds. S&A devices have to be made of extremely high quality components.

So far we have emphasized keeping the weapon safe before it reaches the target. It may be necessary to include an additional S&A device in the firing path to safe the weapon if the intended target is missed. An example would be an anti-aircraft projectile or SAM that could fall on friendly territory if an air-borne target were missed. To preclude this possibility, a surface-to-air missile could have an additional timer S&A device that would normally arm. This would allow firing path continuity until the expected elapsed time from launch to target intercept is exceeded, at which time the device will safe the missile. An alternative to using a timer S&A device would be to place an additional timer firing device in the fuze system in addition to the normal target-sensing device. If the target is missed, then the alternate timer device would send the fire signal for detonation. The timer setting would be based on a time greater than normal intercept, but would detonate while the weapon was still safely airborne. This alternative design method of increasing the safety reliability is found in most antiaircraft projectiles.

(21.3.5) Bomb Stabilizing Devices

Bombs are stabilized in flight by either **fin** or **parachute assemblies**. These assemblies attach to the rear section of the bomb and keep the bomb nose-down during its descent. These assemblies can separate from the bomb after the bomb hits the ground. Two common types of fin assemblies used are the **conical- and box-fin assemblies**. They help stabilize the bomb in flight, allowing the bomb to exhibit the best effects of low drag and stabilization after release. A conical or box-fin mated with a GP bomb results in a low-drag general-purpose bomb. Most of the countries use bombs of this type. The USA also used them for a long time and most of the bombs dropped in Vietnam were of this type only.

Conical- and box-fin assemblies

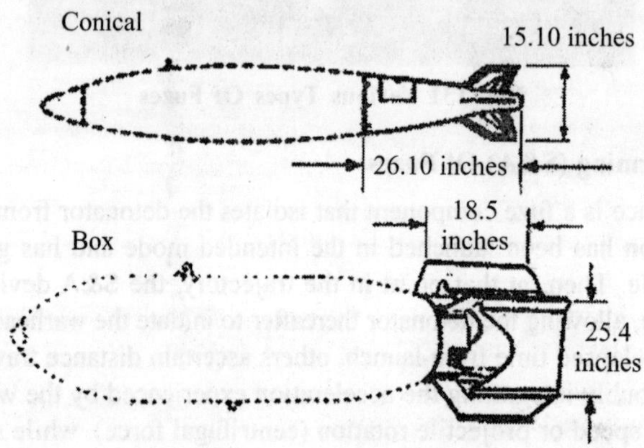

Fig. 132 The Simple Conical- And Box-Fin Stabilizing Assemblies On Bombs Resulting In The So-Called Low-Drag Bombs—Bombs That Fall Quickly

The US now prefers a high-drag **retarding-fin assembly** for most of its general-purpose bombs. In this device, with bombs designed to be dropped at low altitude, the tail section provides some sort of **'retarder'**, a drag-producing mechanism that ensures the bomb falls well behind the launch aircraft before impact and detonation and the aircraft gets safely away. A bomb fitted with a retarder is of course known as a **'retarded bomb'**. (Now this is getting very unfair. So far they were only dumb, now they are retarded too!) While little parachutes were used as retarders in the past, today the mechanism consists either of pop-out fins or, increasingly, a **balloon-parachute hybrid** called a **"ballute"** that resembles a conical or pyramid-shaped inflatable tent. Pop-out fins tend to break at high speeds, possibly pitching the bomb up into the launch aircraft. Ballutes have no such speed restriction. Two types of high-drag retarders are used. The first is the **air- inflatable retarder tail assembly** containing a ballute (combination balloon and parachute) device that deploys shortly after bomb release. There were two types of ballutes, the BSU-49 mated to a 500-pound MK-82 bomb, and the BSU-50 mated to a 2,000-pound MK-84 bomb. The second type of retarding fin is the **Snakeye**, which has four metal vanes that open into the wind stream to slow the bomb after release. These high-drag retarder tail assemblies are used to slow the bomb quickly after a high-speed, low-level release, thereby reducing the chance of an aircraft being damaged by its own bomb fragments. Snakeye fins were used by Navy aircraft to deliver mines into the waters around Iraqi naval bases.

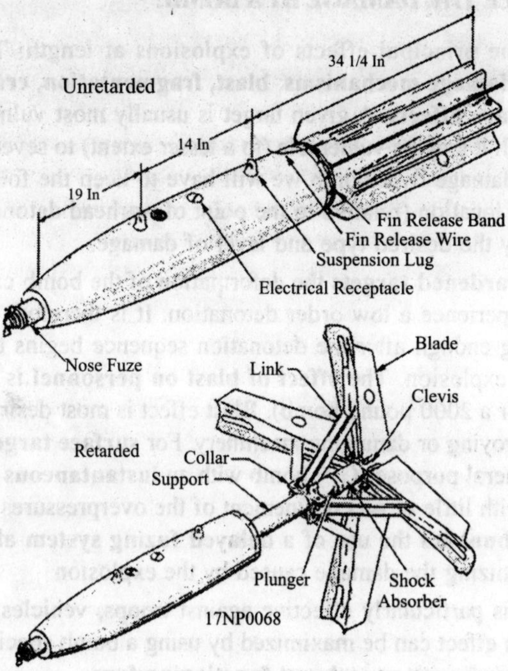

Fig.133 The Mechanics Of Working Of The Retarding Fin Assembly For High-Drag Bombs

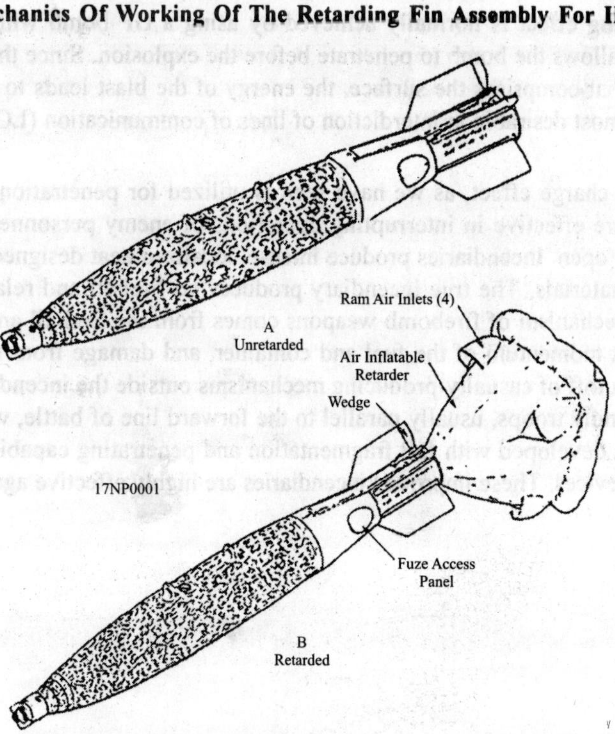

Fig.134 The 'Ballute', A Balloon-Parachute Hybrid Is An Air-Inflatable Retarder, A Clever Idea

(21.4) HOW TO OPTIMIZE THE DAMAGE BY A BOMB?

We have seen the principal effects of explosions at length. There are five **general categories of munitions damage mechanisms: blast, fragmentation, cratering, shaped charge penetration**, and **incendiary effects.** A given target is usually most vulnerable to one particular damage mechanism, though it may be vulnerable (to a lesser extent) to several damage mechanisms. In order to optimize the damage by a bomb we will have to keep the following factors in mind: target construction, target location (relative to the point of warhead detonation), warhead damage effects pattern, and finally the desired type and level of damage.

When **striking hardened targets** the deformation of the bomb casing or fuze may cause the warhead to dud or experience a low order detonation. It is therefore essential that the bomb casing remains intact long enough after the detonation sequence begins to contain the hot gases and achieve a high order explosion. The **effect of blast on personnel** is confined to a relatively short distance (110 feet for a 2000 pound bomb). Blast effect is most desirable for attacking walls, collapsing roofs, and destroying or damaging machinery. For **surface targets the effect of blast** is maximized by using a general purpose (GP) bomb with an **instantaneous fuzing system** that will produce a surface burst with little or no confinement of the overpressures generated by excessive burial. For **buildings or bunkers** the use of a **delayed fuzing system** allows the blast to occur within the structure maximizing the damage caused by the explosion.

Fragmentation is particularly effective against troops, vehicles, aircraft and other soft targets. The fragmentation effect can be maximized by using a bomb specifically designed for this effect, or by using a GP bomb with an **airburst functioning fuze.**

The **cratering** effect is normally achieved by using a GP bomb with a delayed fuzing system. This system allows the bomb to penetrate before the explosion. Since the explosion occurs within the material that comprises the surface, the energy of the blast leads to the formation of a crater. This effect is most desirable in interdiction of lines of communication (LOC) and area denial operations.

The shaped charge effect, as we have seen is utilized for penetration of armor. **Fire** or **incendiary effects** are effective in interrupting operations of enemy personnel and in damaging supplies stored in the open. Incendiaries produce intense, localized heat designed to ignite adjacent combustible target materials. The true incendiary produces no fireball and relatively little flame. The basic damage mechanism of firebomb weapons comes from the fireball and burning residual fuel globules, impact momentum of the fuel and container, and damage from fires started by the weapon. The sharp cutoff of casualty-producing mechanisms outside the incendiary pattern allows delivery close to friendly troops, usually parallel to the forward line of battle, with minimum risk. Munitions have been developed with full fragmentation and penetrating capabilities coupled with reactive incendiary devices. These improved incendiaries are highly effective against fuel and other flammable targets.

□□□

Chapter 22

Some of the Great Bombs in Service Today

(22.1) THE MK-80 SERIES DUMB BOMBS

The MK-80 series of the USA was developed in the 1950s in response to the need for bombs producing low aerodynamic drag. All MK-80 series bombs are similar in construction. The bombs are still going strong and used liberally from all sorts of aircraft against all sorts of targets. They are much cheaper in comparison to other weapon systems and hence used the most. Such LDGP (Low Drag General Purpose) bombs are designed to be aerodynamically streamlined and hence their familiar shape. Their cases are relatively light and 38.5 percent of their complete weight is explosive. They may use nose, tail or radar-proximity air-burst fuzes and conical or retarded tail fins. The 'ballute' air bag, if deployed from the tail provides a high speed, low altitude delivery capability by quickly slowing the bomb and allowing the aircraft to escape the blast pattern.

The MK-82 is a 500 lbs bomb. It is six and a half feet long and 10.75 inches in diameter. The MK-83 is a 1000 lbs bomb. It is ten feet long and 14 inches in diameter.

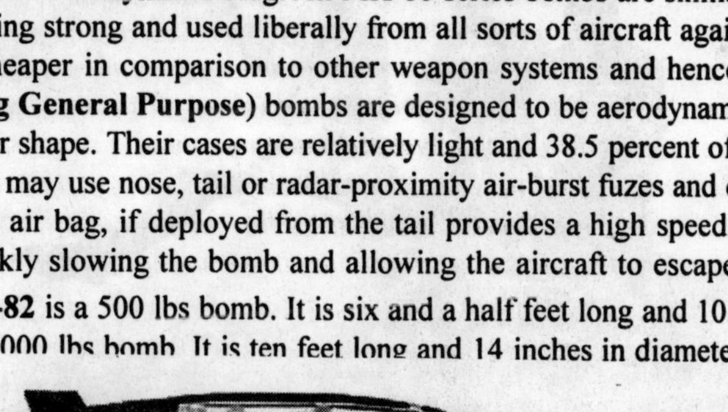

Fig.135 (From Top To Bottom) MK-81, MK-82 And MK-83 Bombs Showing Their Internal Structure

(22.2) THE DAISY CUTTER: THE BIGGEST CONVENTIONAL BOMB USED SINCE THE WWII

The **BLU-82B/C-130** weapon system, nicknamed **Commando Vault** in Vietnam and **Daisy Cutter** in Afghanistan, is a high altitude delivery 15,000 lbs conventional bomb, delivered from a C-130 cargo plane since it is far too heavy and big for the bomb racks on any bomber or attack aircraft. This monstrosity is nearly 12 feet long and four and a half feet in diameter. Originally designed to create an instant clearing in the jungle, it has been used in Afghanistan as an anti-personnel weapon and as an intimidation weapon because of its very large lethal radius (variously reported as 300-900 feet) combined with a tremendous flash and terrible explosion visible and audible at very long distances. The Americans used it to terrorize poor Afghans and the Iraqis. We are emphasizing this terror aspect because no jungles were to be cleared in Afghanistan or Iraq and the utility of the bomb in clearing minefields is highly doubtful. Thus the only possible use could be to terrorize. Next to the MOAB, it is the largest conventional bomb in existence but is still less than one thousandth the power of the Hiroshima nuclear bomb.

Frequent press reports to the contrary, the Daisy Cutter is not a **fuel-air explosive (FAE)**. It is a conventional explosive incorporating both agent and oxidizer. In contrast, an FAE consists only of agent and a dispersing mechanism, and takes its oxidizer from the oxygen in the air. FAEs generally run between 500 and 2000 pounds; it would be difficult to make an FAE the size of Daisy Cutter because the correct uniform mixture of agent with ambient air would be difficult to maintain if the agent were so widely dispersed.

Fig.136 Daisy Cutter, At 15,000 Lbs The Largest Conventional Bomb That Has Been Used Since The WWII

The warhead contains 12,600 pounds of low-cost GSX slurry (ammonium nitrate, aluminum powder, and polystyrene) and is detonated just above ground level by a 38-inch fuze extender, optimized for destruction at ground level without digging a crater. This is an important point because it was originally designed to clear helicopter landing zones and artillery emplacements in Vietnam and you could not afford to have craters. The idea behind an 'air burst' weapon, as opposed to a weapon that explodes on impact with the ground, is to increase its destructive range. A bomb that penetrates the ground and then bursts tends to send all of its energy either down into the ground or straight up into the air. An air burst weapon sends a great deal of its energy out to the side. The weapon produces an overpressure of 1000 psi near ground zero, tapering off as distance increases. This system depends upon the accurate positioning of the aircraft by either a fixed ground radar or onboard navigation equipment. The bomb is parachute launched (one parachute for cargo extraction and one for bomb stabilization) to delay its fall. The blast is so heavy that the minimum altitude from which it can be dropped is 6000 feet. From 6000 feet also it is made to make its descent in 27 seconds so that the plane may safely clear off.

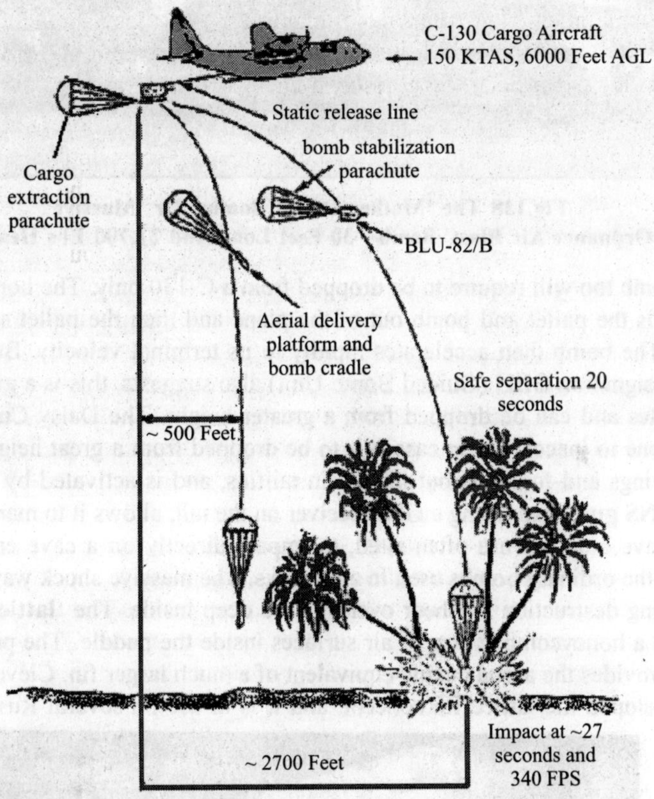

Fig. 137 Deployment Sequence Of The Daisy Cutter

(22.3) THE MOTHER OF ALL BOMBS (MOAB)

A bomb even bigger than the Daisy Cutter, is reported to have been constructed. Developed by **Albert L. Weimorts Jr.** for the US military, it is called the Mother of All Bombs or the **Massive Ordnance Air Blast Bomb** (MOAB). The acronym could also be a childish mockery of Saddam Hussain's reference to the Mother of All Battles. The official designation is **GBU-43/B.** The first test of the MOAB was conducted on March 11, 2003 at the Eglin Air Force Base and then another in November. It is 30 feet long with a diameter of 40.5 inches. The 21,700-pound (9.5 tons) bomb contains 18,700 pounds of high explosive.

Fig.138 The 'Mother Of All Bombs' Or 'Massive Ordnance Air Blast' Bomb—30 Feet Long And 21,700 Lbs Heavy!

This bomb too will require to be dropped from a C-130 only. The bomb rides on a pallet. A parachute pulls the pallet and bomb out of the plane and then the pallet separates so that the bomb can fall. The bomb then accelerates rapidly to its terminal velocity. But, unlike the Daisy Cutter, as the designation GBU (Guided Bomb Unit) also suggests, this is a guided bomb. It does not use parachutes and can be dropped from a greater height. The Daisy Cutter, because of the parachute, is prone to inaccuracy in case it is to be dropped from a great height. The MOAB has two long stub wings and four pop-out lattice-fin tailfins, and is activated by twin nose-mounted fuzes. Its **GPS-INS guidance**, using a GPS receiver on the tail, allows it to maneuver down narrow valleys where cave hideouts are often sited, to impact directly on a cave entrance—something which frustrates the ordinary bombs used in air attacks. The massive shock wave propagates down the tunnel, causing destruction by sheer overpressure deep inside. The **'lattice fins'**, are paddle-shaped fins with a honeycomb lattice of air surfaces inside the paddle. The paddle faces into the air stream and provides the aerodynamic equivalent of a much larger fin. Clever idea, isn't it? The Soviets had developed the lattice fin concept, and it is in use on several Russian missiles.

Fig. 139 The Guidance System Of The MOAB: The Lattice Fins (Folded) Can Also Be Seen

The explosive used in the MOAB is believed to be H6. There are other reports also which stated that the explosive is GSX like that used in the Daisy Cutter or even the old tritonal. It is believed that the bomb used in the first test had tritonal. The blast of the MOAB is strong enough to flatten an area as big as nine city blocks. Except for an all-out war, the utility of the MOAB, however, lies in its immense psychological impact.

It is believed that the McAlester Army Ammunition Plant has made 15 MOABs. The US Air Force Research Laboratory says that an even larger version weighing 30,000 lbs is under development.

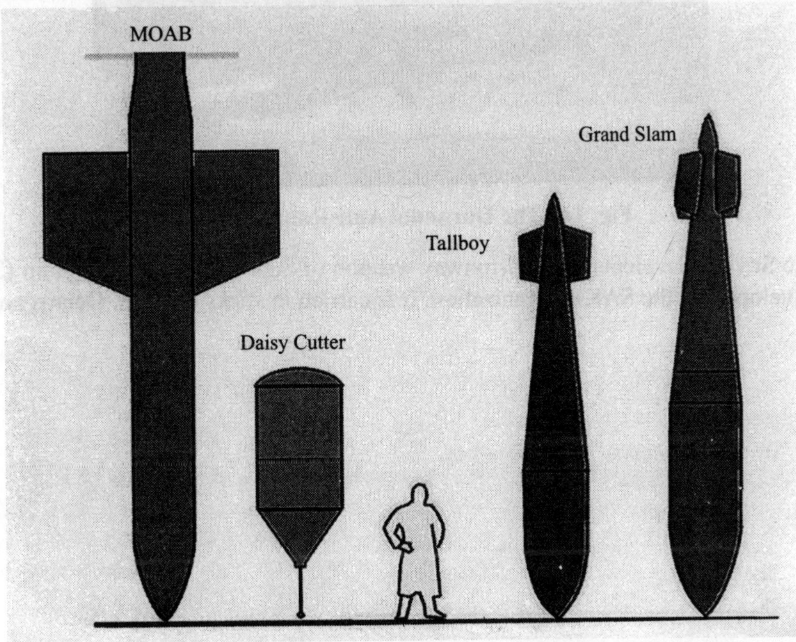

Fig.140 Relative Shapes And Sizes Of The Greatest Bombs In History, Namely The MOAB, The Daisy Cutter, The Tallboy And The Grand Slam With A Man For Comparison

(22.4) THE DURANDAL ANTI-RUNWAY BOMB

The Durandal anti-runway bomb was developed by the French company MATRA, designed solely for the purpose of destroying runways. The purpose of anti-runway bombs is to create such craters on the runway which would be difficult to repair. The Durandal is over eight feet long and weighs 450 lbs of which 330 lbs is explosive. Once the parachute-retarded low-level drop bomb attains a nose-down attitude, it fires a rocket booster that penetrates the runway surface, and a delayed explosion buckles a portion of the runway. It can penetrate up to 40 centimeters of concrete, creating a 200 square meter crater causing damage more difficult to repair than the crater created by a general-purpose bomb of similar weight. USAF F-111s can carry up to 12 of these weapons at a time, while French Mirage 2000s can handle a total of 8.

Fig. 141 The Durandal Anti-Runway Bomb

The Soviets developed an anti-runway weapon of their own, a 100 kilogram (220 pound) munition developed by the SAKR organization. It is carried in sticks of three. Details are, however, unclear.

❑❑❑

Chapter 23

Dig Deep and Hit Hard: The Bunker Busters

(23.1) COME CONCRETE OR SOLID EARTH, IT WOULD GO ALL THE WAY IN!

(23.1.1) Why Do You Need Penetrating Bombs?

What do you do when your targets are burrowed deep inside mountains or simply deep underground in an urban environment? Such hardened facilities are likely to house command centers, ammunition depots and research labs that are either of strategic importance or vital to waging war. Because they are underground as well as fortified, they are hard to find and extremely difficult to strike. Historically there is nothing new about hiding some of your vital centers deep in the bowels of the earth. In the days of the Cold War, the Strategic Air Command of the USA was located inside a mountain and it was believed that it could withstand even a nuclear strike. The Chinese have their highest military command even now inside a mountain only. We learnt about some of the giants like the Tallboy and the Grand Slam that were used in the World War II for this purpose. These bombs could penetrate deep into the earth or right through several feet of reinforced concrete before exploding. Much more sophisticated devices have been built since then. They are popularly known as bunker busters. The Americans had spread the rumor that Saddam Hussein had his command center in such bunkers in Iraq. Earlier for their invasion of Afghanistan also they had spread the rumors that Osama bin Laden was operating from underground bunkers in the caves and mountains of Torah Borah. Neither Saddam nor Osama was ever found in those mythical and exotic bunkers but it gave the Americans an excuse to use their bunker busters—it had great propaganda value, if nothing else.

(23.1.2) How To Make Bunker Busters And How They Work?

The physics of a modern bunker buster is rather simple. For the bomb body you require an extremely strong tube that is very narrow for its weight and is also extremely heavy. In other words, the bomb starts looking like a nail! The bomb is dropped from an airplane at a sufficient height so that it develops a great deal of speed as it falls. When the bomb hits the earth, it is like a massive nail shot from a nail gun and the bomb penetrates deep into the earth. The next question is how you make such a body? During the Gulf War, American scientists solved this problem starting from the drawing board and shipped them to on location in Iraq in 17 days flat! The most expedient and ready to hand material for the casing was to be found in surplus 8 inch artillery barrels. We know that artillery barrels are, in the first place, made of extremely strong hardened steel. Secondly, they are layered structures with the inner layer being under constant compressive stress and the outer layer under constant tensile stress. This enables them to withstand the tremendous pressure repeatedly developed in the chambers and also the shocks generated by the recoil-absorbing mechanism. So they took 16 feet sections of such artillery barrels for making the bomb body. The second problem is that you should have an explosive filling that would not get triggered by the impact of the fall. **Tritonal, AFX 708** and **PBXN-109** have been used. The third problem is that the electronic circuitry of the bomb should also withstand the shock. The fourth problem is that you must have a fuze that will trigger the bomb only when it has penetrated to the desired depth. Attached to the front of the barrel is a laser-guidance assembly. The target is laser-illuminated by a spotter either on the ground or in the bomber aircraft itself and the bomb homes in on to the illuminated spot. The guidance assembly steers the bomb with fins that are part of the assembly. Attached to the end of the barrel are stationary fins that provide stability during flight. In a test drop at the Tonopah Test Range in Utah, the earth penetration was claimed to be about 100 feet! Concrete penetration was claimed o be up to 20 feet!

To make bunker busters that can go even deeper, designers have three choices:

- They can make the weapon heavier. More weight gives the bomb more kinetic energy when it hits the target.
- They can make the weapon smaller in diameter. The smaller cross-sectional area means that the bomb has to move less material (earth or concrete) 'out of the way' as it penetrates.
- They can make the bomb hit faster to increase its kinetic energy. The only practical way to do this is to add some sort of large rocket engine that fires right before impact.

One way to make a bunker buster heavier while maintaining a narrow cross-sectional area is to use a metal that is heavier than steel and nearly as hard or more. One material that is both extremely strong and extremely dense is **depleted uranium (DU)**. Depleted uranium is a by-product of the nuclear power industry. Natural uranium from a mine contains two isotopes: U-235 and U-238. The U-235 is what is needed to produce nuclear power, so the uranium is refined to extract the U-235 and create 'enriched uranium'. The U-238 that is left over is known as 'depleted uranium'.

U-238 is a theoretically a radioactive metal that produces alpha and beta particles. In its solid form, it is not particularly dangerous because its half-life is 4.5 billion years, meaning that the radioactive decay is so slow that for all practical purposes it is stable! DU is the material of choice for penetrating weapons being 2.4 times heavier than steel and 4.9 times harder than iron—its hardness is only a little less than that of tungsten. On the Brinell scale, the hardness of tungsten is 2570 and that of DU is 2400—compare it with the hardness of iron which is just 490! Depleted uranium alloyed with a small amount of titanium is even harder. For these properties, depleted uranium is used to make the 'long rod' penetrator of the Armor Piercing Fin Stabilized Discarding Sabot anti-tank rounds and also in boats and airplanes as ballast. There is an additional benefit of using DU. Heated DU burns like magnesium! Since the process of penetration produces a great deal of heat, it means that once inside the target, the DU could burn adding to the lethality of the bomb.

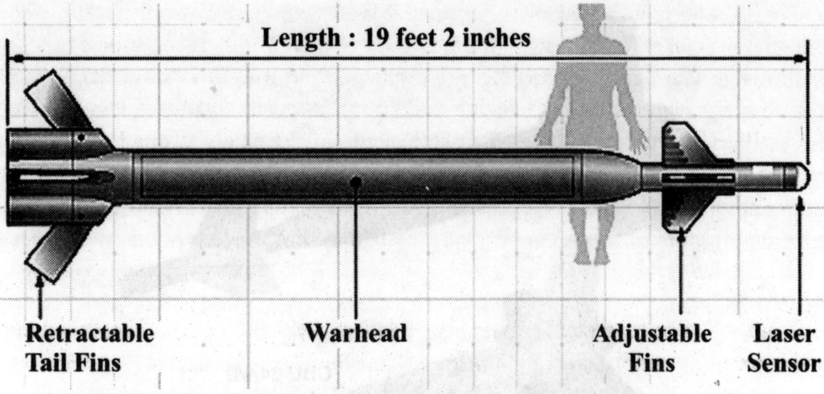

Fig.142 Structure Of The BLU-113/GBU-28 Bunker Buster Bomb And Its Size Compared To A Man

(23.2) BLU-113, BLU-109 AND BLU-116 GUIDED PENETRATION BOMBS

The finished basic bunker buster bomb, known as the **GBU-28** or the **BLU-113**, is 19 feet long, 14.5 inches in diameter and weighs 4,400 lbs. The **GBU-27/GBU-24 (BLU-109)** is nearly identical to the GBU-28, except that it weighs only 2,000 lbs. It is less expensive to manufacture, and a bomber can carry more of them on each mission. The GBU-28 had in the past been fitted with a delay fuze (FMU-143) so that it exploded after sufficient penetration rather than on impact. Then the Hard Target Smart Fuze (HTSF), designated the FMU-157/B, was used.

Fig. 143 A Spectacular Photograph Of The GBU-28 Penetrating Several Meters Of Concrete

One of the newest weapons in the US Air Force's arsenal of hard target penetrating bombs is the AFRL Munitions Directorate-developed **Advanced Unitary Penetrator (AUP)**, made of a special Nickel-Cobalt steel alloy and equipped with the Hard Target Smart Fuze. The bomb and its smart fuze were a big hit during Operation Allied Force. The AUP mimics, in appearance and dimensions, the already combat-proven BLU-109 warhead. The AUP is designated the **BLU-116 (GBU-24 C/B)** and boasts twice the penetration capability of the BLU-109.

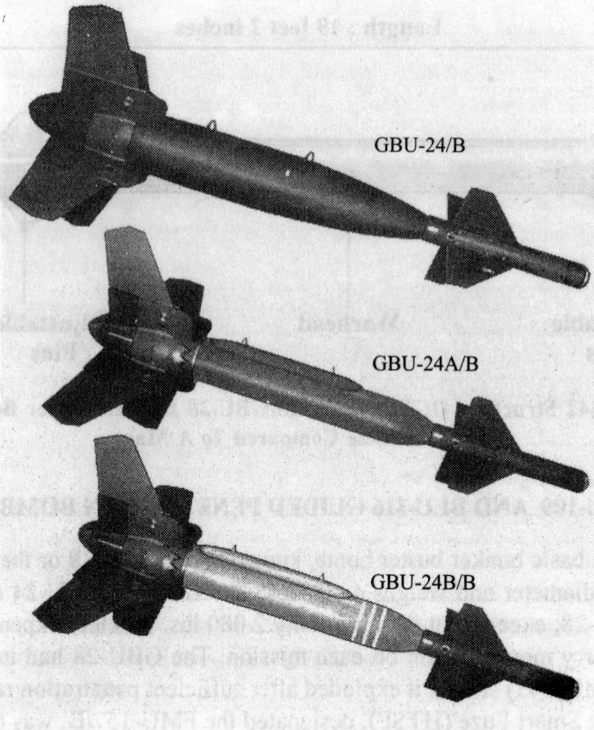

Fig.144 BLU-116/GBU-24 Advanced Unitary Penetrator, A Hard Target Smart Fuze Version Of The BLU-109/GBU-24

(23.3) NUCLEAR BUNKER BUSTERS: WHY THEY DO NOT MAKE SENSE!

(23.3.1) What Makes Some Americans Obsessed With Nuclear Bunker Busters?

Americans have produced some of the greatest weapon systems in the world; yet they are also the most frightened people on the earth, constantly in danger of losing their hegemony. Since the Cold War, the US defense community has become obsessed with the problem of bunkers and how to destroy them. The solution put forward has, of course, been expensive new weaponry. According to the US Defense Department estimates, there are perhaps 10,000 underground military installations in the world. Most of them, no doubt, are crude ammo dumps, but some are literally subterranean fortresses. What the Americans find very disturbing is that most of them are to be found among their potential adversaries! The most dazzling is the complex beneath Russia's Yamantau Mountain, begun under Brezhnev but completed only recently; tunneled sideways into the Urals southeast of Moscow, the complex sits below thousands of feet of quartz, insulated from an American ICBM attack! China, too, has an extensive system of underground bunkers and command shelters, including hundreds of fortified missile silos, complete with living quarters, which are scattered throughout its 3.7 million square miles. In North Korea the reliance on tunnels and bunkers is claimed to be even more obsessive. Underground weapons factories there are believed to employ as many as 20,000 workers; the nation has tunneled under the DMZ and into South Korea, has dug in upward of 10,000 pieces of artillery along the border, and even has built underground airstrips! Iran is said to possess an underground uranium-enrichment plant at Natanz that can be expanded or altered in secrecy. Libya's vast chemical-weapons plant at Tarhunah, though now apparently in disuse, is believed to be still standing, still underground, and could be quietly re-opened by a lapsed Qaddafi. Without satellite surveillance showing its construction and burial, the world might still be ignorant of Syria's As-Safirah chemical-weapons factory. These known installations and may be other unknown ones around the world, constitute the last class of targets that America's current arsenal cannot credibly threaten with swift annihilation. And that scares the shit out of them!

What do you do about such heavily protected installations? The most simplistic answer would be to nuke them. But 'nuking them' is almost a dirty word which the American hypocrites would not like to use for the consumption of the world. They want to use it and yet they do not want to be seen using a weapon of mass destruction! Quite a problem, isn't it? The hypocrites then found a contrived solution. The idea is to marry a small nuclear bomb with a penetrating bomb casing to create a weapon that could penetrate deep into the ground and then explode with nuclear force so that the shock wave may do the job of demolition of whatever had been seeking the protection of Mother Earth. Rocket motors are supposed to accelerate the bomb so that it helps penetration. They have been given the fancy name of **Low Yield Earth Penetrating Nuclear Weapons or Robust Nuclear Earth Penetrator (RNEP).** That way they would be able to use a nuclear weapon and yet proclaim to the world that the nuclear weapon was used on a military target and it was so small and exploded so deep that its use did not really amount to using weapons of mass destruction as there would neither be any radioactive fallout nor any collateral damage!

(23.3.2) Currently What Are The Options For Nuclear Bunker Busters?

The **B61-11**, available since 1997, is the current state of the art in the area of nuclear bunker busters. The warhead is derived from the famous B61 bomb which could be configured to have variable yields in the range of 0.3, 1.5, 5, 10, 45, 60, 60, 80, 170, and 340 kilotons. The hardened steel casing and the nose cone are, however, new creations. The B-61-11 is said to be nearly 12 feet long but only 13 inches in diameter.

(23.3.3) Why A Nuclear Bunker Buster Won't Work The Intended Way?

Even in the USA there are sober scientists who have not been carried away by the jingoistic propaganda. Their research has shown that a nuclear bunker buster is actually a stupid idea and the devices cannot work the intended way. In the first place, it is very much doubtful whether they can really penetrate as much as they are required to penetrate. Secondly, there is little reason to believe that there would not be radioactive fallout. The nuclear bunker buster could work alright but only when the explosion is so big that it would amount to using a conventional nuclear weapon in the conventional way with all its destructive might unleashed in the most ruthless manner!

From the theory of 'long-rod penetration', the maximum possible depth D of penetration even for a penetrator which is much stronger than the target, is proportional to the length and density of the penetrator and inversely proportional to the density of the target. The exact relationship is:

$$D/L = (d_p / d_t) \ln(Y_p / Y_t)$$

Where L is the length of the penetrator, 'd' refers to density, and Y refers to the strength of a material to plastic yielding; the subscripts p and t stand for the penetrator and target.

Robert Nelson of Princeton University has demonstrated that because of the limitations imposed by the yield strength of the steel used in casings, no bunker buster can ever go fast enough to penetrate reinforced concrete deeper than five times its length without destroying itself in the process.

Then there is one more problem. An earth-penetrating nuclear weapon must protect the warhead and its associated delicate electronics while it burrows into the ground. This severely limits the missile to impact velocities of less than about three kilometers per second for missile cases made from the very hardest steels.

In the test drops performed in Alaska, the B61-11 reached speeds of roughly 300 meters per second at impact. In order to penetrate reinforced concrete, it would need to be traveling at approximately 500 meters per second. At around 900 meters per second, the shock wave generated by the missile's slamming into the ground will deform it severely; at 1,200 meters per second, the missile will in most cases break into pieces. To penetrate granite—ubiquitous in mountainous bunkers, and believed to be common above any truly valuable bunker—a penetrator would have to attain upward of 3,000 meters per second, at which speed it would certainly be crushed.

What is more, the length of the bomb cannot be increased much for two reasons: there are no aircraft capable of carrying a weapon much longer than the ones that are currently deployed; and as length increases, so does the tendency of the bomb to snap in two on impact.

Coming to the blast itself, there is no way you can prevent the radioactive debris from going up in the atmosphere or contain the collateral damage. In order to be fully contained, nuclear explosions at the Nevada Test Site had to be buried at a depth of 650 feet for a 5 kiloton explosion—1300 feet for a 100-kiloton explosion! The B61-11 is not going to penetrate to that depth in any case. When dropped from a height of 40,000 feet, the B61-11 was able to penetrate three meters at most into the Alaskan tundra, and not at all into hard rock (that is, without self-destructing). The proponents say that even if the penetration is about 165 feet, it would be able to destroy fortifications buried nearly twice as deep and yet the radioactive fallout would be contained if the yield is about 400 tons. Critics say that this much penetration is just not possible. A more realistic figure would be about 57 feet. And this depth would neither serve the purpose of destroying the bunker nor containing the fallout. Further, even if earth penetrators could be made to perform at their theoretical limit, the only gain would be a temporary advantage over countries that have not yet dug bunkers at a depth that no weapon, no matter how massive, could ever reach. There is serious dispute over whether the bomb, even if it deposits a greater fraction of its energy in shock waves into the ground than in the case of a surface burst, would be able to damage hardened structures made of things like granite.

Even supposing that the missile's point of entry were miraculously neat, a nuclear blast at the depths a real missile could attain would invariably breach the surface of the earth, expelling a hot fallout cloud in what is known as a 'base surge'. Base surges are more dangerous than traditional fallout clouds because they are more toxic, containing irradiated particles of dirt and rock. They also spread more quickly, sweeping across the surface of the earth in every direction, outward rather than upward. The fact is that even the B61-11, at its current, inadequate impact speeds, does not burrow a clean rabbit-hole in the ground but rather kicks up a crater like a meteorite; any faster-moving penetrator would do so to a still greater degree. Even a 1-kiloton bunker buster—a relative firecracker, with a tiny fraction of the explosive power of the high-yield RNEP—detonated at fifty feet under-ground could eject about 1,000,000 cubic meters of radioactive soil.

Chapter 24

Cluster Bombs: Bombs that Breed Bomblets

(24.1) THE RATIONALE OF DEVELOPING CLUSTER BOMBS

Cluster munitions are air-dropped bombs or ground launched shells that eject multiple small **sub-munitions (bomblets)**. Their primary purpose is to kill enemy soldiers, although specialized weapons designed for anti-runway, anti-armor and mine-scattering purposes have also been developed. When they are dropped by aircraft we call them **cluster bombs**. When you fire them from artillery, you call them **ICM (Improved Conventional Munitions)** shells.

Cluster bombs were developed in order to improve the efficiency of aerial attacks, particularly against 'soft' targets like personnel. Single bombs, even though large, are not so useful for this purpose because they cover a smaller area (known as a 'footprint' in military parlance), and their effectiveness is dependent on the accuracy of the bomb's drop. A cluster bomb functions like a shotgun, covering a wider area with a spread of miniature bombs.

Like many other implements of war, the Germans were the first to develop it too during the Second World War though it could not be used effectively. The Germans made what is known as **SD-2** (Sprengbombe Dickwandig 2). It was a two kg device containing about 225 g of TNT. The bomb is more famous as the **'butterfly bomb'**. It was so named because the thin cylindrical metal outer shell hinged open when it was dropped and gave a superficial appearance of a large butterfly. They were packed into canisters containing between 6 and 108 of the bombs, depending upon the canister type, and were released as the canister fell. The bomb was an 8 cm long cylinder which was slightly smaller in diameter before its vanes deployed. A steel cable 15 cm long was attached via a spindle to the fuze screwed into the fuze pocket in the side of the bomb. The outer shell would hinge open as two half-cylinders when it was dropped and spring-loaded vanes at the ends would flip out. These rotated the spindle as the bomb fell and armed the fuze. The bomb could be fitted with an impact fuze, time delay fuze or an anti-disturbance fuze. In October 1940 the British found one butterfly bomb that had failed to explode. This particular piece had an anti-disturbance fuze. How they managed to defuse it is an interesting story. It was disarmed by making a clay wall around the weapon and pouring liquid nitrogen over it, thus freezing the mercury spirit gauge that would activate the bomb once disturbed!

Immediately after the War the Americans made an almost identical copy of the SD-2 and it was known as the **M83 bomb**. It was used both in Korea and Vietnam.

(24.2) DISPENSERS FOR CLUSTER BOMBS

Making bomblets for the cluster bombs is easy. They are just small bombs. The major concern is to make dispensers for the bomblets so that they are dispersed in the desired fashion. Cluster bombs can be made in two ways. You can have a hollow bomb-shaped container in which you can load the bomblets and it can be dropped like a bomb. The sub-munitions themselves may be fitted with small parachute retarders or streamers to slow their descent (allowing the aircraft to escape the blast area in low-altitude attacks). Some people call them **dropped dispensers**. Dropped dispensers fall away from the aircraft and are stabilized in flight by fin assemblies. Dropped dispensers may be in one piece or in multiple pieces. All dropped dispensers use either mechanical time or proximity fuzing. Mechanical time fuzes are activated by a lanyard on release from the aircraft. These fuzes allow the payload to be dispersed at a pre-determined height above the target. Multiple-piece dispensers open up and disperse their payload when the fuze functions. Single-piece dispensers eject their payload out of ports or holes in the body when the fuze functions. The so-called **Rockeye canister** is a **'clamshell canister'** that splits in half to release its sub-munitions. The **TMD (Tactical Munitions Dispenser)** has three panels that are released to disperse the sub-munitions. The three types look almost identical externally but have minor differences to accommodate their specific sub-munition loads. **Attached dispensers,** on the other hand, stay attached to the aircraft and can be reloaded and used again. Their payload is dispersed out the rear or from the bottom of the dispenser.

Fig. 145 The Rockeye Or Clamshell Canister

Fig. 146 An Attached Dispenser (SUU-14/A)

(24.3) VARIOUS TYPES OF SUB-MUNITIONS FOR CLUSTER BOMBS

Sub-munitions are classified as bomblets, grenades, or mines. They are small explosive-filled or chemical-filled items designed for saturation coverage of a large area. They may be **anti-personnel (APERS), anti-materiel (AMAT), anti-tank (AT), dual-purpose (DP), incendiary, or chemical**. As stated above, the sub-munitions themselves may be fitted with small parachutes, drogues, retarders or streamers to slow their descent. Anti-personnel cluster bombs produce shrapnel to kill troops and destroy soft (unarmored) targets. Anti-armor munitions contain hardened spikes with shaped charge warheads to pierce the armor of tanks and armored fighting vehicles. Anti-runway sub-munitions are often designed to penetrate concrete before detonating, allowing them to shatter and crater runway surfaces. Mine-laying weapons do not detonate on contact, but scatter their cargo of land mines for later detonation.

Fig. 147 APERS Sub-Munition BLU-26/B Or Guava

APERS sub-munitions are usually of ball-type. They are very small and are delivered on known concentrations of enemy personnel, scattered across an area. When the APERS sub-munition hits the ground, a small fragmentation ball shoots up and detonates about 6 feet above the ground. The **area-denial APERS sub-munitions (Family Of Scatterable Mines-FASCAM)** are delivered into areas for use as mines. Like a land mine, it will not blow up until pressure is applied to it. When they hit the ground, trip wires kick out up to 20 feet from the mine. All area-denial sub-munitions use anti-disturbance fuzing with self-destruct fuzing as a backup. Most or all of the submunitions used by the SUU-30/B canister are in the form of small spherical ribbed grenades, with the ribbing usually angled to help the sub-munition spin for arming and to aid in dispersal. These sub-munitions come in a range of sizes and can be designed for fragmentation or incendiary effects. The **'BLU-26/B'** is more or less typical, being a fragmentation sub-munition with a weight of 1.2 pounds, and is sometimes called the **'Guava'**, after a small tropical fruit.

The **AMAT** and/or **AT sub-munitions** are designed to destroy hard targets such as vehicles and equipment. They are dispersed from an aircraft-dropped dispenser and function when they hit a target or the ground. A drogue parachute or a fin assembly stabilizes these sub-munitions in flight so they hit their targets straight on. **AT area-denial sub-munitions** can be delivered by aircraft, artillery, and even some engineer vehicles. These FASCAMs all have magnetic fuzing. They will function when they receive a signal from metallic objects. These sub-munitions, similar to the APERS area-denial sub-munitions, also have anti-disturbance and self-destruct fuzing. The **CBU-78/B 'Gator'** cluster bomb contains 45 **'BLU-91/B'** antitank mines plus 15 **'BLU-92/B'** antipersonnel mines. Both these mines look like tuna-fish cans embedded in an open-top box, and measure 5.75 x 5.0 x 2.6 inches. The BLU-91/B anti-tank mine weighs about 4.3 pounds and the BLU-92/B weighs about 3.7 pounds. Both mines have microelectronic fuzes. The BLU-91/B can discriminate between different types of armored vehicles, and detonate when it is likely to have optimum effect against belly armor. The BLU-92/B is activated by wires that it tosses out after landing, and is intended to hinder minefield clearing. Both mines will self-destruct after a preset time period, with the three possible time settings defined by a selector switch on the canister. The **CBU-89/B,** based on the larger TMD canister, disperses the same two types of mines and is also known as a 'Gator'. The **CBU-99/B** cluster bomb contains 247 **'Mark 118'** anti-armor sub-munitions. These sub-munitions are dart-like hollow-charge bomblets with a weight of 1.32 pounds each. Each bomblet is capable of penetrating 7.5 inches of conventional armor. There is also a **CBU-100/B**, with exactly the same configuration but using the SUU-76/B canister instead of the SUU-75/B canister.

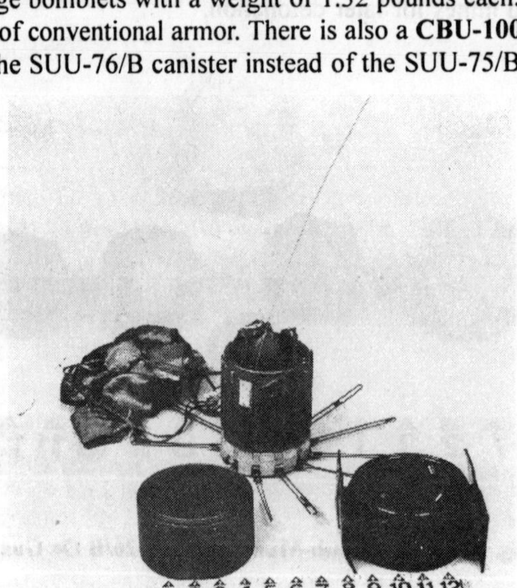

Fig. 148 Some Area Denial Sub-Munitions Including BLU-91/B Anti-Tank Mines (Bottom)

Currently many cluster bombs use sub-munitions that have more than one effect. The **CBU-59/B,** for example, contains 717 '**BLU-77/B**' bomblets, which weigh 1.02 pounds each, and are dart-shaped with a small standoff probe detonator on the nose and fins on the tail. These bomblets have both **antipersonnel** (AP) fragmentation and **anti-materiel** (AM) incendiary effects, and so are referred to as '**APAM sub-munitions**'. The 'BLU-68/B' is an **incendiary sub-munition** with a weight of 0.93 pounds. It is believed to use titanium pellets as the incendiary agent. The **CBU-87/B** contains 202 '**BLU-97A/B' Combined Effects Munitions (CEMs)**". The BLU-97A/B is a can-shaped sub-munition with a weight of 3.4 pounds built around a hollow charge that penetrates up to about 7.1 inches of armor. Once dispensed, it pops out a number of small 'spider' tabs to ensure that it falls nose-down, and then pops out a ballute to retard its fall. It has a scored case that breaks into 300 fragments, as well as an incendiary zirconium ring. Its capability makes it a preferred sub-munition and it is used in a number of guided munitions as well, including a sub-variant of the well-known Tomahawk cruise missile.

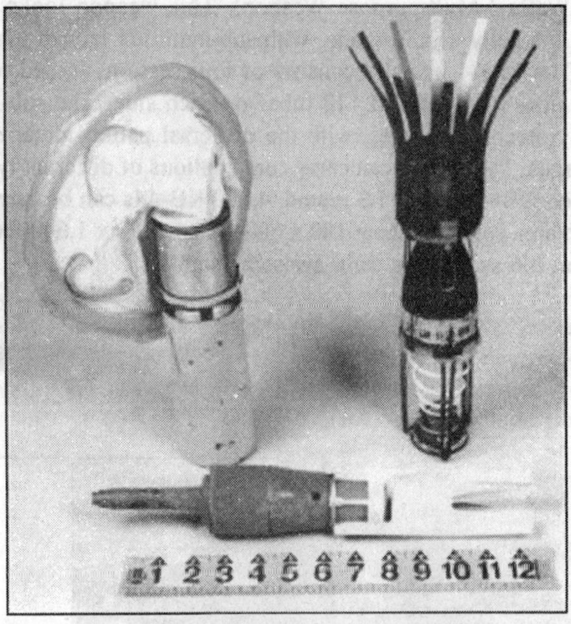

Fig. 149 Various AMAT And AT Sub-Munition Including The BLU-97 A/B Combined Effects Munition (Left)

A growing trend in cluster bomb design is the 'smart' sub-munition, which uses guidance circuitry to locate and attack particular targets, usually armored vehicles.One of the advanced cluster bombs, the **CBU-97/B**, carries the **smart BLU-108 Sensor Fused Weapon (SFW)** sub-munition. The BLU-108 Sensor Fuzed Weapon (SFW) is an un-powered, top attack, wide area, cluster munition, designed to achieve multiple kills per aircraft pass against enemy armor and support vehicles. The BLU-108/B sub-munition contains four armor-penetrating projectiles with infrared sensors to detect armored targets. At a preset altitude sensed by a radar altimeter, a rocket motor fires to spin the sub-munition and initiate an ascent. The sub-munition then releases its four projectiles, which are lofted over the target area. The projectile's sensor detects a vehicle's infrared signature, and an explosively formed penetrator fires at the heat source. The **CBU-94/B** carries sub-munitions with '**soft kill**' capability to attack electrical power stations.

Fig.150 Cutaway Diagram Of The CBU-97 1000 Lbs Cluster Bomb Containing Sensor-Fuzed Sub-Munitions For Attacking Armor

The Germans have developed a unique and highly effective sub-munition dispenser, the 'MW-1' (MehrzweckWaffe / Multipurpose Weapon). This weapon looks something like a big bathtub strapped onto the belly of a Tornado, with sub-munition launch tubes facing outward on each side of the tub. The MW-1 actually consists of four sections joined together, each with 28 launch tubes on each side, for a total of 112 tubes on each side. The sub-munitions are blasted out of the tubes by pyrotechnic charges, with the dispersal pattern determined by the size and firing order of the charges. The MW-1 can carry combinations of different types of sub-munitions. The anti-armor munition KB44 weighs 1.3 pound. 4,500 KB-44s can be carried by a single MW-1 dispenser to saturate an area covering about 180 x 500 meters (590 x 1,640 feet). The entire payload can be ejected in about 0.6 seconds, a truly awesome sight!

Fig. 151 The German MW-1 Dispenser Mounted On A Luftwaffe Tornado (Inset Shows The Sub-Munitions Being Dispersed

(24.4) SOME FAMOUS CLUSTER BOMBS

(24.4.1) CBU-52 And CBU-58

This is just a typical illustration. We have taken two bombs of the USA. Similar things are made by other nations too. American cluster bombs are designed for high-altitude release whereas the British and French designs have to be released at low altitudes. The **CBU-52** is a 93 inches long and 16 inch wide device. It is loaded with 220 anti-material, anti-personnel bomblets, weighs

785 lbs and can be used with a variety of proximity fuzes or the mechanical MK-339 timed fuze. The sub-munition **BLU-61A/B** is a 3.5-inch spherical bomblet weighing 2.7 lbs with a 0.65 lbs high-explosive warhead. A total of 17,831 were expended during the Gulf War. The **CBU-58** has got the same dimensions and weighs 800 lbs. The CBU-58 is loaded with 650 **BLU-63/B** bomblets. These bomblets contain 5-gram titanium pellets, making them incendiary and useful against flammable targets.

(24.4.2) Runway-Cratering Bombs CBU-51/A And JP-233 ETC.

We must make a reference to the **runway-cratering bombs and bomblets** too as they are highly useful in rendering enemy airfields out of action for some time. A single crater can be repaired quickly; repairing a large number of relatively smaller craters is more time and effort consuming.

The American runway-cratering bomb is called **CBU-51/A**. It carries a bomblet called **BLU-67/B** bomblet which weighs 11 lbs.

The British runway-cratering bomb is called **JP-233.** They were extensively used by RAF Tornado strike aircraft early in the Gulf War to disrupt the operation of Iraqi airfields. They need to be released at low altitude and are potentially dangerous operations for the pilots. Two 21 feet 6 inches long JP-233s can be fixed to the bottom of a Tornado. Each of the two dispensers consists of two sections, with the rear section carrying 30 **'SG-357' runway cratering sub-munitions** and the front section carrying 215 'HB-876' anti-personnel mines. Both sub-munitions are parachute-retarded. The SG-357 weighs 75 pounds and is a **'two-stage' munition**, a novel invention of the British, with a hollow charge in the front to blast a hole in a runway, and a cratering charge that falls into the hole and then explodes.

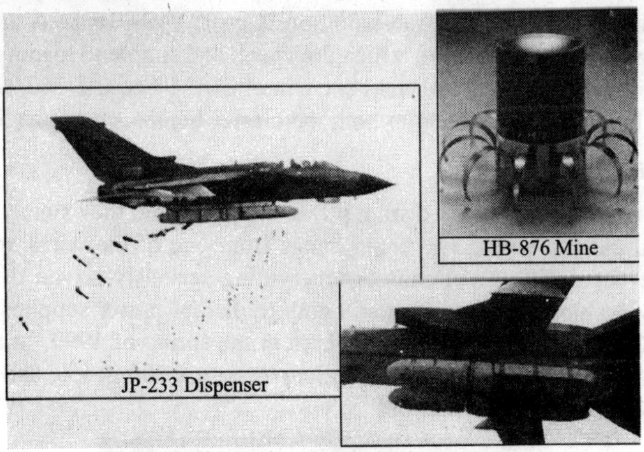

Fig. 152 The JP-233 Runway-Cratering Bomblet And Its Dispenser Mounted On A Tornado Aircraft

The Germans have made the 37.5 pound **"STABO"** runway-cratering sub-munition. STABO is also a two-stage weapon, with a penetrating charge followed by a cratering charge.

(24.4.3) Smart Cluster Bomb For Armor: The CBU-97

The American **CBU-97 sensor-fuzed weapon** was employed during Kosovo War and in Iraq in 2003. The CBU-97 is a 1,000 lbs class weapon containing **sensor-fuzed BLU-108 sub-munitions** for attacking armor. It is intended to be used for engaging an adversary's main armored force in the 'halt' or 'hold' phase. The idea is to disrupt and stop an attack, providing time for other combatant forces to reinforce the theater. After release, the **TMD (Tactical Munitions Dispenser)** opens and dispenses the ten sub-munitions which are parachute stabilized. Munitions specifically intended for anti-tank use may be set to self-destruct if they reach the ground without locating a target, theoretically reducing the risk of collateral damage to civilians and non-military targets. The SFW with **WCMD (Wind Corrected Munitions Dispenser)** is called **CBU-105.**

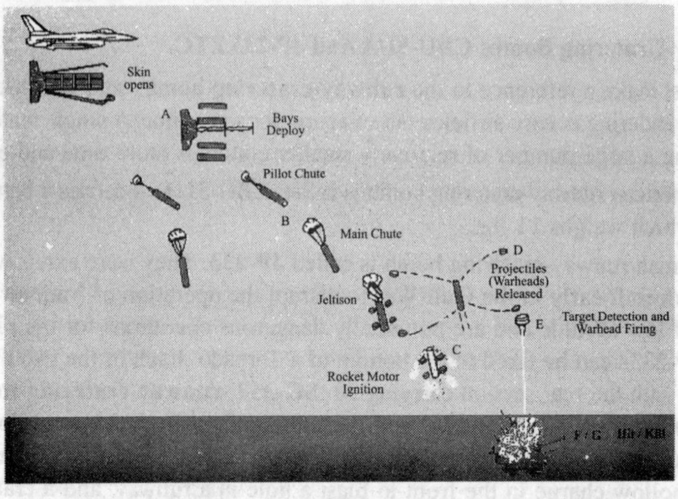

Fig. 153 Deployment Sequence Of The CBU-97 Sensor Fuzed Weapon

The limitation of the smart sub-munition is cost. Such weapons are many times more expensive than standard cluster bombs, which are cheap and simple to manufacture. For example, one unit of CBU-52 costs just $1542 whereas one unit of CBU-97 costs $ 360,000! A MK-82 500 lbs General Purpose bomb, in comparison to both the cluster bombs, costs just $ 249!

(24.4.4) Power Disruption Bombs

How do rioters or saboteurs disrupt power supply? Well, they simply lob a long chain on to some overhead electric wires. The chain hangs from one of the wires and its hanging parts short circuits the wires. The Americans do everything seriously—even the job of a rioter or saboteur! So they developed a sophisticated bomb to disrupt power supplies also. This was first used in the NATO air campaign against the Serbs in the spring of 1999. In the first attack itself they knocked out 70% of the power supply in Serbia. It took 500 people and 15 hours to get just one transformer yard back into operation!

Fig.154 Sub-Munition Of The Power Disruption Bomb Showing The Spools Of Conductive Fiber

This is a cluster bomb called **CBU-94/B**. Its sub-munition is called **BLU-114/B**. They are about the size of two beer cans stacked on top of each other. After the sub-munitions are scattered by the TMD, they pop off a nose cap. 147 spools of conductive fibers are then shot out with a small explosive charge to unravel into a tangle that falls on power lines and other power-distribution equipment, short circuiting them. After the 1999 attacks with the CBU-94/B, the Serbs working at the power plants collected samples of the threads, as well as remnants of the TMD and sub-munition cases, and passed them on to the Russians for analysis. The Russians report that the TMD cases were marked '**CBU-102/B**' instead of 'CBU-94/B', and that the threads were actually made of glass fiber coated with aluminum, and not carbon fiber as the US military had stated. It appears that the CBU-102/B was a minor modification of the CBU-94/B. Presently the wires can be cleaned even if it takes hours. So the Americans applied some more mind to it. An improvement is said to form a sort of conductive cloud that gets into every opening and is very difficult to clean up.

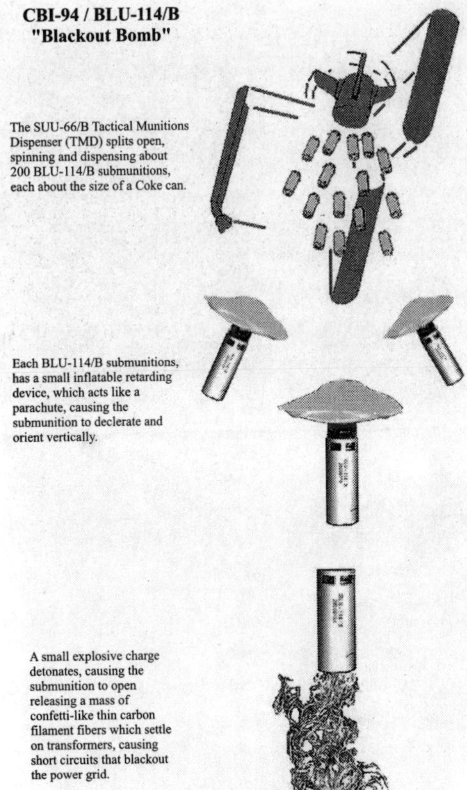

Fig. 155 Deployment Sequence Of The Power Disruption Bomb

(24.5) THE PROBLEM OF UNEXPLODED BOMBLETS

There is a little practical problem in using the cluster bombs. Even with 95 percent sub-munition function reliability, one CBU-58 (with 650 sub-munitions) could produce up to 38 unexploded sub-munitions. A typical B-52 dropping a full load of 45 CBU-58/CBU-71, each containing 650 sub-munitions, could produce an average of some 1700 unexploded sub-munitions. This may be dangerous for own troops as well. They call it potential for fratricide from **unexploded ordnance (UXO)**. Studies show 40 percent of the duds on the ground are hazardous and for each encounter with an unexploded sub-munition there is a 13 percent probability of detonation. Thus, even though an unexploded sub-munition is run over, kicked, stepped on, or otherwise disturbed, and did not detonate, it is not safe. Groups that cry themselves hoarse over landmines, oppose the cluster bombs too for this reason. A solution has been found to this problem also. Israeli Military Industries has developed a spring-loaded pyrotechnic fuze that is activated by the air slipstream after sub-munition deployment, and fires the sub-munition after a few tens of seconds if it isn't detonated by its primary fuze. IMI claims their fuze reduces the dud rate to less than a tenth of a percent.

□□□

Chapter 25

The Much-maligned Napalm, Its Cousins and Fuel-Air Explosives

(25.1) THE USE OF FIRE AS A TOOL OF WAR IN THE ANCIENT PERIOD AND THEREAFTER

The utility of fire in war was recognized quite early in human history. It made sense to throw fire on the enemy as humans have an in-built fear of fire. Flaming arrows, bundles of cloth etc. soaked in oil and firepots (small earthen vessels filled with combustibles) were used all over the world in different forms. Nothing, however, struck so much terror in the hearts of the adversaries as the so-called Greek fire used by the Byzantine Greeks. As one contemporary victim of Greek fire advised his comrades-in-arms, 'Every time they hurl the fire at us, we go down on our elbows and knees, and beseech Our Lord to save us from this danger'. Variants of the Greek fire were used by other people also. The Chinese are known to have used something which they called 'fire oil'. The Byzantines used it in very effectively in naval battles also as it could continue to burn even on water. It was a burning liquid weapon used in the fashion of a flame thrower and it goes to the credit of the Byzantines that they could devise some sort of pumping mechanism to use it that effectively. It would not be an exaggeration to say that the Greek fire was responsible for many important victories of the Byzantines and one of the reasons that the Eastern Roman empire could survive for so long! The Greek fire was such a jealously guarded military secret that its exact formula still remains shrouded in the mist of history. However, it is believed that it was most probably a mixture of naptha, quicklime and sulfur.

In the modern times during the WWI the British made the **Mark IV Incendiary Bomb**. This was just 6.5 oz in weight and was called the **Baby Incendiary Bomb**. It was considered the most effective incendiary bomb of the WWI. The Americans followed suit before the WWII and developed the **Mk II Incendiary Dart**. Italy and Germany also developed incendiary bombs weighing 1 kg and 2 kg each. They were used in Ethiopia and Spain respectively. Military observers were convinced that fire bombs constituted major threat to cities. The next few years were to prove them absolutely right. We have discussed the havoc wrought by the M-69 fire bombs used against the Japanese cities in the WWII.

Napalm is a fire bomb. You can make elementary fire bombs like the famous **Molotov Cocktail** from petrol or gasoline and which we have discussed in connection with IEDs (Improvised Explosive Devices). But there is a little problem with petrol. It burns too quickly. It is fine as long as you are using it for a little terrorist or riotous activity. It is just not good for military use. For military use you require something that would burn slowly for quite some time—at least some minutes. In other words, the problem is to thicken petrol. It is this quest which led to the invention of the thickener called napalm.

(25.2) THE INVENTION OF THE NAPALM THICKENER

During WWI American researchers experimented with as many as 40 potential thickeners including sodium stearate but success could not be obtained. Between the two World Wars there was little research in the USA in this direction. The British had, however, been able to produce a thickened fuel using rubber. It used 8% crepe rubber with gasoline. Other thickeners tried included soap gels and elastomers/certain polymers. One soap gel used was a calcium soap made by the action of cresylic acid in coal tar on lime.

Napalm had to wait for the stroke of a genius to be invented. The word napalm is made from naphthenic and palmitic acids. The original napalm was invented by Dr. **Louis Fieser** of the Harvard University working with the US Army Chemical Warfare Service in 1942. They discovered that the soap based on aluminum naphthenate oleate or oleo-linoleate and aluminum palmitate could be used to gel petrol. The resulting gel was quite effective and was used extensively in the World War II fire bombings, Korea and Vietnam. It was also found that lauric acid could easily replace palmitic acid. The Americans plumped for it because lauric acid can be obtained from coconuts which were available to them in abundance. The 'production models' of napalm were based on lauric acid; the name of napalm, however, stuck. The percentage of the thickener used varied form 6 to 12% depending on how you proposed to use it; in flamethrowers, grenades or aerial bombs.

Early napalm bombs had a bursting charge of TNT or tertyl surrounded by some white phosphorus. The charge scattered both the phosphorus and the gel. It also ignited the phosphorus which, in turn, ignited the gel.

Once the basic principle was discovered, napalm lent itself to great versatility. Besides regular gasoline, people have tried all sorts of fuels ranging from kerosene, benzene, toluene, xylene and unrefined crude oil to refinery residues. For thickeners, they have tried rubber, rubber processing residues and silica gel also. For additives, viscous or solid hydrocarbons like naphthalene or members of the paraffin group, white phosphorous, red phosphorous, magnesium, aluminum, stag formers like calcium silicate, calcium silicide, ferro-silicon, lithium silicide, diatomaceous earth, fine brick dust, asphalt, wood dust, phenolo-formaldehyde resins, benzene sulphohydrazide and carbon disulfide have also been used. Phenolo-formaldehyde resins have the capability of producing water-resistant napalm. The percentage used is up to 10. It forms adhesive crusts, particularly with red hot cinders. Benzene sulphohydrazide converts synthetic resins into foams. It needs to be used only to the extent of 1 to 2%.

(25.3) THE FURY THAT THE NAPALM HAD UNLEASHED IN VIETNAM AND ELSEWHERE

One ton of napalm is enough to incinerate a football field in seconds. Entire villages could be burnt with this. The gel sticks to any surface including human skin and burns there. Napalm sucks oxygen out of the air and can kill people who are not burned to death by asphyxiation. Look at the irony of the whole thing—this invention of Dr. **Fieser** killed hundreds of thousands; his other inventions, this time in the field of medicine, cortisone and synthesized quinones saved hundreds of thousands of lives!

The Americans had used the napalm to their heart's content both in the World War II and in Vietnam without any moral compunctions whatsoever because it was serving their purpose. Between 1964 and 1973 they used nearly 1.47 million napalm bombs (mostly napalm and a limited quantity of other incendiaries like magnesium/thermite, white phosphorus and zirconium) in Vietnam totaling about 420,000 tons in weight! This must be compared with the total tonnage of all sorts of aerial bombs used in the same period. The Americans used about **6.5 million tons of air-dropped munitions in Vietnam.** Thus napalm accounted for a little less than 10% of the total munitions dropped—a very large percentage indeed for those who talk of championing democracy by force throughout the world! It is not known as to what percentage of ground munitions the incendiaries accounted for. It is necessary to keep this in mind because the total amount of all sorts of munitions used in Vietnam comes to a whopping 15 million tons!

The maximum number of napalm bombs was used in 1966 when they used nearly 430,000 napalm bombs of various sizes totaling 124,000 tons in weight. The second highest usage was recorded in 1969 when they used about 228,000 of them weighing over 71,000 tons. The average number of napalm bombs used in the remaining years came to over 100,000 per month weighing nearly 28,000 tons. General Paul Harkins is on record having said that 'it really puts the fear of God into the Viet Cong—and that's what counts'.

They did use **napalm in Iraq** too. In March 2003 it was used to destroy an Iraqi intelligence gathering operation on top of Safwan Hill in southern Iraq. Initially they denied it vehemently and the US Air Force even posted a 'disinformation alert' on its website. However, by August 2003 they accepted that **MK-77 firebombs** were used but maintained that napalm was not used. A very poor effort at a semantic exercise! The MK-77 only uses a different composition but in effect is essentially the same as napalm! The troops call it napalm only. Col. Randolph Alles of the US Marines is on record having said that 'The generals love napalm—it has a big psychological effect'. They used it again in their attack on the town of Fallujah in November 2004 also when the town having a great many civilians too was bombed and raged to the ground.

It may be recalled that the **Geneva Protocol III** (Protocol on Prohibitions or Restrictions on the Use of Incendiary Weapons) of the Convention on Certain Conventional Weapons, adopted on October 10, 1980, made it illegal to use incendiary weapons on civilian populations and restricted the use of these weapons against military targets that are located within a concentration of civilians. Such weapons are considered to be 'excessively injurious or to have indiscriminate effects'. Fifty-one countries initially signed the document and on December 2, 1983, its provisions entered into force. By the end of 2004, 104 countries had signed and 97 had ratified the protocol. The US, the self-proclaimed champion of democracy in the world, is not a party to this protocol and continues to use incendiary weapons in all its major conflicts. It is the only country to do so. Their chutzpah in rushing to any place in the world in force in an ostensible attempt to protect what they perceive as the human rights of some people who kowtow to them and violate the human rights of those who do not is indeed commendable!

The Americans officially claim that they have destroyed their last container of napalm in a public ceremony in 1991! Would you ever trust these hypocrites? They are very clever people and know how to make fools of those who are not knowledgeable. They have only said that they have destroyed napalm bombs. Fine. They have not said that they have destroyed their fire bombs or incendiary bombs or that would never use them. In fact MK-77 incendiary bombs which are better than the old napalm do exist in their inventory. In any case, if they feel like using even the old napalm also again, they will invent some reason for that too. Personally, we don't have any problems with any weapon but we thought we must expose the hypocrisy of those who proclaim themselves to be the only responsible power in the world and who are ever ready to use force against a smaller power for what they perceive to be the violation of civilized conduct of warfare or human rights in their sovereign territory.

(25.4) HOW THEY MADE AERIAL BOMBS OUT OF A GELLED FUEL LIKE NAPALM?

Early napalm bombs were quite ingeniously improvised devices. Small fighters or fighter bombers used in ground attack roles had limited range on their internal fuel tanks and were obliged to use external tanks. These external tanks were mounted under the wings or the fuselage and given a cigar shape too. Made of light aluminum alloy, they could be dropped after use. Hence the name **'drop tank'**. Since early napalm mixture was made by thickening aviation fuel only, it was natural for somebody to think that the drop tanks could be converted to carry napalm instead and used as a bomb! The first such 'bomb' to be standardized was called the **M-116**. It was over 11 feet long and one-and-a-half feet thick. Empty it weighed just 52 lbs. It could be filled with 100 gallons of napalm. The initial intense heat lasted for 6-10 seconds whereas the fire of less intense heat burned for 6-10 minutes. The bomb was good for incinerating an elliptical area 300 feet long and 100 feet wide.

The M-116 was followed by the **BLU-1/B** and the **BLU-27/B**. The un-finned 750 pound / 100 gallon BLU-1/B napalm bomb and its finned variant, the BLU-27/B, were used extensively and with devastating effect in Vietnam. US forces also used the **M35** and **M36 'Funny Bombs'**, which scattered thermite over a target area and proved lethal to trucks and area targets.

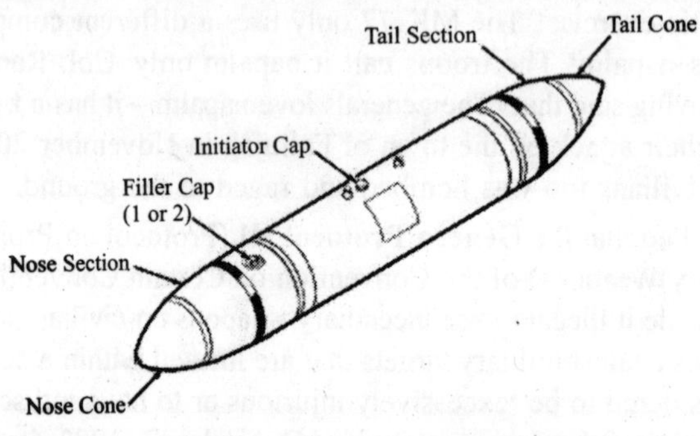

Fig. 156 So Simple And Yet So Deadly: BLU-1/B Napalm Bomb, The Scourge Of Vietnam

(25.5) THE MK-77 BOMB: A DEADLIER COUSIN OF THE NAPALM BUT STILL UN-MALIGNED BECAUSE IT IS NEW

The MK-77 is a 500 lbs bomb which uses a composition radically different from the napalm. It uses a mixture of benzene (21%), petrol (33%), and polystyrene (46%). Some people call it **Napalm-B**. It was developed at the Elgin Air Force Base in association with the polystyrene section of the Dow Chemical Company. The chemistry is simple. Petrol still remains the main combustion agent. Polystyrene is the thickener. It is a synthetic resin made from ethylene and used in great quantities in numerous household and industrial plastic products. Now polystyrene would not dissolve in petrol. But it dissolves in benzene and that's why they use benzene. The fire of this mixture is hotter than the fire of napalm (850 degrees against 760 degrees of the latter) and harder to extinguish. It burns for 10-12 minutes, i.e. nearly two to three times longer. The longer burning time enables it to ignite fuels and lubricants in mechanical equipment also. It is also more stable in storage. In fact it was found feasible to fill the bombs in the factory itself with the fuel and solid pieces of polystyrene in them. They dissolved when the solvent benzene was added to it thereby avoiding the need of special mixing equipment.

The MK-77 bomb ruptures on impact and spreads burning fuel gel on surrounding objects. MK 13 Mod 0 igniters are used to ignite the fuel gel mixture upon impact. The container of this bomb is very light and fabricated of aluminum, with a capacity for about 75 gallons of combustible gel. They are deliberately not provided with stabilizing fins and consequently acquire a tumbling motion on being dropped that contributes to the scattering of the combustible gel over a wide area. Commonsense dictates that they are very effective against dug-in troops, troops in various types of vehicles including armored vehicles and anything that could burn or suffer adversely from fire.

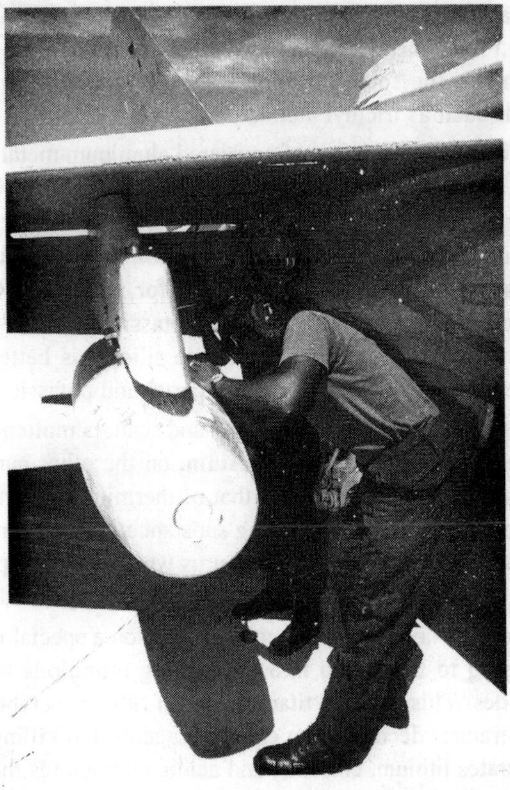

Fig. 157 This Is The Infamous MK-77 Bomb Used In Iraq Even On Civilian Targets: Photo Shows The Bomb Being Loaded On A F/A-18 Hornet In 1993: Note The Fin-Less Design That Facilitates Tumbling And Thus Scatters The Fuel Over A Larger Area

Subsequently it was found that a great many polymers like polyisobutylene, polybutadiene, substituted polybutadiene, polyisoprene and copolymers of hydrocarbon dienes could also be used as thickeners and toluene and methyl-ethyl-ketone could be used as solvents. Some of them give burn time nearly ten times longer than that of conventional napalm. An additional advantage of these thickeners is that they do not separate out in prolonged storage or on high impacts encountered in high altitude droppings as it used to happen sometimes with conventional napalm compositions. They also have much better adhesive properties and can stick to even vertical surfaces without slipping down. In fact they are virtually impossible to remove once they stick.

Addition of naphthalene or similar hydrocarbons to Napalm-B results in copious amounts of black smoke. This increases the psychological impact of the bomb and also hinders the efforts to fight flames.

(25.6) THE INCENDIARIES

(25.6.1) Thermite Incendiaries

Incendiaries produce intense, localized heat designed to ignite adjacent combustible target materials. The true incendiary produces no fireball and relatively little flame. They are highly effective against fuel and other flammable targets. The main incendiary agents are **thermite, magnesium, white phosphorus (WP),** and **oil-based incendiaries** including combustible hydrocarbons and gelled ones. We discussed the last in connection with napalm. Aluminum can be used as a pure metal; as an alloy with magnesium; in the form of napthenate, palmitate or laurate in napalm; or in organometallic compounds such as triethyl aluminum.

Thermite incendiaries are a mixture of powdered aluminum metal and ferric oxide and are used in bombs for attacks on armored fighting vehicles. Another composition consists of 2.75 parts of black iron oxide (ferrosoferric oxide) and 1.0 part of granular aluminum. Yet another composition used was 3 parts of aluminum powder, 6 parts of barium nitrate and 8 parts of ferrous oxide. Interestingly, thermite was used before the WWI for welding railway lines. For military purposes it is necessary to use a binder like sodium or potassium silicate or even sulfur. Sulfur is used when you intend to scatter the thermite; otherwise silicate is better. Thermite requires an igniter. One popular igniter consisted of magnesium (9 parts) and potassium perchlorate (13 parts).

Thermite burns at about 2000 degrees Celsius and scatters molten metal, which may lodge in the skin producing deep multiple burns. **Magnesium,** on the other hand, burns at about 2200 degrees Celsius with a scattering effect similar to that of thermite. Its particles also produce deep burns. Fragments of melted particles of the burning substance may become embedded in the skin of persons close to a bursting projectile, producing burns which are multiple, deep and variable in size. The fragments continue to burn unless oxygen is excluded by flooding or smothering.

The US Air Force and Navy have been collaborating on a special incendiary fill for **'agent defeat' munitions**, designed to be loaded into penetrating munitions to destroy chemical and biological weapon stockpiles. This fill is a 'titanium boron lithium perchlorate intermetallic' that burns at a very high temperature, decomposing chemical agents and killing biological agents. The burning mixture also generates lithium, chlorine, and acidic compounds that kill biological agents. The mixture does not generate high pressures when ignited; reducing the likelihood that it will disperse chemical or biological agents from the target.

Thermite grenades have been made for starting fires and burning through steel plates such as hatches over fortifications. One example is the American grenade AN-M14 which contains 700 g of thermite and burns for 30-45 seconds at a temperature of about 2200 degrees. The German DM-19 grenade consists of 320 g of red phosphorous and aluminum and burns at a temperature of 1200 degrees. Incendiary grenades were made for rifles and AGLs also.

(25.6.2) Pyrogels

The properties of oil-based and metal-based incendiaries are different. Oil-based incendiaries give a spreading effect. Metal-based incendiaries increase the burning temperature and prolong the effect by producing hot slags of solid oxides. It was therefore thought desirable to combine the two.

Pyrogels serve this purpose. They are essentially thickened gasoline in which the thickener is usually isobutyl metacrylate and may include natural rubber also. A metal incendiary is added to it later which increases the burning temperature to up to 1600 degrees. The dissemination of this mixture is similar to conventional napalm compositions when the bursting charge bursts it. Asphalt may, however, be added to the mixture to overcome the shearing strains resulting from high altitude bombing.

Some of the pyrogels used in the US army include:

- **PT1**: Magnesium and an oxidizer, petroleum distillate, asphalt and isobutyl metacrylate
- **PT2**: Uses barium nitrate as the oxidizer, a little of asphalt as binder and 5% isobutyl metacrylate as the thickener
- **PTV**: It contains 60% gasoline with a trace of p-aminophenol, 5% polybutadiene as the thickener, 6% sodium nitrate as the oxidizer and 28% magnesium as the metal incendiary.

(25.6.3) Pyrophoric agents

Pyrophoric agents are those which do not require an external igniter like most of the mixtures described above. They ignite spontaneously in air. White phosphorous, for example, is one such agent. At ordinary temperatures, **white phosphorus (WP)** is a solid which can be handled safely under water. When dry, it burns fiercely in air, producing a dense white smoke. Due to this property it was used extensively in the WWI by almost all the nations involved for making incendiary bullets for small arms and even artillery shells and trench-mortar bombs. As an anti-personnel weapon it was a dreaded thing in the war as large particles of burning phosphorous stuck to clothing and could not even be brushed off easily. It was, however, not really effective for starting fires in wooden structures and the like on account of two reasons; the first is the low temperature of combustion and the second is that the oxides produced by the burning of phosphorous provide some protection against fires. Oil-based targets, however, are liable to catch fire.

Many metals, when very finely divided, exhibit pyrophoric properties. For military purposes zirconium is used in high explosive and armor piercing incendiary shells, the lining of shaped charge rounds and in incendiary cluster bombs. For use in munitions **zirconium** is alloyed with titanium or lead in approximately equal proportions. Fine particles of the metal may be mixed and bound in a synthetic rubber composition. For added durability this incendiary core may be surrounded by a similarly bound composition of iron and nickel carbonate. For that matter even **depleted uranium** (DU) exhibits pyrophoric properties. When a DU dart of an APFSDS (armor piercing fin stabilized discarding sabot) round hits an armor plate, a part of the dart is partially pulverized by the heat of impact and burns spontaneously. Thus the penetrating dart spews flame as well.

Certain organometallic compounds like **TEA** (triethyl aluminum) also exhibit pyrophoric properties. It is a liquid and burns with a temperature of 2300 degrees which is comparable with that of metallic incendiaries. For use in weapons, it is thickened with polyisobutylene. In that condition it is sometimes described as **TPA (thickened polyphoric agent)**. This uses about 6% of polyisobutylene and is used in shoulder-fired rocket launchers etc. It has been shown that if you reduce the percentage of the thickener to about 1%, it becomes a material which can produce a chemical fireball which will radiate so much thermal energy that it will be capable of producing third-degree burns on the occupants of a bunker irrespective of the fact whether the fireball actually hit them or not. Hitherto this capability of inflicting damage by thermal radiation was limited only to nuclear weapons. The problem with TPA is that making it is a complex process demanding a highly developed petrochemical industry. Because of its reactivity with air, the production has to be carried out in an inert atmosphere of nitrogen or helium. Polymerization of isobutylene is also a complex process.

(25.6.4) New Incendiary Agents

A great amount of research has been done to find out newer incendiaries. More interest has been shown for pyrophoric agents because they do away with the need of fuses and can be used profitably with small rockets, artillery projectiles and bomblets to cover a large area. Compounds like aluminum borohydride, pentaborane and boron alkyls like triethylboron have been investigated. An alternate means of achieving **spontaneous ignition** is to use a **hypergolic oxidizing agent**. This agent generates so much heat in the process of oxidizing a fuel that the fuel ignites. Agents which are hypergolic with petrol are interhalogens, chlorine trifluoride, bromine trifluoride and bromine pentafluoride.

(25.7) FUEL-AIR EXPLOSIVES (FAE)

(25.7.1) What Are Fuel-Air Explosives And How They Work?

What makes fuel-air explosives (FAEs) so enigmatic and yet attractive? FAEs, being in aerosol form, literally form a cloud and therefore cover an area much larger in comparison to condensed explosives. Since the cloud can easily envelop structures of awkward shapes also, FAEs do not suffer from any limitations imposed by land contour or face handicaps in front of protective structures. Shock waves from high explosives propagate in straight lines and hence are limited by the contours of the land or obstructions by structures. Then the very size of the explosion of the FAE and the resulting huge ball of fire have a great psychological impact too. The first operational use of FAE was by the US Marine Corps in 1967 in Vietnam. At that time the FAEs consisted of ethylene oxide as fuel and were used for minefield clearance. The technology of FAE weapons was primarily developed at the **US Naval Weapons Center** (NWC) at China Lake, California, and this work led to the standardization of the first real operational FAE weapon, the Cluster Bomb **CBU 55/B**, which saw extensive use in Vietnam.

You can think of FAEs as explosives which dissipate their explosive energy uniformly in a very large area as against conventional explosives which create a very high pressure zone in a small area and then the pressure falls off rapidly as you move away from the point of explosion. Since the explosive substance in a fuel-air explosive is nothing but a fuel, most people are misled into believing that it would be easy to make a fuel-air explosive weapon. It is not so. Fuel-air explosives (FAEs) are highly deceptive things. In fact it is very difficult to make fuel-air explosives work successfully and reliably. All those fools across the world who carry a couple of cylinders of LPG in their cars along with a few jerry cans of petrol and hope to make a devastating fuel-air explosive bomb that would bring down a building as huge as an airport, can only be pitied for their ignorance. Why such devices fail miserably is explained later in the book where we have written at length about LPG-based IEDs and their inherent limitations. Here we will explain the basic mechanism of fuel-air explosives.

Two phenomena similar to fuel-air explosives occur by way of accidents in many chemical industries. They are known as **vapor cloud explosions** and **dust explosions.**

Accidental vapor cloud explosion occasionally take place in refining and chemical processing industry. Such explosions take place when fuel vapor is present in the air in a certain range of concentration and somehow gets ignited. For vapor cloud explosions there is a minimum ratio of fuel vapor-to-air below which ignition will not occur. Alternately, there is also a maximum ratio of fuel vapor-to-air, beyond which ignition will not occur. These limits are termed the lower and upper explosive limits. For gasoline vapor, the explosive range is from 1.3 to 6.0% vapor-to-air, and for methane this range is 5 to 15%. The extent of damage caused by a vapor cloud explosion depends on several factors including the total mass and type of fuel, the strength of the source of ignition, the manner in which the fuel was being released in the air (for example, whether it was in the form of a turbulent jet or a slow leakage), and turbulence induced in the cloud (for example, from ambient obstructions).

Dust cloud explosions are a common phenomenon in the coal mining, grain storage, and the woodworking and paper industries. In all such cases, the dust cloud is made up of material that can easily ignite and explode. Many miners have been killed and injured and massive production losses have resulted from coal dust explosions in underground coal mining operations. In grain storage, for example, billions of tiny, highly combustible particles of grain are generated by grain kernels rubbing together as they move along conveyer belts and shifted between bins. Inside the enclosed chambers, those particles rise in a cloud. When the dust gets in with the right mixture of oxygen and comes in contact with a spark or even an overheated bearing on a conveyer belt, it explodes. In general the smaller the particle, the more reactive the dust is. As the materials become smaller, they disperse and remain suspended more easily, increasing the potential for ignition and propagation of the reaction. In order for a dust explosion to take place, the dust particles must be smaller than a certain size and the amount of finely granulated material per unit of volume must lie within certain critical values as is the case with the concentration of vapor for vapor cloud explosions. Industries adopt simple scientific measures to prevent such explosions. Such measures include, where possible, providing nitrogen gas purging to ensure that the oxygen concentration is kept below that required for combustion.

Now let us see whether the phenomena have any military relevance? There are dramatic differences between explosions involving vapor clouds and high explosives at close distances. For the same amount of energy, the high explosive blast overpressure is much higher than that from a vapor cloud explosion. The shock wave from a TNT explosion is of relatively short duration, while the blast wave produced by an explosion of fuel-air mixture displays a relatively long duration. One of the advantages of the FAEs is that they are more weight efficient, as oxygen is taken from the atmosphere for detonation, and produce high yield due to higher combustion heat of fuel.

The blast effects from vapor cloud explosions are determined not only by the amount of fuel, but more importantly by the combustion mode of the cloud. Most vapor cloud explosions that occur accidentally are deflagrations, not detonations. They would not be of much use as explosives. They can start fires alright but hardly anything else. For the vapor cloud explosion to be militarily useful in the sense that you get higher peak overpressures and more blast energy, we must achieve detonation of the fuel-air mixture. That is the difficult part and a great deal of the research in the field of FAEs by the major military powers has been directed towards solving this problem reliably.

Fuel-air explosions can only occur when the fuel concentration in the air lies within certain limits. For explosion by deflagration these are the Upper and Lower Explosion Limits (UEL and LEL); within this range lie the narrower Upper and Lower Detonation Limits (UDL and LDL).

Most FAE fuels are dispersed by a central explosive burster charge, producing a cloud of droplets and vapour that mix with the surrounding air. The cloud is then detonated after between 0.1 to 5 seconds depending on the system. During this time there is some evaporation of the droplets and further mixing as the cloud expands. This leads to large variations in concentration within the cloud, thereby further reducing efficiency. For this reason only fuels with a wide range of UEL and LEL are used:

SER	FUEL	EXPLOSIVE LIMITS (% Volume in Air)	
		LEL	UEL
(a)	(b)	(c)	(d)
1	Ethyne	2	100
2	Oxirane	3	80
3	Ethene	3	34
4	Methane	5	14
5	Propane	2	10
6	Gasoline	2	8

If the fuel concentration lies within the detonation limits then a transition from deflagration to detonation may occur if the FAE cloud is large enough. However, most systems rely on a high explosive booster to induce detonation.

A Fuel-Air Explosive (FAE) disperses an aerosol cloud of fuel (such as certain hydrocarbons and hydrocarbon oxides, viz., hexane, heptane, ethylene oxide, propylene oxide, etc.). There are **embedded detonators** alright to initiate the explosion but what exactly is the mechanism by which they ensure factors like the critical mixture of fuel with air, take its temperature uniformly beyond the flash point and its detonation instead of a mere deflagration, remains largely a military secret. Usually the FAE bombs consist of a container of a volatile liquid, in some designs including a finely powdered explosive component as slurry, and (typically) two separate explosive charges. After the munition is dropped or fired, the first explosive charge (or some other dispersal mechanism) bursts open the container at a predetermined height and disperses the fuel in a cloud that mixes with oxygen. Once the fuel is appropriately mixed, the second charge detonates, propagating a detonation through the cloud.

The rapidly expanding wave front due to overpressure flattens all objects within close proximity of the epicenter of the aerosol fuel cloud, and produces debilitating damage well beyond the flattened area. The main destructive force of FAE is high overpressure, useful against soft targets such as armored vehicles, aircraft parked in the open, and un-reinforced buildings and equipment and kills and injures personnel. The anti-personnel effect of the blast wave is more severe in foxholes, on personnel with body armor, and in 'stiff' enclosed spaces such as caves, buildings, and bunkers. Their utility against minefields is somewhat doubtful. The overpressure within the detonation can reach 430 psi and the temperature can be 2500-3000 degrees Celsius. Outside the cloud the blast wave travels at over 2 miles per second. This blast wave leaves behind a vacuum. This vacuum then pulls in loose objects. Since blast overpressure for FAEs is much less than that by high explosives, the negative pressure phase here does not play the same role which it plays for high explosives. In this case, it draws in the unexploded burning fuel to create almost complete penetration of all non-airtight objects within the blast radius, which are then incinerated. Asphyxiation and internal damage can also occur to personnel outside the highest blast effect zone, e.g. in deeper tunnels, as a result of the blast wave, the heat, or the following air draw.

(25.7.2) Weaponizing The FAE Systems: How It is Done?

The design and manufacture of a FAE weapon faces two major challenges: achieving the correct fuel-air mixture and then initiating it at the correct time. A typical FAE munition could be a circular cylinder, two or three diameters long, filled with fuel and designed to burst at an optimum height above the ground. A **burster charge** of 1 to 2% of the fuel weight is located in a tube along the central axis of the bomb. The mass of the burster charge-to-fuel ratio has little effect on resultant cloud dimensions once the ratio exceeds 1:40. The purpose of this burster charge is to break open the fuel container and distribute the fuel in a cloud with sufficient volume to contain enough oxygen for complete initiation. This volume is determined by the quantity and reaction chemistry of the fuel. FAEs are different from conventional condensed explosives in that they contain fuel elements only and do not carry their own oxygen. Hence mixing with just the right amount of oxygen is critical.

If the FAE munition contains a liquid fuel, then the burster charge has also to distribute the liquid in an aerosol form so that it can be detonated. This is a critical function as particle size and distribution influences the ability of the fuel to detonate. The fuel is ideally detonated when the cloud reaches the correct diameter for the optimum stoichiometric ratio. Then a second detonator that has been projected into the cloud initiates the explosion. The delay time between dispersion and initiation of detonation is of the order of 150 ms (this could vary from 0.1 to 5 seconds depending upon the system). The weapon system must achieve this delay very reliably. One advantage of this short delay is that local meteorological conditions have little effect on the cloud formation.

Fig. 158 The Explosion Of A Fuel-Air Explosive Bomb: Look At The Sheer Size Of The Area Covered By The Explosive Cloud: The Fireball Is truly Awesome

The expansion of the cloud after initiation is affected by aerodynamic forces acting on the droplets, which are progressively broken up until the cloud expansion ceases, and a distinctive 'pancake' shaped cloud has been formed. The vaporization of these droplets primarily takes place when they have been heated by the arrival of the shock wave or the flame front after initiation.

The following table lists some of the possible FAE fuels that have been experimentally demonstrated to work, and compares their energy output with that of TNT:

Serial	Fuel	Energy/Unit Mass (Kcal/G)	Energy/Unit Volume (Kcal/Cm^3)	Remarks
(a)	(b)	(c)	(d)	(e)
1	Decane	11.3	8.5	
2	Kerosene	10.2	8.2	
3	Propylene Oxide	7.9	6.6	Confirmed Use
4	Aluminium	7.4	11.0	Aluminium Powder
5	Ethylene Oxide	6.9	6.0	Confirmed Use
6	TNT	1.1	1.6	

From the table it should be clear that FAE, such as Propylene or Ethylene Oxide can release over 400% more energy that TNT, although the fact remains that TNT is a high explosive and hence releases its energy in a much shorter time. This distinction should always be kept in mind and the high energy of release should not be confused with destructive potential as we have explained in the first chapter itself. They are useful against 'soft targets' only. Remember that FAEs are highly spectacular weapons; but they are not miraculous weapons as most people who are confused by the total energy release, tend to think.

(25.7.3) Performance Of FAE Systems

The absence of an intrinsic oxidizer in the FAE means that a much greater proportion of the payload of the weapon can be devoted to the fuel. Therefore a greater release of energy can theoretically be achieved at a target for a munition of a given payload. Due to the high heats of explosion of potential fuels, the energy advantage can be as high as ten times that of TNT in ideal conditions. However, the fact that air contains only 21% oxygen, means that the explosive efficiency of FAE systems is less than 40%. The nitrogen in the air dilutes the system and absorbs heat and an even concentration of fuel in the atmosphere is difficult to achieve operationally. Once again this is a complicating factor, something which amateur bombers are not able to understand.

(a) TNT Equivalence: The explosive potential of a FAE system can be expressed in terms of its TNT weight equivalent (W_{TNT}):

$$W_{TNT} = (K \times W_F \times H_F) / H_{TNT}$$

Where, K is the efficiency of the explosion, W_F the weight of the fuel, H_F the heat of explosion of the fuel, and H_{TNT} the heat of explosion of TNT.

(b) Blast Effects: The initial detonation pressures and velocities of FAE are much less than those for an equivalent weight of conventional explosive.

Serial	Type	Detonation Pressure (Bar)	Velocity Of Detonation (m/s)
(a)	(b)	(c)	(d)
1	TNT	190,000	6950
2	FAE	19	1800

Therefore FAEs have little brisance or shattering effect notwithstanding the fact that they do have other advantages in the propagation of the air blast. Condensed explosive detonates as a 'point source' and the blast overpressure falls rapidly with increasing distance from the charge. In contrast, the source for a FAE explosion is a large cloud, (which for a 73 lbs charge can be up to 100 feet in diameter). This can not be regarded as a point source, and therefore the decay in peak overpressure takes place much less rapidly with distance from the edge of the cloud than is the case for condensed explosive:

The following table summarizes the overpressure from the FAE cloud-edge.

Serial	Ethene (1 Ton) / Air Explosion Distance From Cloud Edge (M)	Overpressure (% Of TNT Equivalence)
(a)	(b)	(c)
1	10	50
2	20	139
3	50	374

This low rate of fall of the peak overpressure makes the FAE into a true **'area explosive'**. Although on first sight the lower over-pressure performance of FAE against TNT would seem to be a disadvantage, it is sufficient against personnel, soft skinned vehicles, parked aircraft and antenna arrays.

(25.7.4) Initiation Sytems For FAEs

- **Explosive Initiation:** Present FAE systems are fitted with a two stage fuze. The first to fire the burster charge and the second to initiate the resultant fuel-air cloud by shockwave effect.
- **Chemical Initiation:** The advantage of chemical initiation is that a single event could be used to disperse and initiate the fuel. In this case chemicals that react with one another or the fuel are dispersed with the fuel. Heat from the reaction is then sufficient to initiate the cloud.
- **Hypergolic Reactions:** The hypergolic approach uses bromine trifluoride or chlorine trifluoride that are able to react with the fuel. Perfluorohydrazine is the interacting two component initiator.
- **Pyrophoric Initiation:** It has been tried using boron and aluminium alkyls, but is not known to have been deployed on an operational system.

(25.7.5) Currently Deployed FAE Systems

Examples of FAE systems, together with their current operational status, are summarised below:

SER	SYSTEM	COUNTRY	DELIVERY MEANS	REMARKS
(a)	(b)	(c)	(d)	(e)
1	CBU-55/B	USA	Air dropped	Operational. Contains 3 x BLU-55/B ?

contd. ...

... contd.

(a)	(b)	(c)	(d)	(e)
2	CBU-72	USA	Air dropped	Operational Contains 3 x BLU-73/B Ethylene Oxide filled.
3	BLU-64/B	USA	Air dropped	Hydrocarbon fuel
4	BLU-72/B	Air dropped	Pave Pat - Propane filled Pave Pat 2 - Ethylene oxide filled	
5	BLU-73/B	USA	Bomblet	Ethylene oxide filled
6	BLU-95/B	USA	Air dropped	500 lb Propylene oxide filled
7	BLU-96/B	USA	Air dropped	2000 lb Propylene oxide filled
8	FAACBFuel/Air Aero Cargo Bomb	China	Air dropped	Operational. Similar design to Serial 1.
10	KAB-500KrOD	Russia	Air dropped	TV guided
11	ODAB-500PM	Russia	Air dropped	
12	ODS-OD BLU Dispenser	Russia	Air dropped	Contains 8 x CBU
13	S-8D (S-8DM) 80mm Rocket	Russia	Air delivered	
14	S-13D 122mm Rocket	Russia	Air delivered	
15	AS 11 Rocket	Russia	Air delivered	Unconfirmed
16	AS 12 Rocket	Russia	Air delivered	Unconfirmed
17	SPLAV 220mm BM 9P140Uragan MLRS	Russia	Multiple Launch Rocket System	May be also be known as TOS-1 220mm MLRS?
18	NORINCO 305mm FAE Minesweeping MLRS	China	Multiple Launch Rocket System	Operational.
19	SLUFAESurface Launched Unit FAE	USA	Multiple Launch Rocket System	Trials only, now stillborn.
20	SHTURM ATGM	Russia	Helicopter launched	ATGM
21	KORNET-E LRATGM	Russia	Ground launched	
22	CATFAECatapult Launched FAE	USA	Catapult	Trials and Development. Operational status undefined.

(25.7.6) CBU-72 Fuel-Air Explosive (FAE) Cluster Bomb

The 550 lbs CBU-72 cluster bomb contains three FAE sub-munitions **BLU-73/B**. The sub-munitions weigh approximately 100 lbs each and contain 75 pounds of ethylene oxide with air-burst fuzing set for a height of 30 feet. The container is a little over seven feet long and 14 inches in diameter. An aerosol cloud approximately 66 feet in diameter and 7-10 feet thick is created and ignited by an embedded detonator to produce an explosion. The result is a spectacular fireball that is highly effective against exposed targets.

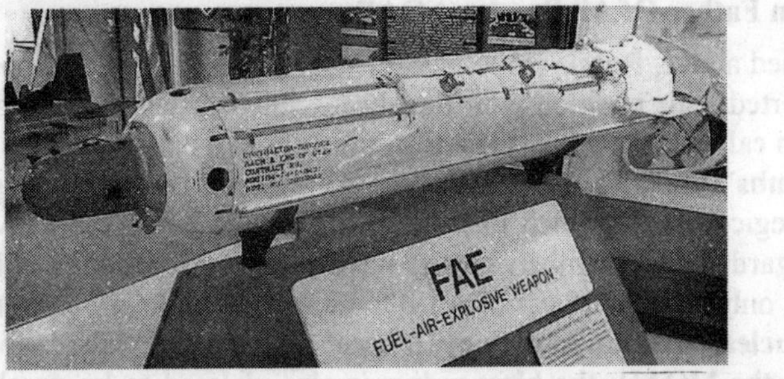

Fig. 159 Fuel-Air Explosive—The 550 Lbs CBU-72 FAE Bomb Containing 3 BLU-73/B Sub-Munitions Of 100 Lbs Each: On Display In The Goleta Air & Space Museum

(25.8) ENHANCED BLAST AND THERMOBARIC MUNITIONS

(25.8.1) Enhanced Blast Munitions (EBM)

Parallel to the development of FAE was the development of Enhanced Blast Munitions (EBM). An EBM is little more than an improved efficiency high explosive, obtained by the addition of metal powders that release energy when they oxidize in the high temperatures of the explosion. We mentioned them here because some people mistakenly associate them with FAEs.

(25.8.2) Thermobaric Munitions

The thermobaric weapon works by propelling a warhead that scatters an aerosol explosive on or before impact with the target and then immediately ignites this to create a high-pressure blast wave. The effect is a much more rapidly expanding blast than obtained in a conventional explosion. Compared with a fuel-air explosive, the thermobaric weapon has a much higher expanding concussion effect and hence they are believed to be more useful against personnel hiding in tunnels etc. They could be using either a finely powdered solid or a volatile liquid.

In contrast with traditional FAEs, which generate a 'big ball of fire', the thermobaric aerosol detonates with explosive shock much more like that of a normal explosive filling. The result is a crushing high-pressure shockwave that propagates through the tunnel system, followed by an asphyxiating wave of flame. The **BLU-118/B**, for example, is described as a thermobaric munition, the term indicating that the bomb produces both thermal and pressure effects, and is intended for attacks on underground tunnel targets. They have been given a popular name of **vacuum bombs** also but we do not consider that to be proper.

Russia is currently the world leader in thermobaric weapons, and has already used them in action, for example, to clear the Mujahideen from caves in Afghanistan and more recently Chechen guerrillas from buildings in Grozny. The Russian **RPG-7 thermobaric warhead**, fired from a man-portable rocket-launcher, is said to produce effects comparable to the detonation of 2 kg of TNT, while the effects of the Russian **RPO-A Shmel Bumblebee rocket-propelled incendiary/blast**

projectile are reportedly similar to those produced by a 152 mm howitzer projectile. Russian variants of this latter weapon include one that combines a thermobaric warhead with a small hollow charge, which is designed to penetrate structures prior to detonation of the main warhead, thus considerably enhancing its effects.

A FAE system from Israel was developed for minefield clearing. The system uses a small rocket-propelled thermobaric charge which explodes over the minefield and activates exposed or buried mines.

(25.8.3) The Russian Father Of All Bombs (FOAB)

As mentioned above, Russia is the world leader in the field of thermobaric weapons. They seem to have reasserted their position with the successful field test of a bomb on September 11, 2007 which has been called the **'Father Of All Bombs'** (FOAB) in an obvious comparison with the **'Mother Of All Bombs'** (MOAB) of the USA. The bomb was dropped with a parachute from a Tu-160 Blackjack strategic bomber which has the distinction of being the heaviest combat aircraft ever built. Data regarding the explosive used were not released but it is believed that it is a thermobaric weapon only. The yield was claimed to be equivalent to about 44 tons of TNT, making it the **largest non-nuclear weapon in history**. In terms of yield it is claimed to be nearly four times more powerful than the MOAB; the blast radius is also claimed to be nearly 1000 feet, that is, about two times higher than that of the MOAB. It is also claimed that the so-called 'highly efficient' explosive used in the FOAB weighs only 17160 lbs in comparison to the 18,700 lbs of high explosive used in the MOAB and yet is much more powerful, producing temperature at its epicenter which was also twice that in the MOAB. There were press reports of the bomb having used nanotechnology also to increase the power of the bomb but how exactly that has been done is not clear. The MOAB was claimed to be capable of destroying nine city blocks in one strike—hence the destructive potential of the FOAB can be easily imagined.

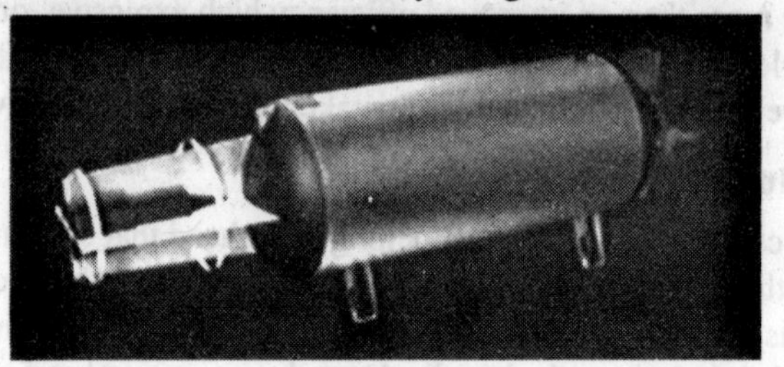

Fig. 160 The Father Of All Bombs: The Biggest Non-Nuclear Bomb Ever Made In History: A Marvel of Russian Expertise In Thermobaric Explosives

(25.8.4) Solid FAEs

The US Air Force has been experimenting with a **'solid FAE (SFAE)'** filling that apparently disperses a fine cloud of aluminum particles, while the Navy has been working on an **'interhalogen oxidizer'** filling, possibly based on fluorine.

In 2003, United States Marines used a thermobaric version of their Shoulder-Launched Multipurpose Assault Weapon, called a **Shoulder-Launched Multipurpose Assault Weapon - Novel Explosion (SMAW-NE)** in Iraq. One team of Marines reported that they had destroyed a large one-story masonry type building with one round from 100 yards. The thermobaric explosive used in this weapon, **PBXIH-135** has been described earlier.

□□□

Chapter 26

Bombs Take A Technological Leap and Become Smart

(26.1) THEY HIT WHERE IT REALLY HURTS

(26.1.1) What Are Smart Bombs Or Precision-Guided Munitions And Why Do We Need Them?

Precision-guided munitions (smart munitions or smart bombs) are self-guiding weapons which achieve a very high degree of accuracy in hitting a target. The idea is to maximize damage to the target while minimizing collateral damage. As we have seen, the explosive effects of a bomb with respect to the distance at which you are measuring the effects follow a power law. This means that if your accuracy is poor then you would require a much bigger bomb or a large number of smaller bombs to do the job. Conversely, it means that if you could improve your accuracy then you can do the job with a smaller bomb or with a smaller number of bombs. This is precisely what the so-called smart bombs do. Bombs falling freely under gravity from a great height were notorious for their inaccuracy. And that was but to be expected. The aircrafts had to fly at considerable heights to save themselves as much from anti-aircraft fire as possible and they had to release their whole load of bombs before they were shot out of the sky. Obviously no one had the time or the foolhardiness to release their bombs in a very calculated manner. In World War II it could take 9,000 bombs to adequately damage a target the size of an aircraft shelter; in Vietnam it took 300; today it can be done with just one laser-guided munition from an F-117, albeit at a high cost! Smart bombs are equipped with guidance technology that actively seeks a target with little or no direction from a remote controller after the munitions are launched.

(26.1.2) How A Smart Bomb Works?

A smart bomb is essentially an ordinary dumb bomb with a few major modifications. In addition to the usual fuze and explosive material, it has:

- an **electronic sensor system**;
- a built-in control system (an **onboard computer**);

- a set of **adjustable flight fins**; and
- a battery

When a plane drops a smart bomb, the bomb becomes a sort of heavy glider. It doesn't have any propulsion system of its own, like a missile does, but it does have forward velocity (by virtue of having been dropped from a speeding plane). It also has flight fins that generate lift and stabilize its flight path. The control system and adjustable fins give the bomb a way to steer itself as it glides through the air. While the bomb is 'in flight', the sensor system and control system track the designated target on the ground. The sensor system feeds the control system the relative position of the target, and the control system processes this information and figures out how the bomb should turn to steer toward the target. To actually turn the bomb, the control system sends a message to actuators that adjust the flight fins. These fins work the same basic way as the various flaps on an airplane. By tilting the fins in a particular direction, the control system increases the drag acting on that side of the bomb. As a result, the bomb turns in that direction. This adjustment process continues until the smart bomb reaches its target, and the fuze mechanism sets off the explosive. Smart bombs generally have proximity fuzes or impact fuzes.

The main difference between the different types of smart bombs is how the sensor system actually 'sees' the target in the first place. In the following we will discuss some of the technologies used and some famous bombs using them.

(26.2) ELECTRO-OPTIC OR TELEVISION-GUIDED BOMBS

(26.2.1) The AGM-62 Walleye

One simple way to solve the problem of 'seeing' the target was to mount a TV camera in the nose of the bomb. Thus you could get a 'bomb's eye' view of the target and transmit it back to the aircraft controlling it. An operator in the aircraft could send controlling signals to the bomb and direct it to the target using steering fins mounted on the bomb. In this mode, the operator may launch the bomb without a specific target and sight, and then pick up the target from the video as the bomb gets closer to the ground. There is no propulsion system in the bomb and it actually glides towards the target, making it a glide bomb. In automatic mode, the pilot locates a target through the bomb's video camera prior to launch and sends a signal to the bomb telling it to lock on to the target. The bomb's control system steers the bomb so that the indicated target image always stays near the center of the video display. In this way, the bomb zeros in on the locked target automatically.

Fig. 161 Walleye: The First Smart Bomb: A Television-Guided Bomb And A Warhorse Of The Vietnam War

The AGM-62 Walleye was introduced in the Vietnam War in 1967 and served till 1995. The US Navy was its main user. The main motivation for developing this bomb was to destroy the bridges which were vital for the VietCong supply lines and extremely difficult to hit by unguided bombs. The Walleye was a purpose-built bomb and not a modified dumb bomb. It was over 11 feet long, one foot thick and weighed 1100 lbs with 825 lbs warhead. It had four large long-chord delta fins for control. Advanced versions of the Walleye called **Walleye II, Walleye ERDL (Extended Range-Data Link)** and **Walleye ERDL DPSK (Digital Phase-Shift Keying)** were also developed and used in that period. The warhead was also increased to up to 2000 lbs.

In operation, the image provided by the Walleye's seeker was displayed on a TV screen in the controlling aircraft. The operator locked the crosshairs on the display on to a high-contrast scene element like a window, door or some prominent structural element, set the fuzing option as needed, and then released the weapon. The weapon would then glide accurately into the target, guided by the target's contrast pattern. The seeker system used an analog pattern-matching scheme. It converted the image into two analog electrical waveforms, one for the vertical axis of the image and one for the horizontal axis, with the amplitude of the waveforms reflecting scene brightness. The seeker control electronics would then detect the transitions in the waveforms corresponding to abrupt changes in scene brightness, and use them as 'goalposts' to guide the missile to target. As such it could be said to be a **'fire and forget' weapon** but the seeker needed to have a stable high-contrast pattern to work properly, and so it sometimes would break lock if the day were gray and hazy, or there were shifting clouds or sources of transient sun glare. However, when the weapon worked, it worked very well, with accuracies of a few meters.

(26.2.2) The HOBOS

The US Air Force developed a guided bomb of its own in 1969. It was called **HOBOS (Homing Bomb System)** and officially known as **GBU-8**. Though the guidance technology was similar, it was an improvement over the Walleye in the sense that it was a kit that could be fitted over an existing MK-84 2000 lbs or MK-118 3000 lbs bomb. The kit included a tail fin section that had four square fins and contained a battery and control and communication electronics, and a nose section that contained an electro-optic seeker system. The two sections were linked by four long slender fins, or **'strakes'**, and an umbilical conduit that ran along the length of the bomb from the tailfin section to the nose section. A HOBOS weapon based on the Mark 84 slick bomb was a little less than 13 feet long, had a span of about four feet, and weighed 2,240 lbs.

(26.2.3) Pinnacle Of Electro-Optic Guidance Technology: The GBU-15

The GBU-15 bomb is an un-powered, glide weapon used to destroy high-value targets and represents the state-of-the-art for this technology. One GBU-15 costs US $ 0.3 million! This highly maneuverable weapon has a low-to-medium altitude delivery capability with high accuracy. It also has a standoff capability with a range of over five nautical miles. The Americans made very effective use of the GBU-15/B during the Gulf War, dropping 71 of the weapons to seal off oil manifolds wrecked by the Iraqis to spew out oil, and other targets.

The GBU-15(V) 1/B is a **TV-guided bomb** and the GBU-15(V) 2/B is an **infrared-guided weapon**. The first was developed in 1983 and the second in 1985. The GBU-15 provides the capability for accurate (automatic or manual) guided delivery of a MK-84 bomb or a BLU-109 penetrator bomb at increased ranges. The GBU-15/B is conceptually similar to HOBOS, with tail and nose sections that can be strapped to a 2,000 lbs MK-84 GP bomb. The standard GBU-15/B with a MK-84 has a length of a little less than 13 feet, a wingspan of about five feet, and a weight of 2,500 lbs. The GBU-15/B had a distinctively different appearance from HOBOS, however, with triangular nose fins on the seeker section and large truncated delta fins on the tail section, with both sets of fins arranged in a cruciform pattern. The larger fins give the GBU-15/B a longer glide distance than HOBOS. When the GBU-15/B is configured to use the BLU-109/B 2,000 lbs penetrator warhead, it features smaller 'short chord' wings.

The GBU-15's effective standoff range is greater than that of laser-guided munitions, since the GBU-15 does not need to have acquired the target before it is released. The weapon is remotely controlled by a datalink system, and the weapon systems operator locates the target area and the specific aim point by observing the video transmitted from the weapon. The weapon's mid-course flight path can be adjusted either automatically or manually. Weapon video is either electro-optical (TV camera) or infrared, and generated in the nose of the weapon. The guidance section is attached to the nose of the weapon and contains a television guidance system for daytime or an imaging infrared system for night or limited, adverse weather operations. A datalink in the tail section sends guidance updates to the control aircraft that enables the weapon systems operator to guide the bomb by remote control to its target. The rear control section consists of four wings that are in an 'X'-like arrangement with trailing edge flap control surfaces for flight maneuvering. The control module of the bomb contains the autopilot, which collects steering data from the guidance section and converts the information into signals that move the wing control surfaces to change the weapon's flight path.

The GBU-15 may be used in direct or indirect attack. In a direct attack, the pilot selects a target before launch, locks the weapon guidance system onto it and launches the weapon. The weapon automatically guides itself to the target, enabling the pilot to leave the area. In an indirect attack, the weapon is guided by remote control after launch. The pilot releases the weapon and, via remote control, searches for the target. Once the target is acquired, the weapon can be locked to the target or manually guided via the datalink system.

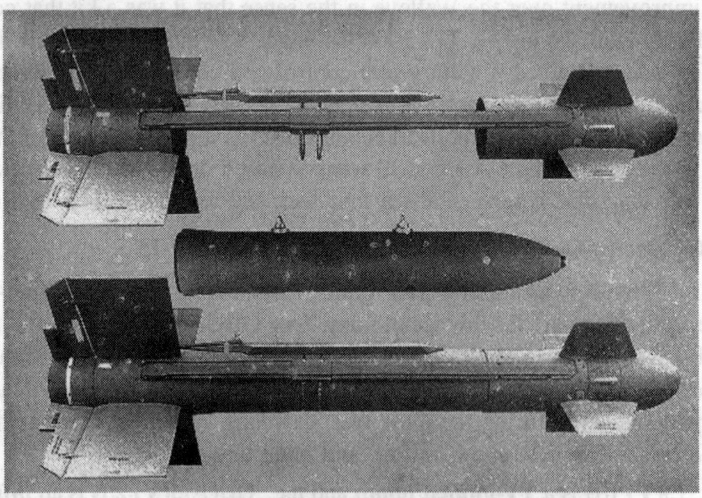

Fig 162. GBU-15: Representing The State-Of-The-Art In Television/Infrared Guided Smart Bombs

More improvements in the GBU-15 are underway. The **Inertial Terrain-Aided Guidance (ITAG) system** is an adverse weather, precision guidance system for the GBU-15. It replaces the guidance system for this weapon. ITAG is an inertial navigator which uses updates from a radar altimeter correlated with terrain elevation maps. GPS is used to initialize the inertial navigator prior to weapon release. Unlike current laser-guided bombs which can only be employed in clear weather conditions, the ITAG will be able to achieve a less than 10 feet circular error probable (CEP) accuracy in adverse weather.

(26.3) LASER-GUIDED BOMBS (LGB)

These are the bombs which were so aggressively advertised during the Operation Desert Storm. You were shown video clippings of LGBs traveling precisely down ventilator shafts and destroying targets with a single strike! The claim was that they were using such accurate bombs so that only the intended targets were destroyed and there was little collateral damage. How clever! The fact is that the number of LGBs used in Iraq was very small in comparison to the dumb bombs. The Americans obviously kept mum about the immense collateral damage caused by the vast numbers of dumb bombs used by them. We talked of the **GBU-28** and the **GBU-24 penetrator bombs** earlier; well, they are laser-guided bombs.

(26.3.1) The Operating Principle Of Laser Guidance

In simple words, lasers have two qualities. The first is that unlike the beam of a flashlight, the laser beam is highly collimated—it does not diverge. You can shoot a laser beam to the moon and get it reflected back with little loss! The second is that it is highly monochromatic, that is, of a single color as against white light which is a mixture of a myriad colors or even ordinary colored light which has a range of wavelengths in it. The fact that a laser beam remains highly collimated over even very long ranges means that it can be used as a precise pointing device. A laser could be strapped to a telescope with cross-hairs so that the beam could be focused to 'illuminate' a particular target to 'mark' or 'designate' it. When used with a bomb, the bomb has only to detect the laser light that is reflected off the target and then adjust its course in such a manner that it keeps on getting the reflected light; if it fails to get the light, it would mean that it is off-course and it has to correct its path. That way it can home in on to the target. Further, the fact that the laser light is highly monochromatic also means that the light reflected off such a target can be easily detected by simple sensors through a filter attached to the lens. The guided bomb could be fitted with such a sensor, with the sensor linked to feedback-control mechanisms so that it would respond only to the laser light reflected off the illuminated target and not get confused with other spurious signals.

In its simplest form, the seeker could have an optical sensor, shielded by a filter lens that would be transparent to laser light but would block light of other wavelengths. The sensor could be a simple array of photodiodes, arranged in quadrants. The quadrant that would pick up the maximum light energy would activate the fins, which could be operated in a simple **'bang-bang' control mode**, that is, they would be either deflected completely to one or the other limit of their range of movement, or would remain straight. The fins could operate in pairs arranged symmetrically around the guided bomb to shift the weapon up and down or left and right. Laser-guided bombs have their own power supply, consisting of a thermal battery for the electronics, along with a hot gas-driven actuator to move the seeker head.

(26.3.2) The Principle Translated Into Practice

In practice, laser designators and seekers use a **pulse coding system** to ensure that a specific seeker and designator combination work in harmony. By setting the same code in both the designator and the seeker, the seeker will track only the target designated by the designator. The pulse coding is based on **Pulse Repetition Frequency (PRF)**. Coding allows simultaneous or nearly simultaneous attacks on multiple targets by a single aircraft, or flights of aircraft, dropping laser guided weapons (LGWs) set on different codes. This tactic may be employed when several high-priority targets need to be expeditiously attacked and can be designated simultaneously by the supported unit(s).

Before dropping the bomb, the bomber aircraft computer tells the missile's control system the specific pulse pattern (via an electronic 'umbilical' connection to the bomb). Once the bomb is in the air, the control system is only interested in laser energy with this pulse pattern. The control system's basic goal is to steer the bomb so that the reflected laser beam is hitting near the center of the photodiode array. This keeps the bomb heading straight toward the target.

Modern versions of laser-guided bombs have a **computer-control group (CCG)**, guidance canards attached to the front of the warhead to provide steering commands, and a wing assembly attached to the aft end to provide lift. The CCG transmits directional command signals to the appropriate pair(s) of canards. The guidance canards are attached to each quadrant of the control unit to change the flight path of the weapon.

An important component of the LGB is the **targeting pod**. Earlier the designating laser was carried in another two-seater aircraft. It consisted of a TV camera bore-sighted with the laser. You needed a two-seater aircraft so that the other man (WSO—the weapons system officer) could keep on viewing the target image provided by the pod, and using a joystick or similar hand controller to keep the target lined up in the display cross-hairs. But TV is limited to daylight and fair weather. One solution was to use a **'forward looking infrared (FLIR)' imager**, which could be used night or day, and also to a degree in murky weather. FLIR generally displaced TV cameras in **'second-generation' targeting pods**. A modern **'third generation' targeting pod** may typically contain a FLIR with multiple FOVs (field of views) for targeting, a solid state charge-coupled device (CCD) visible-light camera, also with multiple FOVs, and a GPS-INS navigation system, or terrain-following radar so that the WSO is able to download the GPS coordinates of a target obtained by the pod into a weapon before launch. Improvements in radar miniaturization have led to a new class of targeting pod carrying **'synthetic aperture radar (SAR)'**. A SAR is a type of imaging radar that uses the motion of the carrier aircraft to give the effect of a larger antenna, improving the resolution of the radar image. SAR can be used day or night, in any weather, to pinpoint targets. The US Navy became interested in acquiring a SAR attack capability after the Kosovo campaign in 1999, when poor weather badly hampered NATO air strikes. The Super Hornet is expected to eventually obtain a SAR capability in any case.

(26.3.3) Limitations Of Laser-Guided Bombs

LGBs are excellent performers in dive deliveries initiated from medium altitude in clear weather. The effects of smoke, dust, and debris can impair the use of laser-guided munitions. The reflective scattering of laser light by smoke particles may present false targets. Rain, snow, fog, and low clouds can prevent effective use of laser-guided munitions. Heavy precipitation can limit the use of laser designators by affecting line-of-sight. Snow on the ground can produce a negative effect on laser-guided munition accuracy. Fog and low clouds will block the laser-guided munition seeker's field of view which reduces the guidance time. This reduction may affect the probability of hit. The LGBs can also miss the target if the laser is turned on too early. During certain delivery profiles where the LGB sees laser energy as soon as it is released, it can turn from its delivery profile too soon and miss by falling short of the target. To prevent this, the laser designator must be turned on at the time that will preclude the bomb from turning down toward the target prematurely.

Thus the manufacturers' claim of '**one target-one bomb**' may be exaggerated. In practice, they had to drop on an average of four LGBs per target.

(26.3.4) The Paveway III (GBU-24) And Its Predecessors

The **Paveway I** was the first laser-guided bomb developed during the Vietnam War. In that era the LGB kits cost only a few thousand dollars, and were produced in great quantity. Tens of thousands of them were used in Vietnam, where 2000 lbs Paveways destroyed bridges that had survived repeated conventional bombing raids.

The Paveway I consisted of a kit that was attached to ordinary MK-82 500 lbs, MK-83 1,000 lbs, and MK-84 2,000 lbs bombs. The kit included a laser seeker head attached to four control fins in a cruciform arrangement, which was attached to the front of the bomb, and a set of four larger fins, also in a cruciform arrangement, which was attached to the rear of the bomb to provide some limited glide capability. All the LGBs used the same seeker head, but had different fin assemblies to accommodate different types of bombs. The LGBs had their own power supply, consisting of a thermal battery for the electronics, along with a hot gas-driven actuator to move the seeker head. They did not require any electrical connection to the aircraft. They could be launched by any aircraft that could carry ordinary bombs of the same size, and could be guided by a laser designator on the launch aircraft, on a spotter aircraft, or operated by ground forces.

Paveway I had fixed wings. The **Paveway II** came to have retractable wings. These weapons featured an enhanced but simpler and cheaper seeker head, and pop-out fins for the rear assembly to improve the weapon's glide characteristics, and make it easier to fit to aircraft. The Paveway II required the launch aircraft to operate from medium altitude so that the bomb would have direct path to the target. This left the aircraft vulnerable to ground defenses, and so in 1976 the USAF issued a requirement for yet another generation of Paveway weapons, the '**Paveway III**'. The Paveway III used larger rear fins and a 'smart' seeker head that incorporated a microprocessor-based digital autopilot, coupled to a more sensitive seeker with a wider field of view.

Fig. 163 Low Level Laser Guided Bomb Paveway III Or The GBU-24/B: One Of the Most Publicized Bombs Of The Gulf War Used To Destroy Hardened Aircraft Hangars And Underground Bunkers

The Paveway II had a simple **'bang-bang' control mode**; the Paveway III has **'proportional guidance'**. Bang-bang control was simple, but wasted much of the bomb's energy and reduced its range. The bomb could be released at low altitude and then sail almost level into the target area, where it could then dive directly into a target. However, tests of the improved LGB kit with 500 lbs bombs showed such weapons to be unstable, so the USAF only fielded the 2,000 lbs version, with test and evaluation completed in 1986.

Now the Paveway III consists of either a 2,000 lbs MK-84 general purpose or BLU-109 penetrator bomb modified with the **low-level laser-guided bomb (LLLGB)** kit with larger wings designed for low altitude delivery and with a capability for improved standoff range of 11.5 miles to reduce exposure. The warheads are now integrated with Hard Target Smart Fuzes also. The 2,000 lbs Paveway III was known as the **'GBU-24/B'**. In the Gulf War they used about 9000 Paveway II and III.

Video recordings taken by targeting systems showed Paveways and other smart munitions dropping into the front doors of aircraft shelters and falling into the ventilation shafts of buildings and blowing them apart. These proved a striking and impressive propaganda tool for US forces and were repeatedly shown by all the channels across the world to claim that their precision strikes were intended to reduce the collateral damage on targets in built-up urban areas. However, videos that showed the effects of high explosive blasts on humans were very cleverly not released to the public by the American propaganda machine and the media controlled by them. One Paveway was even used to blast an Iraqi helicopter out of the sky, an action-movie stunt that amused the adolescent Americans and their toadies immensely. The American propaganda machinery also maintained that in the 2003 Gulf war they had also used the Paveway II as a **'smart rock'** to hit specific targets without explosives so that collateral damage could be minimal. They claimed that they had taken BDU-45/B dummy bombs and filled them with concrete so as to have a 500 lbs non-exploding block. They were fitted with Paveway II guidance kits and used to attack unprotected urban targets. The effect was like a giant 500 lbs hammer hitting the building. The idea was very appealing in theory but unfortunately it was essentially for propaganda purposes only. Whatever numbers of these concrete bombs were used, they neither destroyed anything of consequence nor were of any consequence in shaping the course of war. There is no reason to believe that they can ever replace explosive bombs. **GBU-45/B** is most probably not even an official designation.

The **Enhanced Paveway Laser Guided Bomb**, designated **GBU-24E/B**, is a precision-guided hardened target penetrator used to destroy hardened aircraft hangers and underground bunkers. It integrates a Global Positioning System and a ring laser gyro inertial measuring unit (IMU) to the already fielded GBU-24B/B "Paveway III" with the existing laser guidance.

The famous 'bunker-buster' GBU-28/B bomb also uses the Paveway III kits.

(26.4) GPS-GUIDED AND INERTIAL GUIDANCE BOMBS

As we saw, laser-guided bombs don't work very well in bad weather, or the dusty and smoky conditions that can often occur in combat. The US military had long wanted a true **fire-and-forget guidance system**, where the weapon could simply be given target coordinates and then let it proceed to destroy it. In the early 1990s, the US military completed its GPS (Global Positioning System) navigation satellite constellation. The satellites of the GPS network transmit signals that allow a device with a GPS receiver to determine its own location to within a few tens of meters or better. GPS allows a weapon to be given coordinates and find the target on its own. As GPS signals are faint and in principle can be easily jammed, it is important for a **GPS-guided munition** to have an **INS (Inertial Navigation System)** based on fiber-optic gyroscopes or solid-state accelerometer sensors, to keep the weapon on course if it can no longer acquire GPS signals. The bomb reverts to inertial navigation in the event of losing the GPS signal. Inertial navigation is significantly less accurate; the **JDAM (Joint Direct Attack Munition)** achieves a CEP of about 40 feet under GPS guidance, but typically about 100 feet under inertial guidance. Further, the inertial guidance CEP increases as the dropping altitude increases, while the GPS CEP does not. The precision of satellite-guided weapons is dependent both on the precision of the measurement system used for location determination and the precision in setting the coordinates of the target. The latter critically depends on intelligence information, not all of which is accurate. However, if the targeting information is accurate, satellite-guided weapons are significantly more likely to achieve a successful strike in any given weather conditions than any other type of precision-guided munition. The **MOAB** discussed earlier is also a GPS-INS guided bomb.

The GPS-INS guidance system doesn't render electro-optical or laser guidance systems obsolete, which have higher precision and are usually better suited for attacks on mobile targets whose coordinates can't be easily determined in advance. However, it has proven useful to equip both the older electro-optical and laser-guided systems with backup GPS-INS guidance systems to allow them to stay on course even if seeker lock is lost.

(26.4.1) The JDAM (Joint Direct Attack Munition)

After the Gulf War, the US Air Force decided to develop a GPS-guided bomb known as the 'Joint Direct Attack Munition (JDAM)'. The basic idea behind the JDAM program is to outfit existing 'dumb' bombs with sophisticated rear guidance sections. Actually they wanted to have a precision weapon for their new B-2 stealth bomber to make use of the aircraft's impressive attack capabilities. The B-2 featured a precision, all-weather **'GPS-Aided Targeting System (GATS)'**, and a GPS-guided bomb coupled to GATS was exactly what was needed. This resulted in the JDAM. The Joint Direct Attack Munition (JDAM) is actually a guidance tail kit that converts existing unguided free-fall bombs into accurate, adverse-weather 'smart munitions'. With the addition of a new tail section that contains an inertial guidance system and a Global Positioning System guidance control unit, the JDAM improves the accuracy of unguided, general purpose bombs like the 2,000 lbs BLU-109/MK-84 bomb or the 1,000 lbs BLU-110/MK-83 bomb in any weather condition. The JDAM GPS guidance kit provides a new tail with a **GPS-INS guidance system**, along with a set of strakes that are strapped around the bomb mid-body to control the bomb's fall. The guidance system is programmed by an umbilical connection, though a wireless infrared link is in the works. The stand-off range is up to 24 km. The unit cost per tail kit is about US $ 21,000. The low cost explains the enthusiasm about it. First tests of the JDAM were performed in late 1996. The JDAM was introduced to combat during the 1999 Kosovo campaign, with the B-2 stealth bomber using the weapon to good effect. Incidentally, Kosovo was also the first combat use of the B-2! They also employed the JDAMs in 2001, during the campaign against Afghanistan, dropping them from B-52s and B-1s also. The JDAM was also extensively used during the US invasion of Iraq in the spring of 2003. The JDAM has been marketed so well that there has been talk of ending use of unguided bombs altogether.

The **'JDAM tail kit'** includes adjustable tail fins, a control computer, an inertial guidance system and a GPS receiver. Both the GPS receiver and the inertial guidance system allow the bomb to locate itself in space. The GPS receiver figures out its position by interpreting GPS satellite signals, while the inertial guidance system monitors the bomb's movements, tracking its path from its launch position. Before dropping the bomb, the aircraft uses its own GPS receiver to pinpoint particular targets on the ground. Just before releasing the bomb, the aircraft's computer feeds the bomb's computer its current position and the GPS coordinates of the target. In the air, the JDAM's GPS receiver processes signals from GPS satellites to keep track of its own position. As with other smart bombs, the control system adjusts the flight fins to steer the bomb in the right direction. According to the U.S. Air Force, the system is accurate to within 40 feet. When everything goes exactly right, the bombs generally hit within a few feet of their targets.

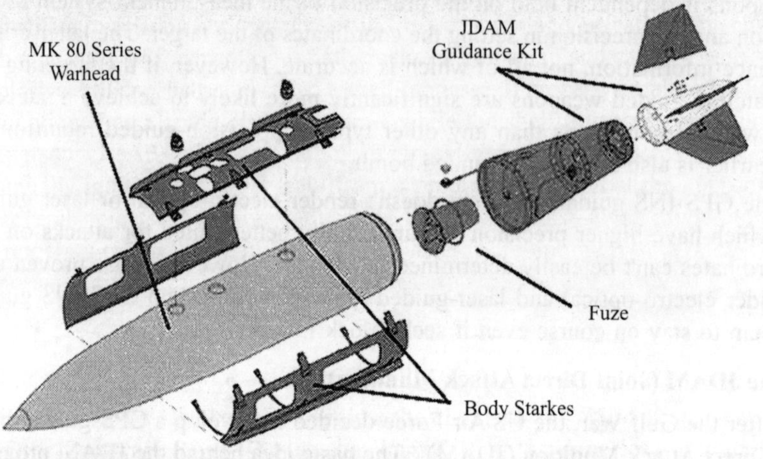

Fig. 164 Structure Of The JDAM, A GPS Guided 1000-2000 Lbs Smart Bomb With An Accuracy Of 40 Feet At A Launch Distance Of 15 Miles

This system works fine even in bad weather, because the JDAM gets all its information from satellite signals, which aren't blocked by cloud cover or obstacles. The bomb doesn't have to see anything at all to find its way to the target. And at around US $20,000 per tail kit (which can be added to an existing warhead), it's much more economical than laser-guided bombs costing over US $120,000!

JDAMs also allowed the heavy bombers to fly all-weather close support missions, a role which their original designers could not have imagined. The bombers take off with a load of the JDAMs without having specific targets and orbit over the battle area, above cloud cover. If ground forces or spotter aircraft find a target, they report the target coordinates through an intelligence network, which passes attack instructions and coordinates on to a bomber. The navigation system is initialized by transfer alignment from the aircraft that provides position and velocity vectors from the aircraft systems. The bomber crew loads the coordinates into the required number of JDAMs, sets them for air, surface, or penetrating detonation, proceeds to the target area, and releases the bombs. Once released from the aircraft, the JDAM autonomously navigates to the designated target coordinates. In its most accurate mode, the JDAM system can provide a weapon circular error probable of 40 feet or so (some Boeing sources report less than 33 feet CEP) during free flight when GPS data is available. If GPS data is denied, the JDAM can still achieve about 100 feet CEP or less for free flight times up to 100 seconds with a GPS quality handoff from the aircraft. Interestingly, Boeing says there is no great interest in improving JDAM's resistance to GPS jamming. They maintain that: "JDAM is not GPS guided, it is GPS aided, and the **IMU (inertial measurement unit**) on it is so good that once it has (target) data transferred from the aircraft, it is extremely accurate with GPS denied." CEP using the IMU acceptable for larger munitions, and an advanced terminal seeker may make GPS something that is nice to have but can be done without.

Fig. 165 The JDAM

The USAF is deeply interested in a 500 lbs JDAM designated **GBU-38/B** based on the MK-82 since it will allow B-1 and B-2 bombers to drop large clusters of precision weapons in a single sortie, each hitting a separate target. Boeing is also performing studies on a winged version using a '**Diamondback pop-out wing kit**' for greater standoff range, as well as a terminal guidance seeker capability. It is expected to more than double the JDAM's stand-off range to about 40 miles. The goal of the terminal guidance seeker is a CEP of 10 feet, and the ability to attack moving targets. Boeing is also performing experiments to add laser guidance to the JDAM. The US Air Force has also begun experiments on the JDAMs to attack moving targets, using JDAMs modified with a datalink to permit course corrections up to weapon impact. Such a datalink would allow a JDAM with switchblade wings to be dropped from a standoff distance and then guided to target by a battlefield surveillance aircraft or potentially even a space-based surveillance platform.

(26.4.2) The AGM-154 JSOW (Joint Stand Off Weapon)

While the JDAM provided the US military with a degree of short range standoff attack capability, other programs have worked on weapons to attack targets at greater ranges. The JSOW project was initiated in 1986 under the initial name of '**Advanced Interdiction Weapon System (AIWS)**', the goals being a lightweight, low cost, fire and forget weapon with medium range and the capability to carry different types of payloads. A development contract was awarded to Texas Instruments in 1992. Texas Instruments was later acquired by Raytheon. Initial tests of the JSOW began in 1994. The 'Joint Stand Off Weapon (JSOW)' was introduced to combat in January 1999, during air strikes by US Navy strike aircraft on Iraqi air-defense sites, and was used in later combat actions also.

The **Joint Stand Off Weapon (JSOW)** is a joint venture of the US Navy and Air Force to deploy a standardized medium range precision guided weapon, especially for engagement of defended targets at ranges outside that of standard anti-aircraft defenses, thereby increasing aircraft survivability and minimizing friendly losses. The JSOW family of kinematically efficient, air-to-surface un-powered glide bombs, in the 1,000 lbs class, provides standoff capabilities from 17 miles low altitude launch to 46 miles high altitude launch. The JSOW has **pop-out switchblade wings** and a **GPS-INS guidance system**, making it a **'launch-and-leave'** weapon. It uses a datalink for terminal homing. The weapon is smart enough to fly a pre-planned path to its target, make turns and hide behind mountains. If the proper pre-planned launch point cannot be reached, the pilot can release it and, as long as the target is within a pie-slice wedge on the cockpit display, let the weapon determine its own flight plan. The JSOW is light enough to be carried by smaller attack aircraft such as the F/A-18, the F-16, or AV-8B Harrier.

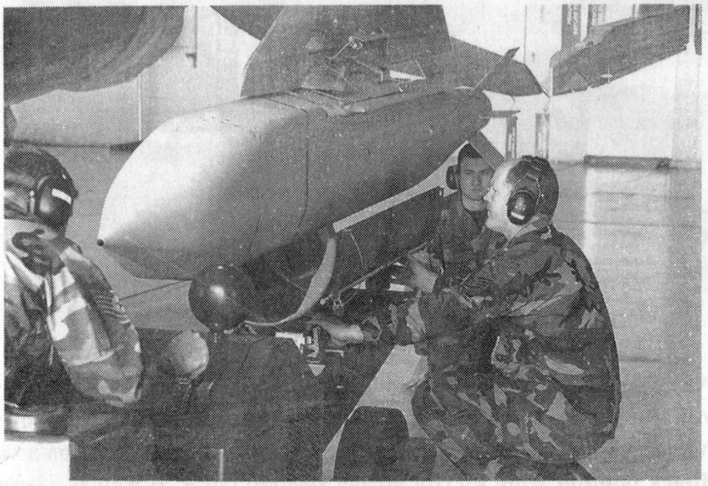

Fig. 166 The AGM-154 JSOW (Joint Stand Off Weapon), A GPS-INS Guided 'Launch-And-Leave' 1000 Lbs Smart Bomb That Can Be Launched From A Distance Of 46 Miles—Shown Here Mounted On An Aircraft

The JSOW comes in three variants, A, B, and C, designed for use against area targets, light-to-medium armored targets, and point targets, respectively. The JSOW is about 13 feet in length and weighs up to 1500 lbs. The wing span is about nine feet.

The **AGM-154A** carries 145 **BLU-97A/B CEMs (Combined Effects Munitions)** for use on 'soft' targets. When the weapon makes its final attack dive on a target, it blows off covers on either side of its boxy fuselage, and a gas-inflated aluminum bladder scatters the sub-munitions out on the sides. The CEM sub-munitions produce armor-piercing, fragmentation, and incendiary effects.

The warhead for the **AGM-154B** is the **BLU-108/B** from the Air Force's **Sensor Fuzed Weapon (SFW) program**. There were reports, however, that the program was cancelled in favor of the **WCMD-ER** which could carry 10 BLU-108/B Sensor Fuzed Weapons instead of the six carried by the AGM-154B.

The **AGM-154C** will carry the **BLU-111/B** variant of the MK-82, 500 lbs general purpose bomb, equipped with the FMU-152 **Joint Programmable Fuse (JPF)** and is designed to attack point targets. For the AGM-154C, British Aerospace **'Broach penetrating warhead'** is also being considered. Broach is a three-stage munition, featuring a small charge to blow away dirt on top of a target, a hollow-charge warhead to punch a hole in the target, and a blast-fragmentation warhead to explode inside the target. The fuzing system is provided by Thales. The AGM-154C has an imaging infrared seeker and datalink, compatible with the existing pylon-mounted control pod for the Navy Walleye glide bomb, to allow precision strikes on a target. The seeker is capable of autonomous operation for 'fire and forget' attacks.

Fig. 167 That's How the JSOW Gets Its Great Range: Note The Extended Wings Of The JSOW

(26.4.3) Famous Glide Bombs Of Other Nations

A European standoff weapon system, developed by DASA of Germany, is available in both glide and powered forms. The best developed of this family of weapons is the **'BombKapsel 90 (BK-90) gliding cluster munition dispenser'**, developed primarily for the Swedish Air Force's JAS-39 Gripen fighter. It has been named 'Mjoelnir', after the thunder god Thor's mighty hammer. The BK-90 contains 12 sub-munitions launch tubes on each side, with multiple sub-munitions contained in each tube. It has a boxy fuselage with very stubby wings, as well as tail fins on each corner of the fuselage box. The BK-90 is intended to be released at high speed and low altitude. Essentially it allows a strike aircraft to 'toss' the weapon onto a target, with the weapon using **INS navigation and terrain-following radar** to navigate into the target area. It can be also released from high altitude for more standoff range, and has an **optional GPS-INS capability** for such long range attacks. Toss range at low altitude is up to about 5 miles, while the range at high altitude is up to 14 miles. Once over the target, the BK-90 blasts its sub-munitions out of the ejector tubes, with the sub-munitions deploying small chutes to stabilize their descent. The dispersal pattern is up to about 800 by 1,300 feet. There are two types of sub-munitions: the 8.8 lbs MJ1 air-bursting fragmentation sub-munition for attacks on soft targets; and the 40 lbs MJ2 anti-armor sub-munition. The BK-90 is being built and marketed in the US as the **'Autonomous Free-Flight Dispenser System (AFDS)'** by an American subsidiary of DASA named CMD. The AFDS has been evaluated by the USAF.

(26.4.4) The Wind Corrected Munitions Dispenser (WCMD)

The **'Wind Corrected Munitions Dispenser (WCMD)'**, provides INS guidance for a cluster munition canister. The WCMD consists of a tail kit that can be attached to a **Tactical Munitions Dispenser**. Unguided cluster bomb units have to be released from low altitude to keep them from being blown off the target by the wind. Adding a guidance system permits such munitions to be dropped from high altitude. The idea is to get a CEP of 85 feet with the INS guidance kit but the tail kit is designed to accommodate a GPS guidance system also should such an upgrade seem necessary.

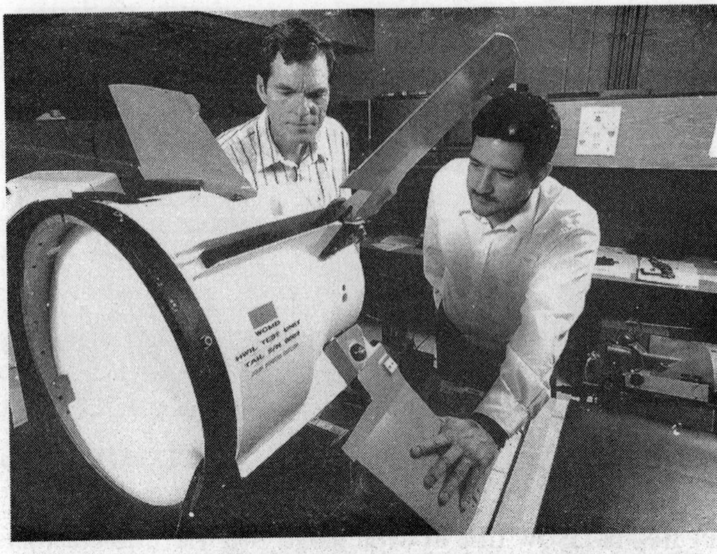

Fig. 168 The WCMD (Wind Corrected Munitions Dispenser) Kit For Attaching To The Tail Of Unguided Cluster Bombs

WCMD configurations include:

- CBU-103/B: Same as CBU-87/B (202 BLU-97/B CEM) with WCMD tail kit.
- CBU-104/B: Same as CBU-89/B (Gator mine unit) with WCMD tail kit.
- CBU-105/B: Same as CBU-97/B (10 BLU-108/B SFW) with WCMD tail kit.

A WCMD with BLU-118 submunitions for power-grid attack has also been introduced, and the Air Force Research Laboratory has developed a new **'agent defeat'** weapon based on the WCMD named 'Stop' to allow it to attack chemical or biological weapon storage facilities using 4,000 titanium flechettes to puncture storage tanks and production facilities.

The Air Force is now working towards an **'extended range'** version of this munition, the **'WCMD-ER'**, which will have pop-out gliding wings to give it a typical range of about 34 miles, and will be fitted with a GPS-guidance upgrade.

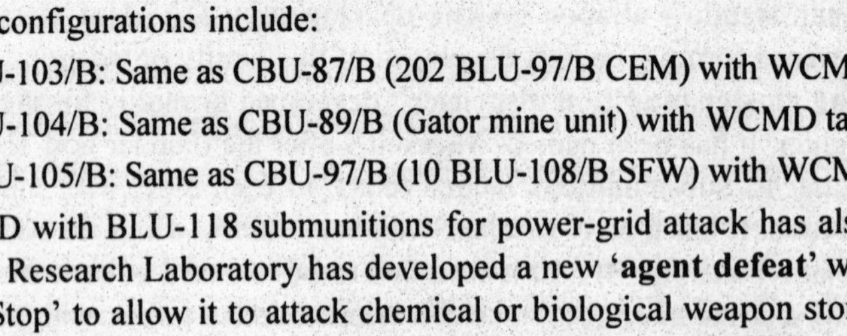

Fig. 169 The WCMD Attached To A Cluster Bomb

(26.4.5) The SDB (Small Diameter Bomb)

The US Air Force is now developing a lightweight penetrating GPS-guided bomb, known as the **'GBU-39/B Small Diameter Bomb (SDB)'**, initially known as the **'Small Bomb System (SBS)'**. It all had started with the so-called **'Miniaturized Munitions Technology Demonstration (MMTD)'** program. The SDB is intended for carriage on new stealthy manned and unmanned strike aircraft to allow them to carry weapons internally, while still giving them enough firepower to destroy a number of separate targets in a single sortie.

Fig. 170 The GBU-39 Small Diameter Bomb, A 250 Lbs Bomb With The Penetrating Capacity Of A 2000 Lbs Bomb

The SDB is a penetrating bomb with weight in the 200 to 250 lbs range, and with a length of about six and a half feet. It is intended to have the same penetrating capability as a 2,000 lbs weapon, but will only carry 50 lbs of explosive. The SDB will incorporate a **'differential GPS guidance system'** to provide a smaller CEP, and an improved penetrating structure to allow the weapon to pierce up to almost six-and-a-half feet of concrete. **Lattice fins** are now being considered to reduce size and improve performance. They are paddle-shaped with a honeycomb lattice of air surfaces inside the paddle. The paddle is faced into the air stream and provides the aerodynamic equivalent of a much larger fin. The Soviets had developed the lattice fin concept, and it is in use on several Russian missiles. The Russian Vympel firm holds a US patent on the technology. You may recall that it is also used on the MOAB. A **'Swing Wing Adapter Kit (SWAK)'** is being built for the SDB to allow an aircraft to drop the weapon from standoff ranges of up to 28 miles at altitude. If the SDB is being launched at a hard target, the wing will be discarded in mid-course to allow the bomb to build up velocity before striking the target. For soft targets, the wing will be retained until the bomb is in range of small arms fire. The SDB will be fitted with the new "Hard Target Smart Fuze (HTSF)". The HTSF is a tail-mounted fuze that incorporates accelerometers and a processor chip. The HTSF can determine if the bomb has struck earth, concrete, rock, or empty space; can count the number of layers it has penetrated; and can compute distance or time to determine the appropriate detonation time for a specific target.

The Russians, not to be outdone, have introduced a 1,100 lbs glide bomb designated the '**KAB-500S-E**' that can use both GPS and the comparable Russian GLONASS satellite system for guidance. **GLONASS-GPS augmentation kits** have also been developed for Russian EOGBs and LGBs.

(26.5) DENSE INERT METAL EXPLOSIVE (DIME)

Dense Inert Metal Explosive (DIME) is an experimental type of explosive that has a relatively small but effective blast radius. We chose to discuss it here because the SDB is supposed to use the DIME. It is intended to be used in urban areas where conventional bombs could result in great collateral damage. DIME produces lower pressure but increased impulse in the near field. Far Field damage is reduced (no fragments / impulse roll-off). The lethal footprint can be tuned to precision footprint. As such they are described as **LCD (Low Collateral Damage)** weapons or **FLM (Focused Lethality Munition)**. The phrase 'inert metal' refers to the metal's non-involvement in producing the blast (as opposed to, for example using, aluminum powder to increase blast strength), rather than the metal being chemically or biologically inert.

The warhead case consists of a low-density, wrapped carbon-fiber/epoxy matrix integrated with a steel nose and base. The low-density composite case can survive penetration into a one-foot hardened concrete wall. Upon detonation, the carbon-fiber warhead case disintegrates into small non-lethal fibers with little or no metallic fragments, thus significantly reducing collateral damage to people and structures. The warhead explosive fill is a dense inert metal explosive containing fine tungsten particles or **HMTA (Heavy Metal Tungsten Alloy)** to provide a ballasted payload with sufficient penetration mass. The tungsten displaces some of the explosive so as to reduce the total explosive used. Two common HMTA used are: rWNiCo: tungsten (91-93%), nickel (3-5%) and cobalt (2-4%); and rWNiFe: tungsten (91-93%), nickel (3-5%) and iron (2-4%). Upon detonation of the explosive the casing disintegrates into extremely small particles instead of the shrapnel which result from the fragmentation of a metal casing. The HMTA powder acts a **micro-shrapnel** which is very lethal at close range (12 feet or so), but loses momentum very quickly due to air resistance coming to a halt within approximately 40x the diameter of the charge. This increases the probability of killing people within a few feet of the explosion while reducing the possibility of causing death/injures or damage further away.

Critics of the concept argue that although the relatively small radius of destruction reduces the area over which casualties may occur, the increased lethality close to the point of explosion and use in urban areas may actually increase the number of unintended casualties (so-called 'collateral damage'). In addition to this the toxic/carcinogenic effects of the HMTA may cause increased deaths in those who survive the initial blast or in people who inhale the dust. As reported in French national newspaper Le Monde, according to a team of journalists from the Italian State radio-television RAI, DIME-type bombs were being used in the Gaza strip by the Israelis against Palestinians during July/August 2006. The investigation was performed by analyzing samples of metals found in the victim's bodies and examining the unusual wounds. Israel, however, denied possessing or using such weapons

(26.6) SENSOR-FUZED WEAPONS: SMART SUB-MUNITIONS

The WCMD can be used to improve the accuracy of cluster bomb canisters carrying a wide variety of sub-munitions. One new sub-munition that can be used in this fashion is the '**CBU-97/B Sensor Fuzed Weapon (SFW)**' cluster bomb. The SFW carries a smart anti-armor sub-munition, designated the **'BLU-108/B'**, that allows a single CBU to engage and destroy multiple armored targets simultaneously. In a test conducted in late 1991, for example, an F-16 dropped four SFW canisters from low altitude, which then dispersed a total of 40 BLU-108/B sub-munitions over a column of 24 vehicles. 17 hits were scored on 11 of the vehicles.

The CBU-97/B SFW is based on the standard TMD canister, and has a loaded weight of 1,000 lbs. The SFW can attack armored vehicles over a wide footprint. The munition can be delivered at altitudes from 200 to 20,000 feet. An F-16 can carry 4 SFWs, with each canister carrying ten BLU-108/B sub-munitions in two rows of 5, and each sub-munition containing four smart hockey-puck shaped 'Skeet' warheads each. An F-15E can carry 10 SFWs.

Once a dispenser is released, a pyrotechnic charge pops open the TMD's three panels to release the ten sub-munitions. Each sub-munition is a cylinder that is decelerated by a small parachute. The parachute also orients the sub-munition vertically over the target area. Chute deployment timing is staggered to allow the sub-munitions to disperse, and the actual area covered is a function of aircraft speed and the timing interval selected. As the cylinder descends, the four Skeet warheads flip out from the body of the sub-munition. Each Skeet warhead consists of a 5.25 inch diameter disk of flat copper backed by an explosive charge that is bore-sighted to a protruding infrared sensor. Once the cylinder is aligned properly over the target area, the parachute is cut loose as a rocket motor fires through two nozzles, which are canted so that they stop the cylinder's fall and start it spinning. Once the cylinder is spinning rapidly, the Skeets are released in pairs to spin away from the cylinder. They wobble in flight to allow the infrared sensor to scan over the ground below in a spiral pattern.

When a Skeet warhead flies over a vehicle, the warhead's infrared sensor identifies it as a target and fires the explosive charge. This creates an explosively formed penetrator (EFP). Explosive reactive armor is ineffective against such a **'kinetic kill projectile'**. If the Skeets don't find a target after a certain length of flight, they explode into fragments as a harassment measure.

The infrared sensor is capable of working through fog, since it is close to the target, and works at night and through electronic countermeasures. Not all the warheads are expected to find targets, but the dispersal pattern of the 40 Skeets carried by an SFW is expected to be effective against concentrated targets such as an armor column.

Further development of the basic CBU-97 SFW has focused on adding the WCMD inertial guidance tail kit, resulting in the **CBU-105/B SFW**, and on improvements to the sub-munition. The **BLU-108/B** sub-munitions are dispensed at higher altitude to cover a bigger footprint of about 600 by 1,800 feet. The new projectile includes a ring of 16 explosively formed penetrators around the central main penetrator, providing an additional shotgun effect.

The Russians have fielded an SFW-class sub-munition, designated **'SPBE-D'** when used with a cluster-bomb unit and **'MOTIV-3M'** when used with the Smerch multiple-launch rocket system. A Smerch rocket can carry five MOTIV-3M sub-munitions. Each MOTIV-3M sub-munition measures 11.2 x 10 x 7.3 inches and weighs 33.1 pounds. When the missile arrives in the target area, the five sub-munitions are dispensed, with each popping out a parachute and then hunting for a target with a dual-band infrared seeker with a field of view of 30 degrees. At about 500 feet above the target, the sub-munition fires on its target, using an explosively-formed penetrator that can punch through 2.75 inches of armor.

(26.7) THE LOCAAS: A REALLY SMART SUB-MUNITION

The SFW is not an entirely smart weapon. It is dropped into a target area, and its warheads will only attack a tank if their paths happen to cross. The US Air Force Research Laboratory (AFRL) has been working with Lockheed Martin for several years on a much smarter sub-munition, the experimental **'Low Cost Autonomous Attack System (LOCAAS)'**.

LOCAAS is designed to fly into a target area, search for targets, and destroy them. Both glider and jet-powered versions have been investigated, though it appears that the glider version was abandoned early in the program. LOCAAS looks something like a big pill with wings and tail. Several different turbojet engines have been evaluated in the program, including the Hamilton Sundstrand TJ-50 and Technical Directions INC TDI-J45G. The sub-munition navigates with a GPS-INS guidance system. The device is just 20 inches long with a wing span of 36 inches and weighs 95 lbs. It carries a warhead of 17 lbs. Flying at a speed of 230 mph, it can go up to 125 miles.

Fig. 171 LOCAAS: The Ultimate Smart Bomb That Will Keep On Loitering In The Battlefield Autonomously To 'Search And Destroy'

LOCAAS sub-munitions could be carried in a TMD containing either six glider sub-munitions or four jet sub-munitions. Four such dispensers can be carried on an F-16 or F/A-22 Raptor strike fighter, for a total of 16 P-LOCAAS submunitions. A B-2 bomber could carry a total of 192 such sub-munitions. LOCAAS is also being considered for an unmanned aerial vehicle due to its small size and lethality. LOCAAS carries a laser radar seeker that sweeps the target area underneath. The seeker is coupled to a target recognition system that compares target features with a library of possible targets and prioritizes them for attack. LOCAAS can be programmed before launch to hunt for specific target types, and will then hunt down and destroy such targets.

The actual attack is performed by firing an explosively formed projectile, similar to that fired by the Skeet warhead used in the SFW, but more sophisticated. The copper slab in the 'multimode' warhead is backed by explosives that can be fired in sequence to either spray the copper in pellets, to attack soft targets like trucks; to wad the copper into a slug, to attack light armored vehicles, or form it into a rod, for attacking very hard targets. LOCAAS will self-destruct if it can't find a valid target. A production LOCAAS will still have a search capability, but it will also have a radio datalink to keep a "man in the loop" who would determine if the target should be attacked.

(26.7.1) The Air Dominator

The US Air Force is now working with Boeing on a new weapon named the 'Air Dominator' that is conceptually similar to LOCAAS, but designed to loiter over a target area for at least 12 hours, searching for a target to attack. The Air Dominator would be dropped in a cluster of three or four munitions by a combat aircraft at high speed. The cluster would deploy a high-speed retarding device, and once it slowed enough, it would break apart and each of the Air Dominator munitions would deploy a parachute, drop down to low altitude, and pop out wings. The proof-of-concept demonstrator had a wingspan of 12 feet and a 4 foot fuselage, with the wing featuring 'wing warping' flight controls, an idea that goes back to the Wright Brothers.

Once fully deployed, the munitions would patrol a 'kill box' for targets, using a laser radar to spot targets and then destroy them with a penetrator or Skeet-type warhead. Each munition would be able to perform three separate attacks, including a final 'kamikaze'-type attack. The weapon could be programmed to obey different rules of engagement, allowing different levels of operator control over attacks.

The weapons would be networked, allowing one to tell the others that a target had been found and that others might be nearby. The cluster would include a 'Gateway' munition to act as a networking hub, carrying a satellite link to provide intelligence or receive commands from anywhere in the world, or a refueling munition to extend the endurance of the attack munitions. The Gateway and refueling munitions would be about three times bigger than the attack munitions, and would operate at higher altitude. An F/A-22 Raptor will be able to carry 24 Dominators and a single Gateway.

(26.8) THE 'BRILLIANT ANTI-ARMOR MUNITION'

The US Army and Northrop Grumman are in advanced stage of development of another smart sub-munition, the **Brilliant Anti-armor Munition (BAT)**. This 44 lbs weapon looks like a cylinder about 5 feet long with a bulbous head when stored in its dispensing system. When released, the BAT pops out four wraparound tailfins and four long, straight cruciform gliding wings around the mid-body. The bulbous nose is fitted with an infrared sensor, and the wingtips are fitted with long spikes fitted with acoustic sensors.

Once deployed, the BAT glides to a preprogrammed target location, with a cluster of BATs dispersing to ensure that they do not attack the same targets, and uses the acoustic sensors to identify the general location of a tank. Once a target has been boxed by the acoustic sensors, the infrared sensor takes over, directing the BAT to hit the target directly from the top, destroying it with a two-stage penetrating warhead. A second-generation BAT with improved seeker system and multimode warhead is being developed as a follow-on improvement.

The BAT is being designed as the primary payload for the Block II Army Tactical Missile System (ATACMS), which will carry 13 BATs. ATACMS is a surface-to-surface battlefield missile and so BAT is not precisely an air-launched munition, but air-launched delivery is perfectly practical. Other possible air-launch schemes include a TMD loaded with four BATs, or a SLAM missile loaded with six or eight BATs.

(26.8.1) The Viper

In the spring of 2003, the US Army performed a demonstration of a laser-guided version of BAT named 'Viper', originally 'Viper Strike'. The weapons were carried on an Army Hunter drone at the end of March 2003 and scored seven hits out of nine drops. Viper is intended for high-precision 'top down' attacks on targets, particularly in urban or other built-up areas where collateral damage is a concern. It was designed as a fast-track modification of the BAT, but for the moment military interest in Viper actually seems to be overshadowing that in BAT.

□□□

Chapter 27

LITTLE BOMBS FOR BIG GUYS: GRENADES

(27.1) HAND GRENADES

(27.1.1) Hand Grenades Reclaim Their Place In Warfare

The inventor of the grenade is not documented, but grenades were first used in the late fifteenth or early sixteenth century. Originally they were just hollow metal balls filled with gunpowder. These early models were ignited with a slow burning fuse and thrown distances over 100 ft. The word 'grenade' has a Spanish origin in the word 'granada' meaning pomegranate in reference to the general size of early grenades, and because their content of pellets reminded them of the seeds in a pomegranate.

As the range and accuracy of firearms increased during the eighteenth century, grenades were virtually abandoned. They were re-introduced to the infantryman's equipment arsenal during the trench warfare of World War I. During World War II over 50 million fragmentation grenades were manufactured in the United States alone.

Hand grenades are the personal bombs of an infantryman and imbue him with a sense of power which guns cannot give. In urban combat they are said to be a soldier's best friend and hence when he has to enter a room where gunmen could be hiding, the grenade precedes him!

(27.1.2) Defensive And Offensive Grenades

You can have defensive as well as offensive grenades. Defensive grenades are the older type. For defensive grenades, the blast radius is greater than the distances up to which they can be thrown. This means that you have to necessarily take cover after throwing them. Offensive grenades were made later. For them the blast radius is less than the distance to which they can be thrown and hence taking cover is not mandatory. In the past one simple technique of increasing the range up to which a grenade could be thrown was to attach it to the end of a stick or wooden handle but that made it bulkier. It was called a **stick grenade**! Offensive grenades contain explosive charge filler of flaked TNT in a body with sheet metal ends and pressed fiber sides.

(27.1.3) Fragmentation Grenades

This is the most common variety of **anti-personnel grenades**. Fragmentation grenades contain an explosive charge in a metal body, designed to break into fragments upon the charge exploding. They typically weigh about 16 - 21 oz, and are constructed of cast iron or steel. A variety of metals and shapes are used for the fragments and the casing. The molding method can also be varied. The explosive charge is about 5 - 6.5 oz. Earlier TNT was used but now Composition B is common. Fragments are provided either by the serrated shell, nicked wire binding, pellets or all of them. When you are using nicked wire for fragments the wire binding is enclosed in a thin sheet metal case. The older '**pineapple grenades**' had the serrations very much visible on the outside. Newer versions have the fragmentation pattern embossed on the inside of the shell and the outer surface is smooth. In any case, the size and number of fragments is pre-determined. Plastic bodies are also being made. There are many polymers that can be used for the fragmentation grenade casing depending on the strength and processing requirements (e.g., polypropylene, polyamides, polyacetals, polycarbonates, polyesters, polyethers, aldehyde/phenolic condensates, melamine resins, and urea resins).

(27.1.4) Stun Grenades

Stun grenades produce a blinding flash of light (up to one million Candela) and a deafening blast (170-180 dB), but they are not lethal as no fragments are produced. The combined effect of the flash and the bang is to concuss, confuse, disorient, or momentarily distract a potential threat for up to five to six seconds. It can seriously degrade the combat effectiveness of affected personnel for up to a minute. They are most useful in confined spaces and thus lend themselves to anti-terrorist operations or law and order situations where you don't want to kill people for whatever reason. The filler is essentially a pyrotechnic mixture.

(27.1.5) Incendiary Grenades

Incendiary grenades have an incendiary mixture in them and they are useful for damaging equipment or starting fires. Some of these grenades—as with the TH M14 thermite—come with metal straps that prevent rolling and an M200A1 igniting fuse with only about two seconds of delay after safety lever release.

(27.2) BASIC MECHANISM OF HAND GRENADES

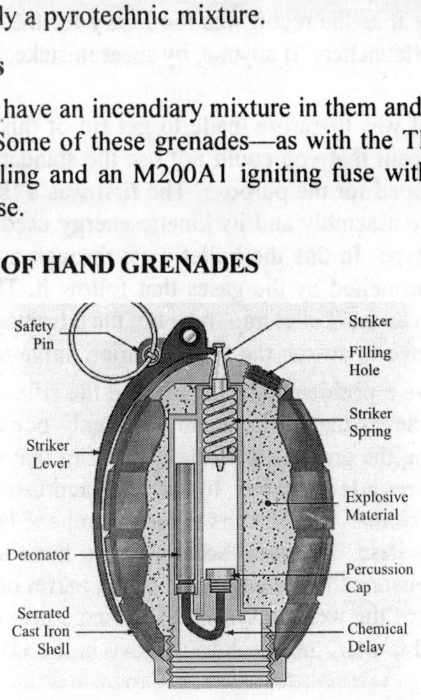

Fig. 172 The Internal Structure And Mechanism Of A Typical Hand Grenade Wit' A Time Delay Fuze

The heart of a grenade is the delay mechanism for the fuze. The most common system is a chemical delay. When the safety pin is removed and the grenade lobbed, the striker lever becomes free and strikes a percussion cap. The percussion cap ignites a chemical delay element. This is nothing but a slow-burning material. One end of this is connected to a detonator which eventually detonates the explosive filling.

(27.3) RIFLE GRENADES AND GRENADES OF AUTOMATIC GRENADE LAUNCHERS (AGLs)

A grenade enables you to engage several persons simultaneously. However, the range to which hand grenades can be thrown is necessarily limited. Hence the desire to have grenades which could be thrown much farther was very natural. After all, the infantryman needed something more than just bullets from his rifle or machine gun! A machine gun kills individuals but the splinters of a grenade create a lethal zone! A grenade gives you 'fire density' which bullets cannot give. Well, the mortar could do the same but there is a fundamental limitation with the mortars. You cannot use the mortar all the time because the mortar cannot be fired effectively at ranges below say, 400 yards. So what do you give to the enemy troops who are lodged between 50 yards and 400 yards? It had to be grenades and that too delivered at that distance.

This led to the development of the rifle grenade. The first development in this regard before the First World War was rifle grenades. The most rudimentary type was a **cup launcher**. In this a special cup containing the grenade was attached to the muzzle and the propulsive force was provided by a special blank cartridge. In the other device, the grenade was attached to a thin rod as a tail and the tail was inserted in the barrel. It acted like a piston and took the force of the gases of the blank cartridge. Both of them used the standard hand grenade. The grenade was primed before fixing in the devices; the clip flew out as the grenade exited. The rifle had to be braced on the ground like a mortar to fire it as the recoil was too heavy for the shoulder to take. There was a serious problem with both the launchers. If anyone, by sheer mistake, used a regular ball cartridge, he would have had it.

The next development was therefore made to get rid of this handicap. Two types came into existence. Obviously it meant that you could not use the standard hand grenade and special rifle grenades had to be developed for the purpose. The first was a **'bullet trap'** type. In this the bullet is trapped in the grenade assembly and its kinetic energy used to propel the grenade. The second is a **'bullet through'** type. In this the bullet exits through a coaxial hole in the grenade assembly and the grenade is propelled by the gases that follow it. There is indeed some loss of propulsive power in comparison to the 'bullet trap' type but the advantage is that in a semi-automatic or automatic rifle you don't have to switch the gas operation valve to 'closed' first.

Both of them still have a problem. You cannot use the rifle in its standard mode as long as the grenade is there. This means that if the enemy suddenly pops up, you will be obliged to waste precious time in removing the grenade or firing it first and that may cost your life! One way out is to develop a **dedicated grenade launcher**. It was done and used for years. That, however, is not a very wise idea because one man is wasted as a purely grenade launcher. Unless it is a multi-shot weapon, it is perfectly useless. The ideal solution is to have a complete-in-itself grenade launcher that could be simply attached to the underside of the barrel of the rifle or that the grenade launcher could be made a part of the weapon system. The first solution is a most popular weapon since the 1960s and the second is what they seek to achieve in the **OICW (Objective Individual Combat Weapon)**.

Numerous rifles are found sporting the under-barrel grenade launcher. They can be fired from the shoulder in the normal fashion. The only problem with the system is that it makes the rifle both bulkier and heavier. Most countries use the 40 mm grenade but 25 mm caliber is also being tried for the OICW.

(27.3.1) The Technology Of The Fuzes Used In Rifle Grenades And AGLs

Grenades launched by rifles or automatic grenade launchers should have impact fuzes for obvious reasons. There is a little problem in designing impact fuzes. They must be 'unarmed' before they are actually fired to rule out accidents. And since they are launched by rifle or AGL, this 'arming system' must be automatic. In some designs, the arming system is triggered by the propellant explosion that drives the grenade out of the launcher. In other designs, the grenade's acceleration or rotation during its flight arms the detonator. The **impact trigger**, at the nose of the grenade, consists of a movable, spring-mounted panel with an attached **firing pin** facing inward. As in the time-delay grenade, the **fuze mechanism** has a **percussion cap** and a **detonator explosive** that ignites the main explosive. But it does not include a chemical delay element. When the grenade is unarmed, the fuze mechanism is positioned toward the tail end, even though it has a spring pushing it toward the nose. It is held in this position by several **spring-mounted, weighted pins**. The firing pin is not long enough to reach the **percussion cap** when the fuze is in this position. The spinning motion of the grenade generates a strong **centrifugal force** that pushes the weighted pins outward. When they move far enough out, the pins release the fuze mechanism, and it springs forward toward the nose of the grenade. When the grenade hits the ground or something else, the nose plate pushes in, driving the firing pin against the percussion cap. The cap explodes, igniting the detonator explosive, which ignites the main explosive. In more sophisticated designs you can have **electronic fuzes** or fuzes with electronic position sensors which can be programmed to explode after the grenade has traveled a certain distance so that you could pinpoint particular targets, even ones behind barriers, with extremely high precision, or fuzes with self-destruct mechanisms.

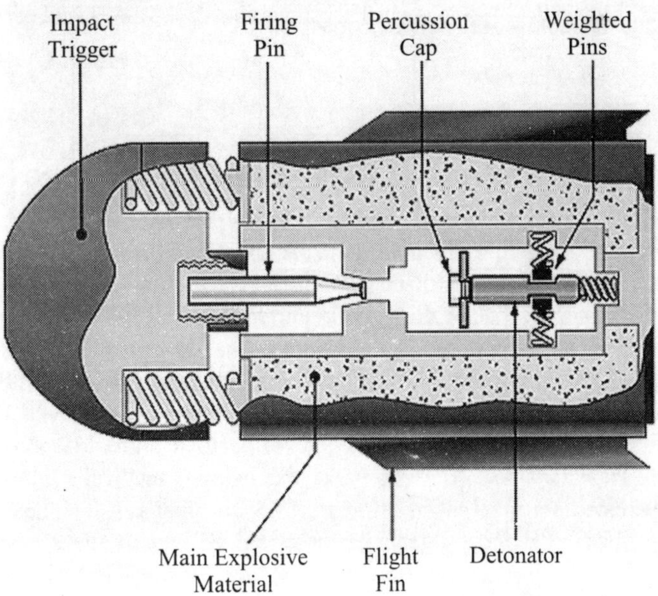

Fig. 173 The Internal Structure And Mechanism Of A 40 Mm Impact Fuze Grenade Fired By The M203 Grenade Launcher And Automatic Grenade Launchers

(27.3.2) Russian Case-Less And Bouncing Grenades

The Russian VOG-25 grenade is case-less and there is nothing to be removed after the grenade is fired and you are ready to fire the next round. Unlike the M203, it is a muzzle-loading weapon with pre-engraved rifling grooves in the grenade to fit in the grooves of the barrel. Grenades are held in the barrel by a spring catch. Unfired grenades can be extracted using a built-in push-out extractor, located at the left side of the weapon. The VOG-25P grenade is a bouncing grenade, that is, upon impact on the ground a special charge in the nose of the grenade is set off and it propels it about one meter in the air when the grenade explodes.

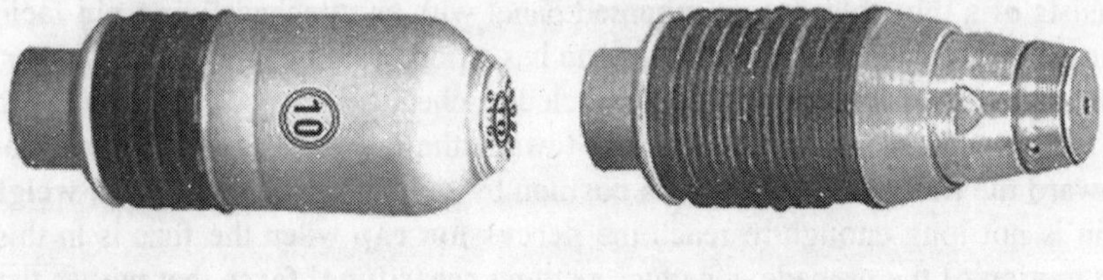

Fig. 174 Russian Beauties: The VOG-25 Caseless Grenade And The VOG-25P Bouncing Grenade Used In The GP-30 Grenade Launcher

□□□

Chapter 28

SPRINGING NASTY SURPRISES: LANDMINES

(28.1) THICK DEFENSIVE WALLS PROMPT EARLY MILITARY MINING

Early in the Bronze Age, walled cities began to appear in the Middle East to protect against raiders and other attackers. Jericho, on the west bank of the Jordan River, just north of the Dead Sea, is the oldest known walled city (dating from approximately 8000 B.C.). The walls at Jericho were about 23 feet high and 13 feet thick and were surrounded by a moat 30 feet wide and 10 feet deep. Later, protective walls developed into huge affairs. Under Nebuchadnezzar II (around 600 B.C.) the walls at Babylon increased to a thickness of about 85 feet.

Early military mining techniques were developed in response to these **walled cities** and probably were devised by impressed civilian miners at the behest of conquerors. Before the invention of military mining, the attackers' options were limited to blockading a city (starving them out), scaling the walls, breaching the walls with a battering ram (which began in Egypt about 2000 B.C.), or by stratagem (such as the Trojan Horse). Although the stone-throwing engine of war was first developed by the Phoenicians, the catapult was one of the first effective missile engines. **Engineers took their name from these "ingenious" devices.**

The Assyrian Army organized the **first known 'corps of engineers'** during the time of **Ashurnasirpal II** (about 850 B.C.). These elite specialists operated siege and bridge trains and provided mobility support for chariots. They were the first soldiers equipped with advanced iron pioneer tools and are credited with the **first known use of offensive mine warfare**. This occurred about 880 B.C. when engineer soldiers drove tunnels (mines) under or through walls and fortifications to gain access to fortified areas or to create a breach large enough for a full-scale attack. These engineers excavated a chamber under the wall and braced the ceiling with timber supports. They were called mines because they were actually dug in the fashion of mines. The supports were then burned, causing the chamber and the structure above it to collapse. Attacking soldiers then assaulted through the breach.

Among the many successful mines throughout history are those used by Alexander the Great and his engineer Diades at the sieges of Halicarnassus (334 B.C.) and Gaza (332 B.C.) and Julius Caesar and his engineer Mamurra during the siege of Marseilles in 49 B.C. Although effective mining and other combat engineering skills were critical to the military successes of both of these great generals, historians have failed to accord them their due place in the history of warfare.

(28.1.1) Powder-Primed Tunnel Mines Take On Massive Fortifications

With the invention of gunpowder, someone thought that it would be much better if the supports of the mines (tunnels or galleries) driven under fortifications that used to be burned till then could be exploded instead. Burning timber was a slow process. Exploding them could increase their effectiveness as well as the element of surprise greatly. The idea caught on and it was soon found out that gunpowder was so potent that powder-primed mines could take on massive fortifications also and not merely walls of cities. The name 'mine', however, for these powder-primed tunnels remained unchanged. Huge quantities of gunpowder came to be used in such mines. The first recorded use of such a mine in Europe was in 1403, during a war between Pisa and Florence, when the Florentines exploded a charge in a forgotten passage in the walls surrounding Pisa. One of the individuals involved with these **early explosive mines** was a military engineer named **Leonardo Da Vinci** (yes, the very same genius), who was working for Ludovico, the Duke of Sforza, around 1500!

In his work on siege warfare published in 1740, Sebastien Le Prestre de **Vauban** (French Marshal, 1630-1707) codified principles of **military mining** that remained valid well into the 19th century. Vauban, in what could be considered the first scientifically based demolitions manual, described a method of charge calculation and placement based on characteristics of the target fortress and the desired effect. He defined these mines by the depth and size of the charge:

- For depths less than 3 meters, it was called a **fougasse** (or **contact mine**).
- For depths greater than 3 meters, it was called a mine.
- When used as a '**countermine**' against an enemy mine, it was called a **camouflet.**
- When intended to destroy an entire fortification (using 5,500 lbs of powder or more), it was called **pressure balls (globes de compression)**.

According to Vauban's tables, explosive charges for mining could go up to 27,000 lbs. The purpose of mines was not only to cause destruction but also, with the rocks and soil ejected, to form an earthen ramp that assault troops could use to gain immediate access to the breach. Because the demolition often came as a surprise to defending forces, it frequently caused panic and confusion among them. Tunnel mines were very time consuming to employ. Typically about 18 miners and 36 unskilled workmen were employed in three 8-hour shifts to construct an assault mine. Military mining during a siege could take 30 days or more, and specialists were required for the job. During the Middle Ages, coal miners were hired. Formal mining units were not formed until standing armies were raised by the absolute monarchs of the 17th century. In Britain the Company of Soldier Artificers was formed in 1772. The miners' work demanded tremendous stamina, perseverance, courage, and special caution—asphyxiation due to lack of oxygen and possible flooding were potentially dangerous hazards.

Against the bastioned fortresses of Vauban's time, mining normally began as soon as sappers (military specialists in attack and defense of fortifications) completed the last parallel in front of the glacis of a fortress or fortified town. Then besieging miners dug galleries a little over four feet high and three feet wide and lined them with wood. You can imagine how difficult it must have been to work in such a cramped space and that too literally back-breaking works like tunneling! We salute the miners who endured such rigors for centuries. Once they reached the site selected for the explosion, they dug the blast hole perpendicular to the previous direction of the gallery.

Then they filled the mine chamber with the amount of black powder determined by the siege engineer.

Fig. 175 That's How Mine Shafts Were Tunneled In The Medieval Days

To ignite the mine, they fed an '**ignition sausage**' out of the mine chamber. This sausage was a tube made of linen and filled with granulated black powder that led back to the point of ignition (minenherd). The ignition sausage, a predecessor of the modern time fuze, was normally laid in a about two-and-a-half inch wide wooden duct and covered with a board to protect it from moisture or other damage. The gallery was finally tamped with sod or earth, over a length of 20 to 33 feet. At the appointed time, the miner ignited the powder in the ignition sausage with an ignition sponge and then scampered back like hell before the sponge burned to the powder. Immediately after the explosion, the besiegers could assault the fortress or extend their sap trenches into the crater and reinforce them with gabions. If necessary, additional mines were used to destroy the palisades of the covered passage and the supporting walls of the counterscarp or scarp, thus facilitating entry into the fortress.

While working in tunnels, miners looked for '**listening tunnels**' and '**countermines**' of the defender. Attackers tried to deceive the defender's listening posts by constructing phony '**noise galleries**', where they intentionally produced a lot of noise.

(28.1.2) Mines Gradually Drift Towards Attacking Personnel Too

The concepts underwent some change with time, however. Frederick the Great, King of Prussia, stated: "Fougasses formed into a T-like mine, in order to blow up the same place three times, can be added to the entrenchments. Their use is admirable; nothing fortifies a position so strongly nor does more to ward off attackers." The fougasses which he talked of were intended to detonate in the face of an enemy assault. A black powder charge was placed in a chamber excavated in the face of a fortification (firing horizontally) or in front of it (firing vertically). The chamber was then packed with a quantity of fragments, normally rocks or iron scrap and called a **stone fougasse**, or filled with explosive artillery shells and called a **shell fougasse**. If properly emplaced, a horizontally fired fougasse functioned as a crude claymore mine, while the shell fougasse could function like a bounding antipersonnel (AP) mine or a simple fragmenting mine. Fougasses were command-detonated by manually igniting a powder train from a protected position at the appropriate time. Fougasses had several defects: they were vulnerable to the elements—even moderate dampness rendered them inoperative—and were difficult to detonate at the optimum time. However, in the right circumstances, fougasses caused heavy casualties, as occurred during the sieges of Ciudad Rodrigo, Badajoz, and Santander in the Duke of Wellington's Peninsular Campaign of the Napoleonic Wars.

Fougasses were employed by one of George Washington's engineers, Francois de Fleury (of de Fleury Medal fame), in October 1777 against the Hessians at Fort Mercer, New Jersey, on the east bank of the Delaware River. During the War of 1812, an American ammunition chest accidentally exploded during a British attack on Fort Erie, Canada. This caused the attack to collapse, and the fear of additional fougasses prevented further British attacks. Incidentally, the only engineer unit in the American Army during this war—the Company of Sappers, Miners and Bombardiers—had fought in this battle. During the Mexican-American War of 1845, the Mexicans attempted to employ fougasses on the approaches to Chapultepec.

(28.1.3) Tunnel Mines In The Modern Era

Tunnel mining continued sporadically into the modern era and was used by Napoleon at Acre (1799), the Crimean War (Sevastopol), General Grant's men in the American Civil War (Vicksburg and Petersburg), and the Russo-Japanese War (Port Arthur). Military engineers incorporated the latest technologies from civilian mining as they became available, including more efficient explosives such as nitrocellulose, dynamite, picric acid, and TNT. Other improvements included electric (galvanic) ignition (1850s) and forced-air ventilation systems. We have seen how the biggest mines in human history were used by the British against the Germans in the World War I on the Western Front and the Isonzo Front. During World War I, both sides employed new mechanical tunnel-boring machines developed for commercial coal mining, as well as traditional techniques. Mines were used in the World War II also on the Russian Front, and in the French-Indochina War, particularly at the historic battle of Dien Bien Phu. Some time back, the Peruvians used tunnel mines to liberate hostages held in the Japanese ambassadors's residence in Lima. The North Koreans may use them in the future—some of their tunnels have been discovered under the demilitarized zone (DMZ) and more are suspected.

(28.1.4) Countermines

The original countermines were simply tunnels dug by the besieged defenders to disrupt enemy mining efforts. A countermine was successful when an enemy tunnel was intercepted. Inevitably, a confused, close-quarters fight in the dark cramped space followed, as the two sides fought to control the tunnel. In the words of a contemporary historian who describes such a fight during the siege of Barca about 510 B.C.: "The Persians excavated underground tunnels that reached the walls. Among the Barcaeans there was a skilled worker in brass who took a brazen shield and, carrying it round within the wall, applied it here and there at places where he thought the workings might be. Where there were no mines the shield was silent, but at places near mining operations the shield made a vibrating sound. By countermining at these points, the Barcaeans broke into the enemy's works and slew the men they found there."

Eventually people realized that using an **explosive countermine** would be better than fighting in a narrow tunnel. The idea was to dig close to the enemy's mine gallery and emplace and detonate a charge that would collapse the tunnel and kill the miners. The first identified use of an explosive countermine was during the siege of Belgrade in 1433, when John Vrano used black powder in a countermine against the Turks. This type of explosive countermine was used up to World War I. During the Thirty Years War in Central Europe (1618-1648), some defenders had released poisonous antimony gas into the tunnels to kill miners.

(28.2) LANDMINES, AS WE KNOW THEM TODAY, MAKE THEIR APPEARANCE

(28.2.1) Early, Medieval And Modern Attempts At Using Charges Against Soldiers

In the Modern Age, as fortifications lost their importance in warfare, someone thought of using a similarly buried charge against humans too. Actually, the idea of using buried powder charges against soldiers had struck some smart people in the ancient period also but somehow or the other it could not become popular. Military engineers in China had employed the **first self-contained explosive anti-personnel (AP) mines** against Kublai Khan's Mongol invaders in 1277. Manufactured in many shapes and sizes, these mines could be command-detonated or activated with either a pressure device (probably based on a match) or a pull-firing device (a forerunner of the flintlock mechanism). However, they were seldom used and were largely forgotten by the time Western explorers arrived in the Orient.

Introduction of the flintlock in 1547 led to the first target-activated AP mine in the West. The device called ***fladdermine***, developed by Samuel Zimmermann of Augsburg in 1573, consisted of one or more pounds of black powder buried at a shallow depth in the glacis of a fortress. It was actuated by stepping on it or by tripping a wire along the ground that released a flintlock igniter to fire the main charge. Like the fougasse, these devices were highly vulnerable to dampness and required frequent maintenance. They were used mainly around fixed fortifications.

Although the Chinese first introduced explosive shells (as opposed to solid shot) in about 1221, they were unreliable and were used mainly with mortars. Reintroduction of explosive shells in the West in the 1700s, combined with the invention of the percussion cap by Reverend Alexander Forsythe of Scotland in 1814, made possible the next important step in the development of reliable mines by greatly improving their resistance to moisture. Confederate soldiers under **General Gabriel Raines** improvised the first of this type of AP mine from artillery shells at Redoubt No. 4 near Yorktown, Virginia, during the campaign of 1862. This became famous as the **'land torpedo'** and was the forerunner of the modern AP landmine. They could be cleared only after several casualties by two companies from the 50th New York Volunteer Engineer Regiment. Land torpedoes were also used against Sherman in Mississippi, by General Raines on roads into Augusta, and by General Wheeler on the roads into Savannah and Pocotaglio. A coal torpedo (a type of booby trap with an irregular sheet-iron case filled with black powder and painted black) was used to destroy General Butler's headquarters steamer Greyhound, and coal torpedoes were implicated in the sinkings of the Chenango and the Sultana. By the end of the Civil War, the Confederates had emplaced thousands of land torpedoes around Richmond, Charleston, Mobile, Savannah, and Wilmington, which produced hundreds of casualties. Generals like Robert E. Lee, John Mosby, and J. E. B. Stuart all advocated the use of AP mines. Union Generals publicly condemned this implement of war! But Union troops of I Company, 3rd U.S. Colored Troops, also used torpedoes near Savannah.

(28.2.2) Modern Anti-Personnel Landmines Are Invented And Leave An Indelible Imprint On Warfare

The British employed mines during the Boer War in 1901 to protect railroads and deny fording sites to the enemy. In the Siege of Port Arthur during the Russo-Japanese War of 1904, the Japanese tried to breach Russian mines with volunteer suicide squads that were to force a passage by sacrificing their own bodies. These guys always found suicide volunteers—the kamikaze pilots of the Second World War were not the first! Upon approaching the minefield, however, the volunteers found that heavy rains had exposed many of the mines!

Between the Civil War and World War I, powerful military explosives were introduced that significantly increased the mines' lethality. Black powder shells of the Civil War period burst into only two to five fragments, while those of the Franco-Prussian War burst into 20 to 30 fragments. By World War I, a 3-inch high-explosive shell produced about 1,000 high-velocity fragments.

The German **tretmine (step-on mine)**, the next high-explosive mine to appear, went into limited production before World War I. Lieutenant Ernst Junger of the 73rd Hanoverian Fusilier Regiment described German improvised mines this way: "These hotheads are forever puzzling out the possible ways of ... making the ground in front of the trench murderous with explosive machines. Perhaps they cut a narrow passage through the wire in front of their posts in order to entice an enemy patrol, by this bait of an easy way through, straight up to their rifles."

The United States also had a fairly advanced concept of landmine warfare, as stated in Engineer Field Manual, Parts I-VII, in 1918. However, all the World War I combatants relied heavily on artillery and machine guns and seldom used AP mines. It was not until the Second World War that AP mines reached full maturity, and they have been an important facet of almost every conflict since. Using mostly plastic in mine construction was the most elementary way of beating mine detectors; now people think of most novel ways to make mine detection difficult. Minefields may be laid by several means. Using humans is the simplest and most time-consuming. Mine-scattering shells may be fired by artillery from a distance of several tens of kilometers; mines may be ejected from cruise missiles, or dropped from helicopters or airplanes. Armored fighting vehicles equipped to lay mines have also been built.

If you have seen the Oscar winning film "No Man's Land" you would very well understand how complex life becomes with AP mines. This film brings out the human tragedy angle of mines extremely well. You could watch another film called "Behind Enemy Lines" too for a 'mine experience'. If you are a man of honor then mines are cowardly weapons. However, if you don't bother about such niceties then you have some very nasty things in your hands which would be even nastier surprises to your enemies. Mines are indeed the deadliest companions of a soldier on a battlefield.

(28.2.3) Invention Of Anti-Tank Mines

Rudimentary anti-tank mines were employed by the German Army in World War I. These were based on shells fitted with super-quick fuses, and proved effective against American-crewed

British-built tanks in 1918. However it was not until World War II that effective anti-tank (AT) mines were developed and used in quantity.

Modern landmines, both AP and AT, can be triggered by a number of things including pressure, movement, sound, magnetism and vibration. The only limitation is that imposed by your imagination. The advantage of a magnetic trigger is that it can be made to go off even if the vehicle's tires or tracks have not made contact with it. It is also possible for the sensor to recognize the magnetic signature of a vehicle and act accordingly.

(28.3) MORE THAN 110 MILLION OF THEM AROUND THE WORLD: YOU MUST TAKE THEM SERIOUSLY

Mines are cheap, costing usually from $3 to $12 only per piece. And that's the main reason that they are made by the millions.

In any case, more than 110 million active mines of over 700 types lie scattered in 70 countries waiting for the unfortunate guys to step on them. So far, at least 300,000 people have lost limbs as a result of anti-personnel mines, 50% or more civilians. You must also note that a buried mine can remain active for 50 to 100 years. For every mine found and cleared, 20 more are planted somewhere in the world. At this rate, it would take 1,100 years to clear the world of mines, assuming no additional mines are put down during that time. On the other hand, the cost of deactivating one land mine is somewhere between $500 and $2,000.

Nearly 144 countries of the world have made it into a fashion statement to chicken out on anti-personnel mines claiming that their victims are often civilians. The Americans (yes, the very same nation where the Union generals had once deplored it publicly) have not chickened out and they euphemistically call them '**area denial munitions**'. Among other countries that have not signed the Ottawa Convention we have China, India, Pakistan, Russia, and Yugoslavia. In September 1977 dozens of nations endorsed the international treaty on a worldwide ban on land mines, but President Bill Clinton said the United States would not sign the treaty. "There is a line I simply cannot cross," Clinton said at a news conference in Washington, several hours after delegates in Oslo adopted the landmine ban by acclamation. Washington had asked for certain exemptions, saying it wanted to use landmines for nine more years on the Korean peninsula, where the US maintains tens of thousands of troops in support of South Korean troops confronting Communist North Korea. Clinton said landmines were an integral part of Washington's defense strategy in that part of the world, and did not pose any threats to civilians since the land mines were in a clearly marked zone. The United States had also asked for an exemption for its '**smart anti-personnel mines**' that self-destruct over a set period of time and a provision that would allow countries to withdraw from the treaty if they come under attack.

(28.4) ANTI-PERSONNEL MINES: ODDS AGAINST THE UNLUCKY MAN!

(28.4.1) Variety Galore

There are several-hundred different kinds of anti-personnel mines in use by many countries. AP mines are devices designed to wound or kill individual soldiers by blast or fragmentation. They are usually small, some no bigger than a soda can or a deck of cards—containing no more than a few ounces of explosives—their effect can, however, be devastating. A mine can be buried, taped to trees, kept underwater, or hidden in any of a hundred places and triggered by a number of means. Mines do their work in a number of ways and the most widely used ones are discussed below.

(28.4.2) Blast Type AP Mines

It is unclear which mine is the first modern **'toe-popper blast AP mine'**, but the Soviet-made **PMK-40** and the British-made **'Ointment Box'** mine are good candidates. Blast type AP mines generally have little to no fragmentation and are just small explosive charges with a method of detonation. Even though these mines can be very small, some no bigger then a golf ball, they are still very effective. The most common type of mine, blast mines are buried no deeper than a few centimeters and are generally triggered by someone stepping on the **pressure plate**, applying about 11 to 35.3 lbs of pressure. These mines are designed to destroy an object in close proximity, such as a person's foot or leg. They are thus designed to maim and not to kill. A blast mine is designed to break the targeted object into fragments, which can cause secondary damage, such as infection and amputation. These mines can be entirely non-metallic and thus very hard to detect. Examples of these types of mines are the Bosnian **PMA-3** and the American **M14**.

(28.4.3) Fragmentation Type AP Mines

Stake-mounted, fragmenting AP mines were introduced in the Russo-Finnish War of 1939, when badly outnumbered Finns improvised them from grenades. When the Finns fought the Russians to a standstill along the Mannerheim Line in November 1939, this setback forced the Russians to organize the first mounted breach of a mined, complex obstacle. In preparation for a deliberate breach, the Russians improvised roller tanks and flamethrower tanks, and conducted extensive rehearsals. The stake mine that emerged from the World War II is still used today without significant changes to its design. The best-known example is the Soviet-made **POMZ-2** mine.

These mines release fragments in all directions, or can be arranged to send fragments in one direction **(directional fragmentation mines)**. These mines can cause injury up to 200 meters away and kill at closer distances. The fragments used in the mines are either metal or glass Fragmentation mines can be bounding or ground-based. These mines are easier to detect because, by nature, they must contain large amounts of metal and usually must be surface laid The **PMR-2A** is an example of this type of mine.

(28.4.4) Bounding Type AP Mines

An 1859 U.S. military engineering manual by **General Halleck** includes the design for an improvised command-detonated bounding AP mine. However, modern manufactured examples of this type did not make their combat debut until early in World War II, when French patrols on the Siegfried Line began to take unexplained casualties. These casualties were attributed to a device the French dubbed 'the silent soldier', the famous German 'S' mine introduced during the 1930s. These mines were commonly called **'Bouncing Bettys'**.

A bounding mine is a fragmentation type mine that contains a small booster charge. The booster charge launches the mine up to a height of 3 to 6 feet where the mine detonates. This increases the fragmentation effect and allows the mine to be better concealed because it can be buried. The **PROM-1** is an example of this type of mine.

Usually buried with only a small part of the **igniter** protruding from the ground, these mines are pressure or tripwire activated. When activated, the igniter sets off a **propelling charge**, lifting the mine over three feet into the air. The mine then ignites a main charge, causing injury to a person's head, chest or groin.

Fig.176 The Most Dreaded Of Them All—The Bounding Mine

(28.4.5) Directional AP Mines

These mines descended from an early directional type of stone fougasse used in Europe. Under the guidance of physicists **Franz Rudolf Tomanek** and **Hubert Schardin**, the Germans developed a directional AP mine, called a **trench mine**, late in World War II. The French fielded a directional AP mine in 1947, but it was the Americans who refined it in response to the human-wave attacks of Chinese Communist forces during the Korean Conflict in the early 1950s. The new mine was developed and placed in production in 1953, too late to see combat in Korea. Called the M18 Claymore after a famous type of Scottish broadsword, it first saw combat in Vietnam in 1961.

Upon detonation a directional mine fires its payload in a fan-shaped pattern into the target area in a certain direction. These mines usually contain several hundred small steel balls like ball bearings as projectiles. Directional mines are primarily used for perimeter defense and command detonated, however they can be rigged with tripwires or other types of fuzes. Like the fragmentation type mines this type of mine must be surface laid. The **M18A1 Claymore** and the **MRUD** are examples of this type of mine.

(28.4.6) Fuel-Air Explosive Mines

A fuel-air mine contains a liquid explosive that is dispensed into a cloud by a small bursting charge then is ignited. The resulting shock wave flattens nearby personnel and lightly armored vehicles. Fuel-air mines are not very common because of their size, expense, and complexity. The Serbian UDAR FEA mine, for example, weighs 88 lbs.

(28.4.7) Chemical Mines

The British-developed **Livens Projector** was first employed in 1917 and is arguably the first chemical mine. The Germans also developed and employed what the Allies dubbed the **'Yperite Mine'** in 1918. It used a delayed action demolition charge containing mustard agent ('Yperite') to deny bunkers that were being abandoned during a withdrawal. The first modern chemical mine, the **Spruh-buchse 37 (Bounding Gas Mine 37)**, was developed and produced by Germany during World War II and normally had a mustard-agent fill. It was never used in combat. Except for the introduction of nerve-agent fills, the design of chemical mines has not changed significantly since the Second World War.

(28.4.8) Flame Mines

The first reported flame mine was improvised by Confederate soldiers near Charleston in 1864, possibly from incendiary shells. During World War II, the Russians used a trip-wire-activated static flamethrower at the Battle of Kursk. These devices were quickly copied by the Germans and used in the Atlantic Wall. The British also employed improvised flame mines during the First Battle of El Alamein in 1942. The United States developed the first modern flame mine, the **XM-55**, for use in Vietnam. It was a pressure- or trip-wire-activated bounding mine. There are no indications that it was ever used in combat. Improvised flame mines, sometimes called **flame fougasse**, are still occasionally used in combat.

(28.5) THE TECHNOLOGY OF THE FUZES OF AP MINES

(28.5.1) Various Types Of Fuzes

The fuze is the part of the mine that initiates the detonation of the main charge. There are many types of fuzes, each having its own use. They are discussed in detail below.

- **Pressure Fuzes**: This is the most common fuze and it is used with both AT and AP mines. It is activated by pressure being applied to a pressure plate or prongs. Anti-tank mines usually require substantially more pressure then AP mines and a man on foot usually won't set a AT mine off. Sometimes though, if a mine has been emplaced for a while, pressure will be applied by natural phenomena such as the ground freezing. If this happens it may only take a little bit more pressure to initiate the mine.
- **Tripwires**: These types of fuzes use thin, almost invisible wires strung across a avenue of approach. When pressure is applied to the wire, either by stepping on it or walking into it, it detonates the mine. Tripwires are primarily used with fragmentation type AP mines. Their drawback is that at least a small portion of the mine must be exposed which makes their detection easier.
- **Tilt Rods**: Tilt rods are normally used with anti-tank mines and consist of a rod 2 to 3 feet long sticking up out of the ground. When pressure is applied to the rod laterally the mine fires. The advantage of this system is that the tracks don't have to run over the mine as is required by a pressure fuze—any part of the vehicle touching the tilt rod will be enough to trip it. Like the tripwire though, it requires part of the mine to be exposed.
- **Magnetic Fuzes**: These fuzes detect a large metallic object and cause the mine to activate. They are normally used in anti-tank and off-route mines. Some of the more advanced versions can even discriminate between tanks, APCs, and other vehicles. Like pressure fuzes, they allow the mine to be buried and completely concealed.
- **Seismic Fuzes**: These fuzes are used primarily used with anti-tank and off-route smart mines. They detect the vibration of passing vehicles. These fuzes are not very common due to their cost and complexity.
- **Thermal Fuzes**: These specialized fuzes are normally used with off-route and other smart mines. They are activated by a vehicle's heat signature. Like magnetic fuzes, they can discriminate between vehicles. They are not very common for the same reasons as the seismic fuze.
- **Command Detonation**: Command detonation is not really a fuze but it provides the same effect. In this system a person manually detonates the mine, usually electrically. This system allows total control over the mine's effects but requires a person to be present. Command detonation is normally used with directional mines such as the MRUD or Claymore.

- **Sympathetic Detonation**: Again, this is not a fuze; however, it is a method of detonating a mine. In this system the explosion of one mine causes one or more other mines to detonate. An example of this would be taking a piece of detonating cord and wrapping it around a small mine such as a PMA-1. The other end is then inserted into the detonator well of a mine such as a MRUD. When the PMA-1 fires, it sets off the detonating cord, which in turn detonates the MRUD. By daisy chaining them together an unlimited amount of mines can be set off at the same time.

(28.5.2) Anti-Lift Fuzes Or Anti-Handling Devices

The first explosive booby traps were employed by the Chinese against the Mongols in 1277. They first appeared in the West during the Seminole War of 1840. During the Civil War, Confederate soldiers employed a variety of these devices—including pull-firing devices, timer-rundown fuzes, and coal or wood 'torpedoes' that exploded when burned in a boiler. Booby traps reached full maturity during World War II, when reliable German mechanical anti-handling devices were introduced, and have been used in almost every conflict since.

Anti-handling devices **(AHDs)** perform the function of a mine fuze if someone attempts to tamper with the mine. They are intended to prevent moving or removing the mine. An AHD usually consists of an explosive charge that is connected to, placed next to, or manufactured in the mine. The device can be attached to the mine body and activated by a wire that is attached to a firing mechanism. Some mines have extra fuse wells that make it easy to install AHDs. An AHD does not have to be attached to the mine; it can be placed underneath the mine. These devices use a spring-loaded striker with a standard base, and they function in one or more modes: pressure, pressure-release, tension, and/or tension-release. Very clever, isn't it? They are primarily used with anti-tank mines.

(28.6) SOME FAMOUS AP MINES

(28.6.1) The M14 Blast Mine

The American M14 is a small, cylindrical, plastic-bodied blast mine designed to incapacitate rather than kill. It is just 1.57 inches tall and 2.2 inches in diameter. It was originally

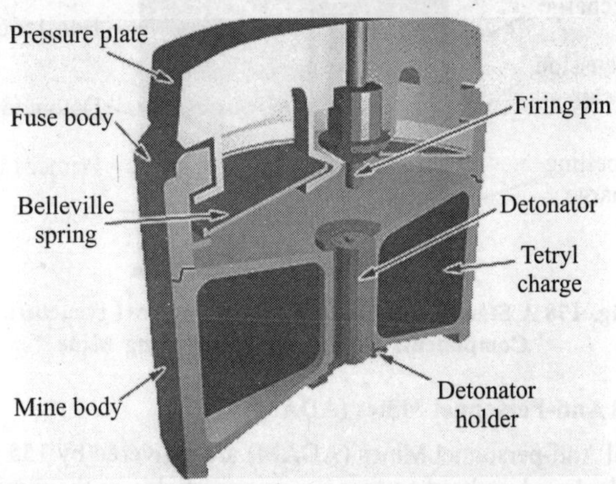

Fig. 177 A Standard Anti-Personnel Blast Mine—Internal Structure And Main Components Of The M14 APM

developed and used by the United States in the 1950s, but it has been used and copied by many nations around the world. Only the firing pin is made of steel; the fuse is made of plastic. This particular anti-personnel mine contains only a small amount of explosive, about 31 grams of **Tetryl.** Once it is armed, any pressure of at least 19.8 pounds can cause the mine to detonate. When the proper amount of pressure is applied it pushes down on the **Belleville spring** underneath the pressure plate. This spring pushes the **firing pin** down on to the detonator.

(28.6.2) The M16 Bounding Mine

The M16 Bounding mine is made of three main parts: a **mine fuse**, a **propelling charge** to lift the mine and a **projectile** contained in a cast-iron housing. It is 7.83 inches tall and 5.24 inches in diameter. The M16 mine contains about 1.15 pounds of TNT explosive. The fuse extends through the center of the mine to the bottom, where the propelling charge is located. To arm the mine, a safety pin is removed from the **striker** on top of the fuse. There are three **prongs** located on top of the fuse, connected to a **spring-loaded wedge**. The fuse encloses a **percussion cap**, a **delay** element and a **black-powder charge**. The M16 can be detonated in two ways: by pressure exerted on the prongs or by pulling the spring-loaded release pin if it is used in a booby trap configuration. Either method causes the pin to pull out of the fuse, releasing the striker and igniting the percussion cap. The percussion cap fires a delay element in the fuse, which fires a detonator after a short delay. The detonator ignites the black powder in the fuse, firing the propelling charge in the bottom of the mine. The mine flies upward to about four feet; the main charge then detonates and releases the fragments. The casualty radius is about 88 feet.

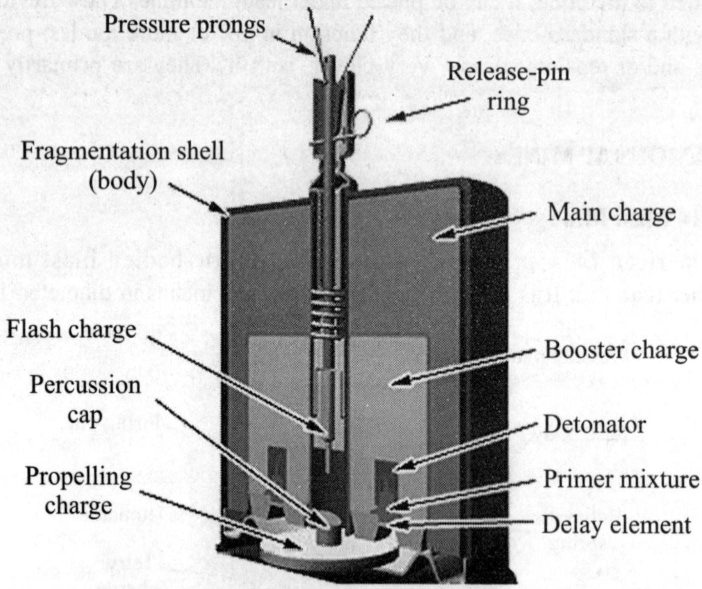

Fig. 178 A Standard Bounding Mine—Internal Structure And Components Of The M16 Bounding Mine

(28.6.3) Area Denial Anti-Personnel Mines (ADAM)

Area Denial Anti-personnel Mines (ADAM) are delivered by 155 mm howitzer. They are not intended to be deployed on hard surface roads or airfields as the mines will shatter or break when hitting asphalt or concrete. There are 36 wedge-shaped mines contained in the 155-mm projectile. Minefield density can be selectively determined by altering the number of rounds applied. The ADAM is an antipersonnel mine activated by deployed trip lines. The mines are expelled from the projectile (approximately 600 meters) over the designated target. Shortly after ground impact, up to seven trip line sensors are released out to a maximum length of 20 feet. The detonators are armed to friction in the event of any small disturbance. The ADAM mine has lethality out to 15 feet. Self-destruct times are 4 hours for short self-destruct (M731) and 48 hours for long self-destruct (M692).

(28.6.4) Delpoyment of AP mines

AP mines are usually placed in a hole and covered with camouflage material. If the hole is only slightly larger than the mine, the weight of the target may be supported by the shoulder of the hole, and the mine will fail to activate. Such bridging action can be avoided if the hole is dug much wider than the mine. Conventional mines are hand-emplaced mines that require manual arming. Personnel place the mines within a defined, marked boundary and lay them individually or in clusters. They record each mine location so that the mines can be recovered if necessary after an area is secured or when some combat action is over.AP mines are generally placed:

- On tracks and pathways frequently used by military forces.
- Beside trees and other spots which are likely to be used by troops as resting places.
- Beside trees and other natural cover near the verges of roads which might be used as cover by troops. This type of mine-laying is frequently used in conjunction with anti-vehicle mines or an ambush, thus causing troops to leave the road in search of cover.
- On new tracks made by troops due to the tendency in thick bush to return by the same route.
- On tracks recently cleared by troops, which leads them to suppose that they are cleared for foot traffic.

(28.7) THE CLAYMORE MINE

(28.7.1) Discovery Of The Misznay-Schardin Effect

The discovery of the Misznay-Schardin effect is very interesting. On November 28, 1935, **Franz Rudolf Tomanek**, a brilliant young Austrian explosives physicist was summoned to Berlin to work on shaped charges—a concept he was responsible for bringing to Hitler's attention. He found himself standing in the Reichsregierung along with Hitler, Hermann Goering, Rudolf Hess and Heinrich Himmler. In as orderly and complete a manner as possible, he presented his concepts for a low-cost, lightweight, effective anti-tank weapon. "This proposal," the Fuehrer declared, "could be the solution. I have always wanted to give the individual soldier a weapon to defeat tanks. Maybe this same concept could be used for bombs and torpedoes." The second lead player was **Dr. Hubert Schardin,** a brilliant German explosives physicist. Tomanek went to work indirectly for Schardin at the Technical Academy of the Luftwaffe, or TAL. Some progress was made, but in order to work properly, the shaped charges required a cavity cone liner made from copper or zinc, the design of which eluded Tomanek and Schardin until well into the 1940s.

It was the spring of 1944 and Dr. Hubert Schardin needed assistance. Under direct orders from Adolf Hitler to develop new super weapons, he needed the Gestapo's help to locate a famous but reclusive Hungarian colonel named **Misznay** who could provide detailed information regarding the complex physics involved in shaped charge explosives. Colonel Misznay was, by all historical indicators, so elusive that even today we are uncertain what his real first name was. After World War II, he dropped out of sight in the Eastern Bloc. The Gestapo headquarters finally managed to get Schardin and Misznay together, but there is no exact record of their conversation or of the mathematical conclusions on which they agreed. Professor Schardin came away from the meeting with enough information relative to direction-effect fragmentation explosives to start development on radically new rocket warheads. He intended to use them to shoot down the American Boeing B-17 bombers then devastating Germany and to thwart Josef Stalin's heavy tanks on the Eastern Front. Schardin knew that the military situation was desperate, and he reacted accordingly.

Misznay's name lives on as a result of a special explosive phenomenon he identified, called the Misznay-Schardin effect—a phenomenon that recognizes that fragments can be thrown from the face of an explosive charge in a predictable pattern, much like a projectile from a rifle barrel. The Misznay-Schardin Effect is a characteristic of the detonation of a broad sheet of explosive. The force produced by an explosive sheet is primarily exerted perpendicular to its surface in both directions. If one side is backed by a heavy or fixed object, the majority of the explosive force will be sent in a single direction. Since the explosive energy is released directly away from (perpendicular to) the surface of an explosive—hence shaping the explosive will focus the blast.

Schardin took that necessary information back to TAL in Berlin. But other pressing developments such as the Messerschmitt Me-262 jet fighter, a 100-mile artillery piece called a Hochdruckpumpe (the V-3 cannon), and numerous rockets occupied Schardin's time. The Panzerschreck, an improvement upon the famous American anti-tank weapon Bazooka was indeed successfully made with a shaped charge that gave armor penetration of 100 mm. He did make an experimental **trench mine** also but could not carry the work on mines far. His anti-tank mine, something that had fascinated the Fuehrer, consisted of a solid slab of explosive with a hollowed front face. The concave surface of the mine was lined with a solid brick of steel and was originally intended as an anti-tank mine. When detonated the explosive force was enough to throw the steel brick forward with enough power and velocity to penetrate a tank's armor plating at ranges up to 50 meters. However the Germans could never fully perfect the weapon.

(28.7.2) Invention Of The Claymore Mine

After World War II was over, **Norman MacLeod** continued work on the Misznay-Schardin effect in the United States. He worked at the China Lake Naval Depot near Bakersfield, California, from 1945 to about 1948. MacLeod was a bit eccentric, a hard-to-get-along-with super patriot who was outraged that American soldiers were forced to endure human wave attacks by Communists during the Korean War. He was especially upset at events north of the Chosin Reservoir in late 1950, when as many as four ranks of Chinese soldiers lined up at arm's length and simply ran, lemming-like, over U.S. Army and Marine troops. The early 1950s found Norman MacLeod managing his own company, the Explosive Research Corporation, in Southern California. In that capacity, he developed a small, anti-personnel mine which he envisioned as a sort of huge shotgun that the man in a foxhole could quickly set up and detonate, taking out any attackers in front of him. That's how the prototype of the Claymore was born. And why did he name it Claymore? Well, Macleod was Scottish and had a great love for the famous Scottish sword Claymore.

The Claymore in its more refined form was made in 1956 by four bright young men--**Don Kennedy, Bill Kincheloe, John Bledsoe** and their chief, **Guy Throner**, working for a munitions builder, Aerojet Corporation. It was this group which made the discovery that hardened steel ball bearings broke up in the device and led to erratic flights. After much experimentation they zeroed in on soft steel. By the time the final design got into production, it was too late for the Korean War but they were used with great effect in Vietnam. Like the original Claymore broadsword swung by a very angry, determined, kilt-wearing Scotsman, Claymore mines cut a large swath in the annals of modern warfare. Grunts in the trenches still use the same 1956 design. Those who have fired Claymores in combat swear by the inexpensive little devices.

(28.7.3) The M18 Claymore Mine

The M18 Claymore is an anti-personnel mine which sprays ball bearings directionally, usually in an arc of 60 degrees up to a distance of over 50-100 meters and at a maximum height of 2 meters. This is essentially a booby trap as the device is not buried into ground; the mine can be detonated on command as well. The planted mine stands on two sets of adjustable legs! The optimum effective range (the range at which the most desirable balance is achieved between lethality and area coverage) is 50 meters. The back blast area is unsafe in unprotected areas 16 meters to the rear and sides of the munition. Friendly personnel within 100 meters to the rear and sides of the munition should be in a covered position to be safe from secondary missiles. The M18 command-detonated mine may be employed with obstacles or on the approaches, forward edges, flanks and rear edges of protective minefields as close-in protection against a dismounted infantry attack. The mine itself is rather small (8 inches long, 3 inches wide and 1.5 inches thick) containing 650 grams of C-4 plastic explosive and 700 ball bearings. The intellectual capacity of the American soldiers is so seriously doubted by their own establishment that the words "Front Toward Enemy" are prominently written on the mine so that in spite of the obvious convex side to face the enemy, some blockhead does not detonate it toward himself!

Fig. 179 The Famous Claymore Anti-Personnel Mine

The M18 A1 version is primarily a defensive weapon. It may be employed to a limited extent in certain phases of offensive operations. The M18 A1 has the same basic capabilities as anti-personnel mines and can be used in most situations where other types of antipersonnel mines are employed. In addition, the M18 A1 has the capability of being sighted directionally to provide fragmentation over a specific area and does not necessarily rely upon chance detonation by the enemy. The M18 A1 is adaptable for covering the ranges between maximum hand grenade throwing distance and the minimum safe distance of mortar and artillery supporting fires.

Fig. 180 Note The Arrangement Of Ball Bearings In A Claymore

The Claymore, in its original form, is basically an anti-personnel weapon. However, inspired by the idea, people are building bigger versions of the device which could be used against vehicles or even aircrafts. The Swedish **FFV-013** weighs 44 lbs and is claimed to **effective against helicopters** at a distance of up to nearly 500 feet!

(28.8) ANTI-TANK MINES HUMBLE THE MIGHTY TANKS

(28.8.1) The First Anti-Vehicle Mines

One of the earliest anti-vehicle 'mines' was described by military engineer Philo of Byzantium around 120 B.C., when he recommended that "in front of the advanced walls (of a city) empty earthenware jars should be buried. These are placed in an upright position with their mouths upward, stopped up with seaweed or imperishable grass, and covered with earth. Troops may then pass over the jars with impunity, (but) the engines and timber towers brought up by the enemy will sink into them." Another early example occurred during the Roman siege of Jerusalem in 70 A.D., when Jewish Zealots dug a tunnel mine under one of the besieger's four massive siege engines (powerful battering rams on wheels), resulting in its destruction and a three-day delay in the battle. During the American Civil War, Confederate soldiers developed and employed pressure-fuzed railroad mines that destroyed at least two heavily loaded trains in Tennessee. To counteract the railroad mines, the Union Army improvised the first mine-clearing roller; a flatcar pushed slowly in front of a locomotive to detonate any mines ahead of it. Though it may seem odd, the explosive anti-vehicle mine predates the appearance of the tank by more than 50 years!

(28.8.2) Proper Anti-Tank Mines Appear On Battlefields To Take The Punch Out Of Tanks

When the British first fielded the tank in the First World War, it was indeed very disturbing for the Germans. The tank was impervious to small arms fire and could easily force its way through obstacles like ditches and barbed-wire entanglements to threaten the infantry. German pioniere (combat engineers) hastily improvised the first anti-tank (AT) mines in response to this menace. Initially, the Germans buried standard artillery and mortar shells with a sensitive fuze pointed up. They also employed command-detonated mines, which are forerunners of full-width-attack AT mines. Later they improvised many types of mines; including a wooden box mine that measured approximately 14 by 16 by 2 inches and weighed about 12 pounds. Twenty 200-gram blocks of explosive were placed in each box, which was normally buried about 10 inches deep. Detonation was initiated by a hand grenade placed inside and against one wall of the box so that the primer passed through the wall. The mines functioned by pressure as tanks passed over them or by command detonation. Electric blasting caps greatly facilitated command detonation. During World War I, Germans scattered their AT mines at random or in locally created patterns to reinforce wire obstacles and AT ditches in front of trench lines.

The Germans began to manufacture standard AT mines in 1916 and produced nearly three million before the Armistice of 1918. Regrettably, no information on the characteristics of these factory-produced mines has come to light. German AT mines had accounted for a significant portion of allied tank losses, including about 15 percent of US tank casualties, during the battles of St. Mihiel, Catalet-Bony, Selle, and Meuse/Argonne. The British also improvised AT mines during World War I. Two varieties have been identified: one based on a pipe bomb and the other on a bombard shell.

In 1929, the Germans introduced the **Tellermine 29**, the first in a series of modern pressure-fuzed AT mines. This series formed the basis for many of the AT mines used to date, including the U.S. M15.

Fig. 181 A Standard Anti-Tank Mine, The American M15 ATM

Modern anti-tank mines are much bigger than anti-personnel mines and have to have triggers that will be activated only by the pressure of an armored vehicle for which they are designed and not by the foot-pressure of a man. AT mines are normally buried with the top of the mine approximately two inches below ground level. These mines are generally larger and much more powerful than AP mines. Weighing several pounds, they are designed to kill or disable main battle tanks (MBTs) and other armored fighting vehicles (AFVs). Generally they are set so that they must have several hundred lbs of pressure, such as a vehicle running over them and not just a man on foot, to set them off.

(28.8.3) Classification Of Anti-Tank Mines By The Effect They Produce

Some mines result in only a **Mobility-Kill (M-Kill)** whereas some result in a **Catastrophic-Kill (K-Kill)**. A M-Kill still leaves the tank armament and its crews able to carry on fighting. A M-Kill destroys one or more of the vehicle's vital drive components (for example, breaks a track on a tank or blows off a wheel) and immobilizes the target. A M-Kill does not always destroy the weapon system and the crew and they may continue to function. In a K-Kill, the weapon system and/or the crew are destroyed.

(28.8.2) Basic Design Types Of Anti-Tank Mines

There are several types of anti-tank mines and they are discussed below.

- **Blast Type**: Blast AT mines derive their effectiveness from the force generated by high-explosive (HE) detonation. Thus they can be thought of as basically a larger version of the blast type AP mine. They rarely kill a heavily armored vehicle like a MBT, instead they disable it by blowing off the track and severely damaging suspension components. These effects can be devastating against lightly armored vehicles. The TMA-1 and M15 mines are examples of blast type AT mines. They usually produce a M-Kill when the blast damages the track or the vehicle, but a K-Kill is also possible.
- **Shaped Charge**: This type of mine can kill a tank by firing a shaped charge through the thin belly armor. They work much like a HEAT round producing a jet that burns through the armor. These mines sometimes have a small initial charge that blows the soil and rocks from the top of the mine before the main charge detonates.
- **Explosively Formed Penetrator**: We have discussed the EFP in detail. Well, these mines fire an EFP. Like the shaped charge mines, they may also have a initial charge to clear away the soil before the main charge detonates.
- **Off-Route Mines**: These **'smart mines'** use thermal, seismic, or magnetic sensors or a pressure sensitive wire laid across the road to detect vehicle movement. Once a vehicle is detected a rocket or other munition hidden alongside the road is fired into the vehicles armor. The more advanced versions can not only differentiate between a tank and other class of armored vehicle, but can even tell friend from foe and between models such as the T-80 and T-72 MBTs.

(28.8.3) Various Types Of Anti-Tank Fuzes

Anti-tank fuzes fall into three categories:

(a) Track-Width Or Pressure Fuzes: They are usually pressure-actuated, requiring contact with the wheels or tracks of a vehicle. Pressure activated fuzes are very simple but people applied quite a bit of ingenuity even for them. For instance, the **double-impulse fuze** works after more than one vehicle has passed over it. This means that a minefield can be designed with several outer lines of AT mines which will only explode when the first wave of enemy tanks are already disabled or damaged deep in the minefield. The double impulse fused mines may therefore damage or destroy valuable recovery vehicles, or the second wave of enemy armor.

(b) Full-Width Or Anti-Hull Or Influence Fuzes: A cheaper form of full width attack mine can be produced by a tilt rod screwed into the fuse well. The rod stands about two-and-a-half feet clear off the ground. If it is pushed out of alignment by the glacis plate, it will detonate the mine underneath the belly of the AFV. Tilt-rod fuses normally require the body of the mine to be buried and the tilt-rod assembly to be clear of the ground. A tilt-rod fuse is preferred in areas where vegetation is sufficient to conceal the extension rod. Camouflage materials are carefully used to prevent premature detonation or interference with the normal functioning of the fuze. Extension rods are camouflaged before the mine is armed.

The Russian **AKS, a tilt-rod-actuated blast AT mine** that appeared on the Russian front during World War II, was probably the first true full-width AT mine. However, the Germans developed the first modern full-width-attack mine toward the end of World War II. Called the **Hohl-Sprung Mine 4672**, it employed a tilt-rod fuze and shaped-charge kill mechanism. Although 59,000 of this mine were produced, there are no reports that it was ever used in combat. Nevertheless, it represented a significant improvement in mine technology. The French probably were first to field a modern full-width-attack mine when the Model 1948 entered service in 1948. Another important advance in the evolution of the full-width AT mine was the Russian development of influence fuzing, both seismic **(VZ-1)** and magnetic, during World War II.

Modern full-width, anti-hull or influence AT mines are activated by several methods: acoustics, magnetic-influence, tilt-rod, radio-frequency, infrared-sensor, command, or vibration. Tilt-rod or magnetic-influence fuzes are the most common. Full-width fuses are designed to be effective over the entire target width and can cause a K-Kill from penetration and spalling metal or from secondary explosions. When a full-width fuse is activated solely by contact with the wheels or tracks of the target vehicle, it usually causes a M-Kill because most of the energy is absorbed by the wheels or tracks. The full-width-attack fuze has proven extremely effective. Its greater coverage enables emplacing units to get the same obstacle effect with significantly fewer mines per kilometer of front. Additionally, this type of mine often produces a K-kill, with fatalities to crew members of all vehicles, including tanks and other tracked vehicles. This lethality definitely decreases the willingness of combat vehicle crews to 'bull through' a mined area.

(c) Off-Route Fuzes: They are designed to produce a M-Kill or a K-Kill, top or side attack, when a target vehicle activates the fuze with acoustic or seismic signals. When activated, an infrared-sensor, explosive-formed penetrating (EFP) or a shaped-charge rocket warhead is launched; and the warhead acquires the target.

(d) Thinking Mines: Some of the 'thinking mines' have magnetic fuzes and will explode only when they detect the specific magnetic signature of a particular vehicle or only when the designated vehicle is fully straddling it so that the blast has maximum effect on the belly of the vehicle. Some of them detect the seismic signature of particular vehicles and can distinguish between say, an Abrams tank or a T-80 tank. Off-route mines, like the French GIAT MIACAH F1 and Swedish FFV 016, fire a shaped charge into the side of an AFV when the vehicle either trips a wire or triggers an electronic sensor.

(28.8.4) Deployment Of AT Mines

The most common areas that anti-vehicle mines are placed are:

- On rises and hills so that vehicles, on detonating the mine, will roll backwards onto other vehicles, thus increasing the damage and number of casualties.

- In rocky areas which will hinder prodding and increase shrapnel effect.
- Next to fords or on tracks running by a river or gorge so that, on detonating the mine, the vehicle will fall into the river or gorge.
- On narrow roads with the aim of blocking certain routes.
- On detours.
- On roads where water has accumulated, making detection difficult.
- In sandy areas where laying and concealment is less difficult.

(28.9) SOME GREAT ANTI-TANK MINES

(28.9.1) The M15 Mine

The American M15 mine is a track-breaker and is activated by 350 pounds of pressure. It is a circular, steel anti-tank mine that contains a main charge of TNT. It has a diameter of about 13 inches and a height of about 5 inches. The main component of the M15 is the 22 lbs of **Composition B** explosive. The M15 is armed by rotating the arming switch so that it is set atop the head of the fuse. The cylindrical fuse is made of iron and is attached to the pressure plate by a copper cover. As a tank rolls over the mine, it pushes down on the **pressure plate**. Underneath the pressure plate is a **Belleville spring** with a **firing pin** affixed to its underside. The firing pin is driven down into the detonator, which detonates and fires the **M120 booster charge** beneath the fuse, which then sets off the **main charge**.

The M-15 can contain a M-603 fuse making it a track-width mine activated by 348 to 744 lbs on the pressure plate, producing a M-Kill. With the M-624 fuse with tilt rod, it's a full-width mine that is activated by a deflection of 20-degrees or 3.7 lbs of pressure to the tilt rod. Depending on the armor, this produces a M-Kill or a K-Kill.

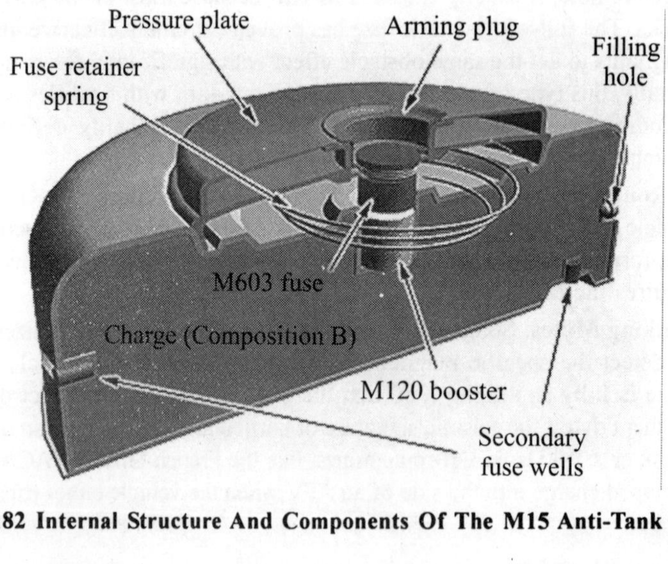

Fig. 182 Internal Structure And Components Of The M15 Anti-Tank Mine

(28.9.2) The M19 Plastic Mine

The M19 is also a track-breaker activated by 400 lbs of pressure. But it is made of plastic and hence difficult to detect. It is square shaped and holds 21 lbs of Composition B explosive.

(28.9.3) The M21 Mine

The M21 is 9 inches in diameter and 4.5 inches in height. It holds 11 lbs of Composition H6 explosive. The mine is activated by 4 pounds of pressure against a 21 inch long extension rod or, without rod, by 290 pounds of vertical pressure on top of its fuze. It uses a Misznay-Schardin plate for providing belly kill and track-breaking capability. The M21 produces a kill against heavy tanks, unless the mine is activated under the track. It can be buried with a tilt rod, or it can be surface-laid with or without a tilt rod. If the M21 is surface-laid with a tilt rod, it must be staked to prevent it from being knocked over and causing the warhead to be directed away from the target. The M-21 with tilt rod must be buried or staked at the 12, 4, and 8 o'clock positions so that enemy vehicles will not tip the mine over. Without the tilt rod, the mine is activated by 287 lbs of pressure on the M-607 fuse and produces a M-Kill by blast effect. The M21 is not compatible with any mechanical mine-dispensing system.

(28.9.4) The M2 Selectable Lightweight Attack Munition (SLAM)

For Special Operations Forces, the M2 SLAM will eventually replace the M15, M19, and M21 mines. The SLAM is an anti-materiel and anti-vehicular (through light armor such as a BMP) munition that is light, compact, effective, and readily usable by Special Operation Forces (SOF) units to destroy enemy vehicles, parked aircraft, and ammunition and petroleum sites while avoiding direct contact with the enemy. The SLAM will have four possible methods of detonation: detonating with a mine function, trip-wire, command detonation function, and a time-delay function (integral timer).

(28.10) FAMILY OF SCATTERABLE MINES (FASCAM)

The Americans are very keen on making life rather easy for their soldiers. Why should you run the risk of emplacing mines yourself when machines can do it for you? The family of scatterable mines (**SCATMINEs**) does precisely this and adds a new dimension to mine warfare. Now you can emplace minefields in enemy held terrain, contaminated territory, or in others areas where it is not possible to emplace conventional minefields. They provide a rapid and flexible means of delaying, harassing, paralyzing, canalizing, or wearing down the enemy forces in both offensive and defensive operations. Thc mines can be delivered by 155 mm shells or a ground dispenser or by air. All FASCAM mines have an active life cycle and self-destruct (SD) time after their active life has expired. They have a safe-arm time from (45 seconds to 2 minutes). When mines fail to arm they will self-destruct. SD times vary from four hours to more than 24 hours.The duration of the active life varies from 4 hours to 15 days depending on the system. The only limitation of scatterable mines is that they cannot be concealed.

There are two general categories of **AP SCATMINEs**—wedge-shaped and cylindrical. The idea of this design is that when they are scattered, as few of them should come to rest on their edge as possible. The M67 and M72 AP SCATMINEs are wedge-shaped and dispensed from an ADAM projectile, which is a special 155-millimeter artillery munition. Each mine weighs 540 grams and is 7 centimeters high containing just 21 grams of explosive Composition A5 (containing RDX with 1.5% stearic acid). The M74, BLU 92/B, M77, and Volcano AP SCATMINEs are all cylindrical in shape. They are 6 centimeters high and 12 centimeters in diameter. Cylindrical AP SCATMINEs kill enemy soldiers through the combined effects of blast and fragmentation. They weigh 1.44 kg and each mine contains 540 grams of composition B4 as its main charge. Each mine has eight trip wires (four on the top and four on the bottom) that deploy after ground impact up to 12 meters from the mine. Trip wires are similar in appearance to very fine thread; they are olive-drab green in color and weighted at the free end. A tension of 0.9 lbs applied to one trip wire is enough to create a break in the electrical circuit and cause the mine to detonate.

All **AT SCATMINEs** whether fired by 155 mm shell or by air have similar functional characteristics. They are cylindrical in shape, weigh approximately 1.8 kilograms, contain 585 grams of RDX as the main charge, and have a magnetically induced fuze. AT SCATMINEs are designed to produce a K-Kill instead of a M-Kill. They produce the K-kill by using an SFF warhead. The EFP is created from a two-sided Misznay-Schardin plate charge to produce a full-width kill.

(28.11) RAAM (REMOTE ANTI-ARMOR MUNITION)

Remote Anti-Armor Munition (RAAM) mines (**M70, M73, M718,** and **M741**) are delivered by 155 mm howitzer. Each projectile contains nine of them. The mines are expelled from the rear of the projectile over the target. After ground impact and roll, the mine is armed and ready to detonate upon sensing a proper armored vehicle signature (electromagnetic). A percentage of the nine RAAM mines are equipped with an anti-disturbance device. You cannot fire them on hard surfaces as they might break on impact.

(28.12) THE MOPMS (MODULAR PACK MINE SYSTEM)

The M131 MOPMS is a man-portable anti-tank (AT) and anti-personnel (AP) mine system. The intended purpose of the MOPMS is to close lanes and gaps in minefields, at chokepoints, to reinforce obstacles, and to emplace point minefields. The M131 module weighs approximately 165 pounds and contains a mix of 17 M78 AT and four M77 AP mines. These 21 mines are launched by 7 launching tubes. Mines are propelled 35 meters to the front and sides. The module may be initiated by hardwire or radio control. The AT mines have a magnetic influence fuze; the AP mines deploy four tripwires upon ejection. Both the AP and AT mines contain a radio frequency receiver which allows the mines to be remotely-controlled (recycle self-destruct time and command destruct) with the RCU (radio control unit) once the mines are dispensed. The RCU can control up to 15 containers or groups of containers via separate pulse-coded frequencies, out to 300 to 1,000 meters. The factory set self-destruct time is four hours. Operators can disarm and recover the container for later use, if mines are not dispensed. The RCU can recycle the 4-hour self-destruct time of the mines three times, for a total duration of 16 hours (4 hours after initial launch and three 4-hour recycles). This feature makes it possible to keep the minefield in place for longer periods, if necessary. The RC can also self-destruct mines on command.

(28.13) FAMILY OF WIDE AREA MUNITIONS (WAM)

The WAM is envisaged to be a smart, autonomous top-attack anti-tank/anti-vehicle munition, designed to defeat armored combat vehicles from a standoff distance. Target vehicles include, but are not limited to, tanks, breachers, and lightly armored tracked vehicles. The **M93 Hornet** consists of three major subsystems: a communications module; a ground platform module, and a smart sub-munition/warhead (sublet) module. Once deployed the WAM uprights itself and autonomously searches for a target vehicle. It uses acoustic and seismic sensors to locate, identify and track armored targets. When a firing solution is satisfied, the WAM launches a sublet in a trajectory over the target. The sublet uses a passive infrared sensor to detect the target and fires an EFP at a vulnerable area. It attacks targets from the top at ranges of up to 100 meters and provides a mobility kill. In addition, the WAM has a command destruct capability for easy battlefield cleanup. In the hand-emplaced version it is about 35 lbs heavy. Deep attack versions would be deployable by missile or aircraft.

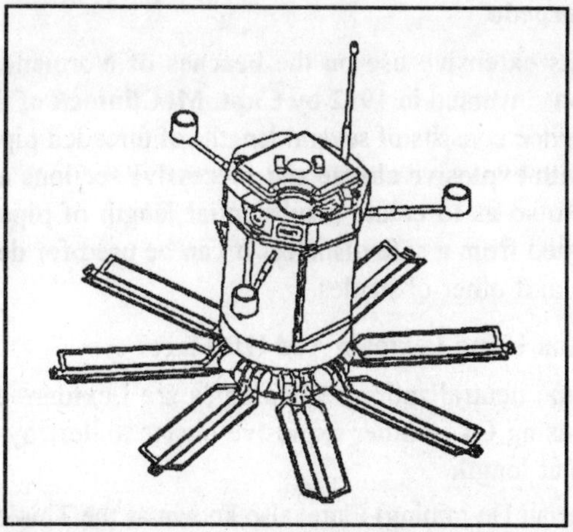

Fig. 183 The M93 Wide Area Munition (WAM)—A Smart Autonomous Top-Attack Munition Designed To Defeat Armored Vehicles From A Standoff Distance

(28.14) BREACHING AND CLEARING MINEFIELDS

(28.14.1) Minefield Breaching Devices

Laying mines is a laborious task but clearing minefields is four times more laborious; the efforts of those who laid the mines are therefore well-rewarded. You can use a mix of both anti-personnel and anti-tank mines and make their job very difficult. A distinction should be made between breaching and clearance. A minefield is breached when lanes are cleared through it to allow men and vehicles to pass through. Normally several lanes are cleared simultaneously so that the enemy cannot concentrate its artillery and air attacks on one vulnerable axis.

Clearing minefields is a painful and risky task. Clearance takes place after the fighting is over, and is a painstaking process in which mines and all types of ordnance are removed from the battlefield. We would mention this topic only in the passing as it does not really fall in the thrust area of the book. Minefield reconnaissance uses both manual and electronic tools.

One of the fastest ways to **breach a minefield** is with explosives. The British system, the Royal Ordnance **Giant Viper**, uses eight rockets to carry a 67 mm diameter, 230-meter flexible hose filled with aluminized PE6/A1 plastic explosive across a minefield. When it has landed, it is detonated in a spectacular explosion, clearing a gap 183 meters long and 7.3 meters wide. It will destroy most mines unless they have double-impulse fuzes or are blast-proofed. Two vehicles towing Giant Vipers working in relays can clear a gap 400 meters deep. As an added precaution an AFV fitted with a mine plough then clears through the gap following the explosion. Similar line breaching systems are in use with the US Army and Marine Corps, who use the **M58A5 Mine Clearing Line Charge (MICLIC)**, and the Chinese Army which uses a **425 mm Type 762** tank chassis mounted twin rocket system. One rocket in the Chinese system will clear a lane 425 feet long and 40 feet to 72 feet wide. The American MICLIC is trailer mounted like the Giant Viper, and will clear a path 45 feet wide and 330 feet long. On a smaller scale, man-packable systems like the British Schermuly Rapid Anti-personnel Minefield Breaching System Mark 2 (**RAMBS-2**) can be fired from a rifle with standard ammunition. It has an explosive line 60 meters long which will clear AP mines to a width of 0.6 meters. Israel Military Industries (IMI) have a portable system (POMINS II) which will clear a path through barbed wire as well as clear AP mines.

(28.14.2) The Bangalore Torpedo

Immortalized by its extensive use on the beaches of Normandy on the D-Day, the so-called Bangalore Torpedo was invented in 1912 by **Capt. McClintock** of the British Army stationed in India. This clever little device consists of several lengths of threaded pipe and connecting sleeves. The leading pipe is filled with explosive charge and successive sections are screwed to the tails of the sections in front of them so as to create a substantial length of pipe. Thus the charge in the pipe in front may be detonated from a safe distance. It can be used for detonating mines or booby traps, clearing barbed wire and other obstacles.

(28.14.3) Mine Neutralization Using Lexfoam And HD Flares

Two successful mine neutralization technologies are Lexfoam and the HD Flare. These systems are alternatives to using C4 or other explosive means to destroy mines in place. We have already discussed Lexfoam at length.

The HD (Humanitarian De-mining) Flare, also known as the Thiokol De-mining Flare, made by the Thiokol Corporation, uses production excess Space Shuttle solid rocket fuel. It is an effective low-order neutralization device against thin-case landmines. It actually burns them using the very high temperatures produced by the flame. After a mine has been detected and partially uncovered, a flare is placed near it and triggered from a safe distance. The intense flame from the flare penetrates the casing of the mine or UXO and ignites the main charge. The mine does not detonate; instead it burns. Even if the mine detonates before the entire explosive has burned, the explosion is significantly reduced.

(28.14.4) Mechanical Clearance Of Mines

One of the most effective manual tools is a **'prodder'**—a non-ferrous metal rod which is pushed into the ground at an angle of about 45 degrees. When it hits something solid the engineer may mark the site for subsequent clearance, or begin the painstaking task himself. This entails digging down around the mine so that a grapnel can be attached and the mine pulled out of the ground by men under cover. If it has anti-handling devices, these will detonate the mine without causing casualties. In Cambodia UN de-mining teams found that the Khmer Rouge had laid some mines at an angle so that a prodder used too vigorously would detonate the mine.

For mechanical clearance of mines, the developments include the **Rhino Earth Tiller**, the **Mine Clearing Cultivator (MCC)**, the **Mine Clearing Sifter**, the **Tempest**, the **Survivable Demining Tractor and Tools (SDTT)**, the **Rotar**, and the **Mini-Mulcher (MAXX).**

The Rhino Earth Tiller is a remote-controlled, tele-operated mechanical system used for large-area AP mine clearance. It neutralizes mines buried up to one foot deep and can also operate in areas of heavy vegetation and small tree cover. It can withstand any AP mine blast and is repairable from AT blasts. The MCC is a remote-controlled mechanical mine clearer for uncovering AT landmines on roads and in large open areas. The MCC is mounted on a modified 200-hp class commercial bulldozer. The Mine Clearing Sifter, which is mounted on the same bulldozer, clears AP and small AT mines from previously scarified soil and loose sand. The Tempest is a small, remote-controlled AP mine blast-protected system designed to clear AP mines from off-road areas inaccessible to large-area mine clearers like the MCC. The SDTT is a modified commercial farm tractor. The system is armored and has mine-survivable steel wheels. The Rotar is an area preparation multi-tool mounted onto an 80-hp Caterpillar backhoe tractor. The MAXX is a small remote-controlled vegetation clearer that can operate in very tight areas. The vegetation clearance attachments fit onto an articulated boom that can be extended and rotated 360 degrees around the unit.

(28.14.5) Supersonic Air Spade

The Air Spade is a tool that allows a de-miner to excavate AP and AT mines in hard soil conditions with increased safety. The system integrates a commercial off-the-shelf air compressor, a gasoline or diesel power engine, and a specially designed air nozzle that creates a focused jet of air that flows at a supersonic speed. It separates and removes the hard-packed soil or clay from around a mine faster than existing digging techniques. Still, the air jet will not detonate the mines because it does not exert enough pressure to do so.

(28.14.6) Novel Technologies For Mine Detection

Novel technologies currently under development for detection of mines involve improved electromagnetic (EM) detection, various forms of **ground penetrating radar (GPR)**, Energy-Focused GPR (EFGPR), infrared cameras and chemical detection of explosive vapor from buried landmines. The Camcopter is a small rotary wing unmanned aerial vehicle that can carry a variety of sensors for minefield detection like IR sensors and an on-board global positioning system (GPS). **Electronic mine detectors** have grown in sophistication as the metal content of mines has diminished. Modern detectors can detect even tiny amounts of metal buried several centimeters in the soil—although the more sensitive the detector, the more false alarms are caused by small pieces of metal debris such as shrapnel fragments.

(28.15) HOW TO MAKE MINE-CLEARING EVEN MORE DIFFICULT

As research progresses in devising better means of detection of mines, people have applied mind to making mine detection even more difficult.

- **Coupling**: Coupling is a method of linking one mine or explosive device to another, usually with detonating cord. When the first device is detonated, it also detonates the linked explosive. This technique is often used to defeat countermine equipment, such as mine rollers.
- **Rolling**: The roller will pass over the initial, un-fuzed device and set off the second fuzed device. This in turn detonates the over-passed device underneath the clearing vehicle. When the linked devices are directional fragmentation mines, they can create a large, lethal engagement area.
- **Boosting**: Buried mines, UXOs, or other explosive devices are stacked on top of one another. The device buried deepest from the surface is fuzed. Fuzing only the deepest ordnance helps mask no-metal and low-metal explosive hazards placed near the surface. This reduces the probability of detection by metal detectors, and it increases the force of the blast.
- **Sensitizing Anti-tank (AT) Mines**: On some nonmetallic AT mines, the pressure plate is cracked and the spring is removed to reduce the pressure required to initiate the mine. Similarly, the pressure plate can be removed from metallic AT mines to create the same effect. A pressure-fuzed AP mine can be placed on the top of an AT mine, thus creating a very large AP mine as an alternative method.
- **Daisy chaining**: A 'daisy chain' of munitions is two or more explosive devices wired together so that a single signal will detonate all the munitions at the same time. AP/AT mines may be used in daisy chains linked with other explosive hazards. Enemy forces may link the mines together with trip wire or detonating cord. When the first mine is detonated, the other mines may detonate. This may also create large, lethal engagement areas.

(28.16) THE SO-CALLED MINES BEING USED BY LEFT-WING AND OTHER EXTREMISTS IN INDIA

The very use of the word mine in connection with the improvised explosive devices that the left wing extremists use is misleading and conveys a wrong impression about the military capability of the extremists. As we have seen at great length above, a landmine refers to devices specifically manufactured in factories for the purpose, as distinguished from IEDs (Improvised Explosive Devices).

The point to be noted is that all these factory-made mines require standard high explosives and fuzes of the type described earlier. They can afford to be small in size only because they use very good explosives. Not a single mine of the above type has been ever used by the extremists. All things that serve the purpose of mines, but do not qualify as above, must be called IEDs only. To call them mine is to convey a wrong impression about the technical and weapons capability of the terrorists. Simply because an explosive device succeeds in inflicting casualties does not mean that the extremists have mastered the technology of landmines or have acquired them. Large quantity of any explosive will do the job. Even gun powder, if used in large quantities will be equally effective. After all, not too long ago great fortifications used to be blown up by something as simple as gun powder. In some instances the extremists have used as much as 80 kg of gunpowder buried in the ground! The point being made is that in order to inflict casualties on unsuspecting men and vehicles, one need not have access to modern technology and resources—even centuries old methods and devices would do. A buried charge of half a quintal of black powder or say, ten kg of gelatin sticks will have greatly devastating effect on even a truck or bus.

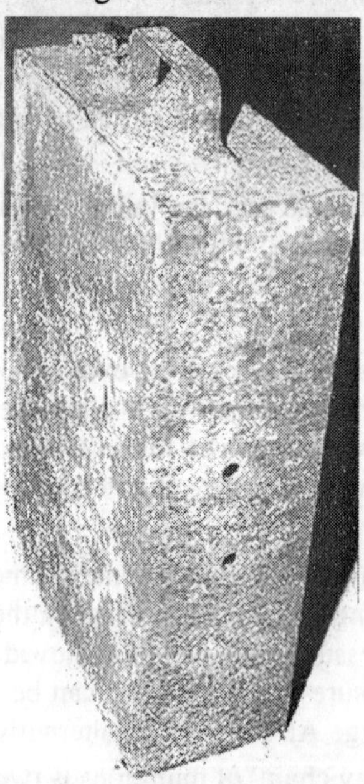

Fig. 184 One Of The So-Called 'Landmines' Recovered In Andhra Pradesh: It Is Essentially Dynamite Sticks Or Some Other Easily Available Explosive Stuffed In A Metal Container With Some Irregular Metal Pieces In It For Shrapnel—An IED Mistakenly Called A 'Landmine' Or Even A 'Claymore Mine' By The Ignoramuses In The Intelligence Agencies And The Security Forces!

For example, in the incident of the attack on the motorcade of Chandra Babu Naidu, the IEDs were not Claymore mines by any stretch of imagination. To call them Claymore mines is nothing but epic ignorance. Eight live devices in the area were also discovered after the incident. While seven of these devices weighed 33 lbs each, the eighth weighed 73 lbs. The recovered devices must have been a backup plan in case the first series were prematurely detonated or did not produce the desired effect. Who has ever heard of Claymores weighing 33 to 73 lbs?

(28.16.1) How The Myth Of The Claymore Mine Got Propagated?

By sheer ignorance! Indian security forces have taken to calling any IED used by the terrorists as a Claymore mine! Most morons do not even know that the Claymore is a small anti-personnel mine. It does not make sense to use a factory-made Claymore against a vehicle—yet the security forces talk of Claymore mines even when the IED has been used against one or more vehicles. They have, of course, never heard of the Misznay-Schardin Effect that is used in the Claymores, and they expect the extremists to be using them!

This brazen ignorance had its origin in some newspapers in Sri Lanka when they started making references to Claymores having been used by the LTTE. The reporting was obviously wrong because they continued to talk of Claymores even when vehicles were targeted. The LTTE, with years of experience of fighting an army, could not have been fool enough to waste Claymores on even unarmored vehicles. In any case, Indian media picked up the word from there. The security forces picked up form them and started using the word indiscriminately.

We will now describe the sort of devices which the left-wing extremists have been found using in India. It is nothing but a metal can like a jerry can in which they just pack gelatin sticks with metal balls or metal pieces of various shapes and sizes. They have used metal pieces as big as cricket balls also, a fact which only illustrates their ignorance in the matter of shrapnel. They have no idea that metal balls as heavy as cricket balls would not fly fast enough to function as shrapnel. It is only an imagination of the media and the security forces that the assorted shrapnel would fly in a directional manner, that is, towards the cars. In the devices made by the extremists, they would fly in all directions. There is no way you can make the shrapnel fly towards the road except by making the IED back against a heavy rock—and in that case it will be too exposed where anyone can see it and thus defeat the very purpose of a surprise attack—the IED would, in any case, not be able to give the shrapnel any significant range! If you happen to bury the IED in the ground then some tamping will be provided by the earth below and the blast would be primarily directed upwards. That is simple common sense and that does not make it into a directional landmine. If you are burying explosive in the ground for blast effect, there is no need to have shrapnel of any size. If one is using shrapnel there also, it would only mean that he does not anything about mines. No AT mine uses shrapnel as the destructive effect is primarily by blast and not fragments; the extremists have not even heard of shaped charges and EFP!

❑❑❑

For example, in the incident of the attack on the motorcade of Chandra Babu Naidu, the IEDs were not Claymore mines by any stretch of imagination. To call them Claymore mines is nothing but crass ignorance. Eight live devices in the area were also discovered after the incident. While seven of these devices weighed 33 lbs each, the eighth weighed 75 lbs. The recovered devices must have been a backup plan in case the first series were prematurely detonated or did not produce the desired effect. Who has ever heard of Claymores weighing 33 to 75 lbs?

(28.16.1) How The Myth Of The Claymore Mine Was Propagated?

By their ignorance, Indian security forces have taken to calling any IED used by the terrorists as a Claymore mine. Most morons do not even know that the Claymore is a small anti-personnel mine. It does not make sense to use a fragmentation-type Claymore against a vehicle—yet the security forces talk of Claymore mines even when the IED has been used against one or more vehicles. They have, of course, never heard of the Misznay–Schardin effect that is used in the Claymores, and they expect the extremists to be using them!

This brazen ignorance had its origin in some newspapers in Sri Lanka when they started making references to Claymores having been used by the LTTE. The reporting was obviously wrong because they continued to talk of Claymores even when vehicles were targeted. The LTTE, with years of experience of fighting an army, could not have been foolish enough to waste Claymores on even unprotected vehicles. In any case, Indian media picked up the word from there. The security forces picked up from them and started using the word indiscriminately.

We will now describe the sort of devices which the left-wing extremists have been found using in India. It is nothing but a metal can like a jerry can in which they just pack gelatin sticks with metal balls or metal pieces of various shapes and sizes. They have used metal pieces as big as cricket balls also, a fact which only illustrates their ignorance in the matter of shrapnel. They have no idea that metal balls as heavy as cricket balls would not fly fast enough to function as shrapnel. It is only in imagination of the media and the security forces that the assorted shrapnel would fly in a directional manner, that is, towards the cars. In the devices made by the extremists, they would fly in all directions. There is no way you can make the shrapnel fly towards the road except by making the IED back against a heavy rock—and in that case it will be too exposed where anyone can see it and thus defeat the very purpose of a surprise attack—the IED would, in any case, not be able to give the shrapnel any significant range. If you happen to bury the IED in the ground then some tamping will be provided by the earth below and the blast would be primarily directed upwards. That is simple common-sense and that does not make it into a directional landmine. If you are burying explosive in the ground for blast effect, there is no need to have shrapnel of any size. If one is using shrapnel there also, it would only mean that he does not know anything about mines. No AT mine uses shrapnel as the destructive effect is primarily by blast and not fragments; the extremists have not even heard of shaped charges and EFP!

PART III
IEDs:
INGENUITY UNLIMITED

Chapter 29

Make it, Set it, Forget it: The Fascinating World of Time Delay and Booby Trap Type IEDs

(29.1) WHAT ARE IEDs AND WHO USES THEM?

As the name suggests, Improvised Explosive Devices are those explosive devices which are not produced in a proper factory which normally produces munitions. Following the American propaganda, people have started taking IEDs to be synonymous with non-state actors at best and terrorists at worst.

(29.1.1) A Revelation For All: The Americans Themselves Have Been Using IEDs For Their Covert Operations

Most people are not aware that the Americans themselves have published entire field manuals for their own army men and Special Forces so that they may learn how to make IEDs. Following the demise of the Soviet Union, the Americans have gone berserk. With no military power capable of challenging them openly, they feel free to send in their army anywhere—if necessary, under the cover of the UN. In other words, they do not feel any necessity of hiding their intentions or activities. This was not the case when the Soviet Union was there. Well, as long as the Soviet Union was there, the Americans were scared. Their intention of establishing their hegemony in the world was still there but they thought it prudent to do many things secretly and in such a manner that they could officially deny their involvement should the necessity arise. The CIA spent several decades and untold sums of money in its covert operations. In the course of such covert and disruptive operations, American Special Forces and intelligence operatives had to operate on their own in hostile territory, whether in the jungles of Latin America or the Far East. That is to say, they could not count on any day-to-day material support of the military and industrial might of the US, something on which ordinary soldiers become so dependent that they take it for granted. They could neither be kept stocked with an endless supply of explosives and ammunition nor could the US air force be sent to their help when they ran into trouble. These Special Forces and operatives had thus to be self-sufficient. They had to learn to fend for themselves without the constant support of the mammoth machinery of the army or the CIA. This meant that they had to improvise their own explosive devices, weapons and ammunition and that too from the material available locally or foraged locally. And they needed special training for this. This was given in all earnestness and with great relish. If you don't believe it, care to have a look at the **CIA Manual 'Explosives For Sabotage'**.

As to the intent and purpose of the book, we reproduce from the Introduction of the manual. Read it very carefully. It is highly illuminating.

"***For quite a few years now, considerable amounts of time, effort, and money have been expended on improving older sabotage devices and accessory gear and in developing new and better items. As a result, a wide variety of manufactured explosive and incendiary items is available for use. At the same time a somewhat smaller effort has been devoted to improvising and testing home-made or field expedient devices and techniques for accomplishing similar results. Since the manufactured, precision devices almost always will be more effective, more reliable, and easier to use, why spend time on improvising field expedients to do the job? For one thing, shelf items will just not be available for certain operations for security or logistical reasons. In these cases the operator will have to rely on materials he can buy in a drug or paint store, find in a junk pile, or scrounge from military stocks. Secondly, may of the ingredients and materials used in fabricating home-made items are so commonplace or innocuous they can be carried without arousing suspicion. The completed item itself often is more easily concealed or camouflaged. In addition, the field expedient item can be tailored for the intended target, thereby providing an advantage over the standard item in flexibility and versatility. While most of the pertinent information on sabotage shelf items is available in catalogs and other publications, much of the improvisation know-how has remained in the minds or files of a few individuals. It is the intent of this manual to consolidate and bring up to date a selected body of this information and make it available for wider use. The devices and techniques included have been selected because they have been well proved out and because they are practical, versatiie, and not too difficult to do. Other techniques have been omitted because they are too unreliable or are just too hazardous to attempt except under the most carefully controlled conditions. The techniques described are not so complex as to require chemical laboratory or machine shop...***"

The Manual also states: "***Commercial and military explosives of one sort or another are in use around the globe. By far the most desirable ones for sabotage purposes are the military plastic explosives and TNT...Plastics have an additional characteristic which makes them ideal for many of the improvisations described later. They can be sliced or molded into any desired shape or pressed snugly against an irregular target...Dynamites also are used commercially world-wide and often may be more available than military explosives. They may be effectively employed in sabotage...If no other sources exist, artillery and mortar shells, bombs, landmines, and other munitions may be cannibalized for the high explosives they contain...***"

The list of topics discussed in the manual is huge, explicit and leaves no scope for any ambiguity whatsoever as to what was the real purpose of the manual. Still, to drive the point home, we will list very few of them:

- Improvised firing systems (including electric pressure switch, water-drip electric delay, expansion electric delay, pocket watch electric delay, seven day electric delay, electric tilting mercury switch, acid and rubber chemical delay, chemical instantaneous initiator, etc.

- Improvised detonators
- Steel cutting techniques
- Derailment techniques
- Improvised cratering explosive
- Improvised shaped charges
- Improvised platter charge
- Improvised counterforce charges
- Improvised diamond charges
- Improvised cable cutter
- Ambush or anti-personnel charges
- Cannibalizing the 3.5" rocket to use it as a shaped charge, landmine, or improvised launching without the launcher, and rigging the rocket as a booby trap, etc
- Improvised incendiaries
- Home-made napalm, and thermate incendiary
- Dust and vapor explosions
- POL charges

It should also be kept in mind that the CIA book is a complete 'how to' manual as against a reference work on the science of explosives and explosive devices. There are other works also of this type, for example, the **CIA Field Expedient Methods For Explosives Preparations** (Desert Publications, P.O. Box 1751, El Dorado, AR 71731-1751, USA, 1977, ISBN 0-87947-207-3). We hope now you realize the utterly shameless hypocrisy of the Americans. You must understand, dear friends; when they commit acts of sabotage or make IEDs, it is quite alright—after all Uncle Sam is the ultimate defender of democracy and human rights in the world. They have got their very special privileges and rights, you see—they are obviously more equal than others! Do you have the gall to expect the same standards from them as you from others?

(29.2) BASIC COMPONENTS OF AN IED

In principle the IED requires only two things; the **explosive charge** and the **firing device.** The bomber has the field wide open for both the things. He can use any explosive he gets. Why? As we saw, military explosives are selected with several factors in mind, such as stability. The IED maker need not bother about such niceties. Even if the explosive is not good in storage, who cares? He does not have to store it for long in any case. Most IEDs are made just hours before the use. If he is prepared to take some risk, he can use even those explosives which are not very safe in handling either. He can even make odd mixtures. All he requires is an explosive substance. He would like to have a powerful one but he would settle for any explosive he is able to get or make. Different explosives have been found to be popular in different regions at different times, depending on the availability.

It is in respect of the firing device that a serious bomber exercises his ingenuity to the fullest extent. Usually he would use a cap-sensitive explosive. Using a non-electrical cap is very much possible but a little cumbersome as some device has to be made to ignite the fuse. Electrical firing devices can be made very neat and accurate.

Let no one be under the impression that IEDs are poorly made bombs and lack sophistication. It all depends on who is making it and how intelligent or dumb he is. The Harvey's

Casino bomb made by John Birges was more sophisticated than any bomb ever made by any nation in any factory and defeated all the attempts of even the scientists from the US Navy's Explosive Ordnance Disposal Facility, the US Army's 34th Explosive Ordnance Disposal Team and Lawrence Livermore Laboratory towards defusing it! By definition, it was also an IED!

Bombing is actually a battle of wits between the bomber and the security forces.

(29.3) IEDs USING NON-ELECTRICAL DETONATORS IN FIRING DEVICES

The basic objective in designing an IED based on non-electrical detonator is to ignite the safety fuse after a delay or when tripped or when a command is given. The safety fuse would, of course, be connected to the detonator.

(29.3.1) Pyrotechnical Firing Devices

For a time delay device, they essentially use a slow-burning fuse or anything that burns slow and eventually ignites a regular safety fuse. For a command-detonated device the safety fuse has to be manually ignited or one has to devise a contraption that would ignite it upon tripping or on command. When it comes to a slow-burning thing, bombers have used even cigarettes and they work fine with very respectable delays obtained. The unlit end of the cigarette leads to say, a box of matches or something similar. The match box ignites the safety fuse. People have used regular pull ring type fuse igniters with trip wires for booby trapped devices.

(29.3.2) Chemical (Straight) Firing Devices

A chemical delay uses some reaction or even a physical phenomenon to start a chemical reaction after a certain length of time. This chemical reaction would then result in production of heat or flame sufficient to ignite a fuse or could result in a small explosion that would set off other explosives. For example, some acid or other corrosive substance could eat through a barrier in

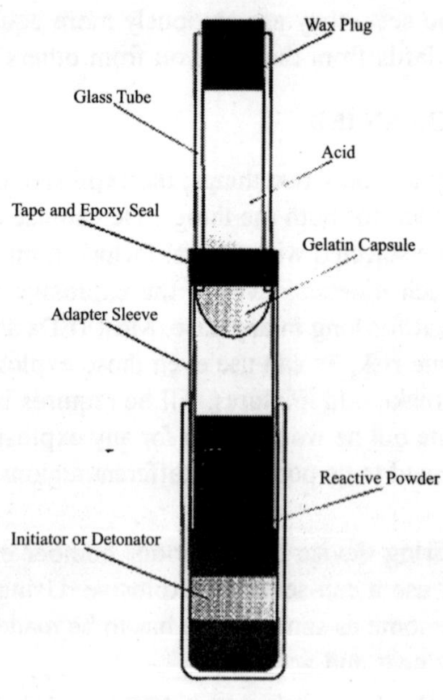

Fig. 185 The Basic Structure Of A Chemical (Straight) Delay Device

some time and then mix with the chemical on the other side of the barrier to start the heat-producing (exothermic) reaction. The most commonly used chemicals are sulfuric acid and a chlorate based composition such as potassium chlorate and sugar or match heads. For barriers, people have used all sorts of things—paper, rubber, cardboard, copper sheets, and even condoms, etc. They have also used simple physical phenomena like dissolving of a gelatin capsule in water for the intended delay. For example, one could keep some sodium in a gelatin capsule of the type commonly used for medicines and let it dissolve in water. Those who are slightly intelligent would put a small ball bearing also in the capsule to ensure that it sinks in the water to facilitate its dissolution. Once it is dissolved, the sodium would react with water. The heat of the reaction would ignite the hydrogen produced or it could be used to ignite other chemicals with greater heat production.

(29.3.3) Chemical (Mechanical) Firing Devices

This type of device does not require a heat-producing or exothermic reaction. All that it requires is that the chemical reaction should release a mechanical device. Then something as simple as a percussion cap would take over. You remember percussion caps? How a pistol cartridge fires? The trigger releases a spring-loaded firing pin. The firing pin strikes the percussion cap. The percussion cap has a primary explosive. It explodes under the impact of the firing pin. Its explosion, in turn, ignites the propellant powder. The same principle is made use of in this type of IED. For example, a taut wire could be holding a firing pin or a similar device against the resistance of a spring. If some acid eats away the wire in some time, the firing pin would be released to strike the percussion cap or something similar. This is the mechanism used in the so-called **'timer pencils'** quite popular with militaries of the world and made in factories. It is difficult to improvise time pencils, however, because it requires some skill and suitable materials. But it can be done

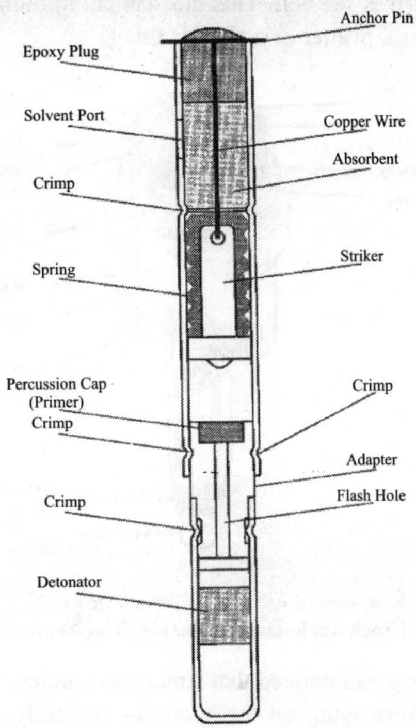

Fig. 186 The Basic Structure Of a Timer Pencil

nevertheless. It requires the machining of a two-part tube—one end of the tube must be of a hard metal to support the spring and the firing pin mechanism; the other end must be of a soft metal which could be crushed to break the sealed glass vial inside so as to release the acid to soak into the absorbent pad and must still be leak proof to prevent any acid from seeping out. Intelligent bombers have successfully circumvented the problem by squirting the acid into the device from outside to activate the device. The acid for this purpose has been carried in a small plastic squeeze spray bottle such as that used for nasal sprays. A hole is provided for this purpose and the acid is made to soak an absorbent pad. The pad then starts attacking the wire. The pad takes sufficient care of any seepage and one could use even cotton for that purpose. Simple, isn't it? Those who find it difficult to work with acid have used even ferric chloride to attack a copper wire. You may recall that ferric chloride is routinely used in the electronics industry to etch copper circuit boards.

(29.3.4) Purely Mechanical Firing Devices

Two types of IEDs are based on purely mechanical firing devices. One could be devices without any delay in them; essentially booby traps. They are devices that use a mechanical motion to trigger a firing mechanism. Just think of a mousetrap and it will become clear to you. The disturbance will release a spring-loaded mechanism. From there the percussion cap could take over. Those who have wanted to make it even simpler though riskier to handle, have used primary explosives mixed with some grit so that the friction generated by say, a kick or a fall on a hard floor, is enough to set it off. That is, the IED would go off if it is kicked or falls to the floor. In the other type there could be a delay incorporated. Such devices use purely mechanical contraptions to introduce delay. Suppose you have an alarm clock. You could set an alarm in it. Once the alarm goes off, considerable mechanical motion is generated in the parts of the clock, such as the movement of the hammer that strikes the bell. This mechanical motion could be tapped or used to make a striker hit a percussion cap, primer or a similar thing.

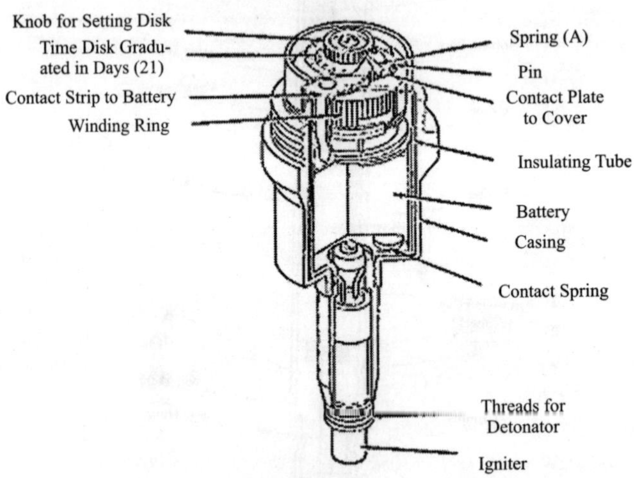

Fig. 187 A German Clockwork Delay (Purely Mechanical) Firing Device

In Vietnam, the Viet Cong had noticed that American soldiers had the compulsive habit of kicking empty soda cans that were lying on the ground—typically adolescent behavior of the retarded Americans. It was not long before the US troops found that many supposedly empty cans lying alongside regular patrol routes exploded when they were kicked. All that the Viet Cong had to do was to rig the empty cans with IEDs of the mechanical firing device type!

(29.4) THE EXTREME VERSATILITY OF IEDs BASED ON ELECTRICAL DETONATORS

An electrical detonator requires a *certain amount of current* to detonate. Hence all that an IED based on electrical detonator requires is that current of a sufficient magnitude should flow through the detonator. What do you require for current to flow in a circuit? The circuit should be complete. That is if all other things like the power source (battery, for example) and wirings, etc. are in place, a circuit requires nothing but a switch to make it complete. That is an **electrical IED** requires nothing but an electrical contact to be made or broken for the device to be triggered. The ways in which it can be done are limited only by imagination. You can construct the most sophisticated devices; you can also construct very simple devices with the help of household things. The basic objective of the device is to send current to the detonator at the desired moment, either after a delay or on command, or on tripping. Did you notice that we emphasized the words 'certain amount of current' above? Why? You would recall from our discussion on detonators that they require a certain 'firing current', say, half an ampere or so. If the electrical IED is so designed that though the detonator gets current upon activation, the current is less than the firing current, it will not work reliably. This is precisely what happens when fools do something. Ask some people to rig a **bomb to a car engine** in such a manner that the bomb explodes the moment the ignition is turned on and chances are very good that 99% of the people will connect the detonator to the **spark plug**! All such morons think that the spark plug generates a spark and hence it must be getting a heavy current! It is not so. The spark plug generates a high voltage alright but at a very low current! Connecting the detonator to the spark plug will not do the job. Current for the detonator has to be drawn from the circuit of the **primary winding of the ignition coil** so that it gets sufficient current! This is not to give any ideas to the terrorists. All terrorists worth their salt know it already. Still, the fact of the necessity of connection of the detonator to the ignition coil should be known to citizens in general and officers of the security forces in particular so that if a car bomb rigged with the spark plug is found, the dumb officers do not take it to be the handiwork of professional terrorists of a foreign country—it could, very well be the stupidity of some small-time local criminal or even a neighborhood boy! Ignorance distorts perceptions in general and intelligence perceptions in particular.

(29.4.1) Mechanical (Electrical) Firing Devices

As we saw, all that the detonator requires is current. It does not matter to it how the circuit is completed. One can make his circuit in such a way that a certain mechanical action directly completes a circuit or closes a switch. After all, most of the switches you have been using all your life are simple mechanical switches. You flip them on or off, isn't it? An IED based on the **mechanical completion of a circuit** is a very simple job. Take a simple press-type switch, for example. Suppose you take a spring-like strip of metal and nail one end of it to a wooden board. The free end would naturally rise above the board. Now connect one wire to the nailed end. Affix another wire to the board where the strip's free end would have touched the board had the other end not been nailed. Thus you have two wires and they could be connected if the free end of the strip is pressed to touch the other wire. This makes a simple but perfect booby trap. Suppose this device is concealed under a sheet of paper or something of that sort. As soon as man steps on the device, the strip would be pressed and when it touches the board, the circuit would be completed and the detonator would fire. The mechanical circuit could be designed in an opposite manner also where pulling instead of pressing does the job. Suppose two ends of a wire are connected to the inner sides of the two arms of a clothespin. That is to say, when the clothespin is in its normal state, the two wires are touching each other and the circuit is complete. Now press the clothespin to open it a little and insert a piece of thin cardboard between the two arms. Cardboard being an insulator will come between the wires and thus would break the circuit. Now one could rig the IED in such a way that a trip wire pulls the piece of cardboard out of the clothespin. The moment it is pulled, the circuit would become complete and current would flow. There are literally millions of ways in which an electrical circuit can be made complete.

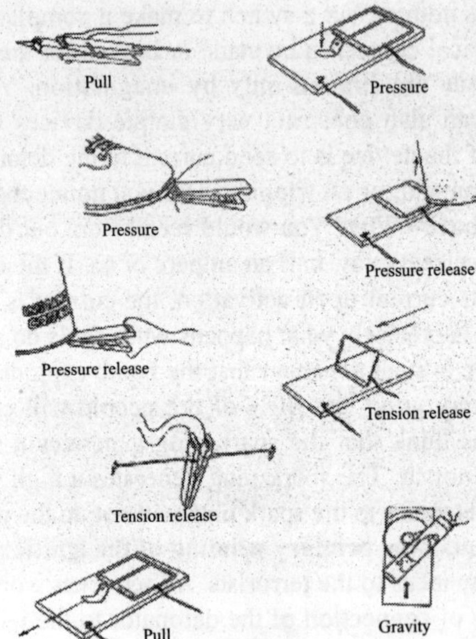

Fig. 188 Some Of The Numerous Improvised Ways In Which An Electrical Firing Circuit Could Be Completed By Mechanical Action

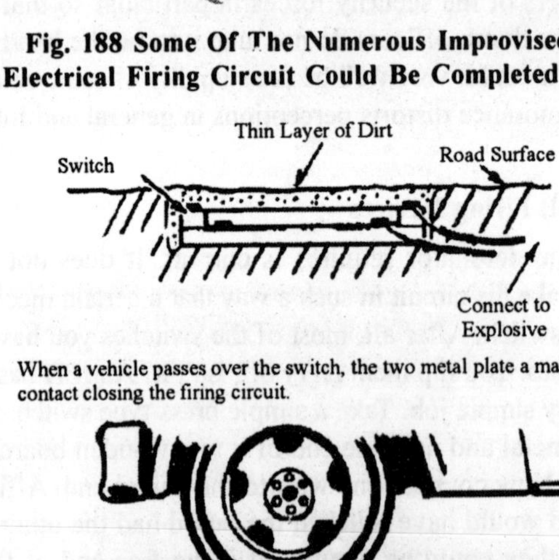

Fig. 189 A Most Basic Pressure Switch For An Improvised Anti-Vehicle Mine

(29.4.2) The Confounding Genius Of Collapsing Circuits

The ingenuity of the designer does not end here. This field has received the loving attention of numerous geniuses. Particularly in the West, bombers have been found to suffer from a massive ego problem and the biggest trip of their life is to make the other guy admit that they are geniuses. They can design the IED in such a way that the mechanical action actually breaks a circuit instead of completing it. Now breaking a circuit would not send current to the detonator. How does it work then? Therein lies the beauty of the mind applied to it. The first circuit is only a trigger for another circuit. The other circuit is so designed that it must get triggered by the breaking of current in the first circuit. It could be connected to the first circuit; it could be completely isolated also. Say for example, if the current flowing in a circuit was making an electromagnet pull a switch to the 'off' position, the breaking of the circuit would demagnetize the electromagnet and make it release the switch to 'on' position. Such circuits are known as **'collapsing circuits'**. Collapsing circuits are extremely clever ideas—they frustrate the personnel who have to defuse the bombs. In technical language we would say that the collapsing of the first circuit would trigger a **relay** in another circuit and that would send current to the detonator. Collapsing circuits may be made to incorporate delays also. One very simple delay would use the life of the battery. As long as the battery is capable of giving the current, it could hold a magnetic relay in the off position. The moment the battery loses power, the switch will be thrown to the on position. The delay time offered would, however, be in a considerable range as you cannot predict the life of a battery accurately.

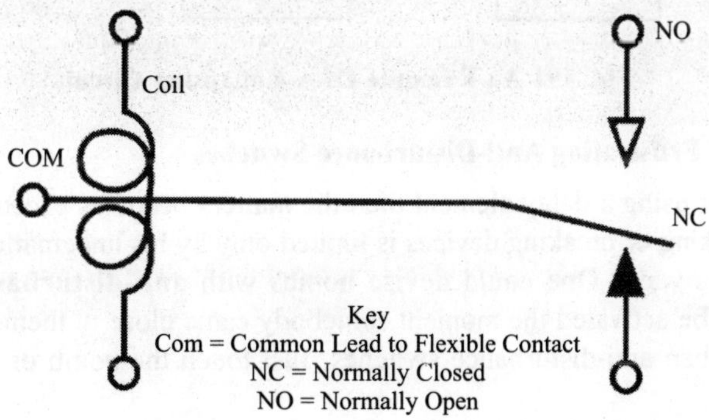

Fig. 190 Schematic Representation Of A DC Relay

Intelligent bombers may easily devise **secondary and tertiary collapsing circuits** also or **various combinations of collapsing and non-collapsing circuits.** You trip a wire and one collapsing circuit is activated. But it does not set off the device as yet. The personnel defusing the bomb may be led into believing that the device has malfunctioned. To confuse them further, the bomber could have very well made dummy connections to a detonator and explosive from the first collapsing circuit. As the personnel disconnect some wire in the secondary circuit, it could activate a third circuit which will actually send current to the detonator.

Every bomb disposal officer is advised to 'attack the cap' when encountering a bomb. This is elementary training. They think that every bomb is eventually exploded by the blasting cap or the detonator. Attacking the blasting cap usually meant to cut the wire leads on the cap, and hoping that it would thus be rendered ineffective. Unfortunately, life is not so simple. Intelligent bombers have turned the tables on this technique by placing a **hidden micro-switch** beneath or to the side of the blasting cap so that when the leads are cut and the cap is pulled out of the explosive device, the micro-switch lever is released and the bomb explodes. The micro-switch is connected to a separate hidden circuit (e.g., in a hollowed-out stick of dynamite, beneath the device in a hollowed-out base, etc.).

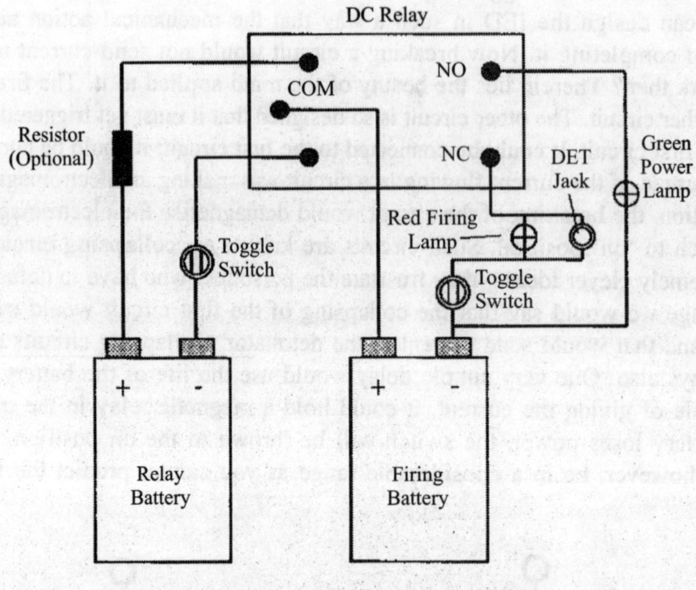

Fig. 191 An Example Of A Collapsing Circuit

(29.4.3) The Utterly Frustrating Anti-Disturbance Switches

If one is not using a delay element then the matters becomes exceedingly simple and the variety of circuit making or breaking devices is limited only by his imagination. One can design his bomb in innumerous ways. One could devise bombs with **anti-disturbance switches** or even switches that would be activated the moment somebody came close to them. Nothing in the world is more frustrating than anti-disturbance switches. Just touch the bomb or even come close to it and it goes off!

Once again anti-disturbance switches can be made in hundreds of ways. The basic thing is that there would be two points or terminals which have to be connected by a conducting medium when the system is disturbed. In general, all one requires for this system is that it something in it should be in a state of unstable equilibrium. Now what is that? Imagine a deep bowl. A ball placed inside it is in a state of stable equilibrium. Even if you disturb it, it will remain inside the bowl only. Now invert the bowl and balance the ball on top of it. If you do it carefully, you can balance it. However, the moment it is slightly disturbed, it will roll down the sides. The ball, when resting on top of the inverted bowl was in a state of unstable equilibrium or, in simple words, was delicately balanced. You can see that you can devise a system where a metal ball is delicately balanced and the moment the system is disturbed, the ball rolls down to a place where it come to touch two terminals, thereby completing their circuit. You can even have a pendulum made of a metal ball. You could fix the two terminals to a side in such a manner that as long as the ball is not oscillating, the terminals are not touched by it. The moment it is set into oscillation due to the disturbance, it will complete the circuit.

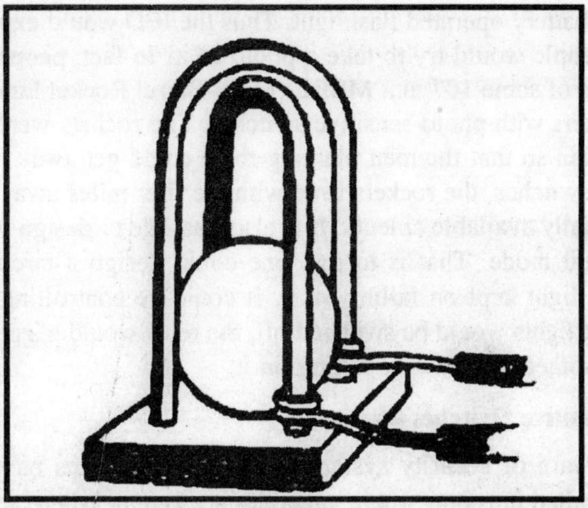

Fig. 192 Illustration Of A Most Elementary Anti-Disturbance Switch

You can also have **mercury anti-disturbance switch.** In this the disturbance could cause the mercury to flow in such a manner as to complete the circuit. In theory, all the devices in the world which can sense vibrations by whatever means can be converted into anti-disturbance switches. All anti-disturbance switches can be made extremely delicate and they can be connected to a relay to trigger a more robust circuit connected with the detonator. You may recall how in October 1940 the British found one 'butterfly cluster bomb' of the Germans that had failed to explode. That particular piece had an anti-disturbance fuze. The British could disarm it only by making a clay wall around the weapon and pouring liquid nitrogen over it, thus freezing the mercury in the mercury anti-disturbance switch! Quite a cumbersome way, isn't it? Clever bombers would not let it be seen that the bomb has a mercury anti-disturbance switch!

Perhaps you might be wondering how they plant a bomb with an anti-disturbance switch under field conditions. Good question! The basic precaution which must be taken with such circuits is that one must provide for an **arming switch**. This switch must be kept off till the very last moment lest there is any mistake in the circuit and the detonator goes off. They also test the circuit with a flashlight bulb in place of the detonator.

(29.4.4) Light Sensitive Switches

This is yet another novel method of completing or activating a circuit. We are all aware of phenomena that deal with light causing some electrical effect on certain materials. Who does not know solar cells? What they do? They convert light into electrical energy. How? They contain certain materials which produce a little voltage when light falls on them. You have photovoltaic cells, photodiodes, and photo-transistors and so on. All of them result in some electrical property being changed as a result of light falling on them. All one has to do is to make these voltages trigger an electronic circuit. The voltages are small, you see and the current may not be enough to trigger a relay directly. Then some electronic circuit is to be used which will respond to the little voltage or current. How this is used for IEDs? The idea is simple. One could make a circuit that would start conducting only when light falls on it. One could keep it in a dark room and hope that when the security forces would open the room and switch on the light or use their flashlights, the device would explode. Clever bombers could do better. They could make the circuit in such a manner that it is triggered only by the bright light of a camera flash and not by the fluorescent light of a room or the light of a battery operated flashlight. Thus the IED would explode only when security forces or the media people would try to take a photo of it! In fact, people have even improvised the ignition mechanism of some 107 mm MBRL (Multi-Barrel Rocket Launcher) rockets launched from makeshift launchers with photo-sensitive switches. The rockets were rigged in darkness and left in the firing position so that the men planting them could get away. With the crack of dawn when light fell on the switches, the rockets fired with the firer miles away from it! Light sensitive switches are commercially available aplenty. It is also possible to design the light sensitive switch in the collapsing circuit mode. That is to say, one could design a circuit which would remain conducting as long as light kept on falling on it. It could be controlling a relay connected to a detonator. The moment lights would be switched off, the relay would trigger the detonator. It could be triggered even by someone's shadow falling on it.

(29.4.5) Infra-Red Sensitive Switches

We are all aware of security systems and intruder alarms based on infra-red beams. Generally they are installed in rooms where valuables are kept or where access is denied for some other reason. The device is simple in principle. There is a source of infra-red light. The source emits the infra-red light in the form of a beam. The light is received by an infra-red sensor at say, the other end of the room. The reception of infra-red light by the sensor keeps the circuit complete. Now if some person or other solid object interrupts the beam, the sensor would not receive any signal. This would result in the circuit getting broken. This breaking of the circuit can be used to trigger an alarm. Now when one has to make an IED out of this, all he has to do is to replace the alarm with the detonator. If the detonator does not get enough current directly, a relay can be used. The infra-red switch lends itself to a good booby trap. The beam would be invisible and the bomb would go off the moment someone crossed the beam. Intelligent bombers could make the circuit in such a manner that it would be triggered only if someone crossed the beam twice or more so as to mislead him into believing that there is no IED.

(29.4.6) Sound Actuated Switches

Perhaps you may recall some toys, particularly cars, which are actuated by sound. The control has a device which produces a sharp clicking sound. The sound is picked up by a receiver in the toy. It can be easily converted into an electrical signal as it does in a microphone. The electrical signal can be used to switch on the motor of the toy car. The same principle is used in sound actuated switches. The circuit of the receiver can be provided with sound filters so that they respond only to certain types of sounds. For example, it could be made to respond to the sound of gunfire or grenade explosions. They could thus be used to booby trap rooms and corridors etc. for urban combat situations. The devices would go off as soldiers would enter there firing and throwing grenades. They may take all the care in the world to not trigger trip wires etc. as they enter the building but the natural soldierly action of shooting may doom them. Now you understand how dangerous this switch could be?

(29.4.7) Electrical Firing Devices With Mechanical Delay

Who does not know the legendary **time bomb with the clock** ticking away? In fact, it is the most enduring symbol of a bomb and immortalized by innumerous films! Even the dumbest of the security forces officers recognize it. The downside is that all the time they keep on looking for a bomb with a clock, little realizing that hardly anybody makes it anymore!

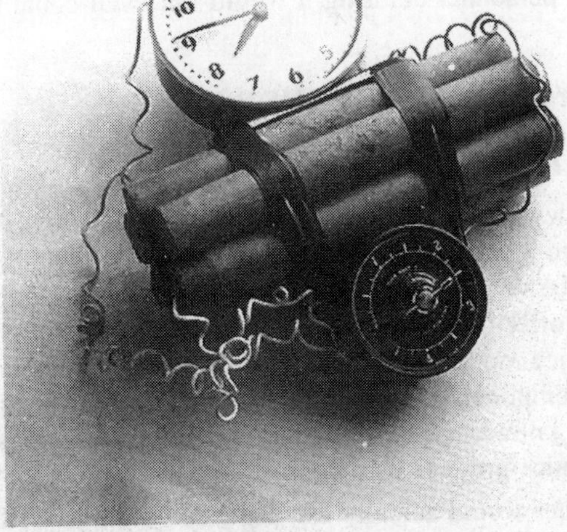

Fig. 193 Who Does Not Recognize This? The Classic Time Bomb With The Alarm Clock Ticking Away

The problem with the mechanical delays are that they are relatively large is size, can be easily detected and the delay offered is necessarily small. The most elementary delay circuit in this category is that based on a clock. Everybody claims to know this but ask them to design it and the dumb heads would flounder. Well, it is not really that easy either and some work is required to make it work reliably. First they remove the plastic or glass cover of the clock. Then they remove the sweep second hand first. Then if the delay required is less than an hour, they remove the hour hand. If the delay required is more than an hour, they remove the minute hand. Then you must provide for an electrical contact at a point on the face of the clock where you want the moving hand to come and complete the circuit. The other wire must be connected to the metal body of the clock. It may require some delicate soldering and fixing of very small screws. They must also replace the plastic over if they want to make the device portable.

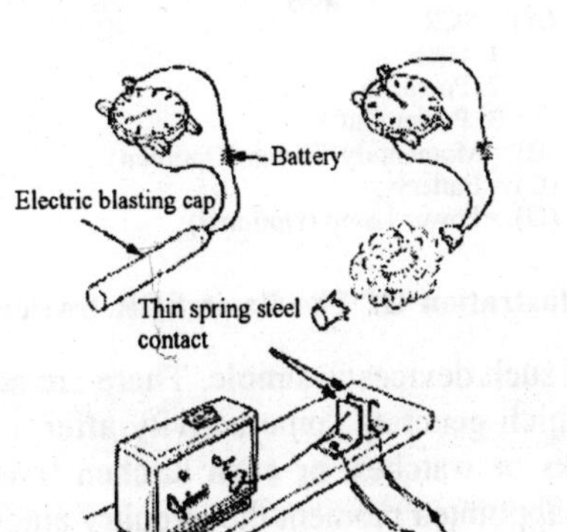

Fig. 194 Improvised Electrical Firing Devices With Mechanical Delays

One can make the bomb with a mechanical delay added to a booby trap so as to make it a source of frustration for the personnel who have to defuse it. For this it is necessary that the mechanical action does not start the electrical activity immediately. The tripping of a wire could, for example, be made to activate a mechanical delay and that could, in turn, activate an electrical circuit after some time. The personnel defusing it would not even come to know that they have started the bomb!

(29.5) ELECTRONIC DELAY CIRCUITS RULE THE ROOST

(29.5.1) The Basic Principle Of IEDs Based On Electronic Delays

They are very neat devices—a bomber's delight in fact! The electronic delay circuits use electronic devices or circuits to initiate current in a circuit immediately upon tripping, on command, or after a delay which could be very long also. They can be made extremely small and hence very easily concealed. They can offer delays of months and that too very precisely. This means that the bomber can hide the device months before the bombing and go home. The best thing about electronic timers is that the components required for most of them are commercially available and all that the bomber needs is knowledge of electronic circuits. Electronics buffs who have been taking up all sorts of electronics projects as a hobby since their adolescence, find it easy.

What device a bomber would actually use depends on a host of factors—availability of technically qualified personnel, electronic parts, chemicals and so on. It also depends on the situation under which the bombing is to be carried out. There are fools who purchase electronic components which could be traced back to them or use the same circuit or mechanism repeatedly which becomes their 'signature'. Intelligent bombers avoid all this.

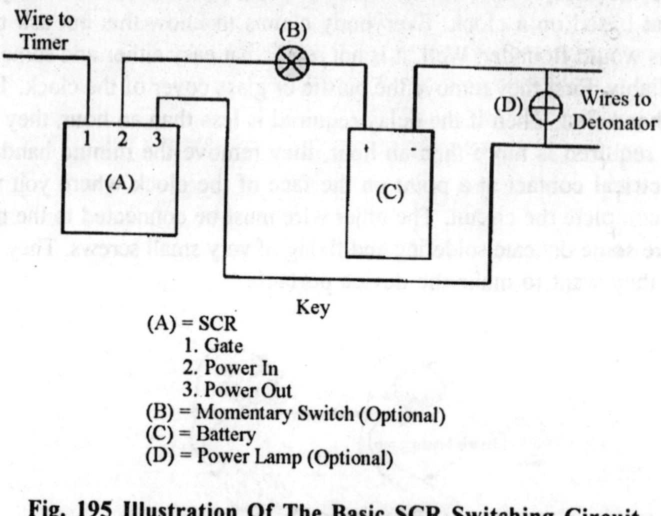

Key

(A) = SCR
1. Gate
2. Power In
3. Power Out
(B) = Momentary Switch (Optional)
(C) = Battery
(D) = Power Lamp (Optional)

Fig. 195 Illustration Of The Basic SCR Switching Circuit

The basic principle of such devices is simple. There are scores of commercially available electronic devices available which generate some activity after a certain time. For example you could have digital alarm clocks or watches, or even kitchen timers. What happens inside such electronic devices is that at the appointed moment the circuitry automatically produces an electrical signal which makes the beeper or bell of the alarm produce the sound, light or whatever. It is this electrical signal which is made use of for the bomb. The designer of the bomb need not tamper with the electronic device—he just has to tap this signal. The signal is used to activate another circuit which will eventually supply current to the detonator. How do you do this? Once again there are commercially available devices like the **SCR (Silicon Controlled Rectifier)** which will do this job easily. A SCR has three terminals. One is called 'power in', the other is called 'power out' and the third is the 'gate'. You give power to the SCR from a battery to the 'power in' terminal'. The tapped signal from the electronic device is fed to the 'gate'. The 'power out' terminal is connected to the detonator. As long as no signal is fed to the 'gate' the SCR does not permit any current to flow in the 'power out' terminal. The moment a signal appears there, the SCR makes the battery send current to the 'power out' terminal. There are other circuits also utilizing devices other than the SCR which accomplish exactly the same thing. One has only to dig through commercially available electronic devices.

(29.5.2) Examples Of Basic Electronic Timer Circuits

A timer circuit is an electronic device for creating a time delay. Timer circuits produce an output or change the output after a pre-determined time delay. You give it an input and it will wait for a certain time before it does the needful on its own. The main component of most electronic timing devices is a capacitor in series with a resistor. Together they create a simple time delay. Nowadays, dedicated integrated circuit chips (ICs) can be used in timing devices. At least 53 companies in the USA alone make **IC (Integrated Circuit) timers**. An intelligent bomber has only to get one and modify it a little for his specific needs.

There are three basic types of **timer circuits**: a **monostable**, an **astable**, and a **bistable**. They are all actually different types of the circuit called **'multivibrator'**. A monostable circuit counts a fixed length of time. It is said to have one stable state. This can be monostable ON, where a device stays on for a fixed time and then stops, or a monostable OFF which stays off for a fixed length of time and then comes on and stays on. Astable circuits are sometimes called square wave timers because they switch on and off repeatedly. The waveform as seen on an oscilloscope has a square shape to it. This repeated on and off switching gives out an accurate pulse or beat that can be used as a digital-type clock pulse. The pulse rate is altered by adjusting capacitor and resistor values. Astable circuits have no stable state because they are constantly switching on and off. A bistable circuit is a timer that has two stable states because it has two outputs and two inputs. Switching the inputs causes alternate outputs to switch on and then off. This is why they are sometimes called **'flip-flops'**. Timers are not only used in timing devices such as clocks, but can also be found in electronic alarms and counting devices. Next we will give examples of two famous timer circuits to illustrate the theory—other commercially available circuits are also possible; they can be entirely fabricated also.

(29.5.3) Electronic Delay Circuits Based On The IC 555

Some of the finest electronic delays are based on the famous IC 555. This integrated circuit chip was introduced in 1971 by the Signetics Corporation and was found to be so versatile that it instantly became famous as **'The IC Time Machine'**. Literally thousands of most novel and unique circuits have been designed over the years using this chip. The bomber would use the IC 555 as an **astable multivibrator circuit** that will produce electrical pulses at a fixed frequency. The frequency will be determined by the values of the components in the RC (resistance-capacitance) network. The pulses are fed to another integrated circuit chip known as the **4017 decade counter**. It takes clock pulses from the clock input and counts them. It is able to count any number from 0 to 9 and has ten outputs, one for each number. One can obtain a great delay by using successive 4017 counters. The circuit functions like this. The IC 555 generates periodic pulses. These pulses are fed to the 4017 counter. After receiving 10 pulses, the first 4017 counter will produce one pulse and automatically reset itself. This is fed to the second 4017 counter. That will also count 10 pulses and produce a pulse and reset itself. That is to say, each successive 4017 counter will multiply the time delay of the previous counter by a factor of 10. The output pulse from the 4017 chip can be used to trigger a circuit that will eventually make current flow in the detonator circuit. One can use as many 4017 counters as he likes. Just five counters will give a delay of over two months. To illustrate by an example, suppose the values of the RC components have been so chosen for the IC 555 chip that it produces a pulse every 69.95 second. This means that using only one 4017 counter will give a delay of 699.5 seconds, two will give a delay of 116.5 minutes, three will give a delay of 19 hours and 25 minutes, four will give a delay of 194 hours and 18 minutes or about 8 days, and five will give a delay of approximately 80 days. One has to ensure, however, that his batteries last that long a time. That's not very difficult as these circuits consume very little current.

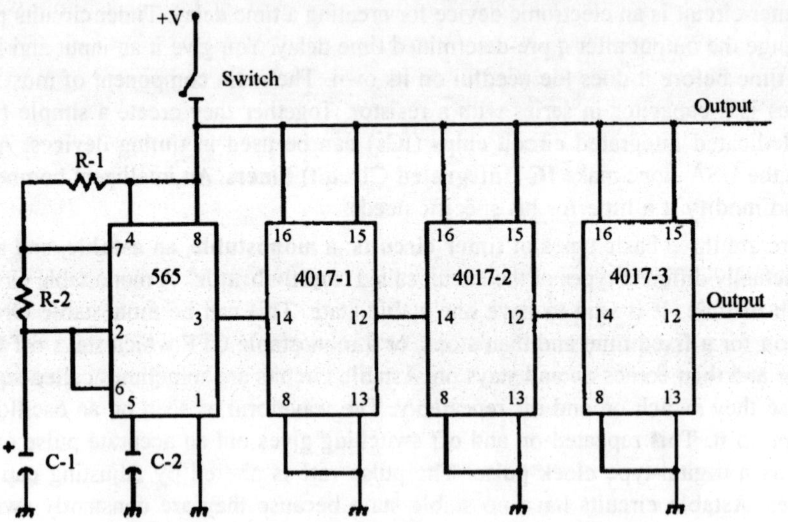

Fig. 196 Illustration Of An Electronic Timer Circuit Based On The IC 555 And Three 4017 Decade Counters Capable Of Giving Over 19 Hours Of Delay

(29.5.4) Electronic Delay Circuits Based On The MC 14536B

Timer circuits based on the Motorola MC 14536B chip are also very versatile. The circuit can provide delays ranging from 8.5 minutes to 194 days. We are making a special reference to it because in the Indian context it seems quite popular with the terrorists. Since 2001 there have been five instances in which this circuit has been recovered from the possession of terrorists. On two occasions it was found with Kashmiri terrorists or those having links with them (December 31, 2001—one Younus arrested at the ISBT, Delhi and February 4, 2006—one Nasir arrested in Defence Colony, New Delhi); on two occasions it was found with Sikh terrorists who were associates of Jagtar Singh Hawara (June 12 and 16, 2005—arrested in Punjab). Once (April 12, 2001) it was found in an explosive device planted in a scooter parked near the North Block, New Delhi. Incidentally, it was also used in the bomb blast at the Maulana Azad Stadium, Jammu in 1995 even as the then Governor was delivering the Republic Day speech there. The same timer was used in the bomb that had been made to explode at the Sawai Man Singh stadium, Jaipur during the Republic Day celebrations in 1996.

The **MC 14536B programmable timer** is a 24-stage binary ripple counter with 16 stages selectable by a binary code. It will take a dc voltage supply of 3 volts to 18 volts. For input it can use either an on-chip RC oscillator or an external clock. By selecting the appropriate counter stage in conjunction with the appropriate input clock frequency, a variety of timing can be achieved. Insurgents are seen using an electronic wrist watch for the input.

Incidentally, the ignoramuses in the Indian media, intelligence agencies and the security forces do not even know that the timer is based on the MC 14536B chip. The popular and pathetically wrong name given to the timer is **ABCD timer**! The insurgents themselves call it so. Their ignorance can be pardoned; the ignorance of the latter cannot be pardoned. Now there is nothing in the world of electronic circuits called an ABCD timer. The insurgents took to calling it the ABCD timer because this was the name told to them by those who designed it for them. The designers did not want to take any chances with the limited intellectual capacity of the insurgents. Four wires come out of the circuit. For the convenience of the users, these wires are marked A, B, C and D so that different combinations could be used to yield different spans of programmable time delays. Had they been marked 1, 2, 3 and 4, the insurgents, the media and the officials would have been calling it the 1234 timer! Schiller had very rightly said: "Against stupidity the gods themselves struggle in vain!"

The bombs used in Jammu and Jaipur were buried in place several months ago when there was no reason for the security to be tight. The bombers took the risk of burying them and leaving them because it could be reasonably well presumed that given the venue being stadiums, the rostrum etc. would be constructed in fixed places only. They escaped detection when anti-sabotage checks were carried out before the Republic Day because in a place like a stadium there is always a considerable quantity of metal at the site on account of the poles and steel used in the reinforced cement concrete—the mine-sweepers therefore get confused.

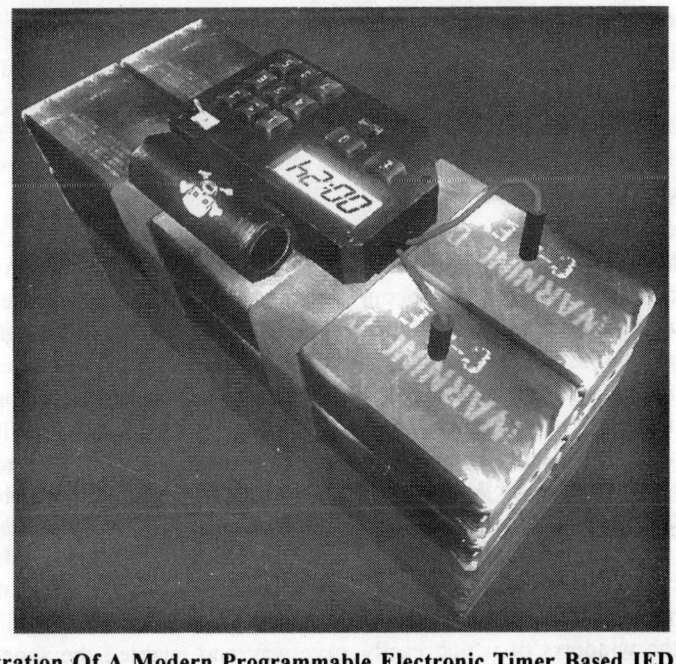

Fig. 197 Illustration Of A Modern Programmable Electronic Timer Based IED In Which The Desired Time Delay Can Be Set By Simply Punching In The Digits

(29.5.5) Powering The Detonators Of Electrical IEDs

As we saw in connection with our discussion on detonators, reliable detonation requires a certain amount of current to the detonator(s). Many IEDs fail simply because sufficient current could not be supplied. One way is to use a fairly powerful battery, such as a 9V alkaline battery. When explosives are being used properly, blasting machines are used. However, a blasting machine is generally ruled out for an IED unless it is a wired device. Blasting machines are capacitor discharge machines. That is, a capacitor stores electrical power and then discharges it in one go to the detonators(s). Some smart guy figured out that the **circuit of a camera flash gun** does nearly the same job. That is also essentially a capacitor discharge circuit. The capacitor stores electrical energy and discharges it to the flash bulb so that it may produce a 'burst' of high intensity light. If you replace the flash bulb with a detonator or several of them, they would get the current discharged by the capacitor. The flash gun has no more significance than a readymade capacitor discharge circuit. Ignorant officers of the security forces, however, think that it is a new or sophisticated invention.

(29.6) SOME HISTORICALLY FAMOUS TIME BOMBS

(29.6.1) The Americans Have Done It Too

Much before terrorism became a fact of life, regular armies had used time bombs in terrorist-like activities. During the American Civil War, Confederate Capt. Thomas Courtney organized what is known as the CSS (Confederate Secret Service). Capt. John Maxwell of the CSS managed to plant a 12-pound black powder device based on clockwork time device on a Union ammunition barge. The bomb inflicted a loss of $ four million in that era besides killing 58 and injuring 126 personnel!

(29.6.2) Stauffenberg's Attempt On Hitler's Life

One of the most famous time bombs in history was made and used by Claus von Stauffenberg in an attempt on Hitler's life in what is known as the Operation Valkyrie on July 20, 1944. The bomb was concealed in a briefcase and used a chemical timer based on acid. The bomb killed four and injured everybody else in that room at the so-called 'Wolfschanze (Wolf's Lair), Hitler's secret headquarters in Bavaria.

The Fuhrer survived because at the critical moment he had moved to a wall map at the end of the table; the heavy oaken table and its supports between him and the bomb shielded him considerably. The briefcase containing the bomb itself was moved to the other side of the table by Colonel Brandt when he had leaned forward to study the map and his foot had caught the briefcase. Secondly, the device had just about 2 pounds of charge. It was made using four captured British 'clam' mines. The mines were in pairs, held together by their magnets and bound one pair side-by-side of the other with tape. The entire bundle was of the size 3 x 3 x 12 inches. Why Stauffenberg used so little a charge? It was later revealed that he had planned another bomb too. The second bomb was to be brought by his aide Werner von Haeften in another briefcase which contained 975 g of British plastic explosive, two detonators and a 30-minute delay device. For some reasons Haeften could not bring the second bomb along with Stauffenberg and Stauffenberg decided to go ahead with his bomb alone. Later, when they were escaping together, Haeften threw the bomb out of their car window. A German Pioneer unit discovered it several days later. Thirdly, it was a conference room the windows of which easily gave way and enabled the shock waves to be dissipated outside. Had it been a proper bunker without windows and where strong walls could have reflected the shock waves, the effect of the even 2 lbs of charge would have been much greater.

Fig. 198 Photograph Of The Wolfschanze (Wolf's Lair), Hitler's Secret Military Headquarters In Bavaria, After The Bomb Blast: Hitler Survived The Attempt

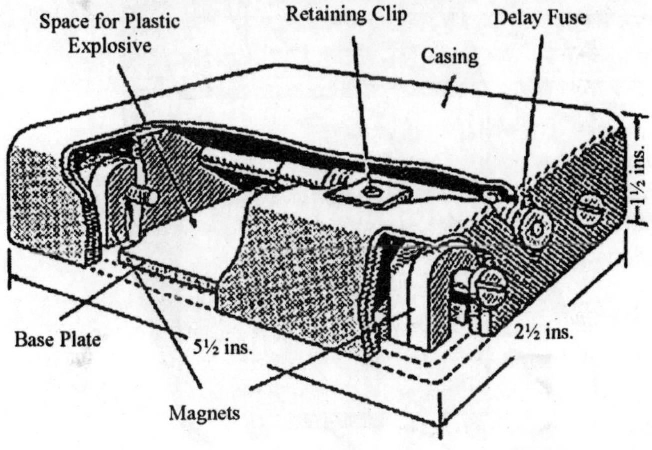

Fig. 199 The British Magnetic 'Clam' Mines Used By Stauffenberg In The Attempt On Hitler's Life: The Bomb Used By Him Carried Four Of Such Mines

(29.6.3) Harvey's Casino Wheel Bomb: A Masterpiece Created By Criminals

Technically one of the most sophisticated bombs ever used on American soil was found in the Harvey's Casino at Stateline, Nevada on August 27, 1980. This was by no means the most powerful bomb used in the USA but it was indeed extremely complex. A pity that it had to be made by criminals who had demanded $ 3 million in extortion! This was one bomb that could not be successfully defused, the great technical expertise of the US government agencies notwithstanding!

The bomb went off as they were trying to defuse it. Though there were no casualties as the building had already been got vacated and they had been struggling with the bomb for days on end, property worth millions was destroyed. Though the criminals were subsequently captured and convicted, their bomb remained undefeated.

The bomb consisted of two metal boxes. The experts had them X-rayed and they had some idea of what was there inside. The smaller box at the top measured about 14 x 14 x 22.5 inches and had essentially electronic circuits in it. There were two banks of toggle switches on its front and back sides and all switches were neatly numbered. The bigger box at the bottom measured about 24 x 26 x 45 inches and contained the charge and some additional circuitry. The boxes were mounted on heavy caster wheels with a dolly handle to facilitate movement. Each corner had adjustable screw legs and there were bubble levels in each corner of the top of the bigger box. It was obvious that the bomb had **anti-disturbance switches** in it and any movement could have set it off. An intermittent whirring sound came from the device indicating that some sort of timer was also at work. X-ray also revealed wired foil linings which meant that one could not drill through the device without **collapsing** some **circuit**. There was indication of a **barometric pressure fuse** also that would have gone off had the boxes been pierced.

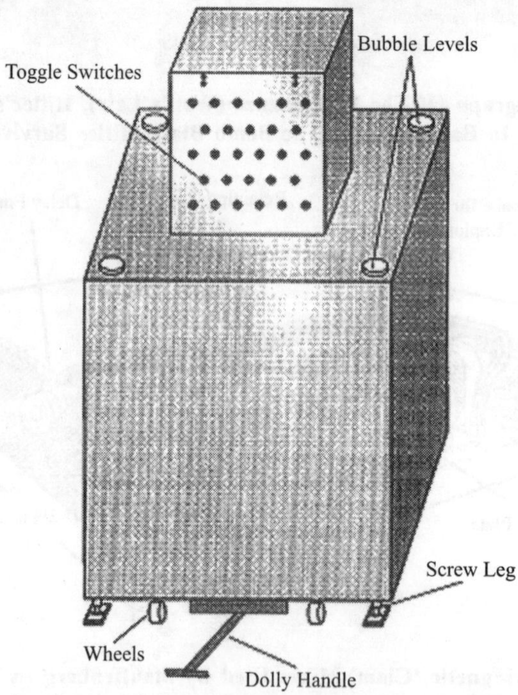

Fig. 200 The Most Famous IED In History: The Bomb That Defeated The Combined Expertise Of The US Navy's Explosive Ordnance Disposal Facility, The US Army's 34th Explosive Ordnance Disposal Team And The Lawrence Livermore Laboratory: A Schematic Representation Of The Harvey's Casino Wheel Bomb

After considerable struggle with the bomb, the experts were of the opinion that only by flipping the appropriate toggle switches in the proper sequence, the anti-disturbance devices could be disengaged to allow the device to be moved to a safer area for further tampering with it. The extortionist himself had given an indication of it when he called the police to tell that switch number 5 was to be flipped first in the first step towards disarming the bomb. The police did not dare—firstly they could not afford to believe in his word and secondly, there were a large number of switches making it impossible to resort to any guesswork in determining the order of flipping the switches. Touching the device was otherwise impossible. One thing was obvious. The electrical signal to the charge in the bigger box was to come from the box at the top. The question was whether this flow could be disrupted quickly enough? Experts of the US Navy's Explosive Ordnance Disposal Facility, the US Army's 34th Explosive Ordnance Disposal Team and the Lawrence Livermore Laboratory argued that one possible way of accomplishing this was to place a linear shaped charge in front of the device and let the extremely hot jet from the shaped charge sever the connections between the two boxes like a very sharp and fast knife. Calculation of the stand off distance was critical. The jet had to be properly formed for it to have maximum cutting effect and yet it had to be as close to the boxes as possible for the cutting to take effect even before the shock of the linear charge itself could set off the anti-disturbance switches in the device. Very elaborate calculations were done and the stand off distance for the C-4 linear shaped charge was fixed at 8.5 inches.

Eventually the genius of the designer proved superior and the bomb went off with an explosive force of about 700 lbs of TNT as the disruptive charge was fired! There is still a considerable difference between theory and real life!

The FBI got a tip off on the bomber after the reward money was raised to half a million dollars. The tip off came from the girlfriend of his son who had dumped her. He was a partner in crime to a lesser extent. John Birges, a 59-year old man got 20 years in prison and died in jail at the age of 74. To neutral observers the FBI story was not very convincing. Birges was a compulsive gambler and certainly had some connection with the casino and the loan sharks. John Birges was convicted basically on the strength of his confession and the testimony of his sons against him and who had apparently tried to talk him out of it. John Birges was a skilled electrician and a welder—it is surprising that a man of his education and training could design such a complex bomb. But committed people do wonderful things. During the search when the police officers used to talk to him, he used to deny everything but at the same time used to boast that if he had designed the bomb he could have bettered it. Since he spoke in specifics of bomb design, it was obvious that he knew a great deal about it.

To the explosive scientist, the incident highlights the fact that it is indeed possible to design a bomb that would beat all attempts to defuse it.

❑❑❑

Chapter 30

Boom at the Press of a Button: All about Command actuated IEDs

(30.1) WHY DOES ONE NEED COMMAND ACTUATED DEVICES?

The great advantage of command actuated devices is that they enable the bomber to select the optimal moment of bombing. Thus they are preferred when a bomb has to be exploded, say only when a car passes a certain point or when a person appears at some place. This requires split-second timing. Even if the movement pattern of the target has a high degree of predictability and even if the time-delay device is very accurate, you can still not depend on the time-delay device because the target is not a machine and can always move a little faster or arrive a little late. It may not make any difference to humans but it makes a great difference to a bomb. A car moving at just 60 mph also covers 88 feet in one second. This means that if there is an error of even one second in the appearance of the target on the designated spot and the time set in the bomb, the target would be off by 88 feet. For a discrepancy of two seconds, it will be 176 feet, a distance at which most IEDs are likely to become ineffective. For such targets human control must take over. The downside of command-actuated devices is that the bomber cannot afford to be very far from the IED whereas for a time-delay device he can plant the device and disappear conveniently. They are therefore not for the chicken-hearted.

Command actuated devices require some command to switch on a circuit to make current flow therein that will eventually fire the detonator. Once again the manner in which the command can be given is limited only by the ingenuity of the bomber. Now commands could be given in two basic manners; wired and wireless. Thus the two most common types of command-actuated IEDs would be: **Command Wire IEDs (CWIED)** and **Remote (or Radio) Control IEDs (RCIED)**.

(30.2) COMMAND WIRE IEDs

If a device is wired you could send an electrical current in it and to the detonator. Nothing could be simpler. The advantage of a wired device is that it is almost foolproof. An electronic circuit may perhaps fail on a rare occasion—there is little which can fail in an electrical circuit. The bomber can keep a powerful battery so that the strength of the current is never an issue. The disadvantage of a wired device is that it is difficult to employ, takes time and even more difficult to conceal. However, where the surroundings and the terrain permit it, people have drawn wires over 600 feet long!

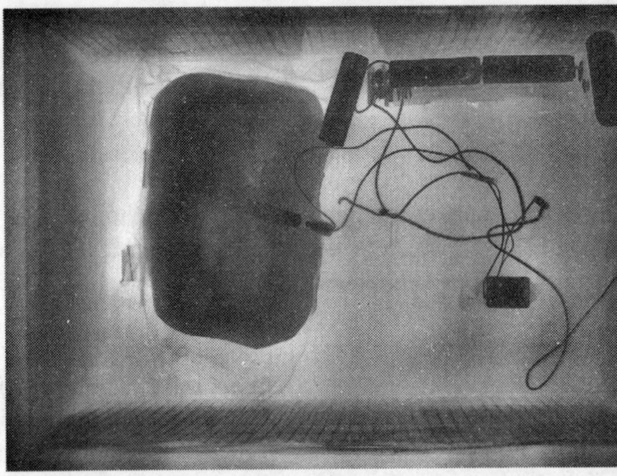

Fig. 201 X-Ray Of A Simple Command Wire IED

(30.3) WIRELESS OR REMOTE CONTROL IEDs

(30.3.1) The Basic Principle Of Remote Control Devices

A remote control or a wireless device must have two things: a transmitting device and a receiving device. The transmitter remains with the bomber. The receiver is attached to the IED. The transmitting device must be capable of sending a signal to the receiver in the IED without wires. The receiver must be able to convert the received signal into say, an electrical signal. This signal triggers an electronic circuit which sends the required current to the detonator. For wireless devices, you could give the command by regular radio transmission in MW/SW or FM, mobile telephone, infrared rays, microwaves, visible light rays, or for that matter even audible sound or ultrasound waves. Subsequent circuitry in a command controlled device is essentially the same as in a regular electronic timer device.

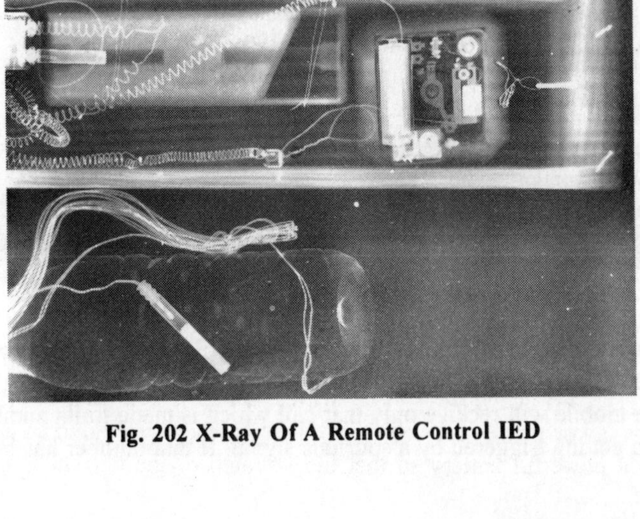

Fig. 202 X-Ray Of A Remote Control IED

Sounds simple? In practice, however, it is not so easy. Why? It is because of the problem of spurious signals. Both the transmitter and the receiver must have a high degree of **selectivity** so that the receiver responds only to the signal from the transmitter. Otherwise the receiver will always be in the danger of triggering the device from spurious signals. For example, if radio transmission is being used, the receiver should not trigger the circuit after picking up regular radio station transmissions, private mobile radio transmission, TV transmission, radar waves or even lightning.

(30.3.2) Mobile Phone / Cordless Phone / Pager Actuated IEDs

IEDs based on mobile phones have become very popular because they are almost readymade and very little modification needs to be done to make the IED functional. The IED requires a minimum of one mobile phone. The mobile phone functions as the receiver in the IED. The phone must, of course, be left 'on'. Unlike other remote control devices, here the bomber does not have to fabricate a transmitter because a call to the mobile set can be made from another mobile or landline. When the mobile phone receives a call, its ringer or vibrator is activated. Both of them require current to work. This means that upon receiving a call, the ringer or vibrator gets a current. This current can be used as the signal for the IED. Fools tend to give the signal directly to the detonator. This current is, however, weak and the device may not function reliably. Intelligent bombers make the output of the ringer or vibrator trigger a relay which then activates the circuit that feeds current to the detonator. They tap the signal from the ringer or vibrator of the phone and feed it to the 'gate' of the SCR or whatever relay one is using. The moment the bomber makes a call to the number of the mobile attached to the IED, the mobile would ring or vibrate. The electrical signal generated therein would trigger the SCR. That requires some knowledge of electronics, however.

Fig. 203 A Mobile Phone Actuated IED

The great advantage of a mobile phone actuated IED is that the bomber can activate it from thousands of miles away from any part in the world. The second advantage is its extremely high selectivity. The mobile will receive only that call which is made to its number. Thus there is no risk of the device getting triggered by a spurious signal. If that number has been kept a secret, it is not likely to receive a 'wrong number' call. The disadvantage of a mobile phone actuated IED is that it can be used only in those areas where mobile network is there. Secondly, from the point of view of the bomber, the problem with the mobile phones is that they can be tracked through the SIM card and the IMEI number (International Mobile Equipment Identity for GSM phones / ESN (Electronic Security Number for CDMA/TDMA phones) / MEID number (Mobile Equipment ID number). People have, however, developed means of **programming new IMEI number** in a stolen set using software like 'Patagonia'or **spoofing** it to confound the investigators. Investigators should, therefore, not become complacent or get overjoyed if they recover a SIM card intact.

Fig. 204 Another Mobile Phone Actuated IED

On exactly the same principle, **IEDs based on pagers** and **cordless phones** can also be made. Pagers are outdated now and leave considerable scope for tracing the caller. For a pager-based device the pager acts as the receiver attached to the IED. The paging signal can be used to activate a relay. Devices based on cordless phones are also very simple to make. In such devices the handset of the cordless phone becomes the receiver of the IED.

Mobile phone actuated IEDs have very serious implications for **aviation security**. There is little proven scientific basis for a highly prevalent myth that puny mobile phones could interfere with the sophisticated and electromagnetically shielded electronic control and navigation equipment of a commercial aircraft—something that is designed to fly through even lightning storms generating tremendous electromagnetic disturbance! We would not be surprised if the two-pronged assault of the public pressure favoring the use of mobile phones on board aircrafts and the resolution of technical issues pertaining to channel re-use etc. eventually leads to mobiles getting permitted in-flight. The authorities could very well remain ignorant unless they read what we are discussing here. As and when it becomes possible for mobiles to be used on board commercial aircrafts, it would be very easy to send up a mobile actuated bomb in the cargo hold and trigger it when the plane is in the air. An IED which is hermetically sealed and has no trace contamination will beat most explosive detectors if they are used at all for items to be loaded in the cargo hold. Further, the device could be cleverly concealed in such packaging or inside such metallic items that it would not arouse suspicion in an X-ray either. For your information, even now the AGRAS network of ground stations, strategically located in the US and Canada is capable of providing seamless coverage to business and private aircraft equipped with compatible telephones. We are emphasizing this point so that if somebody actually succeeds in bringing down a corporate jet or a commercial airliner in future, security forces and intelligence agencies should not be taken by absolute surprise and start wailing like widows. Forewarned is four-armed.

Fig. 205 X-Ray Of A Mobile Phone Actuated IED With The Mobile Phone Concealed Inside A Pipe

(30.3.3) IEDs Based On Walkie-Talkies Or Improvised Radio Systems

Using radio waves is one of the simplest ways of making a remote control device. Radio has been there for a very long time now and is not very complicated either. Then walkie-talkie sets are available aplenty these days. If the bomber acquires two walkie-talkie sets his job is nearly done. All that is required is that the sets must be compatible with each other. Their frequency range is unimportant as far as the basic principle of operation goes. Thus both VHF and UHF sets work fine. One set would serve as the transmitter and the other as the receiver. Devices based on **medium wave, short wave, FM or VHF/UHF** have an advantage over the mobile phone actuated devices that they can be used even in those areas where there is no network of mobile services. To use the system in its simplest mode, all the receiver has to do is to receive some transmission at the tuned frequency. It does not matter whether the transmission is voice or merely noise. The circuit requires just some reception.

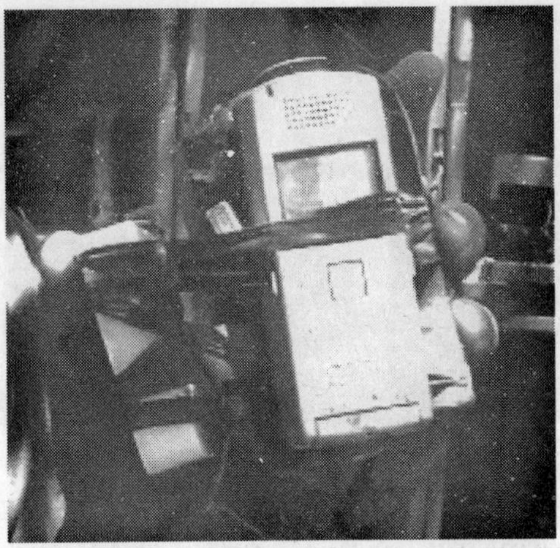

Fig. 206 A VHF Set Actuated IED

Intelligent bombers may go to the extent of designing their own medium wave, short wave or FM transmitters also. It is easy to design radio transmitters of low power in either MW/SW or FM. Circuit designs of both transmitters and receivers are available all over the world since the early days of radio. Electronics buffs take it as a hobby to design transmitters. Then he has either to design a radio receiver if he is using a frequency not used in commercial receivers or use a commercially available receiver if it can receive on his frequency. The receiver has to be kept tuned to exactly the frequency of transmission. The receiver is attached to the IED and signal from its output is tapped in the standard manner. The moment a transmission is made, the bomb will be triggered. In order to prevent the bomb from getting triggered from spurious signals, one may install a circuit in between the 'gate' and the output—the additional circuit will permit current to the gate only if it receives a specific signal from the output, such as a tone or a mixture of tones for a specific length of time. Theoretically a bomb based on **microwaves** can also be made in a similar manner. Designing microwave transmitters is not very difficult. If it comes to that, one could use the microwave generator of a microwave oven also. All he would have to do is to design an antenna so that the microwaves could be properly radiated.

Fig. 207 A Radio Receiver Actuated IED

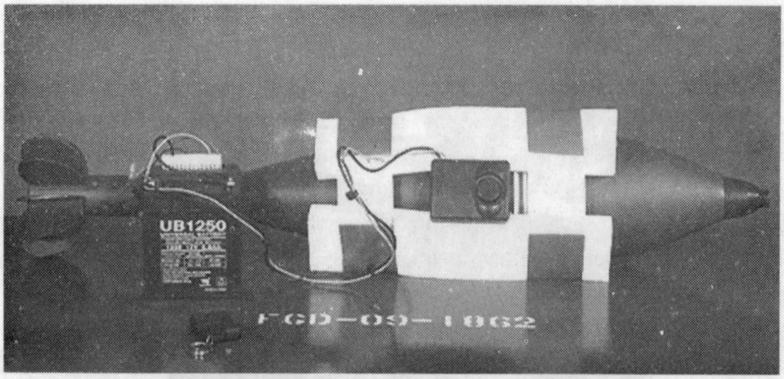

Fig. 208 An IED Using A Mortar Bomb As The Charge And Actuated By The Remote Control Alarm Of A Car

Devices based on VHF/UHF sets have a fundamental limitation. These sets operate in **line-of-sight**. That is, the transmitter and the receiver should be able to see each other. Mobile phones do not suffer from this limitation. That restricts the range of the device and also the places from where the transmitter may be operated. When VHF/UHF sets are used for communication, repeaters installed at sufficient heights are used to increase the range. The bomber does not have this luxury. Then there is another problem. Most of the commercially available sets are made for frequencies, such as the so-called Citizens' Band, which are used by a very large number of users including security forces, private organizations, emergency responders, and even motorists. There is thus considerable crowding of channels. From the point of view of the bomber, this increases two risks. Firstly, unless he has made absolutely sure that the channel selected by him is not at all used by anyone in the vicinity, it is possible that someone could start communicating on his channel and thus lead to an undesirable detonation of the device. Secondly, since the number of channels is limited in any case, it is easy to jam the communications. This is where an improvised MW/SW or FM transmitter/receiver system scores over the VHF/UHF sets. Many of the commercially available jammers are not likely to jam those frequencies. Similarly, a microwave transmitter/receiver system is also not likely to be jammed. These are the possibilities which the authorities must be aware of.

(30.3.4) RCIEDs Triggered By DTMF Codes From Regular VHF Sets

Some bombers have been found triggering their RCIEDs using DTMF codes from VHF sets. Since most officers of the security forces do not have the remotest idea of what DTMF is, they thought that the terrorists have stumbled upon something very sophisticated. That's how ignorance distorts the perceptions regarding the actual technical capabilities of the terrorists. In India this led to the deployment of jammers called 'Ashi' and 'Taurus' for the army. Using DTMF codes as such is a very simple thing and in fact less sophisticated than a mobile triggered IED. Many commercial jammers for VHF sets would not touch the mobile frequency range and they would be functional even with jammers.

(a) What Is DTMF Coding And Why Did We Need it: The Fundamentals Of Selective Calling

In a conventional, analog two-way radio system, a standard radio has noise squelch or carrier squelch which allows a radio to receive all transmissions on a channel. Suppose you don't want to listen to everybody or you want to talk to somebody specific and also want to ensure that not everybody listens to that. What do you do then? The solution is 'selective calling'. **Selective**

calling is used to address a subset of all two-way radios on a single radio frequency channel. Where more than one user is on the same channel (that is, co-channel users,) selective calling can address a subset of all receivers or can direct a call to a single radio. Selective calling features fit into two major categories: individual calling and group calling.

Selective calling is akin to the use of a lock on a door. A radio with carrier squelch is unlocked and will let any signal in. Selective calling locks out all signals except ones with the correct code. Selective calling prevents you from hearing others on a shared channel. It does not eliminate interference from co-channel users. If another user is talking and you try to talk at the same time, your signal will be affected by the other party using the channel.

There are two ways of selective calling. One is called **CTCSS (Continuous Tone Coded Squelch System)**; the other is called **DTMF (Dual Tone Multi Frequency)**. CTCSS is used for group calling whereas DTMF is used for individual calling.

In essence, the DTMF sends out a code. The receiver has to decode them. Once it receives the correct code, it alerts the user. Accurate transmission from the phone and accurate decoding on the receiving end are important.

DTMF was earlier used in telephony. In the time preceding the development of DTMF, telephone systems employed a system commonly referred to as pulse (Dial Pulse or DP in the USA) or loop disconnect (LD) signaling to dial numbers, which functions by rapidly disconnecting and connecting the calling party's telephone line, similar to flicking a light switch on and off. The repeated connection and disconnection, as the dial spins, sounds like a series of clicks. The exchange equipment counts those clicks or dial pulses to determine the called number. LD range was restricted by telegraphic distortion and other technical problems, and placing calls over longer distances required either operator assistance (operators used an earlier kind of multi-frequency dial) or the provision of subscriber trunk dialing equipment. DTMF was developed at Bell Labs in order to allow dialing signals to dial long-distance numbers, potentially over non-wire links such as microwave radio relay links or satellites.

DTMF is a signal to the phone company that is generated when you press an ordinary telephone's touch keys. In the United States and perhaps elsewhere, it's known as **'Touchtone phone'** (formerly a registered trademark of AT&T). With DTMF, each key you press on your phone generates two tones of specific frequencies. Why two tones are generated? This is to ensure that the tones are not imitated by a voice—one tone is generated from a high-frequency group of tones and the other from a low frequency group. They may sound rather musical when dialed (and representations of many popular tunes are possible), but they are not intended to be so.

The tones are divided into two groups (low and high), and each DTMF signal uses one from each group. The tone frequencies, as defined by the Precise Tone Plan, are selected such that harmonics and inter-modulation products will not cause an unreliable signal. No frequency is a multiple of another, the difference between any two frequencies does not equal any of the frequencies, and the sum of any two frequencies does not equal any of the frequencies. All this prevents any harmonics from being misinterpreted as part of the signal.

The DTMF keypad is laid out in a 4 x 4 matrix, with each row representing a low frequency, and each column representing a high frequency. The original keypads had levers inside, so each button activated two contacts. The multiple tones are the reason for calling the system multi-frequency.

There are twelve DTMF signals, each of which is made up of two tones from the following selection: 697 Hz, 770 Hz, 852 Hz, 941 Hz, 1209 Hz, 1336 Hz, and 1477 Hz. The additional four DTMF tones A, B, C & D are seldom used. These are extensions to the standard touch-tones (0-9, *, #) which originated with the U.S. military's Autovon phone network. The original names of these keys were FO (Flash Override), F (Flash), I (Immediate), and P (Priority) which represented priority levels that could establish a phone connection with varying degrees of immediacy, killing other conversations on the network if necessary with FO being the greatest priority, down to P being of lesser priority. Nowadays, these keys/tones are mainly used in special applications such as amateur radio repeaters for their signaling/control. Modems and touch tone circuits tend to include the A, B, C and D tones as well.

The following table shows the DTMF frequencies and the corresponding keys:

		Upper Band			
		1209 Hz	1336 Hz	1477 Hz	1633 Hz
	697 Hz	1	2	3	A
Lower	770 Hz	4	5	6	B
Band	852 Hz	7	8	9	C
	941 Hz	*	0	#	D

For example, when the key '1' is pressed in order to generate the DTMF tone corresponding to '1', the circuit mixes a pure sinusoidal 697 Hz signal with a pure sinusoidal 1209 Hz signal. The amplitudes of the two tones must be similar to enable the decoder to decode them reliably. The net signal of the two tones mixed with each other is also known as a DTMF digit. Digits are usually sent at a minimum of 55 milliseconds (ms) in length with at least 55 ms of silence between each digit.

The relevant code consisting of one or more digits is sent and the station programmed to respond to the number, typically one or two digits, responds by opening the squelch on the receiver to let the audio through. The disadvantage of this system is that if the receiver does not pick up the code at the instant the DTMF signaling takes place then it will not respond to any of the message. This can be a significant disadvantage because mobile stations often loose the signal for short periods as they are on the move.

A number of companies make microchips that encode and decode DTMF signals. The Telephony Application Program Interface (TAPI) provides a way for a program to detect DTMF digits. Many companies have trademarked names for their DTMF features. For example, Motorola calls their DTMF options, *Touch Call.*

Radios with DTMF decoders may monitor all system traffic or remain muted until called, depending on the system design. When the radio receives the correct digit string, it may momentarily buzz or sound a Sonalert. An indicator light may turn on and remain latched on. In most systems, the radio's receive audio would latch on after receiving a valid digit string if normally muted.

Some systems use DTMF for push-to-talk unit ID. Each time the push-to-talk is pressed, the radio sends a string of DTMF digits. Each radio has a unique string of digits. This allows the base station to know who last called or who last pressed the push-to-talk.

In some organizations radios are required to work alongside and with the telephone system. It is quite simple to integrate a phone patch to a radio system, so that office workers are able to make internal telephone calls to workers who carry radios, and the workers can call the internal phones.

(b) How DTMF Coding Is Used For Triggering IEDs?

Many companies such as Kenwood, ICOM, Alinco, Yaesu, Vertex, and Hy-Com make VHF/UHF sets from which DTMF calls may be made. To make a DTMF call all one has to do is to press the PTT (press to talk) button and press the numeric keys. One could use one, two digits or three digits. The risk with two digits is that of accidental pressing of the keys. That risk is reduced with three digits. One could theoretically use even more digits though it does not offer any great advantage. The set mutes the microphone and transmits the DTMF tones. The sets can be so programmed that if you are pressing the keys within an interval of two seconds then you don't have to hold down the PTT button; the set would remain in the transmit state for two seconds after releasing a key. The sets can also be programmed so that after entering and storing a DTMF number with up to 16 digits maximum in the dedicated memory channel, one can transmit the stored number by simply pressing the 'dial' button.

The receiver with DTMF decoder will be with the IED. The moment it receives the signal, it will generate an alert signal. This alert signal can be used to activate a relay circuit in the standard manner and that can be connected to the detonator.

Since DTMF uses standard frequencies throughout the world, jamming IEDs that get triggered by DTMF calls is easy.

(30.3.5) Infrared Rays Actuated Devices

With infra-red sources and detectors being commonplace, it was but natural that someone smart would think of using them for remote triggering. You may recall that such devices are commercially available—the remote control of the TV, for example, works on this. The bomber does not have to fabricate anything. All he has to do is to rig the receiver or detector in such a manner that the output is used to trigger a relay. The basic idea of an infra-red actuated device is simple. If the bomber has concealed himself in a place from where he has a direct line-of-sight to the bomb, he can use infrared rays to start the device. Small range is a problem with infrared rays but the bomber can overcome it by installing a delay circuit in the bomb. Thus he can activate the bomb and then slip out for the bomb to explode later. If he is intelligent, he could provide for a collapsing circuit or anti-disturbance switch also with the device. The main motivation for making such devices is to circumvent the jammer. Just think over it, it is an unbeatable combination; a cursory reading will not enable you to grasp the full potential of this devastating technique.

(30.3.6) Laser Actuated Devices

Lasers are commercially available. Even in the visible range and at relatively low power levels, laser light travels far. This can be easily used for actuating a device. All one requires at the other end is a light sensitive system. Many systems are commercially available. They include photovoltaic cells, charge-coupled devices and numerous other photosensitive devices. One has only to aim and fire the laser, in much the same way one aims the laser light used for aiming a gun. In fact people have used rifles and the aiming laser lights mounted on them to take aim. Even the detectors of laser are commercially available.

(30.3.7) Sound Actuated Devices

If the bomber apprehends that jammers could be in use, he could beat them all by selecting frequencies which are beyond the jammer's frequency range. Jammers are commercial devices and their frequency ranges are well-known. Hence selecting a frequency outside its range is not a great problem. And if push comes to shove, one can devise an IED which is triggered by sound waves of a specific frequency. No electromagnetic waves, no jamming! The basic principle is simple. All he has to do is to install a microphone-like receiver which converts sound into electrical signal. The rest follows. Once again, in order to prevent the device from going off by spurious sounds, he can install such filter circuits which will send a current to the gate only when a specific signal is produced in the microphone. Or one could use ultrasound. Ultrasound travels much farther and can be made to propagate in the form of a focused beam also. One can design an ultrasound transducer or use even simple ultrasound whistles which are commercially available and are used for calling dogs because dogs can hear ultrasound. A very clever idea, indeed!

❑❑❑

Chapter 31

Designs of the Time Delay and Remote Control IEDs Found in Recent Use in the Middle East

(31.1) WHY WE MUST LEARN ABOUT THE IEDs BEING USED IN THE MIDDLE EAST?

One reason is fairly straightforward. As far as bombings are concerned Iraq and Lebanon etc. are the happening places. A large number of improvised explosive devices are being exploded there by the Iraqi resistance. Hardly a day passes when a bomb is not exploded at some place there. Thus, in a way, the bombs in use there represent the most prevalent trends in bomb design. Many bombers across the world take inspiration from them. Many of them could be in communication with them for ideological reasons or whatever. The world is after all a small place and people do have a natural tendency to look up to the leader or whoever has displayed a remarkable degree of courage and resolve. Hence it is quite likely that similar bomb designs would be subsequently found in use elsewhere in the world also. We must therefore know the designs of the IEDs being used there. A study of the devices being used in the Middle East would show what sorts of devices are being used there. This information would help one to make a correct assessment of the technical capabilities and resources of those who have designed the bombs there. It will also be useful to assess the capabilities of those who might try devising similar improvised explosive devices elsewhere in the world. Some difference is inevitable. And that may well give vital clues about the people who might have designed the devices. It happens—at some places you find people who are good in devising electronic circuits but not so good in cannibalizing high explosives; at some other place you may find people who are good at improvising explosives but are flummoxed by electronic circuits. At some places, you would find many people who are very good computer buffs; at some other places you would find people who can make quite complicated mechanical devices but are unable to exploit the computer.

This chapter is intended to educate the people in general and officers of the security forces and intelligence agencies in particular about the most prevalent bomb designs. It will help them cope with such devices better should they encounter them in their parts of the world. If they remain ignorant as they are, they are likely to describe any new device as a most sophisticated one, and thus distort the intelligence perception. Even a design that could be made by a bright high school student would be attributed to the ISI or the Al Qaeda. Secondly, having knowledge of what has been developed and has been in use in some parts of the world will preclude people and officers from jumping to the conclusion that the terrorists have great scientists or technically qualified personnel among their ranks or that their technology is imported. The truth is that most people and officers are flummoxed by the IEDs simply because they do not know anything about them. We have seen instances where even totally dysfunctional devices were described by the intelligence agencies and security forces as highly sophisticated devices. In one case the prankster had stuffed the parts of a broken transistor radio in a geometry box such that the antenna was protruding out. He then proceeded to place the remote control switch of a Sharp call bell also there. There was no way the remote switch would have functioned on its own. This was described as a highly sophisticated device! In another case, the prankster simply placed an alarm clock besides a packet which was made to look like a bomb. The alarm clock was in the shape of a mosque. It was not even connected to the packet. In both the instances the prankster had taken care to stuff some old Urdu newspaper also along with the packets. Both the dysfunctional devices were attributed to Islamic terrorists and both times it was proudly proclaimed that a major tragedy which could have taken the lives of hundreds was averted by their discovery! Schiller could not have said something truer: "Against stupidity gods themselves struggle in vain." In fact many of the officers of the security forces and intelligence agencies secretly harbor a notion that the Islamic terrorists can design such great things which cannot be beaten. Thirdly, it also serves to bust the myth propagated by the Americans about the technical capability of the so-called Islamic terrorists. Many of the IEDs found in the Middle East are based upon commercial delay elements or remote control devices. The overall impression one gains is that the bombers are adept at very cleverly modifying commercially available devices to their use but would not bother to fabricate any complicated circuits on their own.

(31.2) AN ANALYSIS OF THE FIRING DEVICES OR TRIGGERS USED IN THE IEDs

(31.2.1) Qunxing RC Toy Transmitter And Receiver

Remote control toys provide the simplest and most easily available source of remote control devices. Qunxing QX-2218 TX RC Toy Transmitter has found application numerous times.

Fig. 209 A Simple RC Toy Transmitter

Unmodified RC toy controllers are used to send the command signal. Several models of the Qunxing RC toy receiver have been used in various devices. Recoveries have been made for model H20101RC toy car receiver and a newer model H20101RD receiver. All of them are commercial receivers operating in the RC hobbyist frequency range of 26 to 28 MHz. The toys, in any case, are provided with pulse width modulation in the transmitter to control forward/reverse and left/right movement of the toys. This is used to provide the control code for the IED. At times they have made and attached relays also and used a separate 9 volt battery for the detonator. Some devices have had provisions for 'safing and arming' also. Some had time delays also incorporated into them.

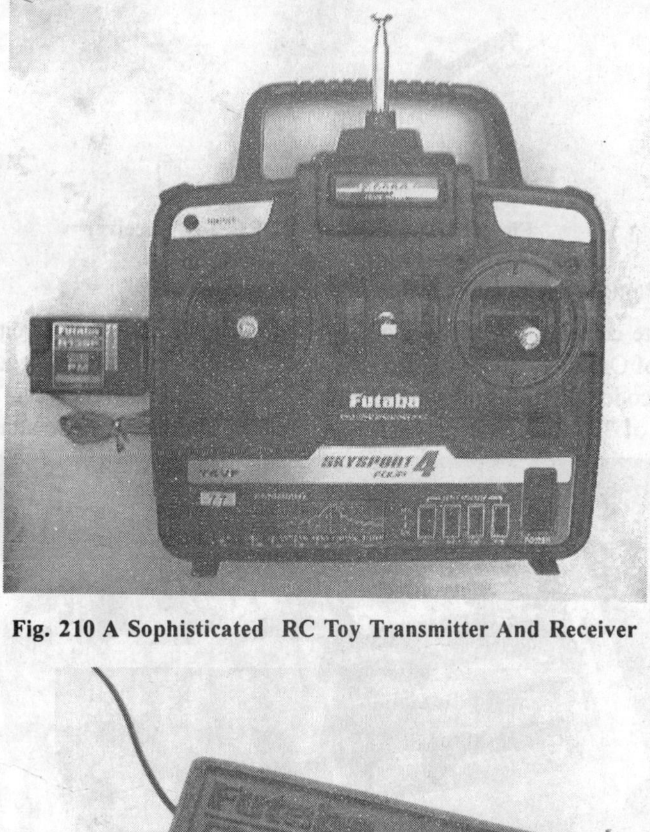

Fig. 210 A Sophisticated RC Toy Transmitter And Receiver

Fig. 211 A Simple RC Toy Receiver

Fig. 212 The Crystals Of RC Toy Receiver

(31.2.2) Qusan Remote Doorbell Transmitter And Receiver

A remote control doorbell is another easily available source of a remote control device. Several models of Qusan wireless doorbells have been employed as IED triggers. The circuit base is Type 2262 encoder IC (transmitter) and type 2272 decoder IC (receiver). They operate in the frequency range of 307 to 319 MHz. Once again relay and delays can be attached.

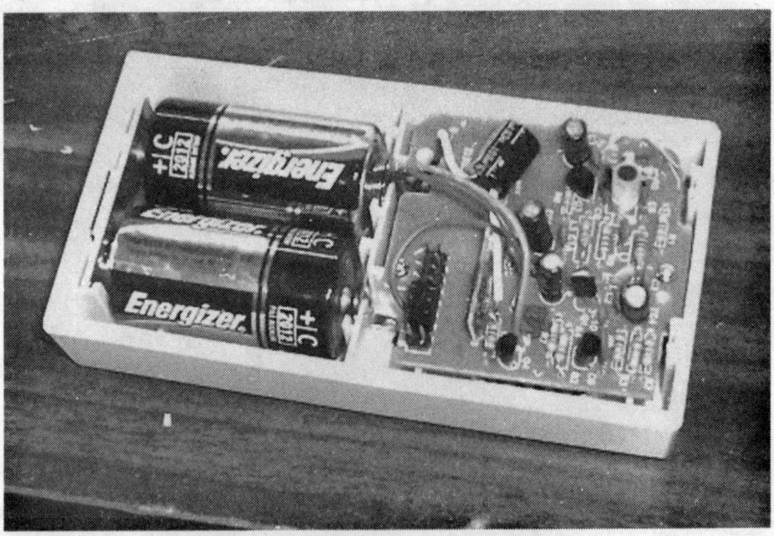

Fig. 213 The Receiver Of A Wireless Doorbell Modified To Activate a Relay

(31.2.3) Red Scorpio & Falcon Car Alarm Pager

A large number of Red Scorpio and Falcon car alarm pager receivers have been recovered. The pager receiver works with a transmitter in the vehicle that sends a signal when the vehicle is tampered with. They operate in the frequency range of 433 to 436 MHz. Relays have been also used with them. The car alarm pager is a much more sensible device than an ordinary pager.

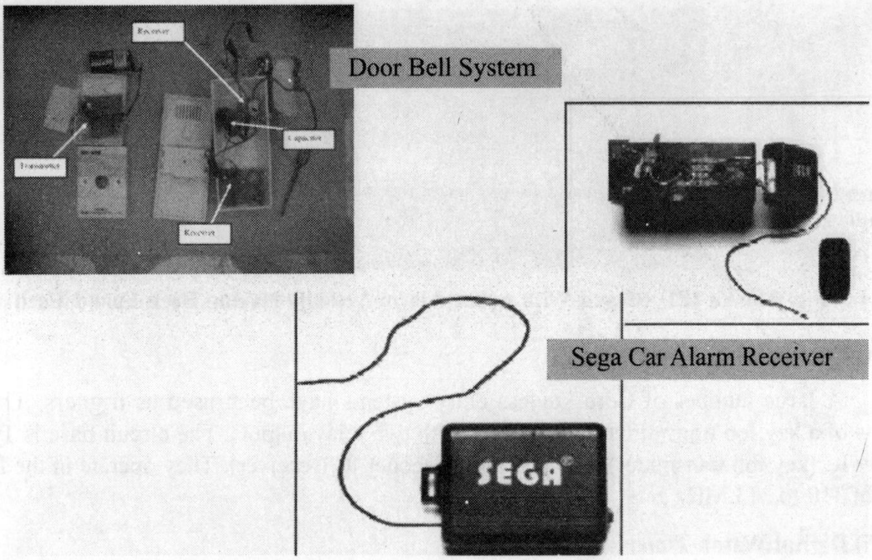

Fig. 214 Doorbell And Car Alarm Actuated IEDs

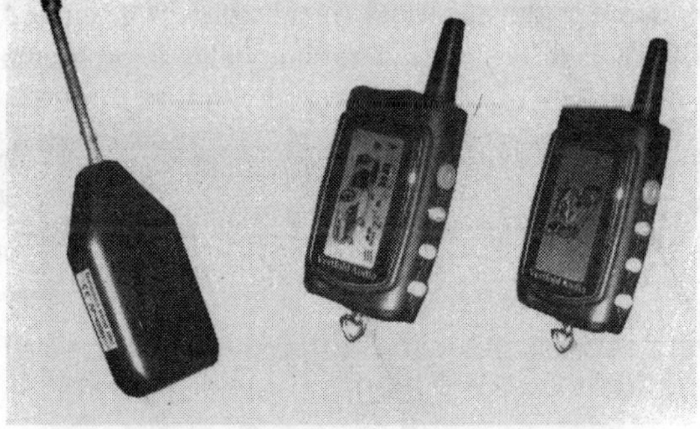

Fig. 215 A Typical Car Alarm Pager

Fig. 216 Another Car Alarm Pager Actually Having Been Found Used

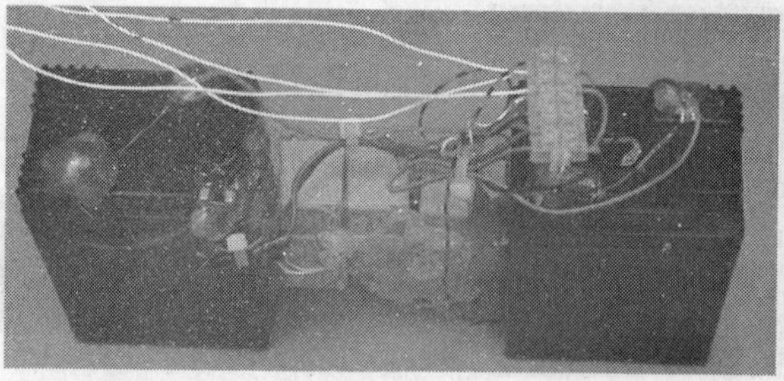

Fig. 217 An IED Rigged With A Car Alarm Actually Having Been Found Used

(31.2.4) Cora Keyless Entry System

A large number of Cora keyless entry systems have been used as triggers. The system consists of a key fob transmitter and receiver with two relay outputs. The circuit base is Type 2262 encoder IC (key fob transmitter) and type 2272 decoder IC (receiver). They operate in the frequency range of 310 to 315 MHz.

(31.2.5) Digital Watch Timer

Timers built using a digital watch are merely modern day versions of the old alarm clock-based timers. In the old alarm clocks, one tapped the mechanical motion generated in the alarms. In a digital alarm clock one taps the electrical signal which is generated at the appointed hour to activate the alarm and feeds it to the relay. The time delay is programmable up to 23 hours, 59 minutes in one minute increments.

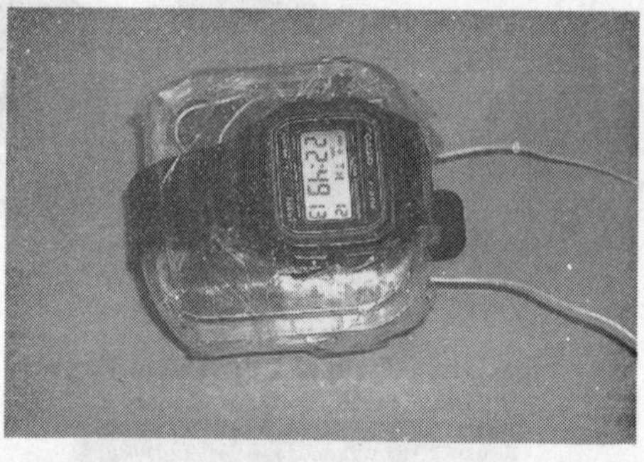

Fig. 218 Digital Watch Timer Based IED: Complete Package As Recovered

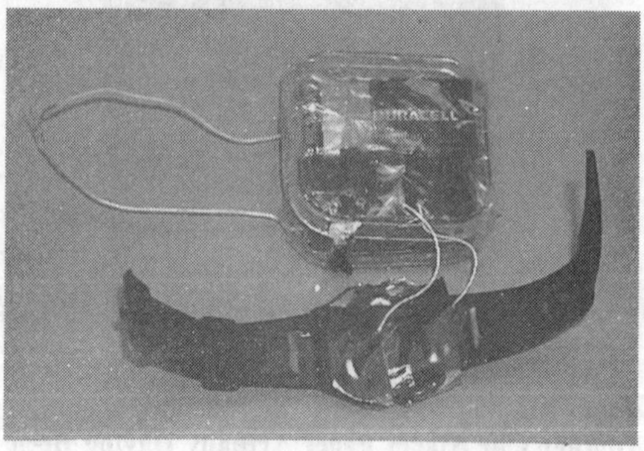

Fig. 219 Digital Watch Timer Based IED: The Watch Separated

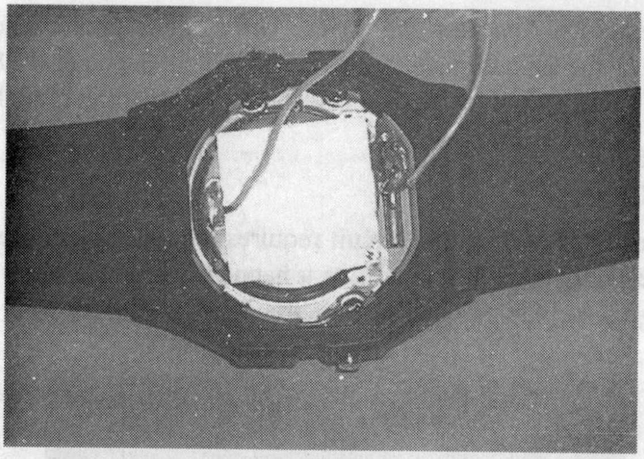

Fig. 220 Digital Watch Timer Based IED: Details Of Wiring From The Watch

(31.2.6) Digital Watch Timer With Opto-Isolator

This device has been found both in Afghanistan and Pakistan. It uses an opto-isolator to interface with the Casio digital watch alarm and a safe-arm delay timer. This is a much safer circuit to use than other versions of the Casio watch timer. Generally Casio F-91W watches are found to have been used but other digital alarm watches could also be utilized. The time delay is programmable up to 23 hours, 59 minutes in one minute increments. The circuit base is a Type 4060 Oscillator/ Divider IC for the safe-arm delay timer, and a Type 4N25 opto-isolator for the watch interface.

Fig. 221 Digital Watch Timer With Opto-Isolator

(31.2.7) NT-R12A/SB1000 Wireless Remote Control Based On RX-3400 IC

Several commercial digital remote control systems can be used for the IEDs. These systems, however, generally operate over short distances, since the transmitter is powered by a single 1.5 volt battery. The circuit base is Type 2262 encoder IC (transmitter) and type 2272 decoder IC (receiver). They operate in the frequency range of 315 to 433.92 MHz. A relay can be provided.

(31.2.8) Long Range Cordless Phone Handset

One of the readymade and very easily available remotely operated devices is a cordless phone. All the bomber has to do is to get a cordless phone which has sufficient range so that the base unit of the phone could be kept in a safe place. The cordless phone models used include the model EB268 produced by the Chinese manufacturer East Bright, Senao (SN-288D) and Sinao (358-CID) though one could always do a market survey and find numerous models. **DTMF codes** can be used for triggering.

(31.2.9) Interrupted IR Trigger

As we have discussed earlier, this circuit requires modulated infra-red (IR) signal to hold the output off. When the IR signal is interrupted, as it happens in commercially available IR security alarms, the output triggers the device. One of the circuit bases used is Sony IR receiver module, model SBX1610-52.

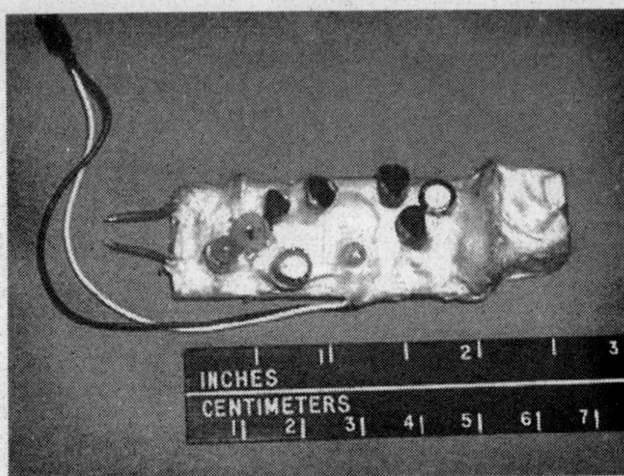

Fig. 222 Interrupted IR Trigger

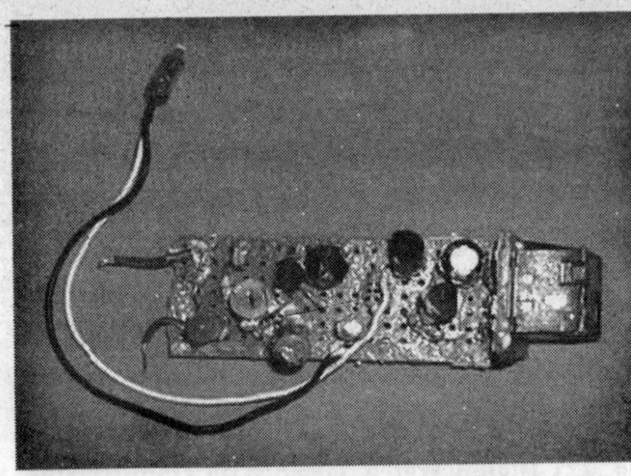

Fig. 223 Circuitry Inside The Interrupted IR Trigger

(31.2.10) Receiver with Ten Addressable Outputs

The device is of considerable importance because it is capable of independent control of ten outputs. This system consists of two assemblies, a model RX46 receiver and a model M461 sequencer. It uses a modified commercial pager as the RF receiver with additional circuits to decode the received non-standard DTMF codes. It operates on 156.500 MHz with narrowband FM modulation. The command signal is single non-standard DTMF tone pair for each of the ten outputs. Ten of sixteen possible codes are used. The battery saver cycle in the receiver assembly requires the DTMF code to be received for at least two seconds.

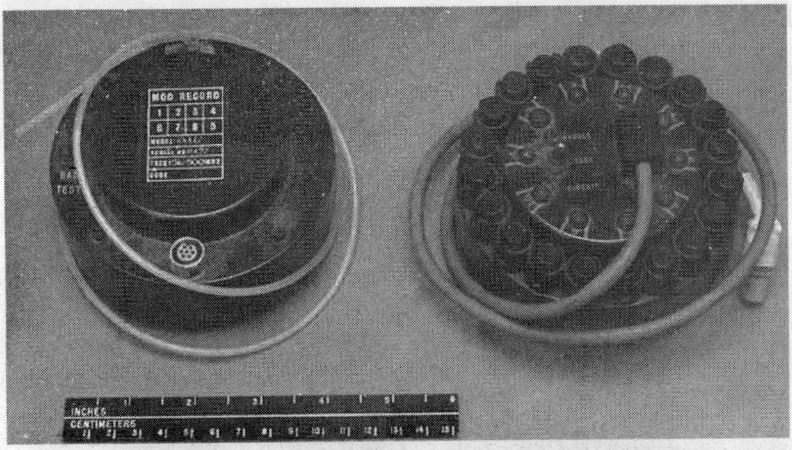

Fig. 224 Receiver With Ten Addressable Outputs

Fig. 225 Circuitry Of The Receiver With Ten Addressable Outputs

(31.2.11) Radiation Trigger As An Anti-Disturbance Switch

This circuit triggers when the radiation source is separated from this device. The detector portion is a commercial radiation monitor like the Monitor 4 from S.E. International Inc.The meter signal is processed on a circuit fabricated for the purpose. When the radiation source is removed, the circuit triggers a SCR to fire the detonator.

(31.2.12) Light / Break-Wire Trigger With Safe-Arm Delay

We have discussed the basic principle of operation of light-sensitive switches. These devices have improved a little upon the basic device. The first improvement is to provide for a safe-arm delay. After the safe-arm delay, light falling on the IED will trigger the device. The second improvement is the provision of a booby trap in the device. This booby trap is in the form of a simple break-wire. Breaking of this wire, even in the dark, trips a circuit and triggers the device. The idea is that normally the device would be triggered by incident light; however, should there be no light, the device would not be wasted and would function like a simple booby trap to be triggered by a tripwire.

(31.2.13) VHF Transceiver Using DTMF

Standard VHF sets with the DTMF decoder operating in the frequency range of 433 to 434 MHz is used with three-digit sequential DTMF codes. They have also provided for a relay and a LED and a buzzer for testing. Decoders have at times been fabricated also. The circuit base for single-digit DTMF code is Type MT8870 DTMF decoder; that for two-digit DTMF is MT 8870 DTMF tone decoder IC.

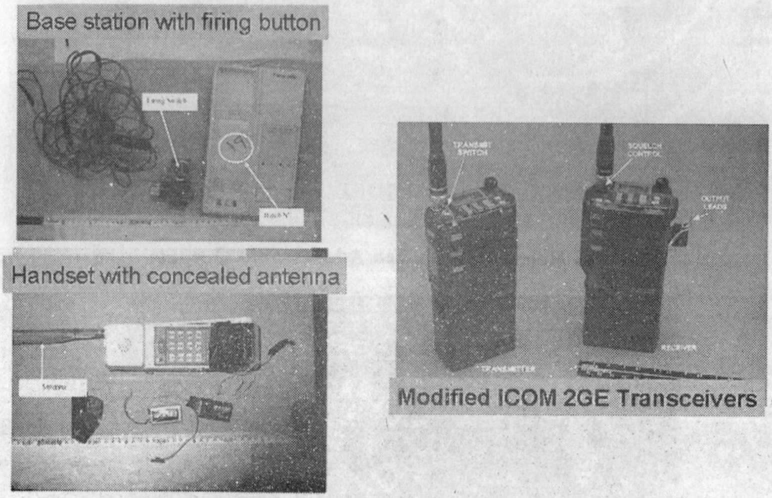

Fig. 226 VHF Sets Modified For Triggering IEDs

(31.2.14) MC 14536 Based Programmable Timer

The device is similar to what has been discussed in the theory. They have, however, provided for a sixteen-position (hexadecimal - base 16) thumbwheel switch (marked 0-9, A-F) to select the number of divider stages used in establishing the time delay. With the switch set to '0', the base time delay is generally set in the range of 30-45 seconds. Each click of the switch doubles the delay, to a maximum, at 'F', in the range of 11-17 days.

(31.2.15) 'Mushroom Timer' For Shaped Charges

This timer is designed for about 18 lbs heavy shaped charges that could have been intended for sabotage in oil pipelines. The charge is triggered by a 0-39 day delay timer assembled in a section of pipe. Increments of one day can be had in the delay. The shape of the timer assembly has earned it the nickname 'Mushroom'. A magnetic plug must be installed at the end of the timer assembly for power to be applied and the timer to function. Removing this plug disconnects power from the timer and the detonator.

(31.2.16) Programmable Q&Q Timer

Based on a commercial Q&Q Timer and concealed in a standard jewelry box, this circuit can produce a maximum delay of 99 minutes 59 seconds in one second increments. In the count-down mode, it will produce alarm when the counter reaches zero, triggering the output relay. The circuit base is Q&Q 'Timer Count Up/Down' DT26-501-GJB manufactured by Japan CBM (Citizens Business Machine) Corporation. Safe-arm switch is provided but that undoes to some extent what the jewelry box seeks to do by way of concealment.

(31.2.17) 1-3-5-7 Hour Timer Using PIC Microprocessor

This timer can be set to one of four different delays. Cutting one of the four jumper wires on the end of the package will set the delay time to 1, 3, 5, or 7 hours. A crystal-controlled PIC 16F84A microprocessor sets the delay time. Timing begins when a battery is attached to the clip and the desired jumper wire is cut. This can be done in either order.

(31.2.18) Mechanical (Washing Machine) Timer

This is a very good example of ingenuity. They have simply taken unmodified mechanical timers from old washing machines and other home appliances like morning-walker machines, and heaters-cum-massagers etc. and adapted them for use as IED time-delay triggers. The timer produces contact closure across a wire pair after a time period ranging normally from 35 seconds to 14 minutes 30 seconds.

❑❑❑

Chapter 32

The Last Resort: Suicide Bombs or Victim Operated IEDs

(32.1) SUICIDE OR 'FIDAYEEN' ATTACKS: HISTORICAL BACKGROUND AND DEFINITIONAL ASPECTS

(32.1.1) Who Resorts To Suicidal Attacks?

Suicidal attacks are not a new concept in warfare. There have been very many suicidal charges in the history of warfare of all races and religions; some of them heroic, some desperate; some intended, some unintended; and some simply foolish. From the almost ritualistic and highly deliberate post-Jauhar suicidal charges of the Rajputs in medieval India to the unintended suicidal charge of the Light Brigade at Balaclava; from the Japanese kamikaze pilots to the 'human wave' attacks of the Chinese, history is replete with examples of attacks which were suicidal in nature. Suicidal attacks are indeed the last resort of a warrior or an army. You may try to glorify them as much as you wish and write epics on whom you laud as heroes; deep down your heart you cannot shake off the feeling that they did it when they were left with no other option. Unintended suicidal attacks are a different thing, however; they are actually military stupidities of a high order. Tennyson was essentially trying to cover up a terrible military mistake when he wrote that poem on the charge of the Light Brigade. For all the deliberate suicidal attacks, the fact remains that when a fighting group, organization or nation adopts suicidal operations consciously as a battlefield tactic, it indicates nothing but desperation; a classic example being the 'kamikaze' aircraft strikes resorted to by the Japanese in World War II when the tide had turned against them.

Suicide terrorism is the use of suicidal operations by terrorist groups. It is not unique to Muslim terrorists as some ill-informed journalists in the West would have you believe. It has been witnessed in Sri Lanka, India, Lebanon, Kuwait, Israel, Panama, Algeria, Pakistan, Argentina, Croatia, Turkey, Tanzania, Kenya and the U.S.A. The LTTE in fact has conducted the highest number of suicide operations in the world, followed by the Hizbullah, Hamas and the PKK (Kurdistan Workers' Party, Turkey).

(32.1.2) Who Are The Fidayeen?

Of late, the media has started making indiscriminate use of the word 'fidayeen', particularly in respect of Isalmic terrorists and those in the Middle East. The Indian media uses this very frequently for Kashmiri terrorists. This word is being used to describe those terrorists who are supposed to be members of mysterious things called '**suicide squads**'. The word is being used out of context and the usage is not correct.

Etymologically, the word 'fidayeen' has a fairly respectable origin. Its root word in Arabic is 'fida' which means enamored, in love, captivated, or totally devoted. From 'fida' we get 'fidayee' which means a lover, devotee, faithful or one who would even give up his life for another person or for a cause. 'Fidayeen' is the plural of 'fidayee'. In a political context, the word 'fidayeen' has been in use since the late 11th century when Hasan Sabbah had divided his fraternity into murshids, da'is and fidayeen. However, it drifted into popular usage in the 1950s when it was used to describe Arab guerillas, especially those operating against Israel. It must be kept in mind that a fidayee should be prepared to sacrifice even his life for his love. If his love is for a cause, he should be prepared to lay down his life, if it comes to that, for the cause. The emphasis is on his willingness to sacrifice his life; it does not mean that he should necessarily sacrifice his life. The simple idea is that his devotion ought to be of such a high degree that he should not hesitate from paying the highest price for it. It does not say that he has to commit suicide in the process. Islam values martyrdom but there is no religious approbation for achieving martyrdom in a suicidal fashion. Thus, one who goes into an attack that could be very dangerous or even suicidal, is merely a daredevil there is no need to exalt him to any superior status by calling him a fidayeen. He is merely more motivated; he is not more skilled!

The decision to commit suicide for a cause or thought is something which the Western mind-set is never able to comprehend. And that's why they think that the suicide bombers are either crazy in an inexplicable manner or so vehemently fired by some obscure religious philosophy that they are able to do so. The very idea of giving up one's life voluntarily for a cause is abhorrent to their deep-rooted cultural and societal mores. They are wrong on both counts. The suicide bomber is neither crazy nor he is necessarily steeped in Wahabi philosophy. There are things which the Westerners would never understand because one has to belong to a certain society to understand them. Since they are not able to understand them, they find simplistic explanations for them. Wahabi philosophy is one of them. Little do they realize that there are any number of Muslim suicide bombers who have not had even heard of Wahabi philosophy, not to speak of Hindu suicide bombers amongst the LTTE and suicide attackers amongst the followers of other religions. Many of the Islamic suicide bombers are well-educated and even technically qualified too. The Glasgow airport bomber (2007) Kafeel Ahmed had an M. Phil. in aeronautical engineering from the Queen's University, Belfast and was pursuing Ph.D. in computational fluid dynamics. The Haifa bomber (2003) Hanadi Taysser Darajat was a law graduate from Jareesh University, Jordan—a rather unusual distinction for a Palestinian girl. It is utterly simplistic to say that: "The profile of a suicide-bomber is that of a young man from a religious home who has been approached by a mosque, then recruited and finally trained". It can be easily seen that such simplistic theories would fail to explain the use of suicide terrorism by groups like the LTTE. The **role of religion** in suicide terrorism has obviously been greatly overstated with vested political interests in mind and to propagate the current American world-view.

Fig. 227 A Palestinian Woman Suicide Bomber

(31.1.3) The Socio-Cultural Roots Of Suicidal Attacks

This socio-cultural handicap to understand a phenomenon is found in other spheres of life also. For the westerners, surrendering before a victorious enemy is not dishonor—for the Oriental warrior it is the ultimate dishonor, even if it is guaranteed that the victor shall not ill-treat him. In many ancient societies including India, one whose life was spared by the enemy lived in eternal disgrace thereafter. The westerners used to get utterly surprised when many Japanese generals and officers preferred to commit suicide rather than surrender. Or for that matter consider a princess, the estranged wife of a king-to-be, who was so promiscuous that even lowly persons like horse trainers, junior army officers, football players, doctors and businessmen had been freely enjoying her sexually. Had it been the case in the Orient, her sons would have died of sheer shame and would never have been able to show their faces in public. To the Orientals, the mere thought of their mothers having behaved like lowly whores is the ultimate disgrace imaginable. And yet in a Western society, the sons are not only able to lead a normal and exuberant public life, nobody even comments that their mother was a slut of the worst kind. It is precisely for such handicaps that they have never been able to understand the lofted philosophy of 'seppuku' by the samurai as an act of atonement. Rather they have debased it by using a wrong term 'hara-kiri' for it which simply means 'belly cutting' (that is, the physical mechanism of seppuku) and has nothing to with the philosophical content of it. That's why the very concept of suicide terrorism confounds the West as something utterly crazy. To the Orient, there is nothing crazy about it—after all, have we not been taking great pride in the tradition of the Rajputs who, once they determined that a military defeat was certain, decided that death was preferable to dishonor. Had they merely wanted death to escape capture, they could have simply consumed poison. But no, that would have been utterly cowardly. They had to die fighting. And what would have happened to their womenfolk once they were slain in battle? The women could have been turned into sex slaves, another grave dishonor. Hence they consigned their womenfolk to the flames of Jauhar so that their minds could be freed from the worries of protecting their honor and then, wearing the sacred saffron headgear, sallied forth in their last suicidal charge from the fort upon the enemy. When the West cannot understand the utter dishonor of surrender in a war, how can they understand the desire to sacrifice one's life in a suicidal act rather than live with what is considered as the utter ignominy of subjugation, having surrendered to or having lost to a power militarily stronger than them? It is the weapon of a man driven to the wall—he is left with no option but to offer the supreme sacrifice in an attempt to prove that he is prepared to offer the maximum that he can do for the cause under the circumstances—beaten he may be, down and out he may be, but his devotion to the cause remains unmoved.

Fig. 228 Details Of the Suicide Bomb Worn By A Palestinian Woman

(32.2) HOW SUICIDE TERRORISM HELPS TERRORIST ORGANIZATIONS?

In conventional warfare, suicidal attacks have never paid off. No decisive victory could ever be clinched by a suicidal charge. The French Marshal Pierre Bosquet had very rightly observed for the charge of the Light Brigade at Balaclava: "It is magnificent, but it is not war!" The charge of the Light Brigade was suicidal only in its content—it was never intended to be a suicidal charge! Then we know how the tactic of kamikaze aircrafts had failed. The situation, however, is materially different for unconventional warfare or terrorist strikes. Suicidal attacks do pay off for terrorists.

Terrorists resort to suicide terrorism for a variety of reasons. Agreed, that all of them stem from an implicit admission that they are desperate. Nobody loves to blow himself up. They are obliged to do so when, at a military or operational level, they find that their usual repertoire of numerous terror tactics or other tactics of unconventional warfare is not proving sufficiently effective. One should always keep in mind that suicidal attacks are normally against the basic tenets of warfare. It they are forced to adopt unconventional means to keep the group or the movement going, it can only mean that they are desperate.

Terrorists who are using suicidal techniques are doing it simply because they have, at that particular juncture in their struggle, found it expedient to do so. It is incorrect to say, as most people seem to believe, that they do it under the influence of some fundamentalist indoctrination or some sort of religious brainwashing. The cause comes first, the religion later. The point being made is that suicide terrorism is both a tactical as well as emotional decision. While the West can possibly understand the former, it is difficult for them to understand the latter.

In the following we list the reasons due to which a person or group decides that it would be expedient to resort to suicide attacks.

- Suicide attacks have got a certain **shock value** attached to them. They send the target into a state of shock much beyond the actual damage inflicted upon him.
- Suicide attacks are **sure-fire methods**. The chances of the attacks succeeding and inflicting considerable damage on the enemy are quite high.
- As such they are very **cost-effective**. The same amount of damage to the target's military resources (such as aircrafts and ammunition dumps), men (such as the US Marines in Beirut) and economic resources (such as fuel depots) cannot be easily inflicted through conventional means of attack. Hence it makes them particularly attractive to the non-state actors. When non-state actors find that militarily they cannot take on the military might of a state, they resort to suicide attacks.
- They are a cost effective **means of countering sophisticated weapons** at the disposal of the state, particularly air force, armor and heavy artillery, things which non-state actors lack, either in the combat zone or outside it. The destruction of more than a dozen aircraft including eight military and some civilian aircrafts by the LTTE in a single suicidal attack on the Katunayake military airbase is a good example.
- As the LTTE Black Sea Tigers have demonstrated, considerable **damage** can be inflicted on **small navies** equipped with not-so-sophisticated vessels or vessels that are incapable of defending themselves adequately from such threats. Their small explosive-laden boats and divers become naval versions of the car bombs and have taken a heavy toll of naval vessels of Sri Lanka, which is way disproportionate to the LTTE's actual strength on the seas.
- Suicide attacks are much **simpler to plan**. They do not have to concern themselves with making an escape plan, often the most difficult part of an operation. They have only to worry about penetration.
- They do not have to **worry about** some member getting **caught** and revealing sensitive information in the course of interrogation or torture.
- It has **great psychological impact** on the enemy.
- At times, if properly executed it can even result in **strategic advantage** to the non-state actor particularly if they have taken out some important leaders, vital for the conduct of operations against them. While a state can recover from any such elimination and introduce substitutes, there is undoubtedly some dislocation both in terms of time and planning of subsequent moves.
- It also makes the **state waste its manpower and material resources** in strengthening the security of its top functionaries, vital installations and public places. Indians have

only to recall how the administration had reacted after the bomb blasts in Sarojini Nagar in Delhi which were not even suicide attacks.

- In fact the non-state actors can force the state to squander its resources merely by **propagating a 'hit list'** of their real or imaginary targets. Once that list is out, there are few governments which can afford to think rationally and most of them are forced to protect even those things which they would have never protected otherwise.
- Not only that, it forcefully introduces **widespread panic** and feeling of insecurity amongst the people and towards this end they are aided by the media which is keen only on sensationalization. Further, over reaction from the government aggravates the feeling of insecurity and indirectly fosters the image of the group being almost invincible if not into an all-pervading force. They get a **larger-than-life image** at a very low cost.
- They can deliver **maximum shock** to the enemy in a single attack. That is to say, they do not have to worry about conserving their resources for some other day.
- The fact that, in a suicidal attack, even high security places can be attacked, serves as a **great morale booster** to them. It has **great propaganda value too**.
- Suicide attacks are living examples of the supreme sacrifice. They constitute the most convincing proof of the dedication and devotion of those who took part in it for their 'cause'. Thus they act as a **source of inspiration** to all those who find their resolve wavering and that they are lacking in the fervor which they had once.
- They can hope to gain some sympathy amongst their covert or half-hearted supporters by such acts of daring and 'bravery'. Every movement looks for and **eulogizes its martyrs.**

Terrorists know that this is a technique that cannot be used indiscriminately; otherwise the movement will soon run out of members. The idea is to keep a sufficiently motivated cadre ready for the act and use them whenever the leadership feels like it, keeping the above-mentioned reasons in mind. These days we find the Palestinians resorting frequently to suicide attacks. This is because militarily the Israeli army has attained such overwhelming superiority that the terrorists are not really left with many other viable options. Any group placed in their position would have been obliged to do the same. It is unfair to blame Islamic fundamentalism per se for that – in their particular circumstances they might be telling their cadres that they would go to Paradise after attaining martyrdom and that houris would be waiting for them there; some other group might have given some other inducement such as establishing their credentials as heroes, or even heavy monetary compensation to their families.

(32.3) THE TECHNOLOGY OF SUICIDE BOMBS OR VICTIM OPERATED IEDs

(32.3.1) Explosive Belts Or Jackets

What is popularly known as an explosive belt or jacket is simply an improvised belt or jacket in which a complete IED has been concealed. A serious suicide bomber would make the explosive belt or jacket with plastic explosives so that they may be given the desired shape. The idea is that it should blend into the clothing as much as possible. Sophisticated explosive belts or jackets should have the explosive in the form of sheets or plates. Otherwise, they have to go for what amounts to grenades as used in the case of Rajiv Gandhi. If one can get plastic explosives, they are the best. Otherwise, one could use any other HE like TNT or for that matter even dynamite.

The most popular shrapnel still remains ball bearings and with good reason. Those who use nails, nuts and bolts deserve to be beaten up first.

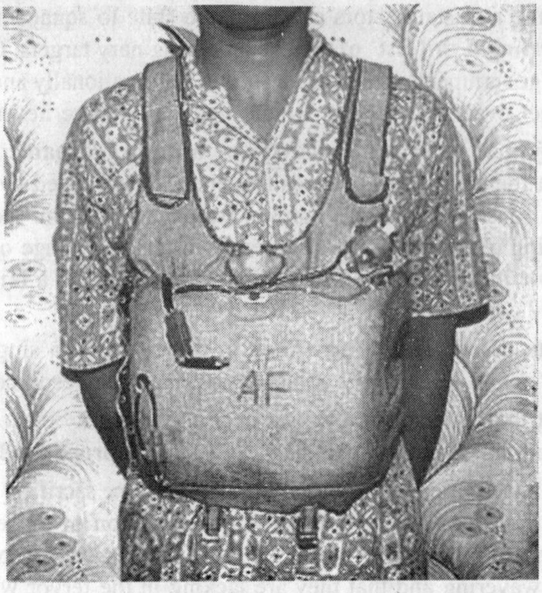

Fig. 229 A Suicide Bomb Made With Two Pull Cords: One To Arm And The Other To Explode The Device

Those who have not been able to make concealable shapes have also resorted to the time-honored method of the bomber pretending to be a pregnant woman. That's precisely what the West German RAF did in the 1970s. The designers had, however, no intention of getting the bomber woman blown up there. She removed the bomb from her abdomen and emplaced it at the target site; the bladder of a soccer football was provided to inflate her belly again!

The use of the words **'human bomb'** for a suicide bomber is so atrocious English that we would not even mention it again.

(32.3.2) The Bomb That Killed Rajiv Gandhi

In the assassination of Rajiv Gandhi, it is believed that the bomb had six charges of 80 g each of C-4 encased in castings of TNT and they were fired by a 9 volt alkaline battery. The shrapnel are believed to be 2800 ball bearings of 2 mm diameter each on all the charges. From the description available it appears that the charges were in the nature of six modified fragmentation grenades. They took special care to provide sufficient shrapnel so that even if the blast by itself proved ineffective for whatever reason, the shrapnel would do the job. They had made a vest or wide belt of denim which could support nearly one kg weight. There were two toggle switches provided in the front; one to arm the device and one to activate. The bomb was designed by a man called Arivu and the design was approved by Shivrasan. It stands to reason that the bomber must have provided for a good capacitor discharge circuit for the single battery to fire all the detonators simultaneously. This is supported by the fact that Arivu was supposed to be an expert in designing electronic circuits and Shivrasan also dabbled in electronics. As such it was not a very sophisticated design but was adequate for its intended purpose. It is not known whether the charges were interconnected by detonating cord or not. The bomber Dhanu detonated the bomb as she bowed to touch Rajiv Gandhi's feet after garlanding him with a sandalwood garland and as he bent to lift her up. The explosive material was on her back side. Since her body provided some tamping effect, the explosive force was directed predominantly upwards. Thus the target's frontal torso got the full force of the blast. With such a close distance, there was no chance of failure with that quantity of charge. Six modified grenades at that close a distance became a case of overkill.

Fig. 230 A Typical Suicide Belt Made Of Denim

(32.4) THE LTTE AND THE CULT OF SUICIDE BOMBING

The LTTE is the biggest proponent of the cult of suicide bombing. It meshes well with their psyche also. Suicide bombing has been raised to the status of a highly respected cult within their ranks. The LTTE suicide squad is known as **Black Tigers**. According to their own data released in 2006, 273 of the Black Tigers had given up their lives till then in suicide attacks. Of them 194 have been the so-called Black Sea Tigers, mainly divers and sailors of small explosive-laden boats who have blown themselves up to take down Sri Lankan navy vessels. A substantial number of them have been women. In fact women make up 29% of the Black Sea Tigers and 23% of the Black Tigers killed so far. Forty-nine more have been killed in the year following July 2006, taking the total to 322 in July 2007 since 1987. Black sea Tigers account for 241 out of them.

The LTTE's first ever suicide attack was made on July 5, 1987 when Vallipuram Vasanthan (better known by his LTTE name as 'Capt. Miller'), a 21 year-old Jaffna youth, had rammed a truck filled with explosives into a building housing Sri Lanka Army soldiers, killing 40 soldiers at Nelliady. The LTTE in fact celebrated July 5, 2006 as the Black Tigers Day in the memory of those cadres who have died in suicide attacks. Their chief Prabhakaran also made a rare appearance at Vanni on the occasion and that in itself should indicate how much they value suicide attacks. One of the speakers on the occasion was a man called Yogaratnam Yogi. He is a close associate of a man called Mahathiya who was once the deputy leader of the LTTE. Yogi announced to the world: "War itself is violent. There are no soft methods in it. Weren't bombs made to blow up and kill men? So why there is such a cry if only a man becomes a human bomb?" The chilling words speak for themselves.

The LTTE is one organization that has been selecting its targets very carefully for their value. They sure know their opponents and their news value. They have been able to kill President Premadasa and Rajiv Gandhi. They made an attempt on Chandrika Kumaratunga also but she survived, losing only an eye in the process. The number of ministers, MPs and senior army officers attacked is very large. They made attempt at Lt. Gen. Fonseka, successfully killed Maj. Gen. Parami Kulatunge, Deputy Chief of Sri Lanka Army and made an attempt to land frogmen to destroy ships in Colombo harbor. By selecting prominent persons or apparently high value targets, they force an ignorant and undiscerning media to launch into a description of the capability of the organization in a disproportionately high manner. Attack on a fuel dump may neither cause many casualties nor result in any great loss to the nation, yet the burning dump makes for excellent TV coverage and the media goes after it in full force, making you believe that the attack has either incinerated the entire island or has left Sri Lanka without any fuel!

Fig. 231 The LTTE Black Tigers

(32.5) DEALING WITH A SUICIDE BOMBER

(32.5.1) What Devious Ways The Suicide Bomber Can Devise?

There is nothing you can do if the bomber does his job undetected. Then you are taken by absolute surprise and cannot do anything. However, if you spot him then don't pounce on him in some mistaken notion of bravery. If the bomber has any brains he can even pretend to surrender so as to draw security personnel closer to him. Worst, the bomb could be under the remote control of somebody else! They invented this mechanism so as to counter the Western security forces who insisted that the bomber wishing to surrender must remove himself first to an isolated place, keep the stand off distance and then remove the bomb from his person. His remote controller would activate the bomb the moment he found or suspected that the suicide bomber is losing resolve!

Western forces have developed a notion that one should shoot or nab the bomber. Little do they realize that the bomber could have easily made the device in such a way that it went off with an audio command which he could give when he got nabbed or could be linked to his pulse so that it went off the moment he died and so on! These are known as **'dead man's switches'**. The phrase comes from **'dead man's handle'**. This was a device used in the early days of steam locomotives. People had a morbid fear as to what would happen to a train if the driver suddenly died of say, cardiac arrest. Hence this device was invented. The driver had to keep his hand on it all the time. If his hand were to be removed by any chance, the handle would activate an emergency brake. Western security personnel without brains who try to shoot him dead don't realize that the bullet itself may trigger the device! Finally, an intelligent suicide bomber could have easily provided for **secondary and tertiary devices**; or for that matter there could be **multiple suicide bombers** in one place!

A somewhat more intelligent way of dealing with a suicide bomber is to fire a TASER at him. That would leave him immobilized and yet leave him alive so that a dead man's switch is not activated. It could be countered only by making a spring loaded switch which would require to be kept pressed as long as the bomb were to remain un-triggered. The moment the bomber would be immobilized, paralyzed or become unconscious; he could lose muscle control over the switch and set it off. More intelligent bombers could make his body sensitive to the receipt of electrical current also.

(32.5.2) Safe Distances To Be Observed

Device carried	TNT equivalent explosive	Building evacuation distance	Outdoor evacuation distance
Suicide belt	10 lbs	90 ft	1080 ft
Suicide Vest	20 lbs	110 ft	1,360 ft
Briefcase/Suitcase Bomb	50 lbs	150 ft	1,850 ft

(32.5.2) Systemic Measures To Deal With Suicide Terrorism

Theoretically, the number of ways in which suicide attacks may be planned is limited only by the ingenuity of the terrorist. Hence it must be admitted that it is rather hard to fight suicide terrorism. Still, the fact remains that there are not very many terrorists who are sufficiently intelligent and who possess a rare blend of a high level of technical knowledge with field expedient methods. Most terrorists are therefore likely to stick to the usual bombings or shootings. We can expect some application of mind in these methods. For example, it was learnt that one terrorist group had hired the services of a dog handler from France to monitor the ability of sniffer dogs. It is likely that this group could develop a suicide body suit with a repellent to evade the attention of sniffer dogs or use a substance that could instantaneously reduce the olfactory capacity of the dog. Further, as terrorists know that security personnel frisking people for suicide devices generally concentrate on the abdominal area, they might incorporate plastic explosives into brassieres which could, in a cursory examination, easily pass off as padded or sports bras.

The first thing that must be admitted in planning security measures against suicide attacks is that terrorists would not go for a suicide attack unless they have adequate information about the target. If they have that intelligence it can only mean that there has been a breach of security. Security managers should constantly search for such present or likely breaches. In other words, a greater emphasis on **OPSEC (Operations Security)** is required to deny vital information to terrorists.

Secondly, those who are interested in protecting certain facilities and premises should be prepared to introduce **fundamental and scientific changes in the physical security measures**. A hackneyed bureaucratic statement that perimeter security ought to be strengthened is not enough. Such pious wishes are expressed only by intellectually bankrupt senior officers. Perhaps they do not realize that most barriers can be broken by heavy vehicles—one has to think in terms of **small concrete ditches** and **drawbridges** (or for that matter, any **removable platform**) on them. The ditches, in our considered opinion, should have a rectangular cross section, and a width large enough to trap the wheels of a typical heavy vehicle only would be adequate. The removable platform can be configured in any way that is found suitable there. Anybody who still thinks in terms of drop gates and caltrops, is not fit to deal with the problem of suicide terrorism.

It must be kept in mind that intelligent terrorists would attack a campus, premises or installation only with explosives carried in vehicles. Only fools resort to shooting. But the world has a great number of fools both amongst the terrorists and the security forces. Hence we will devote some space to the eventuality of the terrorists attacking a place with guns and trying to enter inside with the intent of attacking a protected person inside or just carrying on shooting randomly.

As far as the layout of the structures and buildings inside the outer perimeter is concerned, they may have to be modified with the following objectives in mind. Assuming that somehow the terrorists are able to gain access inside, the first objective should be impeding the progress of the terrorists towards the primary target. This requires two things. Firstly, the access to the primary target should include more than one '**killing zone**', that is, spaces which are covered by the overlapping arcs of fire of more than one gun firing from vantage positions. Secondly, the layout must **deny undefended points** to the terrorists, which they could use as vantage points. This introduces a fundamental conceptual change. Presently when terrorists manage to enter a place, security forces are thrown into utter confusion and purely ad hoc gun battles are fought from whatever cover they happen to get at that time. One has only to recall the instances when terrorists entered Indian army camps or camps of the paramilitary forces. On all such occasions utter confusion has prevailed. Gun battles have lasted for day's altogether because the terrorists take up positions in rooms etc. and keep on firing at the rate of a few rounds per day just to keep the forces engaged. The terrorists have nearly always succeeded in inflicting considerable damage before getting killed. And when they eventually get killed it is not because of any effective security system or tactic—it is only because the terrorists have got limited ammunition whereas the forces have virtually unlimited ammunition. We insist that for places under threat of suicide attacks, there ought to be pre-selected strong points inside the premises (besides the outer perimeter) from where the intruders may be engaged or, at least, held at bay till reinforcements arrive. Elementary things like access control, CCTVs, scanners, metal detectors, proper location of reception counter, etc., need not be emphasized here.

❑❑❑

Chapter 33

Car Bombs: The Terrorist's Most Devastating Weapon

(33.1) CAR BOMBS OR VBIEDs VEHICLE BORNE IEDs

(33.1.1) History's First Car Bomb Was Made By An American!

The resistance fighters in Iraq are the current hate symbol of the Americans and the American propaganda machine would have you believe that the Iraqi fighters are synonymous with the car bomb. Few people would know that it was an American who had ushered in the era of car bombs by using the precursor of a car bomb way back in the 1920s.

On September 16, 1920, a few months after the arrest of his comrades Sacco and Vanzetti, a vengeful Italian anarchist named Mario Buda parked his horse-drawn wagon near the corner of Wall and Broad Streets, directly across the J. P. Morgan Company, the world famous financial firm. It was a delivery wagon common in those days and hence did not arouse any suspicion. Buda nonchalantly climbed down and disappeared, unnoticed, into the lunchtime crowd. A few blocks away, a startled postal worker found strange leaflets warning: "Free the Political Prisoners or it will be Sure Death for All of You!" They were signed: "American Anarchist Fighters". Meanwhile, the bells of nearby Trinity Church had begun to toll at noon. When they stopped, the wagon packed with an estimated 100 lbs of dynamite and a much larger load of shrapnel exploded. The shrapnel were made by carefully cutting lead sash weights in half.

The Americans were used to small bundles of dynamite that had been liberally used in the Wild West and criminals elsewhere. But this was a different ball game altogether. Nobody outside the military had ever seen anything of this type. As Paul Avrich, the celebrated historian of American anarchism who uncovered the true story describes: "Glass showered down from office windows, and awnings twelve stories above the street burst into flames. People fled in terror as a great cloud of dust enveloped the area. In Morgan's offices, Thomas Joyce of the securities department fell dead on his desk amidst rubble of plaster and walls. Scores of bodies littered the streets outside."

Buda must have been utterly disappointed when he learnt that J.P. Morgan himself was not among the 39 dead and more than 200 injured—Morgan was away in Scotland at his hunting lodge. Nonetheless, a poor immigrant with some stolen dynamite, a pile of scrap metal, and an old horse had managed to bring unprecedented terror to the sanctum sanctorum of American capitalism.

Fig. 232 History's First Car Bomb: The Scene At The Site Of Explosion In Front Of The J. P. Morgan Company At The Wall Street, New York

(33.1.2) And Then Others Took Over To Create The Poor Man's Air Force!

Buda's wagon was, in essence, the prototype of a car bomb: the first use of an inconspicuous vehicle, anonymous in almost any urban setting, to transport large quantities of high explosive into precise range of a high-value target. After Buda the world had to wait till January 12, 1947 when the Stern Gang drove a truckload of explosives into a British police station in Haifa, Palestine, killing 4 and injuring 140. The Stern Gang (a pro-fascist splinter group led by Avraham Stern that broke away from the right-wing Zionist paramilitary Irgun) would soon use truck and car bombs to kill Palestinians as well, a creative atrocity immediately reciprocated by British deserters fighting on the side of Palestinian nationalists.

There was a lull after that in the use of vehicle bombs, we don't really know why. There were a few instances like Saigon (1952), Algiers (1962), and Palermo (1963) — but the thing got a revival in a big way in 1972, when the Provisional **Irish Republican Army (IRA)** accidentally, so the legend goes, improvised the first **ANFO car bomb**. These new-generation car bombs, requiring freely available fertilizer and fuel, were very cheap to fabricate. They elevated urban terrorism to an industrial level from the level of an artisan!

The car bomb thus became a semi-strategic weapon that quite like aerial bombs was able to bring down huge structures. After all it was the truck bomb attack on the US Marine Landing Team barracks in Beirut in 1983 that eventually forced the Americans to get the hell out of Lebanon, their fighters and warships of the Sixth Fleet notwithstanding. Two-hundred-and-forty-one Marines were killed in the attack! It is estimated that in this attack the truck was carrying about 12,000 to

18,000 lbs of high explosive—almost as much as a Daisy Cutter bomb (or perhaps even more) that only a nation like the US possesses. Precisely for this reason car bombs have been described as the **poor man's air force**!

(33.1.3) The Technology Of Car Bombs

There are basically two types of car bombs. They are differentiated by the intended mode of use and not so much by the design of the initiating devices. Car bombs have been made using all the initiating devices that we have discussed so far. As you may recall, Timothy McVeigh used something as simple as safety fuse to trigger his bomb used in the Oklahoma City Center bombing. No electrical or electronic device was used in his bomb! In the first type the explosive-laden vehicle is left **stationary** at some place and then detonated either by command or by a time delay device. In this case the car bomb is just a large bomb with the car or the vehicle providing a large and yet inconspicuous container for the huge quantity of explosives. People have used such diverse vehicles as even a concrete mixer! In the other type, the **moving car bomb** is used as the **tool of a suicide attack to force its way** in a protected location such as in the attack on the US Marine barracks in Beirut. In this type, use is made of the strength of the body of the car and its momentum in crashing through barriers. For this type of attack, heavy vehicles are naturally preferred.

Fig. 233 Backscatter Image Of A Car: The Bomb Can Be Seen In The Boot

When a car bomb is being used in the suicide mode, then the terrorist has to contend with the possibility of getting shot even before he has the time to command detonate the bomb. To counter this, they have devised several means. One of the simplest means is to provide a sound-actuated switch which triggers by a certain sound which the terrorist can emit even as he is dying. The bomb can be given a time delay or a 'dead man's switch' also. Alternatively these switches can be provided as a back up to the command actuation so that they swing into operation only when command detonation fails or when the engine of the vehicle stops due to bullets or grenades fired on it. There is one more way of ensuring detonation. The detonation can be made remote controlled and the remote can be operated by a man outside the area of operation but in sight of the proceedings. In fact several American authorities suspected that this last method was used in the attack on the Marines' barracks. What is of more concern to the terrorist is that even if the driver is shot up, the vehicle should keep on moving in a certain direction for it to reach where it originally intended to reach. This is not very difficult because if the vehicle intends to ram into a barrier or barriers, it must be moving at a high speed in a straight line. Some armor plating on the vehicle has been tried to protect the driver but unless done in a professional manner, that makes the vehicle conspicuous and thus takes away the element of surprise. To ensure that the vehicle keeps on moving irrespective of the condition of the driver, they have devised mechanisms by which both the steering and the accelerator pedal can be jammed so that even the death of the driver does not affect the motion of the vehicle. Security forces which think that the best way of dealing with a car boring down on them is to shoot at the driver, may be in for some rude surprise!

A third use of large truck bombs is also possible. As we saw, the truck that attacked the US Marines' barracks was carrying 12,000 to 18,000 lbs of high explosive. Used properly, this kind of charge can be extremely effective **for sabotage** of a very serious kind. A truck full of explosives, if exploded inside an important tunnel could inflict irreparable damage to it and may well put it out of action for long. Remember that the tunnel could be an arterial one linking two important areas and its closure or loss could have catastrophic consequences even during normal times, not to speak of during a war. The 2.5 mile long Salang tunnel in Afghanistan, the world's highest road tunnel, is one such arterial tunnel that connects northern Afghanistan with southern Afghanistan. During the Soviet occupation, there was a devastating fire in this tunnel in 1982 in which 1000 to 2000 people had perished. A Soviet military convoy was passing through the tunnel at that time on its way to the south. The exact cause of the fire could never be ascertained given the cult of secrecy amongst the Soviets. Several theories were advanced. One was that a fuel tanker had exploded; the other was that a munitions truck had exploded as a result of another vehicle crashing into it. Yet another theory was that the munitions truck was actually attacked by the mujahideen and that's why it exploded. The tunnel's entrances and ventilation system were destroyed again in 1996 in the course of the fighting between the Taliban and the Northern Alliance and it could be cleared for traffic only in 2002. Whether the mujahideen had a hand in the Salang tunnel fire of 1986 remains uncertain but it is certain that an attack of that sort can be very much made with catastrophic consequences. Security planners must therefore be ever alive to the possibility of such an attack.

Fig. 234 A Car Bomb Recovered Intact In Iraq: The High Explosive Charge Of The Bomb Is Made Of Unexploded Artillery Shells: Cans Of Fuel Can Also Be Seen— They Are Intended To Build Up A Fire: The Fire Makes It Look Impressive Too

(33.2) WHAT MAKES THE CAR BOMB SUCH AN ATTRACTIVE OPTION?

What makes the car bomb such a favorite weapon of the bomber? Well, from the point of view of the bomber, there are several reasons.

- The first is the **enormous load of explosives** that can be used in it. Trucks, vans, or even SUVs can easily transport the equivalent of several conventional 1,000-pound bombs to the doorstep of a prime target. There is no other way in which the bomber can use this quantity of explosives.
- The second is **stealth**. Who knows what you are carrying in a vehicle? No government can check all the vehicles plying on the roads. And nothing prevents the bomber from using even a trailer full of explosives either.
- Thirdly, it is **dirt cheap,** thanks to the introduction of the ANFO car bomb. Fertilizer and fuel worth just $400 are enough to take care of 40-50 people. Ramzi Yousef, the mastermind of the 1993 attack on the World Trade Center, bragged that his most expensive outlay was in long-distance phone calls. The explosive itself (one half ton of urea) cost $3,615 plus the $59 per day rental for a ten-foot-long Ryder van. Compare it with the Tomahawk cruise missiles that have become the classic American weapon for striking what they call terrorist targets abroad with pinpoint accuracy—they cost $1.5 million each!
- Fourthly, car bombings are **operationally simple** to organize. You just have to recall what just two individuals Timothy McVeigh and Terry Nichols could do in the Oklahoma City bombing and with what resources!
- Fifthly, if the purpose is not very specific and it is **just a terror weapon**, even then it is the ideal choice. Explode a parked car in a market and you inflict casualties indiscriminately.
- Sixthly, car bombs are **highly anonymous** and leave practically no forensic evidence unless the bomber is fool enough to use a certain pattern of cars or explosives. Even in the 1920s, Buda had managed to quietly slip out to home in Italy, leaving William Burns, J. Edgar Hoover, and the Bureau of Investigation (later, to be renamed the FBI) to make fools of themselves as they chased one false lead after another for a decade.
- Seventhly, the car bomb lends itself to **safe and assured delivery**. The IRA had made a habit of kidnapping the family of one of the employees of the target and then forcing him to take the car to the target. This also ensures that the car will get past the usual checks.
- Eighth, car bombs have the great advantage of having one of the most **sophisticated guidance systems** available: human beings. This allows for last-minute changes to the location and detonation of the explosive as conditions dictate.
- Ninth, the mobility of car bombs also affords certain **flexibility in operation** to groups that employ them; if the location or circumstances of the target change prior to the execution of the operation, car bombs grant their user an ability to adjust and develop alternate plans in a way that stationary IEDs do not.
- Tenth, they allow the bomb to be **assembled at a safe location**, for example, and then driven to the target area, as opposed to 'roadside bombs' that have to be placed on routes routinely traveled by enemy personnel.

Fig. 235 The Scene Of Destruction At The Site Of A Car Bomb Explosion In Iraq

The **effectivity of the car bomb** can be judged from the following figures. In **Iraq** alone, in the first seven-and-a-half months of the year 2007 (up to August 15) there have been 36 major incidents of car bombings in which they have inflicted a whopping 2082 **casualties** in dead; an average of nearly 58 persons per attack! No other means of attack could be so consistently rewarding in terms of the casualties inflicted. Out of these 36 attacks, 11 were **truck bombs** which inflicted a casualty of 1067, that is, at an average of 97 casualties per attack. This also shows what can be done with a large amount of explosives in a truck.

The details are given below:

Date	Location	Deaths	Type
August 14, 2007	Yazidi compounds in Kahtaniya, Iraq	500	truck bombs (4)
August 1, 2007	Baghdad, Iraq	74	truck bomb, car bombs (2)
July 26, 2007	Baghdad, Iraq	92	car bomb
July 25, 2007	Baghdad, Iraq	50	car bombs (2)
July 16, 2007	Kirkuk, Iraq	86	car bombs (3)
July 7, 2007	Armili, Iraq	156	truck bomb
June 19, 2007	Baghdad, Iraq	75	car bomb
May 13, 2007	Makhmur, Iraq	50	truck bomb
May 6, 2007	Baghdad, Iraq	35	car bomb
April 28, 2007	Karbala, Iraq	68	car bomb
April 18, 2007	Baghdad, Iraq	198	car bombs (4)
April 14, 2007	Karbala, Iraq	65	car bomb

contd. ...

... contd.

Date	Location	Deaths	Type
March 29, 2007	Khalis, Iraq	47	car bombs (3)
March 27, 2007	Tal Afar, Iraq	152	truck bombs (2)
February 18, 2007	Baghdad, Iraq	62	car bombs (2)
February 12, 2007	Baghdad, Iraq	79	car bombs (3)
February 3, 2007	Baghdad, Iraq	135	truck bomb
January 22, 2007	Baghdad, Iraq	88	car bombs (2)
January 16, 2007	Baghdad, Iraq	70	car bomb

Fig. 236 American Soldiers Examine The Wreckage Of A Car In Which A Bomb Had Exploded

(33.3) DEFENSE AGAINST CAR BOMBS

The only surefire defense against car bombs is to not allow cars at all in the vicinity. Nations usually do it for the residences of their highly protected persons or for very important offices. The next best option is to increase the standoff distance as much as possible under the given circumstances. Standoff distance is the distance that is allowed to be maintained between the target and the last point accessible to a potential vehicle bomb. This depends on where your building is located. If you are erecting a new building then this can, of course, be incorporated in the site layout itself. The options get limited if the building is located in a congested area. The basic principles of security in respect of car bombs for a target having a campus of its own pertain to strength and negotiability of barriers and to restricting the speed of incoming vehicles. We have described the sort of barriers that are needed in terms of strength and difficulty of negotiating in connection with suicide bombs earlier. Restricting the speed is a relatively simpler affair. You can provide for narrow approach roads which would permit only vehicles of less than a certain width, or design them in such a way that they force the vehicles to take sharp turns, thus reducing their speed. Road obstacles too are useful if they have been intelligently emplaced and can withstand assault by heavy vehicles. For example, if obstacles like concrete blocks are placed in a serpentine fashion, they can easily frustrate a speeding vehicle.

If you find a car bomb that has not yet exploded or is about to explode, try to keep a safe distance. The very minimum distance you should find is 600 feet.

Vehicle type	TNT equivalent explosive	Building evacuation distance	Outdoor evacuation distance
Compact Sedan	500 lbs	320 feet	1500 feet
Sedan	1,000 lbs	400 ft	1,750 ft
Passenger/Cargo Van	4,000 lbs	640 ft	2,750 ft
Small Moving Van/ Delivery Truck	10,000 lbs	860 ft	3,750 ft
Moving Van/Water Truck	30,000 lbs	1,240 ft	6,500 ft
Semi trailer	60,000 lbs	1,570 ft	7,000 ft

(33.4) BOOBY TRAPPED CAR BOMBS: NO LIMITS TO INGENUITY

Booby trapped car bombs have been around since the gangland slayings of the 1930s. They are useful for criminals as tools of assassination but the terrorists may find it difficult to get access to the cars of targets of their interest, such targets being usually protected persons. Terrorists use booby trapped cars mostly for planting bombs in cars which are left at some place to be picked up by the police or other people. The technology of booby trapped cars has come a long way since the days of the gangsters planting six sticks of dynamite under the hood of the rival gang's car and connecting the charge to the ignition. Now a modern car has an electrical/ electronic 'nervous' system that can be tapped along many points. A basic knowledge of electronics is required to access it and then the bomb can be planted in many ways.

(33.4.1) Elimination Of Wiring

They have applied mind to devising methods by which the bomb could be planted in the shortest time and without the tampering becoming all-too-evident. For example, they have discovered a method in which no wiring is used at all with the bomb. Now there is a glue available in the market that conducts electricity. The bomber need only affix the circuit components to a board and run the glue to the components, thus creating the circuit. If the device is X-rayed, the bomb technician would be, in keeping with his training and conditioning of the thought process, would be looking for the tell-tale wiring of the bomb but would not find any. Many bomb disposal experts don't even know that such a glue exists.

(33.4.2) Bypassing The Ignition Coil And Using Other Sources

Most of the intelligent bombers work on the habits and psychological make-up of the target with respect to car bombs. Intelligent bombers know that most targets expect the bomb to be wired with the ignition that is designed to explode the moment the ignition is turned on. Hence if they have to plant a bomb they would not connect it to the ignition. They would rather connect it to the **headlamps or the horn** so that the car may get started and going but the bomb would explode the moment the headlights were switched on or the horn used. In fact they need not tamper with the wires at the lamp or the horn. They can take the connection from the fuse corresponding to the headlamps or the horn from the fuse box itself. An unwary target, even if he carried out a cursory check would find that the ignition coil is unmolested and that the car is starting normally.

Few people are likely to check the wires from the headlamps, the horn or the fuse box! He would, in all probability, abandon further check the moment the engine started.

Another technique is to use the explosive charge with a non-electrical detonator. Some match heads could be glued to the other end of the fuse. This could be then taped to the exhaust manifold. When the manifold will get hot enough, the match heads would ignite and in turn ignite the fuse.

(33.4.3) Discouraging Police Investigation Of A Suspect Car

Terrorists are known to have devised several methods for discouraging police investigation of a car suspected to have a bomb. The simplest way is to install an anti-disturbance switch. They have also used proximity sensors—the bomb goes off the moment somebody comes close to it.

Fig. 237 This Is One Car Bomb Which Had More Of Fuel And Less Of High Explosive: The Burning Car May Look Terrifying But The Actual Damage Caused Is Limited To The Car Itself And Little Damage To The Surroundings Results: More Suited For Assassination By Criminals Than For a Militarily Useful Strike

(33.5) DEALING WITH LPG TANKER FIRES

When LPG (Liquid Petroleum Gas) containers are engulfed in fire, they can explode. Though there is nothing like the detonation of a high explosive in it, the sheer volume of the LPG is so high that a devastating fireball results. The phenomenon is known as **BLEVE (boiling liquid expanding vapor explosion)**. As the liquid burns the top of the tank becomes very hot and is weakened. Eventually the high pressure liquid causes it to rupture and explode. This could happen accidentally as well as deliberately by attaching a sufficient amount of bursting charge with incendiaries to it. Then it becomes an act of sabotage. Fools would continue to shoot cylinders and tankers hoping that they would explode; they never do. The subject has been discussed at greater length later.

Vehicle	LPG Mass or volume	Fireball diameter	Safe distance
Small LPG Tank	20 lbs/5 gal	40 ft	160 ft
Large LPG Tank	100 lbs/25 gal	69 ft	276 ft
Commercial/Residential LPG Tank	2,000 lbs/500 gal	184 ft	736 ft
Small LPG Truck	8,000 lbs/2,000 gal	292 ft	1,168 ft
Semi tanker LPG	40,000 lbs/10,000 gal	499 ft	1,996 ft

Prevention of BLEVE accidents is, however, possible. It is based upon the principle that the liquid at the bottom of the tank is cooler than the liquid/gas at the top. There is a pressure relief valve at the top. A pump is fitted to that and it is driven by the pressure of the vapor being released. It sprays cold liquid LPG from the bottom of the tank onto the inner surface of the top side. That cools it down and reduces the chances of the top rupturing and exploding.

❑❑❑

Chapter 34

Getting into the War Mode: Terrorists' Use of IEDs Capable of Being Used in a Military Context

(34.1) IEDs MADE WITH A MILITARY PURPOSE IN MIND

IEDs made with a military purpose in mind must be distinguished from IEDs intended for purely terrorist activities with essentially 'soft' targets in mind. To most morons it comes as a surprise. We thought of including this chapter to educate the people, the intelligence community and the security forces that the making of even those IEDs which can be used quite effectively in a military context or for fighting a guerilla war is not at all difficult and people have already done it. Should the security forces ever come to face such devices during counter-insurgency operations they should not panic over the discovery and start imagining that they have been attacked by regular soldiers! Ignorance breeds fear. Fear leads to debacles. We therefore wish to clear the ignorance.

(34.2) IMPROVISED CLAYMORE MINE

A Claymore mine, for example, has little use for say, urban terrorism. The terrorist would be happier exploding a bomb at a bus stand which he could conceal in any ordinary bag and just leave it behind unnoticed. It would be impossible for him to properly site a Claymore mine there. The Claymore has to be exposed and erected on its legs—a luxury which an urban terrorist cannot have. On the other hand, as we have seen, the Claymore is of great use to those who are fighting a guerilla war. They may find themselves in a jungle laying an ambush where they have all the time to plant it. Given the intricate nature of terrorism in the modern era, it does not really follow a set pattern of progression first from the organization phase to terrorism phase and then from the guerilla warfare phase to the mobile warfare phase. These days several phases are found to exist simultaneously. The LTTE for example, is waging a mobile warfare for all practical purposes. And yet they frequently resort to acts of pure terrorism. So do the Kashmiri terrorists. They are basically in the terrorism phase in the sense that most of their attacks include throwing of grenades in markets, and the like. Yet at times they have laid ambushes on army convoys and attacked army camps very much in the manner of guerilla warfare. It is all a matter of expedience. Since they are not proper ideological insurgencies, they do whatever is expedient for them at that moment.

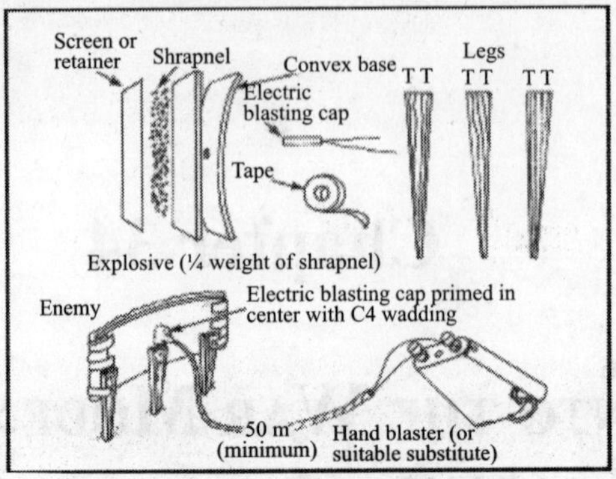

Fig. 238 The Basic Principle Of Constructing An Improvised Claymore Mine

It is not at all difficult to make an improvised claymore mine. People have written entire books on it. For the improvised claymore device or more precisely a grapeshot anti-personnel device with or without a directional effect, a plexiglass sheet (4" x 8.5" in size and ¼ or 1/8 inches thick) is used for making the slightly concave (or convex, whichever way you look at it) shaped mold for the shrapnel. The sheet is first given the desired shape in a clamp and then heated gently to make it retain the shape. Double-stick tape mounted on the mold is used for placing the layer of the rows of ball bearings (say 702 balls of 7/32 inch size) on the mold. Then the bearings are bound in place by epoxy material like Devcon Plastic Steel Liquid. Once the epoxy sets, the tape and the plexiglass mold both can be peeled off to yield a properly shaped layer of ball bearings that is encased in a steel-hard matrix of epoxy. The back of the epoxy matrix is covered with a heavy duty tin foil. If they do not have plastic explosive then three sticks of 80% commercial dynamite are slightly warmed and then pressed on the frame so that they conform to its shape. They are held in place using the same plexiglass sheet and the whole thing is bound with duct tape. Fixing 5/16 inch thick and 10 inches long improvised legs by tape does not present a problem. The detonator is inserted in the middle of the 8.5" side of the explosive.

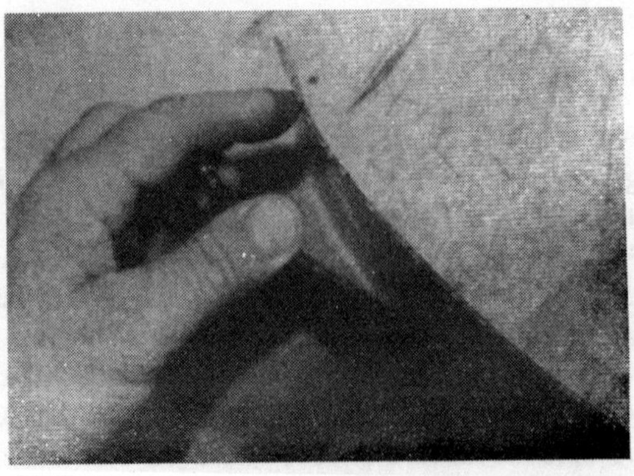

Fig. 239 Improvised Claymore Mine: Ball Bearings Encased In A Steel-Hard Matrix Of Epoxy Are Being Removed From The Plexiglass Frame

Fig. 240 Improvised Claymore Mine: Three Sticks Of Dynamite Pressed On The Ball Bearing Matrix To Conform To Its Shape

Fairly simple, isn't it? They have applied some more technical mind to it. Normal ball bearings are made of hardened steel which tend to either break or deform and thus adversely affect the flight characteristics. Intelligent bombers therefore prefer to anneal them so as to make them a little milder. In a pinch one could use any shrapnel but that makes the device less effective. Some directional effect can be achieved by heavy tamping on one side when it is sited.

(34.3) IEDs MADE BY CANNIBALIZING ARTILLERY SHELLS AND OTHER MILITARY WARHEADS

(34.3.1) Anti-Tank And Anti-Personnel IEDs

An artillery shell, mortar bomb or the warhead of a shoulder-fired rocket provides a readymade source of factory-made high explosive of excellent quality. Not only that, the explosive is packed in a case which is designed for proper fragmentation! Thus two of the biggest worries of making IEDs are taken care of. What more he could ever ask for? You have high explosive and you have proper containment also. A large number of unexploded ordnance is easily found in a war zone. Even unexploded aerial bombs can be found. Non-state actors look for them and make use of them. The Iraqi resistance fighters, for example, are making good use of them. The Americans are discovering scores of them. The Iraqis have been using even anti-tank mines as IEDs because they can convert them into command-actuated devices, a role in which they would be more useful than as a merely dumb anti-tank mine waiting for some tank to come up, and which may never come up in an urban situation. When command-detonated, it can be effectively used against other types of vehicles also.

Fig. 241 IEDs Made From Artillery HE Shells And Anti-Tank Mines Recovered In Iraq

The catch lies in removing the fuze of the warhead safely. That demands some specialized tools and some mechanical skill. Even those tools can be improvised, however. Once that is done, priming the shell with one's own detonators is easy and one has his own IED. He can make a booby trap device, time delay device, command actuated device or whatever he wants out of that. However, suitable detonators are required as the explosives in the shell are rather insensitive. The bomber must know the strength of detonator or the amount of detonating cord required or else he would be wasting that precious warhead!

The intended use of the IED will depend on the size of the warhead. If the bomber has got just the bomb of a 61 mm or 81 mm mortar or has taken out the warhead of a shoulder-fired rocket, he can make only an AP device. If he has got bigger things such as a 155 mm artillery shell or aerial bombs he can make anti-tank devices also. They would be great as AP devices and would work very satisfactorily as AT devices.

Fig. 242 IED Made Using Mortar Bombs Recovered In Iraq

Fig. 243 Command Detonated IED Made Using Artillery Shell And A Cordless Phone

(34.4) IEDs USING EXPLOSIVELY FORMED PENETRATORS (EFP)

EFPs have been used in improvised explosive devices against armored cars, for example in the 1989 assassination of the German banker Alfred Herrhausen (attributed to the Red Army Faction), and by the Hezbollah in the 2006 Israel-Lebanon conflict. A recent development is their widespread use in Iraq.

The charges are generally cylindrical, fabricated from commonly available metal pipe, with the forward end closed by a concave copper or steel disk-shaped liner. The explosive is loaded behind the metal liner to fill the pipe. Upon detonation, the explosive projects the liner to form a projectile at a velocity of well over 1 km/s, depending on the design and type of explosive used.

Often mounted on crash barriers at window level, they are placed along roadsides at choke points where vehicles must slow down, such as intersections and junctions. This gives the operator time to judge the moment to fire, when the vehicle is moving more slowly. Detonation is controlled by cable, radio control (RC), or remote arming with passive IR (PIR) trigger. EFPs can be deployed singly, in pairs, or in arrays, depending on the tactical situation.

(34.5) IMPROVISED PLATTER CHARGE

The platter charge consists of a suitable container that is filled with uniformly packed explosive and placed behind a platter. The platter is metal (preferably round, but square is also satisfactory) and weighs 2 to 6 lbs. The explosive required is equal to the weight of the platter. The container may not be necessary if the explosive can be held firmly against the platter—they have used even tape for the purpose. The charge needs to be primed from the exact rear center, and the blasting cap should be completely covered with a small amount of C4 to ensure proper detonation. The platter charge can be used as an AT or an AP device. When used as an AP device, it must be command-detonated.

The charge should be aimed at the direct center of the target. The effective range (primarily a matter of aim) is approximately 115 feet for a small target. With practice, experienced personnel can hit a 55-gallon drum (a relatively small target) at 82 feet with about 90 percent accuracy.

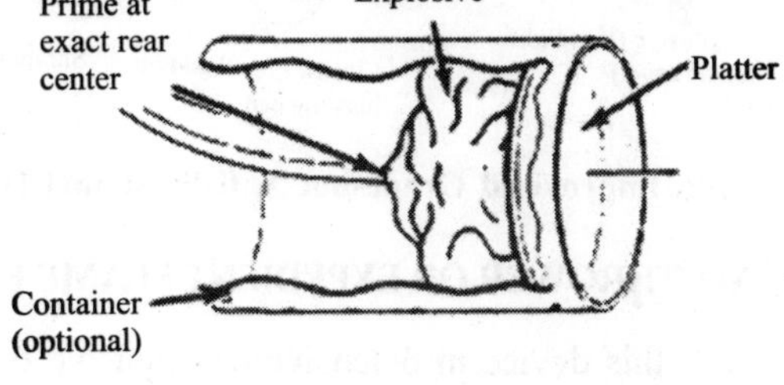

Fig. 244 Improvised Platter Charge

(34.6) IMPROVISED BARBED WIRE ANTI-PERSONNEL DEVICE

This takes the cake amongst the IEDs. A brilliant idea! One can make an excellent anti-personnel device by this stratagem. One roll of standard barbwire (weight about 50 lbs and of the size of a basketball) is placed into position, and 1.25 lbs of C4 or an equivalent quantity of some other high explosive is placed in the center of the roll and primed. This device must be command-detonated. The barbwire AP device can be made somewhat directional by placing the wire against an embankment or a fixed object. This causes the force of the explosion to expel the barbwire fragments in the desired direction.

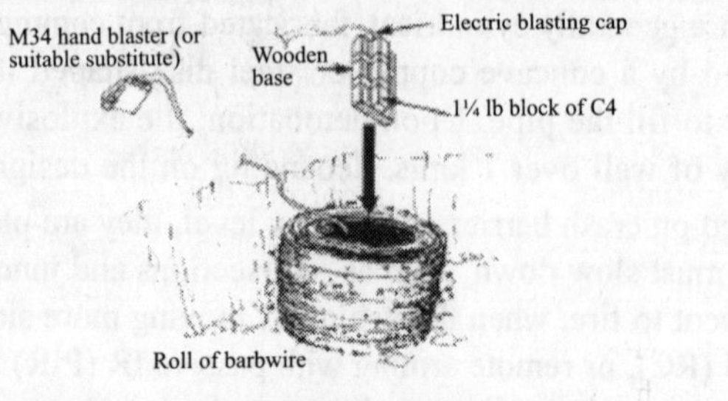

Fig. 245 One Of The Most Ingenious Inventions: The Barbed Wire Anti-Personnel Device: Incidentally An Invention Of The American Army Itself

One can also make an **improvised grapeshot anti-personnel device** by taking a metal canister, packing some shrapnel in it, using explosive charge of one-fourth of its weight packed on top of it and priming it in the center.

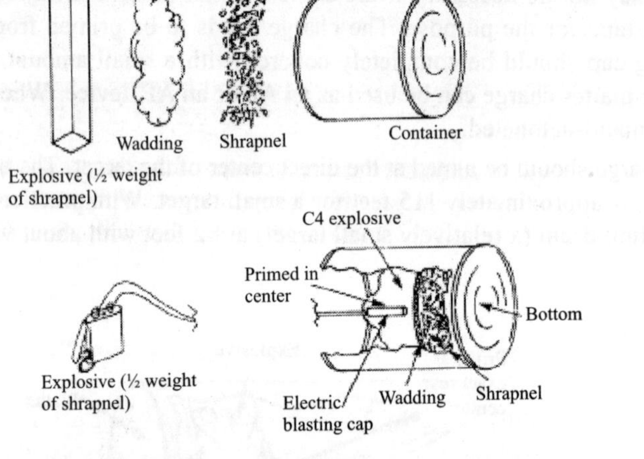

Fig. 246 Improvised Grapeshot Anti-Personnel Device

(34.7) IMPROVISED FLAME THROWER OR EXPEDIENT FLAME FOUGASSE

The terrorist can use this device in defensive or offensive operations for its incendiary effects. Uninitiated soldiers or those who have never even heard of flame throwers might get petrified at the very sight of these devices. The device is not as complicated as it would seem to be. In fact the US Army itself has worked out the problem in one of its publications. The terrorists have learnt from them and gone a step ahead in improvisation. The operation of the device is broken down to steps and then handled accordingly. The charge would consist of a 55-gallon or so drum of thickened fuel. There are two problems in using this as a flame thrower. The first is how you break open the drum so that the thickened fuel is available for dispersal. This is solved by using a detonating cord. They wrap the detonating cord several times around the top of the drum and it breaks it open cleanly without unnecessary spilling of the fuel. The next problem is to ignite the fuel. The fuel is of no use unless it is burning. Ignition can be achieved using something as simple as a trip flare. The next problem is to propel the burning fuel. It can be propelled using a kicker charge of about 2 lbs of TNT or so. Then comes the question of propelling it in a certain direction. A trench angled at less than 45 degrees is dug for the 55-gallon drum that will allow the best coverage and dispersion of the flame fougasse. They make a small cutout area in the back of the trench for the kicker charge. For using it effectively so that all these operations take place almost simultaneously, the detonating cord comes in handy. It has to be used in a ring main. When initiated, the ring main initiates the detonating cord to the trip flare, the drum top, and the kicker charge. The wraps cut the top of the drum off, the kicker charge propels the thickened fuel outward, and the trip flare ignites the thickened fuel as it travels downrange. The result is a flash of flame that spreads downrange for approximately 100 meters.

For the terrorist, the real problem lies in thickening the fuel and it demands some knowledge of chemistry. It is not really difficult and has been discussed later in connection with improvised firebombs.

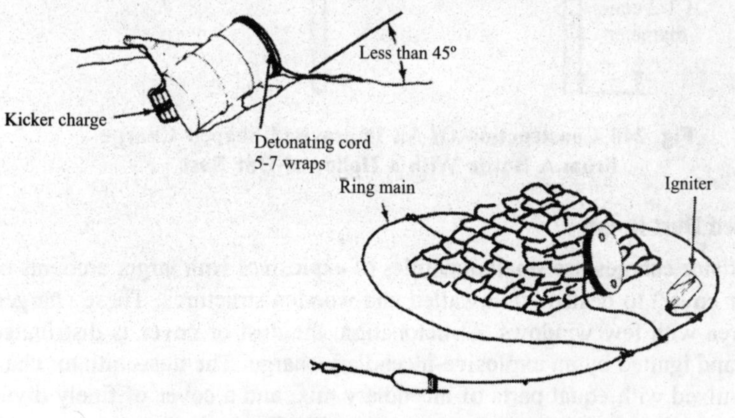

Fig. 247 Construction And Emplacement Of The Improvised Flame Thrower Or Expedient Flame Fougasse

(34.8) IMPROVISED CHARGES SUITABLE FOR DEMOLITIONS

(34.8.1) Improvised Shaped Charge

Shaped charges can be improvised but because of the many variables (configuration, explosive density, liner cavity density, and so forth), consistent results are impossible to obtain. The most popular item for making a shaped charge are bottles that come equipped with built-in cavity liners, such as champagne or cognac bottles. With the ends removed, the container is ready for a cavity liner and explosive. People have used even dynamite and molten TNT in the absence of plastic explosive. Liners are made from copper, tin or zinc. Bombers have worked out some expedient formulas for using the shaped charge. Cavity angle is sought to be 42-45 degrees but anything between 30 to 60 degrees does the job in a pinch. The explosive height is two times the cone height, measured from the base of the cone to the top of the explosive. The normal standoff distance is one and one-half cone diameters.

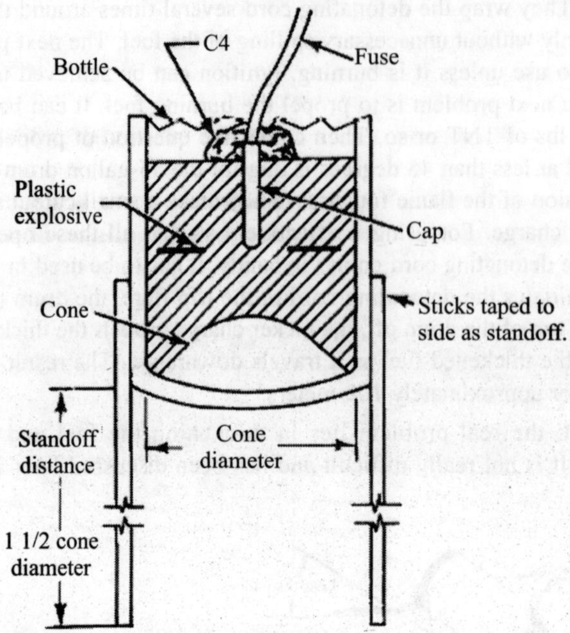

Fig. 248 Construction Of An Improvised Shaped Charge From A Bottle With a Hollowed-Out Base

(34.8.2) Improvised Dust Bomb

Dust-initiator charges use small quantities of explosives with larger amounts of powdered materials (dust or cover) to destroy thin-walled and wooden structures. These charges work best in an enclosed area with few windows. At detonation, the dust or cover is distributed in the air within the target and ignited by an explosive-incendiary charge. The dust-initiator charge consists of an explosive, mixed with equal parts of incendiary mix, and a cover of finely divided organic material. One pound of explosive-incendiary mixture will effectively detonate up to 40 pounds of cover. For making a pound of explosive-incendiary mixture, half a pound of TNT is mixed with half a pound of incendiary mix (two parts aluminum powder or magnesium powder and three parts ferric oxide). C4, in fact, cannot be used here because the explosive component in C4 will not combine properly with the incendiary mixture. The rule of thumb is that 3 to 5 pounds of cover are used for each 1,000 cubic feet of target (3 pounds for enclosed buildings, 5 pounds for partially enclosed buildings). The cover can consist of coal dust, cocoa, powdered coffee, confectioner's sugar, tapioca, wheat flour, corn starch, hard-rubber dust, aluminum powder, magnesium powder, or powdered soap.

(34.8.3) Improvised Plastic Explosives

Improvised plastic explosives have been made in a pinch using any regular high explosive and mixing it with potassium chlorate and Vaseline.

(34.9) IMPROVISED FIREBOMBS AND INCENDIARIES

In spite of what they have been showing you in the films, the traditional **Molotov Cocktail (or Molly** as the West Germans called them) consisting of a glass bottle filled with petrol, rags and a burning wick is a fairly useless device in the open. It is of use only when it is used in a confined space. Then it inflicts burn injuries and can also ignite other inflammable materials present there. For the serious terrorist, firebombs are adolescent pranks. As such we should not have even discussed them in this book. However, in deference to the immense popularity of firebombs in films and fiction, it became necessary for us to discuss the subject so that the myths are cleared.

Throwing a petrol bottle with the wick burning is always dangerous. The wick may even fall out or if you have delayed throwing, the gasoline may start burning in your hands only! The Soviets devised a little more scientific way in the WWII. It consisted of potassium permanganate added to the petrol. Just before use a glass vial containing sulfuric acid was added to the bottle. Upon impact both the vial and the bottle broke. The acid reacted with the permanganate and generated enough heat to ignite the petrol.

There is a fundamental problem, however, with all such devices. The basic Molotov Cocktail was supposed to be a simple device which could be made instantly. Using chemicals in it made it cumbersome, thus defeating its very purpose. Carrying chemicals like sulfuric acid was another problem. Chemical added devices could be made and kept, but it was difficult to use them in large numbers. With these issues in mind, Special Forces of the world and anarchists have devoted a great deal of mind to improvising the firebomb.

The first step towards improving upon the basic gasoline firebomb is to somehow **thicken the gasoline** without napalm. One simple way is to add half a bottle of engine oil to half a bottle of gasoline. The second step is to get rid of the burning rag. Then they need something which will ignite by itself once the bottle breaks. For this they have to use some sort of **hypergolic mixture**. One standard composition used throughout the world is a mixture of potassium chlorate and sugar that is made to react with sulfuric acid. It is taped in a paper packet outside the bottle. Inside the bottle they fill one-fourth part with acid and the remaining with gasoline. When the bottle breaks, the acid will react with the mixture and produce enough heat to ignite the petrol. Another mixture uses glycerin in a capsule and calcium hypochlorite—the glycerin capsule has to be crushed before throwing.

The third step towards improving the Molly is to make **gels**. The advantage of making gels is that one can revert to the old wick which makes life much simpler. Improvised gels can be made from gasoline and kerosene. Diesel is difficult to gel. Gels can be made from soap flakes; lye and castor oil; and egg whites and salt. Enterprising bombers have been able to make even the Napaim B gels using polystyrene taken from Styrofoam cups! Half a cup of soap flakes would do for a quart of gasoline. Soap flakes may require a little gentle heating if stirring alone does not work. A little alcohol (half a tsp) also helps. For the other recipe, one needs two tsp of castor oil, half tsp of alcohol and one-fourth tsp of lye flakes. It takes about two days for the mixture to get the consistency of thick honey. The gel based on two egg whites and one tsp of table salt forms within 10 minutes but does not keep for more than a day. When one is trying to make a gel from Styrofoam, he should see whether the pieces show any tendency of dissolving in the gasoline—if they don't, benzene would have to be used.

As long as someone is trying to throw a bottle of thickened fuel or gel, it is no better than a nuisance. A much more sensible idea in the military context was the invention of the **incendiary mines**. This is something the security forces should be really concerned about. They take a petrol drum and fill it with napalm mixture. Then they either bury it in the path pf advancing troops or hang them from the trees at a height of about 2 meters. You can make a chain of them and trigger them remotely. You will have to give them an igniter of white phosphorous or thermite and a high explosive bursting charge. **US forces in Korea** had made them even better by tightly winding barbed wire over the drums and hanging the mines from the trees. When the drum exploded, the barbed wire also got broken into fragments making up a large amount of shrapnel to boot.

Incendiaries, however, lend themselves eminently to sabotage operations of Special Forces and commandoes. The idea is to make a **small incendiary device** that could be surreptitiously placed on the target. Obviously this should have some delay element in it. Fieser, the inventor of napalm, himself had made such a device for the US Army in the WWII. It was known as the **Harvard Candle**. It consisted of a cylinder of celluloid filled with napalm. It was ignited by a red phosphorous composition in a metal pull cap which served as a scratching disc.

□□□

Chapter 35

Busting the Myth of the Mysterious Liquid Explosives Called TATP and HMTD

(35.1) THE VICIOUS AND MISLEADING PROPAGANDA OF THE WESTERN MEDIA

The western media has been full of utterly irresponsible and speculative reporting about two explosives, namely, TATP and HMTD following the July 2005 bombings in London. The purpose of this chapter is to bust those myths for once and all. The impression sought to be generated was that the terrorists have hit upon a secret formula using which they can make **deadly**

Fig. 249 Paranoia Of The West At Its Best: Notices Displayed Banning Liquids From Aircrafts

liquid explosives from everyday chemicals. Not only that those everyday chemicals can be carried easily on board aircrafts because they are not explosives to start with, the actual explosive would be made on board by simply mixing them there and then the aircraft would be blown up. Such was the paranoia that the morons had banned all sorts of liquids on flights for quite some time and their imbecile copycat followers back home, having no knowledge or intellect of their own, had done the same. Mothers were made to taste baby food in front of airport security guards to prove that it contained no liquid explosives. Liquids and gels were banned from flights. Travelers repacked their luggage in airports, stowing all but the most necessary items in the hold.

Fig. 250 All The Liquid Items Were Collected From Passengers And Dumped Like This

It would be interesting as well as enlightening to quote from some of the western media reports. The reports will show not only how widespread is the realm of ignorance, but they also highlight the fact that practically entire nations can choose to close their eyes to reason when it comes to even a vague reference to terrorism and that too at the hands of Islamic terrorists. It should serve as an eye-opener to those who take the words of the western media as the gospel truth.

"British authorities said Thursday they thwarted a terrorist plot to simultaneously blow up several aircraft heading to the United States using explosives smuggled in hand luggage, averting what police described as 'mass murder on an unimaginable scale'... Police arrested 21 people, saying they were confident they captured the main suspects in what U.S. officials said was a plot in its final phases that had all the earmarks of an Al-Qaida operation. U.S. President George W. Bush called it a "stark reminder" of the continued threat to the United States from extremist Muslims."

"U.S. Homeland Security Secretary Michael Chertoff said the terrorists planned to use liquid explosives disguised as beverages and other common products and set them off with detonators disguised as electronic devices... The bombs were to be assembled on the aircraft apparently using a peroxide-based solution detonated by such devices as a disposable camera or a music player, two American law enforcement officials told The Associated Press..."

"This is a new approach to destroying our aviation system using what I call clean bombs or explosive components to take down multiple aircraft," Rep. John Mica, R-Fla., chairman of the House Aviation Subcommittee says. "We face a very serious challenge in that regard. If they get to the checkpoint, our chances of detecting them are limited."

"Common liquids found in the home, including hair bleach and food preservatives, could be processed, then combined on board a plane after takeoff, to make HMTD or TATP... Hydrogen peroxide, an ingredient in many homemade bombs and the one the would-be British bombers planned to use, is found in many medicine cabinets."

"Liquid explosives are the oldest kind and can be made of ingredients found under the sink or in a bathroom cabinet—acetone found in nail polish, acids in drain cleaner or hydrogen peroxide in hair bleach—common materials that don't set off alarms at the airport."

"As many homemakers know from experience, mixing bleach (hypochlorite) with ammonia in your washer may give you cleaner clothes, but may also blow the cover off your washing machine. Mixing household bleach or chlorinated reagents used in swimming pools with many other consumer chemicals may be capable of producing a sizable explosion. In an e-mail received from Dr. Jay Siegel, Director of the Forensic and Investigative Science Program at Indiana University, he states that their laboratory mixed a small amount of the commercially available hair product, Brylcreem, with an amount of hypochlorite, and "got quite an explosion." Years earlier in Virginia, a similar experiment with pool chemicals (hypochlorite) and Brylcreem had produced a 'nasty explosion'."

"An anonymous source close to the British investigation has presumably leaked to the media that the London bombs were to be made from peroxide-based material. The most likely candidate here is TATP, triacetone triperoxide, the same material used in the London subway bombs, the shoe bomb, and by many suicide bombers. TATP is made from acetone (nail polish remover), hydrogen peroxide (disinfectant from a drug store or hospital) and sulfuric acid (hardware store cleaner). All of these would go undetected by any of the current sniffer devices used for explosives at airports. Combined in a rest room, they would undergo the reaction to form TATP."

We will show that it is all bullshit. Our analysis will expose the extremely high level of general ignorance, paranoia and above all a preconceived bias that pervades the West and of course, their copycats back home.

(35.2) THE TRUTH ABOUT TATP: WHAT IS IT AND WHAT IT CAN DO?

(35.2.1) How The Western Media And Security Agencies Latched On To TATP?

TATP is triacetone triperoxide; HMTD is hexamethylene triperoxide diamine. In the first place, we must state that **liquid explosive** is actually a misnomer given by the media. The explosives are not liquids. The ingredients used to make them are liquids.

According to Western agencies, investigation into the July 7, 2005, London mass transit attack revealed that the four suicide bombers used peroxide-based improvised explosive devices (IEDs), concealed inside rucksacks. Investigation into the attempted bombings on the London mass transit system on July 21, 2005, again indicated that the perpetrators had used rucksacks to conceal their explosive devices, and which might also have contained peroxide-based explosives. The suspicion was apparently based on the smell they detected after the tube railway blast! Since TATP has a strong smell, they concluded that if there was smell, it must have been TATP! So much so for their investigative skills! Independent experts hotly contested this claim and described it as a concoction of the western propaganda machinery. The smell could have come from many sources, such as failure in properly detonating an ammonium nitrate based composition. Either way, it becomes imperative for us to elaborate on the subject for educating the people.

Both TATP and HMTD appear as white crystalline substances, and may be indistinguishable from one another without laboratory analysis. TATP is formulated from acetone, hydrogen peroxide, and a strong acid. It is a primary explosive. HMTD is formulated from hydrogen peroxide, a weak acid (such as citric acid), and hexamine solid fuel tablets. HMTD is also a primary explosive. It is slightly more stable than TATP. The misconceptions took hold quickly because people did not care to verify facts. They thought that since acetone is a common solvent found in hardware stores, hydrogen peroxide is also easily available, and the acid used in car batteries could serve the purpose of the strong acid, there was nothing easier than making TATP. So was the case with HMTD. They thought that since hexamine fuel tablets, commonly used to fuel camp stoves, could be found in many outdoor recreation stores and thus making HMTD was also child's play.

It never occurred to the morons that had it really been so easy to make high explosives little quantities of which could bring down aircrafts, had the world to wait centuries for those four guys to discover that? They should have been rewarded for making a discovery that had eluded the galaxy of brilliant scientists who spent their lives doing research in the field of explosives!

(35.2.2) Why An Improvised Preparation Of TATP Is An Impractical Proposition?

We will first describe the **laboratory method of preparation** of TATP. TATP is a highly sensitive primary explosive and needs to be prepared very carefully. It is particularly sensitive to friction—in fact so sensitive that if a few small crystals are lodged in the cap of a bottle and if you try to screw the cap back, the resulting friction is enough to explode it.

For a proper laboratory preparation of TATP both the hydrogen peroxide (35%) and the acetone must be cooled for several hours in a refrigerator. Then you must prepare an ice bath for the glass beaker. If the quantity of acetone in your possession is sufficient, it may be desirable to utilize it in a more effective form of cold bath for the reaction. Instead of an ice bath, one may substitute a container filled with acetone, cooled by small amounts of dry ice. This mixture can be cooled to temperatures very well below the normal freezing point of water, thus making it possible to more easily restrict the temperature of the reaction. The cooling is achieved by placing the dry ice in small amounts at a time into the solution and allowing them to sublime, thus cooling the acetone. Next you must put 0.68 ml hydrogen peroxide into the beaker in the ice bath, and add 0.95 ml chilled acetone. Measure the temperature of the liquids with a thermometer and wait for it to drop to around 4.4 degrees. Then slowly add 0.24 ml sulfuric acid. The respective amounts may, of course, be changed proportionately provided you can handle them safely. Monitor the temperature closely while adding the acid as the temperature rises. If the temperature gets around 10 degrees, quit adding the sulfuric acid. Failure to quit adding the sulfuric acid when the temperature rises may result in an explosion. If the acid is added too quickly, it will not have time to create the proper peroxysulfuric acid and this is what leads to the explosion. When all of the acid has been added, stir the mixture for about 15 minutes. After 15 minutes, place the reaction mixture into a refrigerator for about 24 hours. A white crystalline solid will result. This crystalline mass can then be filtered out with filter paper. When the crystalline mass has been filtered, pour 6.3 ml of water over it to wash the acid residue away. Set the crystalline mass out to dry. The crystalline mass is tricycloacetone peroxide. If you carried out the reaction at above 10 degrees, you will get the much unpleasant dicycloacetone peroxide. The acid-catalyzed peroxidation of acetone leads to up to three forms of acetone peroxide, including a monomer, cyclic dimer, and a cyclic trimer. The trimer is generally thought to be marginally more stable than the other two and is formed preferentially when the temperature of the reaction and subsequent product formation is kept below 10 degrees Celsius. Some of the more unstable dimer is always found associated with the trimer, and TATP degrades over long-term storage, making it unsuitable for conventional explosive ordnance.

It should be obvious to even a layman from the above that an improvised method of preparation is an impractical proposition for the simple reason that the proper procedure cannot be resorted to under the field conditions, that is, on board an aircraft! Further, it would result in a product that would, in all probability, explode right there in the preparation and handling process. By an improvised method we mean hurrying through the reactions and that too at room temperature. After all if, as the western media and the security agencies claimed, the bombers intended to make it on board aircrafts—they would certainly not get laboratory equipment or refrigerators there! And more importantly, how much ingredients they could smuggle on board and how much TATP they would get out of it? Making even a few grams would be extremely difficult. And they suspected that aircrafts would be blown up! No example could be more illustrative of the absurdly biased, irrational, unscientific and paranoid attitude of the West.

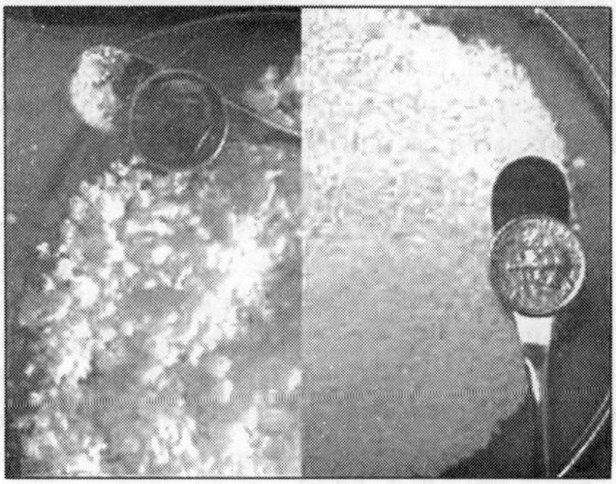

Fig. 251 The Crude Form Of TATP Contrasted With Its Pure Form

(35.2.3) Why Even Pure And Laboratory-Prepared TATP Cannot Be Used To Blow Up An Aircraft?

Even if you prepare TATP carefully in the laboratory, this explosive is not suitable for blowing up an aircraft! Why? The morons in the West and their imbecile copycats back home never knew that TATP is an **unusual explosive**. We normally think of explosive materials, like fuels, as highly energetic compounds, releasing their energy content in fast, exothermic reactions; that is, in the form of heat and gases. The heat makes the gases develop the high pressure that results in the blast. Thus the key property of all explosives, high or low, is the production of heat. Although TATP does burn when it is set alight, releasing carbon dioxide and water, very little heat is created when it explosively decomposes. You get the catch? TATP is an explosive alright in the sense that it is a highly unstable compound, but it does not produce heat when it decomposes explosively. That's why we called it an unusual explosive.

The explosion process of TATP is a slightly complicated thing for those who do not have a background of physical chemistry. Still, we would try to explain it in simple terms. TATP actually undergoes an **entropic explosion**, each molecule of TATP collapsing to produce three molecules of acetone and one of ozone. An entropic explosion is an explosion in which the reactants undergo a large change in volume without releasing a large amount of heat. It is somewhat like the rapid reaction that produces gas in the safety air-bags of cars during accidents.

The three isopropylidene units of the TATP molecule do not play the role of fuel that can be oxidized and release energy during the explosion. Instead, these units function as molecular scaffolds that hold the three peroxide units close together spatially in the appropriate orientation for the decomposition chain reaction. One proof of this unusual mechanism of explosion is the post-detonation residue. You analyze it and you will find large quantities of acetone. Now, had the decomposition of TATP been similar to the heat-releasing (exothermic) decomposition of other explosives we should have got oxidation products. Here you don't get any oxidation products. Rather you get the parent chemical acetone! Scientists have confirmed it in another way also. It has been found that the heat released by breaking of the peroxide bonds is just not enough to drive a chain reaction that could lead to explosion. The methyl groups of TATP do not act as a fuel that could get oxidized and release thermal energy in the process. Instead, the methyl groups act as a scaffold, keeping the peroxide units close to one another and oriented properly for the explosive decomposition reaction.

So what the British were afraid of? The main fear of British security agencies, as the Americans maintained, was that the terrorists planned to hide **micro-bombs** in false bottoms built into opaque energy drink bottles, enabling them to still drink the contents and thus convince a bystander that the bottle was genuine. Then the device would have been exploded using a detonator concealed in the flash mechanism of a disposable camera to puncture a hole in the aircraft skin. MP3 players or electronic key fobs could also have been used to trigger an explosion.

We have shown above that TATP is useless as an explosive for bringing down an aircraft because it does not release heat. And who requires a detonator for a primary explosive like this? Did they not know even that much? Setting it alight with the batteries of mobile phones, iPods, etc. (as the western security agencies suspected) is very much possible, given the extreme sensitivity of TATP but it is utterly puerile to even suggest that it could burn through the skin of a plane and bring it down. The heat produced is simply not enough for the job! The skin of the aircraft is able to bear the stresses of flying at speeds close to the speed of sound, take off and landing with its own enormous weight and the load of the passengers and the cargo. And someone thinks the skin of an aircraft is so weak that it would give way to what these ignoramuses describe as micro-bombs concealed in the false bottoms of bottles of energy drinks! They should give the bombers the Nobel Prize in chemistry for making such a wonderful discovery! If anybody is claimed to have tried that, as the western media and security agencies say, we can only lament that both are fools—the bomber for trying to use an ineffective thing and the target for believing in the stupidity of the bomber! Stupidity of the bomber can be understood for a moment—after all they were ordinary half-educated guys with no knowledge of chemistry. Stupidity of the security agencies is unpardonable. There are infinitely better ways of bringing down an aircraft than sneaking in a bottle of the ingredients and trying to make TATP in the toilet of the aircraft. In all likelihood the bomber would injure himself and nothing else.

This is precisely what is believed to have happened to **Henry Hinrichs**, a mechanical engineering student at the **University of Oklahoma** in October 2005. According to investigating agencies, Hinrichs was killed when about two-three lbs of TATP exploded in his lap. Incidentally, he was about 200 yards away from the Oklahoma Memorial Stadium where over 84,000 spectators were watching a football game. The paranoid Americans claimed that the explosion could be heard five miles away and there was considerable panic thinking it to be a terrorist attack. Since his roommate was a Pakistani called Fazal Cheema the Americans immediately jumped to conclusions and claimed that there had to be a terrorist connection and that Hinrichs had secretly converted to Islam and had been going to a mosque also. Subsequently, all these turned out to be absolute falsehoods. Hinrichs was found to be a depressed and lonely individual who loved to dabble, like many deranged American adolescents, in amateur bomb-making and had been going to Red Rock Canyon in Caddo County to test his bombs made in plastic soda bottles! Seventy-five minutes before his death he had typed on his computer "None of you are worth living with. You can all kiss my ass". The message was still on the screen when agents entered his apartment. The presence of the TATP in his backpack was confirmed by detailed tests—initially the imbeciles thought that it was hydrogen peroxide! TATP was found in his apartment also. The bomb was detonated by a battery. The remains of the backpack contained a circuit board, wires, and a battery. Other items at the bomb site included a crescent wrench, a white sock with protruding wires, a screwdriver, unused wooden matches, and a chemistry book. The incident is highly illustrative; both of the paranoia of the Americans and also of the fate which befalls those amateur bombers who venture into a field they know only little about. Hinrichs was incidentally a National Merit Scholar. You should be convinced now that explosives are a different ball game altogether and engineering or a few books on chemistry do not make you into an explosive expert! There was no doubt about the intentions of Hinrichs; the poor fool did not know, however, that with two-three pounds of TATP all he could accomplish was to kill himself—and he had thought that he could blow up a stadium! Look at the similarity with the foolishness of the Glasgow bomber Kafeel and you would be amazed at the universality of ignorance—higher education notwithstanding! Hinrichs got killed because he had over two pounds of TATP—an amount which is not possible to be made in liquids smuggled aboard aircrafts. And yet they think that it could blow up aircrafts!

Modern aircrafts do not fall down by some small hole punched in the sides. And even that is impossible with the few ounces of **wet TATP** that can be made in the toilet. The whole idea is utterly childish! Filtration takes a long time and how could one possibly dry the TATP even if he succeeded in making a little of it in the toilet? Wet TATP would not be of much use in any case.

Thus the fears were not only unfounded, they betrayed a deeply ingrained hostility which precluded entire nations from seeing reason. How it came about? There are several issues involved in it. The first is plain and simple ignorance. Ignorance is far more widespread than you can imagine. The media people and the security agencies had no idea of the unusual explosive property of TATP and also the impracticality of using it as a high explosive. Opinions these days are made by populist writings, TV and the like—people have neither the intellectual capability nor the patience to grasp sober scientific opinion. Then everybody loves sensation. People love to panic. It really thrilled them to no end that the Islamic terrorists had hit upon a secret formula using which they could make deadly explosives from daily household items. The pursuit of sensation effectively buries any sane voice. As for the intelligence and security agencies back home, they were simply and utterly stupid.

Imbecility knows no bounds—the same security agencies had shot dead an absolutely innocent Muslim a few days after the tube railway bombing when, from his puffy jacket, it appeared to them that he was a suicide bomber!

The stupidity of the Western media and the so-called experts did not end there. In a search of the suspects' homes following the 7/7 attack, British police claimed to have found low-concentration hydrogen peroxide, sugar solution, peanut butter, and in one instance, nine tins of baked beans under a suspect's bed. They had to concoct some theory to make the discovery fit into their general story, that is., these ingredients were intended for bomb-making purposes. It was, of course, not even considered that perhaps the guy just liked peanut butter! Somebody had in fact gone to the extent of saying that common household ingredients such as peanut butter and sugar solution could be modified to provide the fuel and oxygen needed to make explosives. Later somebody realized that it was indeed crossing the limits of commonsense. Then another theory was advanced that the suspects might have intended to use the containers of peanut butter and baked beans to make the bombs look innocuous. The tins of baked beans were X-rayed, and said one official, "By golly, they all had baked beans!" Incidentally, the German police also arrested some people in September 2007 on the charge of keeping hydrogen peroxide!

(35.2.4) TATP Is Not Undetectable Either

One great fear associated with the rumor of TATP was that it was undetectable. We agree that since peroxide-based explosives are so materially different from the whole line-up of high explosives in use that most of the common explosive detectors are unable to detect them. Moreover, if the TATP was to be prepared on board aircrafts then the chemicals required to make it, namely, acetone, hydrogen peroxide and sulfuric acid are not even explosives. We also agree that post-blast analysis of these explosives has proven problematic, as the residues left behind are short-lived. Is that a cause for alarm? No!

TATP can be detected and it should not be flaunted as an undetectable explosive. **Desorption electrospray ionization (DESI)** technique, for example, can be used to detect trace amounts of TATP on paper, metal and brick, and in complex matrices such as methanol, vinegar, and diesel. No sample preparation is required, it gives low nanogram detection limits and it takes less than five seconds to obtain spectra. DESI works by directing a pneumatically-assisted electrospray (cloud of tiny, highly charged droplets) onto a surface. The charged micro-droplets interact with analyte molecules on the surface, generating secondary ions that are collected and analyzed. Other detection technologies based on chemical reactions are also available. A trace analysis can also be done by **High Performance Liquid Chromatography-Atmospheric Pressure Chemical Ionization-Tandem Spectrometry (HPLC-APCI-TS)**.

(35.3) WHAT IS HMTD AND IS IT ANY BETTER?

Hexamethylene triperoxide diamine, or HMTD is also a primary explosive, first synthesized in 1885 by **Legler**. The theorized structure lent itself well to acting as an initiating or primary explosive. While still quite sensitive to shock and friction, it was relatively stable compared to other initiating explosives of the time, such as mercury fulminate, and proved to be relatively inexpensive and easy to synthesize. As such, it was quickly taken up as a primary explosive in mining applications. However, it has since been superseded by much better compounds. HMTD degrades too quickly for modern commercial and industrial applications, becoming useless in a matter of weeks.

We are, however, mentioning it because of late it has caught the fancy of the West as an explosive popular with the terrorists. Why is it so? The reason is simple. It is the ease of manufacture lending itself to making at home. The easiest synthesis of HMTD involves unrestricted ingredients including hydrogen peroxide, citric acid or dilute sulfuric acid as a catalyst and hexamine fuel tablets, which are available from camping supply stores or with military rations.

As is the case with TATP, the laboratory method of preparation of HMTD also is rather cumbersome. Fourteen grams of hexamethylenetetramine is dissolved in 45 grams of 30% hydrogen peroxide and stirred mechanically at zero degrees. Twenty-one grams of powdered citric acid is slowly added to it while stirring continuously. The mixture is stirred for 3 hours at zero degrees and then allowed to warm to room temperature and stand for 2 hours. The white crystalline product is filtered off, thoroughly washed with water, rinsed with methanol, and air dried. If more than double the recommended amount is made at once, the product will decompose exothermally while the solution warms. HMTD may be stored under water with no explosion risk and no increase in the rate of decomposition. It can be seen that hasty preparation in the field would not be possible. Hence all the objections which are there for TATP's use as explosive against aircrafts apply to HMTD also.

HMTD is slightly more powerful than TATP and its behavior is a little more predictable than TATP because it has got only one isomer against three of TATP. When one makes TATP, different batches will have different ratios of isomers and hence they will behave differently. Still, that does not make it into an explosive ideal for bringing down aircrafts. HMTD is not to be mixed with many other substances such as metals as it will react to form highly unstable products that may lead to detonation with out any aggravation. HMTD is also UV sensitive and should avoid sunlight.

❑❑❑

Chapter 36

Explosives and IEDs Made by Amateurs

(36.1) IMPROVISED LOW EXPLOSIVES USED BY AMATEUR BOMBERS

Amateur bombers who cannot make even military grade gunpowder are obliged to make improvisations for getting their explosives. We are mentioning them for the sake of completeness so that the reader who happens to find them somewhere is not flummoxed. It has been our personal experience that the level of ignorance to be found in intelligence agencies and security forces is so high that many of such explosives have been described as powerful explosives, radically new explosives or explosives made with chemical warfare angle in mind. In our considered opinion, anyone who uses such mixtures deserves to be beaten up first for degrading the art of bomb-making and insulting all those geniuses who spent their lives researching explosives. Their basic chemical idea is common—there is an oxidizer and some fuel or combustible substance. (All proportions are by weight.)

(36.1.1) Chlorate Mixtures

They consist of 1 part potassium chlorate, 1 part sugar or 90% of potassium chlorate and 10% of hydrocarbons such as 7% of wax and 3% of Vaseline. The latter composition was called **Minelite B** by the French a century ago and when pressed to a density of 1.3 g/c.c., it was capable of giving a fairly respectable detonation velocity of about 12,000 feet per second. In the so-called **Composition M**, they use 9 parts of potassium perchlorate and 1 part of Vaseline. The secret lay in compressing the mixture, something beyond the capacity of the amateur bomber.

(36.1.2) Flash Powders

They consist of 4 parts potassium perchlorate, 1 part aluminum powder, 1 part sulfur or 7 part potassium perchlorate and 5 parts aluminum powder. In a pinch, people have used potassium chlorate also instead of potassium perchlorate but that's a poor substitute.

(36.1.3) Bangor (Firecracker Powder)

It consists of 6 parts potassium nitrate, 3 parts aluminum powder, 1 part sulfur or 4 parts potassium nitrate, 1 part aluminum powder, and 1 part sulfur.

(36.1.4) Permanganate Powder

It consists of 3 parts potassium permanganate and 2 parts aluminum powder.

(36.1.5) Ammonpulver

This is a poor cousin of the ANFO explosives. It consists of 85% ammonium nitrate and 15% charcoal powder, both easily available items.

(36.1.6) Ammonium Nitrate And Aluminum Powder

This is a clever idea and perhaps the best bet of the amateur bomber. 82% ammonium nitrate and 18% aluminum powder makes it a fairly good explosive in terms of power. Amateur bombers use a 75/25 mixture also. The ammonium nitrate must, however, be absolutely dry. It is hygroscopic and the problem is that moist ammonium nitrate in the presence of aluminum tends to produce hydrogen which can easily cause accidents. It is therefore difficult to store. Ammonium nitrate in this has to be used in powder form and not in the prill form people are used to. This mixture lends itself to detonation by a blasting cap. It is believed that this mixture was used in rather large quantities in three incidents of bombings of apartment complexes in Moscow in 1997.

(36.1.7) Ammonium Nitrate Gel Explosive

Essentially a modification of the above, it is made by taking 55% of ammonium nitrate, 15% of aluminum powder, 10% of potassium nitrate, 5% of sugar, 2.5% of guar gum, 1% of borax and 11.5% of water. The storage quality of the gel improves by adding small quantities of urea, monobasic potassium phosphate and lye.

(36.1.8) Arsenic Sulfide And Potassium Chlorate Mixture

Those who want to make a bomb-like thing which can be thrown without lighting anything and which will explode on impact, use mixtures based on **tetraarsenic tetrasulfide** (red orpiment or realgar). The device is popular because it can be hurled and it will explode if it strikes a hard surface—a matter of great convenience! Arsenic sulfide is mixed with potassium chlorate and some grit for providing the necessary friction. The grit doubles up as improvised shrapnel also. The mixture is wrapped first in paper and then bound in layers of twine. Sometimes some red phosphorus, sugar, kerosene or sulfur are also added for more heat and smoke should they so desire for psychological effect, though they are not absolutely necessary. This mixture has been particularly popular with small-time criminals in India for very long. Even the so-called terrorists in India have been found using it but being absolutely ignorant of the history of this mixture, the intelligence agencies thought that the terrorists have invented this mixture so that the so-called bomb could be used to produce toxic fumes of the oxides of arsenic upon explosion and thus have a **'chemical warfare effect'**! The intelligence community and the security forces, out of their sheer ignorance, attributed a devious design to the terrorists which has never existed in the first place! The terrorists merely copied an age-old practice of the criminals because they are really dumb and the intelligence community thought they have invented a new thing! Arsenic sulfide is freely available in this part having been used as a pigment. It is used even in Ayurvedic medicine! The Sanskrit name for the compound is Manahshila and in Hindi it is known as Mainsil or Manseel. The device is quite risky to use and innumerous accidents have occurred with such country-made bombs. The binding of the paper packet with a couple of layers of twine serves two purposes. Firstly it provides for containment of the mixture so as to increase the explosive effect. Secondly, it reduces the impact sensitivity of the device to some extent. Still, it has been frequently seen that if someone is carrying a couple of them in a cloth bag hung from the handle of a bicycle, the devices explode merely by the banging of the bag with the frame of the bicycle!

(36.2) SOMEWHAT BETTER EXPLOSIVES USED BY A TAD BETTER BOMBERS

(36.2.1) CO-OP Sugar Or Rack-a-Rock Explosives

The Rack-a-Rock explosive mixtures were discovered in the 19th century by a man called **Silas Divine**. He patented a number of mixtures. Some of them are:

- 3-4 parts potassium chlorate, 1 part nitrobenzene
- 3-4 parts potassium chlorate, 1 part nitrobenzene, 1-3% sulfur added after mixing the two
- 8 parts potassium chlorate, 1 part turpentine
- 9 parts potassium chlorate, 1 part kerosene

The Provisional Irish Republican Army made a slight variation of Rack-a-Rock and called it CO-OP Sugar. This is 10 parts sodium chlorate and 1 part nitrobenzene. This turns out to be more powerful than ANFO and equally easy to make. It also, of course, requires a detonator. The IRA used it for their large car bombs and culvert bombs. Sometimes they laced detonating cord through the mixture to boost the detonation velocity. Since nitrobenzene is quite volatile, this mixture needs to be sealed in air-tight containers.

(36.3) THE PIPE BOMB

We are mentioning the pipe bomb in view of its historical importance and immense popularity. Amateur bombers all over the world have used them. The most elementary of the IEDs,

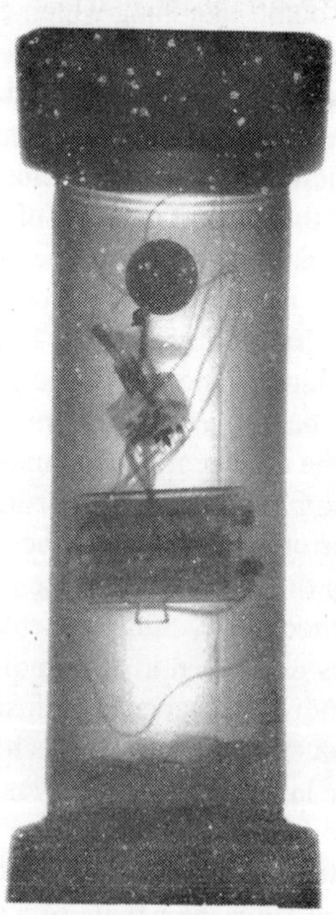

Fig. 253 X-ray Of An Electrically Detonated Pipe Bomb

a pipe bomb is nothing but a piece of GI pipe the ends of which are closed by regular pipe plugs or threaded caps. Bondo (auto body filler) plugs can also be used instead of the caps. Those who are so poor or unskilled that they have to resort to making a pipe bomb, fill it with gunpowder. A hole is drilled in one of the caps to insert the fuse. If the bomber manages some HE, he has to procure a detonator as well to go inside. Pipe bombs can be made better. One can provide for a squib and a timer device also. A squib, as you would recall, is a small metal tube which is filed with say, gunpowder. The powder is ignited by a filament such as that of a flashlight bulb. The squib provides a jet of flame to ignite the main charge. Current to the filament can be given by a command circuit or by a timer device. But one who takes all this trouble could as well take some more trouble and get some HE to make a proper bomb; for serious bombers making a pipe bomb is a matter of grave insult.

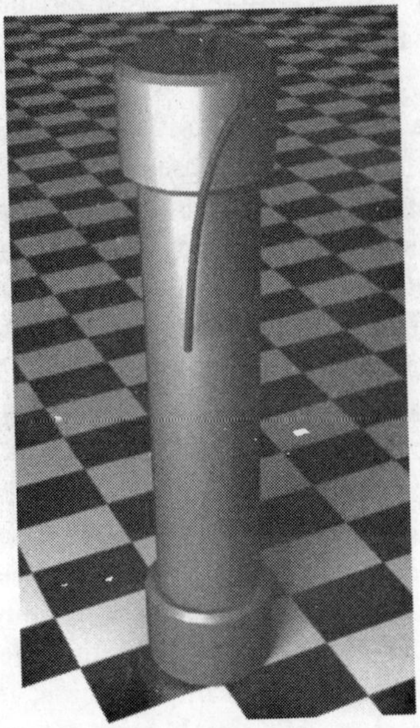

Fig. 252 The Most Elementary Fuse Lit Pipe Bomb

Still there are people who, for reasons of expedience, fill their HE also in a metal pipe. Actually making a fairly strong container this way is so easy and hence the choice, particularly if the amount of explosive is small. Such bombers can use all sorts of actuation devices with the bombs.

(36.3.1) Pressure Cooker / Lunch Box Bomb

It is this quest of a strong container which makes some people use the pressure cooker or a steel lunch box for their HE, gunpowder or whatever. Another advantage is that the containers are inconspicuous.

(36.4) IEDs BASED ON LPG CYLINDERS: WHY THEY FAIL TO WORK?

This is a dream which fools all over the world have been chasing with regularity and, of course, without any success. Had they been wiser, they would have spared themselves the trouble.

Anyone who thinks that he could create a fuel-air explosive by taping a stick of dynamite to the LPG cylinder is sadly mistaken. Fuel-air explosives, as we have seen, are a different ball game altogether. All such people need to be pitied for their ignorance and beaten up for their audacity. Contrary to popular perception and the ignorance of the security forces and the intelligence agencies, it is neither easy nor efficient to make a good FAE bomb out of a LPG cylinder. The onslaught of ignorance is universal; be it the matter of TATP or LPG cylinder bomb, the ignoramuses are too eager to believe that the terrorists have stumbled upon a cheap, simple and yet absolutely devastating formula. How people love to delude themselves!

In such devices, the LPG cylinder is supposed to provide the explosive and it can be rigged in numerous ways for command actuation. The perceived advantage of the LPG cylinder is that one gets about 14 kg of highly inflammable gas. One requires a charge to break the cylinder and ignite the gas. But will it really explode? All of us are aware that if a cylinder has been leaking, it results in a fire accident. Can that be replicated in a bomb?

Accidents involving LPG cylinders are of two types. In most of the accidents that take place in households, the defective cylinder keeps on leaking for quite some time say, overnight. This allows the gas to mix with air comfortably. Subsequently when the mixture catches fire either as a result of someone trying to light the burner in the morning or some spark produced in a switch when it is turned on in the morning, the already prepared gas-air mixture catches fire. Since a large volume is already prepared and catches fire, the gases produced by combustion may be large enough in volume to result in an explosion which is particularly damaging to weak structures like wooden frames, partitions and tin/asbestos roofing etc.

Another type of accident takes place when the cylinder itself is exposed to fire from outside. What happens is that fire reaches the cylinder and starts heating it. As a result, the pressure in the gas cylinder rises in direct proportion to its temperature. If the internal pressure exceeds the mechanical limitations of the cylinder and there are no means to safely vent the pressurized gas to the atmosphere, the vessel will fail mechanically. If the vessel contents are ignitable, this event may result in a 'fireball'. This type of mechanical failure is commonly referred to as a **Boiling Liquid Expanding Vapour Explosion** (BLEVE). To cope with this they are provided with valves that allow the contents of the cylinder to escape in a controlled fashion, lessening the risk of explosion. Many common pressure relief devices are of a simple burst disc type. In these, a small burst disc is installed in the back of the valve. A burst disc is a small metal gasket engineered to rupture at a pre-determined pressure. Some of these burst disc are backed with a low-melting-point metal, so that the valve must be exposed to excessive heat before the burst disc can rupture. In other words, the process of rupture is spread over some time. That naturally reduces the violence of the explosion.

For the LPG cylinder to be converted into a bomb all the gas must burn simultaneously. This would require that all of the gas be released in one go, allowed to mix with the proper amount of oxygen, and then the mixture ignited simultaneously throughout its volume. The entire process must take place very rapidly as against accidents which take quite some time to develop. Is it possible to achieve this? For all practical purposes and within the limitations of amateur bombers, no!

To understand it properly, you must first understand how a flammable gas like LPG burns. The gas requires oxygen to burn. How important it is? Let us first have a look at the physics of the Bunsen burner used in laboratories all over the world or even at the gas cooking ranges in your kitchens. Have you ever thought why you require the gas burner at all? Could they have not simply attached a metal tube to the cylinder and you could have lighted at the other end? Would such a device work satisfactorily? No! Why? The answer is to be found in the working of the Bunsen burner. In a Bunsen burner so commonly found in the laboratories and their everyday versions to be found in the kitchens in the form of gas burner cooking ranges, the amount of air (or rather oxygen) mixed with the gas stream affects the completeness of the combustion reaction in the flame and it can be regulated. Less air yields an incomplete and thus cooler reaction, while a gas stream well mixed with air provides oxygen in an equimolar amount and thus a complete and hotter reaction. You can see it yourself from the color of the flame. When the combustion is incomplete, the flame is yellow and sooty; when the combustion is complete, the flame is blue. If more air can mix with the gas before combustion, the flame will burn hotter, appearing blue as a result. If the holes are closed, the gas will only mix with the air available at the point of combustion, that is, only after it has exited the tube at the top. The reduced amount of oxygen produces an incomplete reaction, producing a cooler but brighter yellow flame which is often called the 'safety flame'. The yellow flame is luminous due to small soot particles in the flame which are heated to incandescence. When the holes are opened to allow more air to mix with the gas, the burner produces a hot, blue flame that can be nearly invisible against some backgrounds. In the cooking ranges of kitchens, however, the vent of air is kept permanently fixed at an optimal level so that the cooking range offers the best combustion all the time. One cannot afford to do experiments in the kitchen. The laboratory Bunsen burner has however, provision for controlling the amount of air that is mixed with the gas and hence the effect of differing amounts can be seen easily. The Bunsen burner has, of course, one more control, common with the kitchen cooking range which controls the rate of flow of the gas. Increasing the amount of fuel gas flow through the tube by opening the needle valve will increase the size of the flame as long as you are not increasing it greatly. If you increase it substantially then unless the airflow is adjusted as well, the flame temperature will decrease because an increased amount of gas is now mixed with the same amount of air, starving the flame of oxygen. Cooking ranges keep the variation within reasonable limits.

For a LPG cylinder bomb you cannot have the luxury of a mixture that has already been prepared and kept ready for ignition as it happens with the LPG cylinder in the kitchen that has been leaking all night. For a bomb you require the mixture to be produced instantaneously on command. How can you do this? You can rupture the cylinder with a bursting charge of high explosive. Still, the problem is that for a sudden rupture and release of the gas in a large volume, this desired mixing with oxygen does not take place efficiently. In a Bunsen burner or the burner of your cooking range, when the gas exits from the nozzle, it comes out in the form of a jet. Since the jet has gas moving at a high speed, this makes its pressure low in accordance with Bernoulli's Principle. The air from the sides, being at a higher pressure than the gas jet, rushes in through the holes to the area of low pressure and is automatically mixed with the gas. How can such a mixing take place when a large amount of gas is suddenly released at one place by bursting open a cylinder? Then there is something known as flash point. The fuel-air mixture needs to be heated to above the flash point of the fuel-air mixture. If you burst the cylinder then the sudden release interferes with the heating of the gas-air mixture above the flash point. You see, the bursting charge, being a high explosive, would release its heat suddenly and it may not be enough to heat up the entire mixture dispersed in a large volume. To maintain the temperature above the flash point some incendiary may also be necessary. Propane, for example, has ignition temperatures in the range of 493 to 604 degrees Celsius. This means that a cloud of propane would ignite simultaneously everywhere only if temperatures in this range could be ensured everywhere. Accomplishing this is not easy. Now you understand why all the nations have not been able to make FAEs.

The mythical LPG cylinder bomb would therefore fail to explode. Yes, the gas will burn and there will be a lot of fire. But there will not be any explosion of the type you get with a FAE bomb. This is precisely what happened at the Glasgow airport in 2007. All that the bomber succeeded in was burning himself to death. It also shows that a degree in aeronautical engineering or pursuing a Ph.D. in computational fluid dynamics does not imbue one with knowledge of explosives! Security forces and intelligence agencies should therefore not get shit scared if they find that somebody holding a bachelor's degree in science, some certificate in electronics, some degree in engineering or for that matter even a Ph.D. degree holder has joined the ranks of the terrorists.

Fig. 254 Some Fools Persist In Rigging A Charge To A LPG Cylinder Hoping To Make A Devastating Fuel-Air Explosive Bomb Cheaply—Unfortunately Life Is Not That Simple And The Device Does Not Work As Expected

(36.4.1) Would LPG / Gasoline Tankers Explode If They Are Shot Through ?

Precisely for the reasons discussed above, and against what they have been showing you in films, even if somebody shoots through LPG / Propane cylinders, LPG tankers, or gasoline tankers, they would not explode. A bullet will result in only a small leakage in the first place. It is not that the contents of the tanker are going to get spilled out as a result of just one or even a couple of shots. Secondly, the bullet is so small that the total amount of heat carried by it is very low. How much heat do you think the bullet could be carrying in its body? After all it was just a small amount of powder that propelled it and the contact with the hot gases produced by the powder was for a very short time—the time it takes for a bullet to exit the barrel! Thirdly, the bullet is moving at a very high speed. Hence if it passes through the tanker's skin, it will be in contact with the gas for a very short time. In that short time of contact, it can deposit only a small fraction of the little heat it is carrying. Thus a bullet or several bullets also will not be able to heat up the great volume of gas substantially. A small tanker also carries about 12,000 liters of gas! Even if the bullet gets lodged, there is simply not enough heat in it. If the bullet is an incendiary or a tracer which gets lodged with its tracer composition still burning then the chances of a fire are increased. But that too only for the fuel which is leaking. You see, oxygen is available only to the gas which is leaking from the holes created and not to the entire tank! If a tank is full, there is simply no oxygen in it. Even if is not full, the amount of oxygen available is not enough for the fuel present. You cannot fire so many bullets into the tank that it practically gives way. Thus it is only the leaking gas which could burn if the incendiaries or tracers have deposited enough heat at the leakage point. This in itself is not a very likely event because unless the tanker wall is of such a critical thickness that it permits the incendiary or tracer bullet to just penetrate and then be there; the bullet will go deep inside, losing heat as it paves its way through the gas, and not deposit enough heat at the hole it has punctured in the skin. Thus all you can get is fire or fires at the punctured holes. There is no way the entire tank is going to catch fire, not to speak of exploding. It can happen only if there is fire at several holes simultaneously and the fire from the punctured holes is allowed to be there for such a long time that a BLEVE can take place. Tankers, as we discussed above, have technical means to deal with BLEVE and don't you think someone would be dousing the fire in the meantime?

Had it been so easy to make gas tanks explode by shooting and considering that there are hundreds of thousands of tankers that ply across the length and breadth of any country absolutely unprotected, the terrorists would have been utter fools to not shoot them from hundreds of yards away and watch the flames leaping? And yet those who close their eyes to reason think that this is indeed possible.

Courtesy the world of films, the myth of a gas tank exploding as a result of a bullet shot in it or a charge strapped to it is so deep-rooted and so highly prevalent that even premier agencies believe in it. The FBI suspected that the Al Qaeda could infiltrate its members as fuel tanker drivers and explode them in populated areas to cause mass casualties. CBS News had reported the following in August, 2005: "FBI Warns Oil/Fuel Trucks May Be Bombs On Wheels: The FBI warned police on Thursday that Al Qaeda cells may use fuel tankers as weapons on Los Angeles, NY and Chicago. The warning first went out Aug. 9 and officials stated tanker drivers may be pulled over today by police to check licenses and registrations. The Transportation Security Administration has considered fingerprinting all truckers who drive hazardous materials, but with 2.7 million drivers under that category, the cost to fingerprint them all would be astronomical—roughly $540 million."

The report is highly illustrative. It shows not only pitiable ignorance but also the terrible paranoia that has consumed the Americans and, of course, their ignorant followers back home. Had they been watching even 'The Mythbusters' on the Discovery Channel in their own country, probably they would not have issued such a stupid alert!

□□□

Appendix A

JAMMERS

(A.1) WHAT ARE JAMMERS?

Jammers are devices that jam radio communications in an area and hence the name. In simple words, jammers are themselves radio devices that produce a wide range of radio frequencies. When you are making radio communication, your signals are ordered in such a way that they carry some meaning. Jammers, on the other hand, produce radio noise. That is to say, their signals do not have any meaning. The noise drowns the meaningful signals. This is known as jamming of the signals. Cell phone jammers are now common things. Since irresponsible people or people lacking elementary courtesies refuse to switch off their mobiles even at places where a call would be a nuisance, many organizations use cell phone jammers to jam cell phone communications during meeting or some other important work. Corporations use jammers to stop corporate espionage by blocking voice transmissions and photo transmissions from camera phones. In technical terms, jammers form a part of **ECM (Electronic Counter Measures)**.

Jammers create their noise in a wide range of radio frequencies, such as from 20 MHz to 522 MHz. Since they are commercially available, hence the frequency range in which they emit the noise is not a secret. Security forces which jealously guard the frequencies of their commercial jammers as if they are a state secret are simply intellectually bankrupt. They are deluding themselves with a false sense of security. Devices for 20 to 3000 MHz are also available and they announce so, such as the **Spot-barrage RCIED Jammer of Shoghi Communicatins Ltd**. The **TRC 274 jamming system of Thales Communcations** covers the frequency spectrum from 1.5 MHz to 3000 MHz! In the first place, the frequency range cannot be kept secret for a commercial product. Then perhaps the security forces don't realize that if it comes to that, one can arrange for a jammer to be purchased and analyze its output electronically to determine its frequency range. Many companies restrict their sales to law enforcement agencies but fooling them is not a problem for a determined group. Hoping that one could save himself from a bomber merely by keeping his jammer's frequency range a secret is puerile. The cult of secrecy is therefore nothing but the result of a clerical mind-set which is neither developed nor amenable to reason.

Fig. 255 The TRC 274 Jamming System Of Thales Communcations

Why do the jammers have to produce a wide range of frequencies? A wide range of frequencies needs to be produced if you do not know the exact frequencies which may have to be jammed. If you know the frequencies correctly then, of course, you have to produce only those frequencies to jam them. For example, since the frequencies of cell phone transmission and reception are known, a cell phone jammer becomes a simple device.

By way of illustration we are giving the technical specifications of the Spot-Barrage RCIED Jammer of Shoghi Communications Ltd. from their own literature available in the public domain.

Frequency Range	**20 - 3000 MHz**
Power Output	
20-500 MHz Frequency	Upto 50/200/500 W
500-3000 MHz Frequency	Upto 50/200 W
Jamming Modulation	Digital Noise, FM
Barrage Jamming Bandwidth	Interference Spectrum band can be set by operator at a rate of 25KHz to a barrage signal in the range of 100-3000MHz with resolution of 100KHz
Jamming Method	Interference generator frequency synthesizer variation setup(not more than 20KHz)
Type of modulation	AM, FSK, PSK, PSK2, PSK4, and Noise
Frequency change time	Time to generate an interference signal from the moment of receiving control command is <100ms
Programmable Channels	Simultaneous generation upto 4 interference signals

contd. ...

... contd.

Frequency Range	20 - 3000 MHz
MTBF	20,000 hours
Harmonic Attenuation	>40 dB
Operating Temperature	-10°C to +55°C
Remote Control	From PC
Built-in Test	Yes
MIL-Specifications	Yes
Weight	Not more than 70kg
Power Supply	24V DC

Fig. 256 The EADS SGS 2000 Jamming System Of Norway Called 'Tor', Capable Of jamming From 1.5 MHz To 1 GHz (Or 3 GHz If Required)

(A.2) WHY JAMMERS ARE OF INTEREST TO SECURITY PLANNERS?

Jammers are of interest to security planners because potential bombers might be using radio-controlled devices to trigger bombs. By using jammers, security planners hope to achieve two things. The jammer will either be able to jam the communication to such an extent that the receiver of the IED will not get any meaningful signal at all. However, it may so happen that the IED is designed to get triggered by any signal on the correct frequency, irrespective of whether it is meaningful or not. In such an eventuality, it could be hoped that one of the frequencies of the jammer would match with the frequency of the IED and thus it would trigger the IED prematurely.

(A.3) HOW JAMMERS WORK AND WHAT ARE THEIR LIMITATIONS?

Jammers could work in two ways. Firstly, they could simply drown the bomber's radio signal in their own signals of different frequencies and higher signal strength. The phenomenon is very simple. Try tuning your radio receiver to some weak and remote station during a thunderstorm. You will not be able to get clear reception even if you sit in a soundproof room. Why? The transmission of that remote and small station is weak in any case. Then the transmission is drowned in the background radio noise created by the lightning. What is noise? As we stated above, noise is a meaningless signal. That is to say, in technical terms it is a random signal. Noise has to be sufficiently strong to create a problem. Any spark for that matter would produce radio noise. Even the spark plugs of cars produce radio noise. That noise, however, is not strong enough to cause serious disturbance. Lightning is a much bigger spark and its radio noise is very strong. In technical terms the noise-to-signal ratio at the input level (that is the antenna of the receiver) becomes so high that the receiver cannot extract any meaningful intelligence from it. Thus if the noise signal of the jammer is strong enough, the receiver of the IED would not be able to selectively pick up sufficient signals of the desired frequency so as to result in an adequate meaningful output. For this purpose it is necessary that the power output of the jammer be in the range of say, 300 W per frequency band. If your power is low then the effective radius in which jamming is found, may be reduced to a few feet. Such jammers are used, for example, to jam the cell phones in a conference room only. You need a much larger radius if you are trying to protect say, a military convoy. Theoretically one could design excellent noise filtering circuits to filter out the noise. But they are not easy to make. It does not cost much to the security planners to take a chance and hope that the bomber has not designed some very good noise filtering circuits.

The other eventuality is to expect that even if the IED receiver does not require any meaningful output to trigger the bomb, the jammer would trigger off any radio-controlled RCIED because the frequency used by the IED transmitter and the receiver could exactly match with one of the frequencies emitted by the jammer. It would therefore result in a premature detonation of the IED and defeat the purpose of the bomber. A jammer is therefore to be used by an advance pilot vehicle that might get sacrificed in the process.

Jammers would obviously work only against those radio-controlled RCIEDs that use the frequency in the region emitted by the jammer. If someone makes a radio-controlled RCIED which uses a frequency below or above the range of the jammer's spectrum the jammer would be a total waste. This may happen at times because many cheap commercial jammers cover only the VHF spectrum hoping that the bomber would be using only VHF sets for triggering his devices. They leave the HF or the UHF spectra open in the hope that the bomber would not have enough knowledge or technical skill to himself fabricate HF or UHF sets. Having a very wide range of jamming frequencies has problems of its own. Then your own radio communications including cellular phones could get jammed. Having no radio communication at all is yet another security risk which makes security forces uncomfortable. It is for this reason that some frequency band is left open for their own communication. For example, if you jam VHF then you can use UHF for your own communication. The moment they leave some narrow band open, the bomber too can use that band or channel. All he requires is scanning of the radio spectrum. Alternatively, if the RCIED is designed to be triggered by things other than radio waves for remote control (such as visible light, infrared rays or even ultrasound); the jammer would not once again be of any use. The jammer should therefore be not thought of as th ultimate protection against RCIEDs.

(A.4) BASIC TECHNIQUES OF JAMMING

(A.4.1) Barrage Jamming

Barrage jamming is the jamming of multiple frequencies at once by a single jammer. The barrage jammer is based on the argument that it is nearly impossible to match exactly a jammer frequency to that of a radiating source; therefore, it is usually necessary to broaden the bandwidth of the noise so that it is greater than the source bandwidth. A barrage jammer has therefore a very wide bandwidth to cover all sources with frequencies in that band. But this broadening of the jammer bandwidth causes the jammer to require more power than one that is exactly matched, because the power that matters for any source is the power that is accepted by the receiver.

Barrage jamming may be accomplished by presetting multiple jammers on adjacent frequencies, by using a single wideband transmitter, or by using a transmitter capable of frequency sweep fast enough to appear radiating simultaneously over wide band (for example, a carcinotron). Barrage jamming makes it possible to jam emitters on different frequencies simultaneously and reduces the need for operator assistance or complex control equipment. These advantages are, however, gained at the expense of reduced jamming power at any given frequency. When jamming radars, barrage jamming is the method of choice particularly against frequency-agile radars, which change frequencies too quickly to follow them in a conventional way.

Fig. 257 The VIP-300 U/O Barrage Jammer Of Homeland Security Strategies, Inc. Designed Specifically For RCIEDs

(A.4.2) Sweep Jamming

In sweep jamming the jammer's full power is shifted from one frequency to another using time division multiplexing. While this has the advantage of being able to jam multiple frequencies in quick succession, it does not affect them all at the same time. Yet in practice the fact remains that any electronic device will take some time to recover from a burst of jamming and by that time the next burst may hit it, thus rendering it useless. The effect is like that of trying to have a conversation with a blacksmith who is hammering a piece of iron. Before the ear has recovered from the noise of one hammer blow, it will be exposed to the next.

Barrage and sweep jamming spread their energy over a bandwidth much wider than that of the source signal. Both barrage and sweep jamming can be used against any number of sources and only require a receiver to tell them that there is a source present. The difference between barrage and sweep jamming lies in the modulation techniques and size of the frequency band covered. Barrage jamming often uses an amplitude-modulated signal covering a 10-per cent frequency band (bandwidth equal to 10 per cent of the center frequency). Sweep jamming often uses a frequency-modulated signal, and the frequency is swept back and forth over a very wide bandwidth.

In practice, the sweep is momentarily stopped in time, at random frequencies within the band, to provide a holding period at a fixed frequency. The sweep rate is also varied in a random fashion to prevent any possible synchronization of the receiver with the jammer for the purposes of defeating the jamming, i.e. anti-jamming or **ECCM (Electronic Counter-Counter Measures).**

(A.4.3) Spot Jamming

Spot jamming is the jamming of a specific channel or frequency. In spot jamming all the power output of the jammer is concentrated in a very narrow bandwidth, ideally identical to that of the triggering device. Spot jamming is usually directed against a specific source and requires a panoramic receiver to match the jamming signal to the source signal.

(A.4.4) Factors On Which The Effectiveness Of Jamming Depends

The effectiveness of a jammer depends, among other things, upon the power output of the transmitter, losses in the transmission line, antenna gain in the direction of the victim receiver, and transmitter bandwidth. In addition, the amount of the jammer transmitter emission delivered into the victim receiver is a function of the receiver bandwidth, its antenna gain, and the radar cross-sectional area of the target. In order to be effective, the jammer transmitter must be capable of emitting enough power in the bandwidth of the victim receiver to jam its intended signal or to simulate a deceptive signal realistically.

Covert jammers have also been designed which can be conceal in the trunk of a car. One of them has the unique feature that it does not break the squelch of a VHF set. Jammers which break off the squelch give away their presence.

Fig. 258 The VIP-300S Vehicle Mounted Jammer Of Homeland Security Strategies, Inc. Designed Specifically For RCIEDs

(A.5) BASIC COMPONENTS OF A JAMMER

The main electronic components of a jammer are:

- Voltage-controlled oscillator: It generates the radio signal that will interfere with the signal to be jammed.
- Tuning circuit: It controls the frequency at which the jammer broadcasts its signal by sending a particular voltage to the oscillator.
- Noise generator: It produces random electronic output in a specified frequency range to jam the target signal (part of the tuning circuit).
- RF amplification (gain stage): It boosts the power of the radio frequency output to high enough levels to jam a signal.

Every jamming device has an antenna to radiate its noise. Some are contained within an electrical cabinet. On stronger devices, antennas are external to provide longer range and may be tuned for individual frequencies. Good jammers consist of a matching array of broadband antennas which work independently of each other. The modules jam frequency bands with differing bandwidths, which are located in different areas of the frequency spectrum. All the modules can be mounted on one vehicle.

(A.6) BEYOND THE JAMMERS AND INTO THE REALM OF ALMOST SCIENCE FICTION TO COMBAT IEDs

Intelligent people realize the limitations of jammers and are trying to explore alternate means also. After more than three years of war in Iraq, roadside bombs remain the deadliest single threat to US troops, and countering them has emerged as one of the chief technological problems of the conflict. The Pentagon has spent tens of millions of dollars on the most obvious fixes—adding armor to vehicles and deploying jammers to block radio signals used to explode the devices—only to see the bombers develop larger, better-concealed and more complicated explosives triggered by cell phones, garage-door openers, pressure hoses and other methods. Soldiers have even developed solutions of their own. Many Humvees in Iraq are outfitted with metal devices the size of a hockey stick that can catch tripwires or detect heat-sensitive triggers on roadside bombs.

Now, a Pentagon agency with a $3.3 billion budget and a staff of 300 has a mandate to focus the defense industry on the problem. The undertaking has attracted not only the country's top weapons makers but also dozens of small businesses like AMTI, all pitching a science-fiction gallery of possible solutions. Lockheed Martin Corp. has established a corporate team with $22 million in internal funding that is looking for 'best of breed' technology, including ways to study attack patterns. International Business Machines Corp. has a system it says will create a digital image of often-traveled roads and alert soldiers to changes that could indicate bombs hidden in trash, rocks or animal carcasses. General Dynamics Corp. is pitching a laser-based system adapted from Israeli technology that it says could burn away trash often used to conceal bombs and disable the devices. The Massachusetts Institute of Technology is studying whether there is a way to sniff out bombs with electronic polymers that mimic a dog's ability to smell. Lawrence Livermore National Laboratory has a toy car of its own. After hearing complaints from soldiers that robots operated by wireless controllers were unreliable and subject to radio interference, Livermore came up with one attached to a 1,000-foot tether.

❑❑❑

Appendix B

DETECTION OF EXPLOSIVES

(B.1) A REVIEW OF EXPLOSIVE DETECTION TECHNOLOGIES

Explosive detection techniques can be divided generally into two categories: **bulk detection and trace detection**. In bulk detection, a macroscopic mass of explosive material is detected directly, usually by viewing images made by X-ray scanners or similar equipment. In trace detection, the explosive is detected by chemical identification of microscopic residues of the explosive compound. Trace detection looks for residues after an explosion or contamination of any surface that might have come in contact with the explosive before explosion. These residues can be in either or both of two forms: vapor and particulate. Vapor refers to the gas-phase molecules that are emitted from a solid or liquid explosive because of its finite vapor pressure. Particulate contamination refers to microscopic particles of solid material that adhere to surfaces that have, directly or indirectly, come into contact with an explosive material.

It should be kept in mind that one consequence of using trace detection is that a valid alarm may be recorded for the object (person, package, vehicle, etc.) being screened, even if the object does not contain a concealed bomb. This can happen if the object has been contaminated with trace explosive material for any number of reasons, legitimate or otherwise. For example, when screening people it is possible to record a positive alarm for nitroglycerin from a heart patient using nitroglycerin tablets for medication purposes. He might even be working in a college laboratory. Security forces should neither get unduly alarmed at trace detection nor harass such persons.

(B.2) THE BASIC CONCEPTS USED IN TRACE DETECTION

(B.2.1) Vapor Pressures of Explosives

To have a good understanding of the trace detection of explosives, it is important to understand the concept of vapor pressure of solids and liquids. The vapor pressure of a solid or liquid substance at a given temperature is the pressure of the gaseous phase of the substance that exists at equilibrium above the surface of the solid or liquid. Contrary to popular perception, all solids and liquids emit a certain amount of vapor at all temperatures above absolute zero (-

273°C), and at a given temperature the amount of vapor emitted is characteristic of the particular substance. For solids this may be very low but it is there nevertheless. Don't solids have smell? It is not only liquids that vaporize. Now you get the picture?

Coming to solid or liquid explosives, all of them have characteristic vapor pressures. For some explosives like EGDN, it is quite high; for some explosives like HMX, it is a billion times less than the vapor pressure of EGDN. TNT, for example, comes in between the two extremes.

The relative values of the vapor pressures mentioned above have important implications for the trace detection of explosives. The high-vapor-pressure explosives are relatively easy to detect from their vapor using detectors such as ion mobility spectrometers or electron capture detectors. The medium-vapor-pressure explosives can sometimes be detected from their vapor, but in many cases this will test the limits of sensitivity for gas phase detection, and particle detection based on surface swiping is usually preferred. That is also the preferred way for low vapor pressure explosives including all the plastic explosives. Ammonium nitrate is an exception. Though it is a medium vapor pressure substance, it is almost always likely to be used in rather large quantities. Hence it is more amenable to bulk detection techniques.

A universal technique to deal with low vapor pressure explosives is to spike them with a high vapor pressure taggant during the manufacturing stage itself so that they may be detected with vapor detection technologies. Common taggants include ortho-mononitrotoluene (o-MNT), paramononitrotoluene (p-MNT), and dimethyldinitrobutane (DMDNB). But you cannot depend on taggants—somebody could have older stock of explosives when taggants were not used and somebody could make them himself!

Vapor detection has a fundamental limitation. It pre-supposes that the vapor is allowed to come out. If the bomber is intelligent he can seal the material hermetically so that no vapor at all comes out. Detection by vapor then becomes impossible.

(B.2.2) Particulate Contamination

Particulate contamination, which can also be referred to simply as particle contamination, consists of microscopic solid particles, often with masses of the order of a few micrograms. Explosives in general tend to be rather sticky, and a person handling a macroscopic piece of the solid material with bare hands will quickly transfer small, unnoticeable amounts of such contamination to his or her hands. This material will then be transferred to any additional surfaces that are touched by the hands, including parts of the person's clothing, doorknobs, tabletops, and other objects. He may not be aware of such contamination as it is microscopic but, quite like the way forensic scientists work, the contamination can be detected. While it is hard to make generalizations, a typical fingerprint may contain many particles, often with a total mass in the range of even 100 microgram or so.

Particulate contamination is usually sampled by wiping the surface to be screened with a swipe pad provided by the manufacturer of the trace detection system being used. Once this is done, the swipe pad can be analyzed for the presence of explosives. This works best with briefcases and similar small packages. When screening people, this sort of surface swiping will necessitate physical contact with the test subject, and this may well be objected to.

Please remember that if the bomber is intelligent, he will not allow particulate contamination to take place at all. This is not difficult. Have not people been dealing with hazardous chemicals, toxins and even dangerous biological agents? The safe practices have been standardized since long. The bomber too can handle the material with great care thus rendering particulate detection more or less impossible.

(B.3) TRACE DETECTION TECHNOLOGIES

(B.3.1) Ion Mobility Spectrometry (IMS)

Ion mobility spectrometry (IMS) is one of the most widely used techniques for the trace detection of explosives. The spectrometer consists of two main sections: the ionization region and the drift region. In a typical IMS, ambient air is drawn into an inlet port at the rate of a few hundred cubic centimeters per minute (ccm/min). The purpose of the instrument is to analyze this air for explosives or other compounds of interest, which may be present in the air in the form of vapor or airborne particulate matter. The air first enters the ionization region, where electrons hit the incoming molecules to form positive or negative ions. In the case of explosives, it is negative ions that are formed. The source of the ionizing electrons is a small, sealed piece of metal that has been coated with a radioactive material, usually Nickel-63. Once ions are formed, they are periodically admitted into the drift region through an electronically shuttered gate. The ions are drawn through the gate by a static electric field, which pulls them towards a metal collection plate at the far end of the drift region. This 'drift' of the ions from one end of the drift region to the other occurs at atmospheric pressure, with many collisions between the ions and the various molecules present. The time that it takes the ions to travel the length of the drift region is called the drift time, and is a complex function of the charge, mass, and size of the ion. Typical drift times are on the order of a few milliseconds. The current collected at the metal plate is measured as a function of time, and an IMS spectrum is a plot of ion current versus time, with different peaks representing different specific ions. Sometimes an additional gas called the dopant or carrier gas is admitted into the IMS to aid in the ionization process; very commonly methylene chloride or some other gas that easily forms chloride ions is used. Ions from this gas usually form the largest peak in the IMS spectrum, commonly known as the reactant ion peak or RIP, which serves as a reference peak.

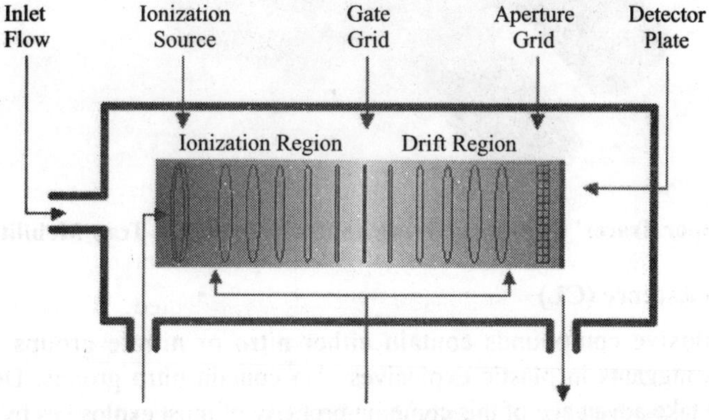

Fig. 259 The Basic Principle Of Operation Of An Explosive Detector Based Upon Ion Mobility Spectrometry

There are a number of features of IMS that make it attractive for the trace detection of explosives. This technique has probably been more widely developed for commercial applications of trace explosives detection than any other. Many of these systems are portable. These instruments have response times of only a few seconds, the proven ability to detect a number of key explosives, subparts per billion sensitivity in some cases, and audio and visual alarms that tell the operator when an explosive has been detected, and its type.

Like all detection techniques, IMS also has certain weaknesses and drawbacks. The warm-up time is considerable. The drift time associated with a given ion is dependent on atmospheric pressure and can thus change during inclement weather or when the spectrometer is moved more than a few hundred feet in elevation. This requires little more than routine, periodic recalibration, but users need to be aware of this potential problem.

A slight variation of the standard ion mobility spectrometry is the so-called **Ion Trap Mobility Spectrometry (ITMS)**. In this the use of an "ion trap" mechanism allows ions to "build up" in a trapped area and then be periodically released into the drift tube for measurement.

Fig. 260 The 'Vapor Tracer' Of GE Ion Track Is Based Upon Ion Trap Mobility Spectrometry

(B.3.2) Chemiluminescence (CL)

Most explosive compounds contain either nitro or nitrate groups. The compounds commonly used as taggants in plastic explosives also contain nitro groups. Detectors based on chemiluminescence take advantage of this common property of most explosives by detecting infrared light that is emitted from electronically excited nitro molecules, denoted as NO_2^*. In a chemiluminescence system, explosive molecules are first pyrolyzed to produce nitric oxide. The nitric oxide molecules are then reacted with ozone in an evacuated reaction chamber maintained at a pressure of about 3 torr. This reaction produces the excited state molecules, NO_2^*. A photomultiplier situated behind a red light filter is used to detect the infrared photons of a characteristic frequency that are emitted when the excited molecules decay to form unexcited molecules. The signal output measured by the photomultiplier is directly proportional to the amount of nitric oxide present in the reaction chamber, and this signal is thus used to detect the presence of explosives in a chemiluminescence system.

Used alone, chemiluminescence is not capable of identifying the type of explosive molecule present. Indeed, all that can be said is that a molecule must initially have been present that decomposed to yield nitric oxide molecules, and such molecules include not only explosives and taggants but also species found in fertilizers, some perfumes, and other potential interferents. For this reason, chemiluminescence detectors are not used alone but are fitted with a front-end gas chromatograph (GC) and then they are known as **GC/CL detectors**. The major drawback of chemiluminescence systems is their inability to detect explosives that are not nitro-based.

(B.3.3) Electron Capture Detectors (ECD)

An electron capture detector or **ECD** detects explosives and other types of molecules having high electron affinities. In an ECD, a vapor sample is drawn into an inlet port, and this vapor mixes with a stream of inert carrier gas (usually helium or argon). The gas flow then travels through an ionization region to an exhaust line. In transit, the gas flow passes through a chamber with a radioactive material that acts as an electron source, as in an IMS. The source material is usually Nickel-63 or tritium. The emitted electrons become thermalized through collisions with the gas in the chamber, and eventually are collected at an anode. Under equilibrium conditions, there is thus a constant standing current at the anode. The basic principle behind an ECD is that this standing current is characteristic of the gas mixture being drawn into the system. If the gas mixture originally consists, for example, of helium and room air, the standing current will be reduced if the vapor of an explosive enters the chamber. This happens because the explosive molecules have a high electron affinity and thus a tendency to capture free electrons and form stable negative ions, leaving fewer electrons to reach the anode. Thus, a reduction of the measured standing current is evidence that an explosive or some similar species is present. As with a chemiluminescence detector, the ECD by itself cannot distinguish individual types of explosives from each other or certain interferents. If you have to identify the explosive then a gas chromatograph is placed on the front end to allow temporal identification of different explosives and then they are known as **GC/ECD detectors**.

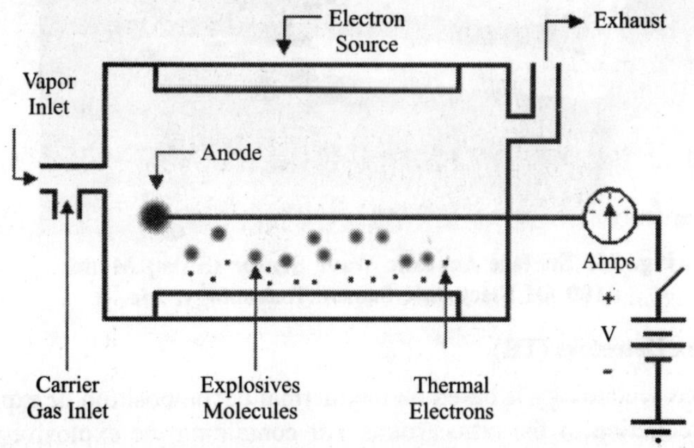

Fig. 261 The Basic Principle Of Operation Of An Explosive Detector Based Upon Electron Capture

This type of detector has a rapid response and typical sensitivities of about 1 ppb for most electron-capturing compounds. This sensitivity is somewhat less than the sensitivity of a typical IMS or chemiluminescence system, but it is still adequate for some applications. However,

GC/ECD detectors usually cost less than IMS or chemiluminescence systems. These systems also tend to be smaller, lighter, and more easily portable.

(B.3.4) Surface Acoustic Wave Sensors (SAW)

Surface acoustic wave (SAW) sensors are usually coupled with a front end GC, as is the case with ECD and chemiluminescence detectors and are known as **GC/SAW detectors.** The principal component of a SAW sensor is a piezoelectric crystal that resonates at a specific, measurable frequency. When molecules condense on the surface of this crystal, the resonant frequency shifts in proportion to the mass of material condensed. The frequency shift also depends upon the properties of the material being deposited, the surface temperature, and the chemical nature of the crystal surface. In a typical system, the exit gas from the GC is focused onto the SAW resonator crystal using a carefully positioned and temperature controlled nozzle. A thermoelectric cooler maintains the SAW surface at sufficiently low temperatures to ensure efficient trapping of the molecules of interest. The crystal surface can also be heated in order to desorb vapors and thus clean the surface. The temperature of the surface allows control of sensor specificity, by preventing adsorption of species with vapor pressures above a certain level. This feature is useful in distinguishing between high and low vapor pressure explosives. During sampling, vapors are concentrated in a cryo-trap before being desorbed into the GC for temporal separation.

Fig. 262 Surface Acoustic Wave Sensor (SAW) Model 4100 Of Electronic Sensor Technology, Inc.

(B.3.5) Thermo-Redox Detectors (TR)

Thermo-redox technology is based on the thermal decomposition of explosive molecules and the subsequent reduction of the nitro groups. Air containing the explosive sample is drawn into a system inlet at a rate of approximately 1.5 l/min. The air is next passed through a concentrator tube, which selectively adsorbs explosive vapor using a proprietary coating on the tube's coils. The sample is then pyrolyzed to liberate nitro molecules, and these molecules are detected using a proprietary technology developed by Intelligent Detection Systems. Since only the presence of nitro groups is detected, this technology cannot distinguish among different explosives and potential interferents that contain nitro groups. Thus, the system identifies the presence of an "explosive-like" material, without identifying a specific explosive. Furthermore, the technology cannot detect explosives that do not contain nitro groups.

Fig. 263 Model EVD-3000 Of Scintrex/IDS Using Thermo-Redox Technology

(B.3.6) Field Ion Spectrometry (FIS)

Field ion spectrometry (FIS) is a relatively new (1994) trace detection technology that is related to IMS. It incorporates a unique ion filter using dual transverse fields, which allows interferences to be eliminated electronically, without the use of GC columns, membranes, or other physical separation methods. FIS is similar to IMS in that it involves separating and quantifying ions while they are carried in a gas at atmospheric pressure. Furthermore, both systems utilize soft ionization methods that yield spectra where the species of interest produce the main features (i.e., under proper conditions there is little decomposition of the analyte). In FIS, ions enter an analytical volume defined by a pair of parallel conducting plates where they execute two motions. The first is a longitudinal drift between the plates due to the bulk motion of a clean, dry carrier stream of air. The second is an oscillating motion transverse to the bulk flow velocity that occurs as the ions respond to an asymmetric, time-varying electric field imposed between the two plates. In response to the asymmetric field, the ions tend to migrate towards one of the plates where they will be neutralized. A second DC field is simultaneously established across the plates and can be used to balance or compensate for the drift introduced by the primary field. The DC field intensity needed to compensate for the AC field induced drift depends on the mobility of the particular ion under investigation, so that only specific ions can pass completely through the analytical volume and into the collection area for detection. Therefore, the device can be tuned to selectively pass only the ions of interest. Scanning the DC field intensity produces a spectrum of ion current versus field intensity that is known as anionogram.

(B.3.7) Mass Spectrometry (MS)

Mass spectrometry (MS) has long been one of the most powerful techniques available for laboratory chemical analysis. Although it is rarely used in routine field applications and may thus be of little interest to law enforcement agencies doing explosive detection, it is discussed briefly here for completeness and because of this widespread laboratory use. While there are different types of mass spectrometers, this is basically a magnetic filtering technique. Molecules are first ionized and then passed through a magnetic filter, which allows ions to be identified based on their charge-to-mass ratio. In some systems, the MS is connected to a front-end GC. Mass spectrometers have excellent specificity for identifying different ions, and some systems have sub-picogram sensitivity. Mass spectrometers tend to be expensive. The advertised sensitivity is in the low parts per billion range for some analytes.

(B.3.8) EXPRAY Field Test Kit

EXPRAY is a unique, aerosol-based field test kit for the detection of what the manufacturer refers to as Group A explosives (TNT, DNT, picric acid, etc.), Group B explosives (Semtex H, RDX, PETN, NG, smokeless powder, etc.), and compounds that contain nitrates that are used in improvised explosives. Detection of explosive residue is made by observing a color change of the test paper.

Initially, a suspected surface (of a package, a person's clothing, etc.) is wiped with the special test paper. The paper is then sprayed with the can provided for Group A explosives. The appearance of a dark violet-brown color, for example, indicates the presence of TNT, a blue-green color indicates the presence of DNT, and an orange color indicates the presence of other Group A explosives. If there is no reaction, the same piece of test paper is then sprayed with the can for Group B explosives. The appearance of a pink color indicates the presence of Group B explosives, a group which includes most plastic explosives. If there is still no reaction, the same paper is then sprayed with the third can. The appearance of pink then indicates the presence of nitrates, which could be part of an improvised explosive. If the second can is applied after a positive result with the first can, a change to pink color indicates a double or triple base explosive. The chemistry associated with these color changes is proprietary. The order of spraying is critical, and all three sprays should be used in order to perform a complete test.

Fig. 264 The EXPRAY Field Test Kit

The EXPRAY system has certain limitations. Obviously, it is not always possible to identify the specific type of explosive present when a positive result occurs. Sampling is entirely by surface swiping; there is no method for obtaining a vapor sample. Furthermore, not all types of explosives can be detected. Some nitrate esters and the chlorate group give a negative result. These include mixtures of potassium chlorate, sodium chlorate, and potassium nitrate with sugar, sulfur, and/or carbon. In addition, only the specific colors mentioned above can be judged a positive detection. Other discoloration is possible, but should be judged negative. The main attraction of the system is its low cost.

(B.4) BULK DETECTION TECHNOLOGIES BASED ON X-RAYS

X-ray technologies are used to search for explosives in a variety of situations. X-ray systems are most often used for screening luggage, packages, mail, and other relatively small items. All of the systems involve irradiation of a target item with X-rays, followed by detection of an image created by X-rays that are either transmitted or backscattered by the item. Vehicle screening systems are also available, though these are large, expensive, and require passengers to exit the vehicle while it is being screened.

The absorption of the X-rays in a material depends mainly on the atomic numbers of the elements present in it. The black and white images must be viewed and subjectively interpreted by an operator. Systems measuring effective atomic number can automatically alarm in the presence of materials that have an effective atomic number that is in the correct range for explosives.

X-ray imaging systems can be divided into two broad types of categories:

- Systems that simply produce an image.
- Systems that detect explosive-like materials and may generate automatic alarms. The simple black and white images from fluoroscopic imaging, image storage panel, and standard transmission systems cannot identify the actual explosive material, but do allow the operator to see wires, batteries, detonators and other bomb components.

(B.4.1) Fluoroscopic Imaging (FI)

FI refers to a transmission X-ray system, where the transmitted x-rays form an image of the objects investigated on a fluorescent screen. This is the simplest type of system, exposing the entire object with a cone of X-ray energy, often for extended periods. The fluorescent screen can be viewed in real time using a mirror angled at 45 degrees (to avoid standing in direct line with the x-rays), or can be used in conjunction with a video camera and a video image storage panel. Use of the camera and image storage panel minimizes the X-ray dose to the object being viewed, by allowing the image to be captured during a brief exposure. Even with a camera and storage device, this type of system is not film safe. The main reasons for the continued use of fluoroscopic imaging and image-storage panel systems are low cost and portability.

(B.4.2) Standard Transmission X-Ray (ST)

A higher level of technology is the standard transmission X-ray. These systems use a fan or flying-spot of X-rays to scan the object as it is carried past the scanner by a conveyor belt. A black and white image is produced directly with a linear array of sensing diodes. The resulting image is stored in digital memory and displayed on a TV or computer screen for operator viewing. These systems are often found at airport checkpoints, where carry-on bags are screened. While standard transmission systems deliver a smaller X-ray dose to the object being screened (they are film safe), they produce black and white transmission images similar to the fluoroscopic systems and are subject to the same detection limitations. Specifically, standard transmission systems cannot identify the actual explosive material, but do allow the operator to see wires, batteries, detonators and other bomb components.

(B.4.3) Backscatter Imaging (BI)

Backscatter systems produce an image from X-rays that are scattered from the screened object. You see, this uses a fundamentally different phenomenon. So far we had dealt with penetration of objects by X-rays. Here we deal with scattering of X-rays by objects. Because low atomic number materials are more efficient at scattering X-rays, explosive-like materials are imaged as bright in the backscatter image, while they are barely visible in the transmitted image. The backscatter system produces two images and both the backscatter and transmission images are displayed. The backscatter image is usually most effective for the detection of low atomic number materials such as explosives, narcotics, and plastic handguns while the transmission image is most useful for viewing metals. Backscatter systems cannot discriminate between various low atomic number materials, for example, between C-4 military plastic explosive and harmless plastic. This is a very serious limitation of a very costly technology.

(B.4.4) Dual-Energy Systems (DE)

Dual-energy X-ray systems yield superior material discrimination through comparison of the attenuation of x-ray beams at two energies. Thus, identification of low atomic number materials can be achieved by using incident X-ray beams of two distinct energies. Materials of specific atomic numbers (the same effective atomic number as that for explosives) can be clearly highlighted for the operator by adding color to the image. A material that has a high atomic number (for example, metals) is often colored green, while low atomic number materials are colored orange, and materials with the same atomic number as explosives are shown red.

Some standard transmission systems may have color instead of black and white displays. Do not confuse dual-energy systems with colorized single energy transmission systems, which do not color the image on the basis of atomic number, but rather color the image based on the level of X-ray energy that is transmitted. These simple single-energy colorized systems have never been proven more effective than the single-energy black and white systems, although their displays may look similar to a dual-energy display.

(B.4.5) Computed Tomography (CT)

Computed tomography (CT) is an even more sophisticated X-ray technique in which two-dimensional images ('slices') through an object are added together to produce a three dimensional image. Along with the three-dimensional image, the effective atomic number is calculated and materials with the same atomic number as explosives can be identified. This is essentially the same technique that may be familiar to the reader from medical CT scans. Development of a dual-energy CT scanner is underway. This system may be capable of detection with lower false alarms than the single-energy CT scanners. Any system that can determine that low-atomic number materials are present can have an **automated alarm** function added. Systems that have been converted into automated detection systems include backscatter, dual energy, and the CT. In fact, CT systems are most useful as automated detection systems because the image they produce may be difficult for the operator to interpret (the image is often heavily distorted).

(B.5) CANINE DETECTION

Trained canines provide a reliable and time-proven method for detecting concealed explosives. There is in principle no explosive compound that dogs cannot be trained to sniff out, and dogs used in the US military are typically trained to detect nine different explosives. Compared to technology-based 'sniffer' systems, dogs have the advantages of superior mobility and the ability to rapidly follow a scent directly to its source. Because of these advantages, canines are an excellent choice for explosives detection applications that involve a significant search component. Such applications include the search of vehicles, warehouses, luggage and cargo, aircraft, buildings, offices, and areas exterior to buildings such as parking lots, property perimeters, etc. Disadvantages of canines compared to trace technologies include limited duty cycle (i.e., a dog works about 1 h before requiring a break), the need for regular retraining, and the inability to communicate to the handler the type of explosive that is detected.

(B.6) NOVEL DETECTION TECHNIQUES

(B.6.1) Thermal Neutron Activation (TNA)

This technique is based on the interaction of neutrons with the nitrogen atoms contained in an explosive compound and analysis of the resulting gamma rays. Thermal neutrons have to be generated by slowing down (moderating) fast neutrons, as apparently there are no portable means of producing them directly. In summary, TNA is able to characterize high explosives by their nitrogen and hydrogen signature, possibly exploiting the absence of other elements as well. It is probably the 'easiest' among the neutron-based techniques, apart from neutron backscatter. On the other hand it is relatively slow; typical response times range from minutes to tens of minutes, depending on the material being investigated. TNA it is not capable of detecting, in practical terms, neither oxygen nor carbon.

(B.6.2) Fast Neutron Analysis (FNA)

The Fast Neutron Analysis (FNA) technique measures the results of the interaction of fast neutrons with suspect matter, in particular inelastic scattering and the production of characteristic gamma rays. The neutron energy has therefore to be above a given threshold for each of the elements involved. The energies of the gamma rays emitted indicate the elements present in the material. The intensity of the gamma rays indicates the relative amount of material present. It is therefore in principle possible to calculate the elemental ratios—how much of each element is present with respect to the others—in order to determine the type of substance under analysis. In the instance of explosives, carbon, nitrogen, and oxygen ratios are considered. FNA has therefore the potential of delivering better results than TNA because it is sensitive to nearly all elements in explosives and opens the possibility of identifying the substance under analysis, but is usually far more complex and expensive.

(B.6.3) Pulsed Fast Neutron Analysis (PFNA)

It is also based on the interaction of neutrons with the explosive material, which results in the emission of gamma rays with characteristics that are indicative of the specific material. Pulsed operations allow the use of timing information (e.g. coincidence or anti-coincidence measurements), and can be very useful for background reduction. Pulsed operations are particularly interesting when using very short fast neutron pulses (typically nanosecond wide)—short compared to the flight time across the object to be analyzed. In addition, the neutrons have to be as mono-energetic as possible (they have to travel at roughly the same speed). Given these conditions, **Time-Of-Flight (TOF)** techniques can be used to determine the location of the detected material—the measurement start time is given when the neutron pulse is created, and the stop time when the gamma rays are recorded (the gamma rays travel at the speed of light, much faster than the neutrons). When combined, for example, with the vertical scanning of the neutron source and the horizontal movement of the object relative to the source/detector, pulsing provides a three-dimensional spatial resolution capability (still rather coarse but potentially very useful). The nature of the material is again provided by gamma spectroscopy.

(B.6.4) Pulsed Fast-Thermal Neutron Analysis (PFTNA)

In a typical Pulsed Fast-Thermal Neutron Analysis (PFTNA) setup a neutron generator produces microsecond wide fast neutron pulses, for example, 14 MeV neutrons from a D-T generator. During these pulses, and possibly also shortly thereafter, prompt gamma rays resulting from fast neutron inelastic scattering reactions (and nuclear reactions) are measured, in particular to identify carbon and oxygen. The accelerator is then kept off for a time of about 100 microseconds, and during this interval the neutrons thermalize and prompt gamma rays resulting from neutron capture reactions can be measured (in particular for the detection of nitrogen). The cycle then starts again. A longer pause can also be exploited (a few ms), in order to measure delayed gamma rays due to activation of one of the elements (such as oxygen). In a pulsed approach the signal-to-noise ratio improves, but the overall signal strength is reduced by the duty factor of pulsing, requiring much stronger peak neutron intensity (to compensate for the low duty factor). This in turn can make the spectrum measurement during the pulse very difficult.

(B.6.5) Nuclear Quadrupole Resonance (NQR)

Nuclear quadrupole resonance (NQR), a derivative of nuclear magnetic resonance (NMR), is a bulk inspection technology for detecting crystalline explosive solids containing Nitrogen-14 nuclei such as RDX, TNT, and nitrates. NQR has been described as 'an electromagnetic resonance screening technique with the specificity of chemical spectroscopy'. Unlike NMR where an external (static) magnetic field is needed, quadrupole resonance takes advantage of the material's natural crystalline electric field gradient, that is, the electrical gradients available within the molecule itself. These gradients are due to the distribution of the electrical charge within the molecule and do therefore depend on the chemical structure (they will be different for different materials). The electrical field gradients align the electric quadrupole moments, which are a physical property due to a non-spherical (say ellipsoidal) nuclear charge distribution, of the Nitrogen-14 nuclei. As a result, the material being diagnosed need not be contained within large magnetic field-producing coils. NQR therefore amounts to NMR without a magnet.

When a low-intensity RF signal is applied to the material at certain frequencies, usually in the range 0.5 to 6 MHz, the alignment of the Nitrogen-14 nuclei is altered. As the RF is removed, the nuclei precess to their original state (actually a transition between the energy states resulting from the previously described interaction), producing a characteristic radio signal. The signal can then be measured for analysis. Detecting the presence of explosives becomes similar to tuning a radio to a particular station, and the uniqueness of a molecule's electric field allows NQR technology to be highly compound specific. This high selectivity is partly a disadvantage, as it is apparently not that easy to build a multi-channel system necessary to cover a wide range of target substances.

The impossibility of detecting substances fully screened by metallic enclosures (also foils, depending on their thickness) is a serious problem for NQR as it is for NMR also. What can possibly be done is that the presence of such objects would throw the NQR probe out of tune, in which case the operator would come to know that something is wrong.

(B.6.6) Portable Isotopic Neutron Spectroscopy (PINS)

It is also based on irradiation of the explosive material with neutrons to produce gamma rays of a characteristic energy which are used to identify the element. The neutrons are obtained from a Californium-252 source. It is designed primarily for field use in military applications, where it identifies explosives, nerve agents, and blister agents associated with firing ranges and munitions burial sites. The presence of phosphorus, for example, would indicate nerve gas. The relative heights of the peaks are related to the ratios of the elements inside the container. These energies and ratios are in themselves uniquely characteristic of the contents of the container, whether high explosive, nerve gas, or something else. In this way, with a 'library' of known signatures, it is possible to determine unambiguously the contents of the container.

(B.6.7) Associated (Alpha) Particle Time-Of-Flight Analysis (API)

The Associated (Alpha) Particle Technique (or API, **Associated Particle Imaging**) uses an interesting property of the deuterium-tritium (D-T) reaction. What happens is that the source emits a low energy alpha particle simultaneously with, and at exactly at 180 degrees ('back-to-back') to, each fast 14 MeV neutron produced. In a modified form of the D-T generator these alpha particles can be detected internally with a position sensitive detector, measuring their time of generation (and therefore that of the accompanying neutron) as well as their direction relative to the source (and therefore also the direction of the accompanying neutron, since its line of travel is opposite to that of the alpha particle). The fast neutrons thus produced are therefore also defined as 'tagged' (by the associated particle), in time as well as in direction, which can lead to a significant reduction in background signals. The neutron then collides with a nucleus and produces a gamma ray, whose time of arrival at the detector can be precisely measured. This is therefore again a **Time-of-Flight (TOF)** technique as is the case with the PFNA, allowing determining the distance traveled by the neutron (as both the speed of the neutron and of the gamma ray are in principle known). As its direction is also known, three-dimensional spatial resolution of targets can in principle be provided, without the need for scanning. The nature of the volume element of the material being analyzed is again provided by gamma spectroscopy. Associated Particle Imaging is therefore a technique which can potentially provide three-dimensional (3D) information (i.e. voxel by voxel) on the chemical nature of the object under analysis, ideally stoichiometric information.

(B.6.8) Prompt Gamma Spectroscopy (PGS)

In a number of cases it is actually not necessary or appropriate to discriminate between TNA and FNA, stressing instead that prompt gamma rays be measured, independently of them being due to thermal neutron capture or fast neutron inelastic scattering. For example, in practice the neutrons used for TNA will also contain a fast neutron component, and therefore gamma rays due to their inelastic scattering reactions will be present in the spectrum together with the lines due to neutron capture (which is what one is looking for in pure TNA). Pulsed operations allow some degree of discrimination.

(B.6.9) Neutron Backscatter (NB)

Neutron backscatter is probably the simplest neutron based technique. It relies on fast neutrons emitted into the target (as we have seen they can penetrate easily several cm of steel if necessary), where they are slowed down by collisions with hydrogen nuclei. The number of detected backscattered thermalized neutrons, i.e. slow neutrons coming back in the direction of the source, provides a measure of the hydrogen content of the material. These devices are therefore also called 'Neutron Thermalization Gauges' (NTG), and are in current use in a number of other fields (for example, petroleum industry). Current applications of neutron backscatter systems include for example the discrimination of ordnance containing explosives, inert substances and chemical warfare agents based on the considerable differences in their hydrogen content. A system has also been reported to have been in use for quite some time to detect explosives hidden in car doors, tyres, etc. For the detection of buried objects (for example, landmines) the system has problems in presence of too much water and seems therefore likely to only work in dry or slightly humid environments.

❑❑❑

Appendix C

High Explosives at a Glance

CLASSIFI-CATION	EXPLOSIVE	COLOR	USES	RATE OF DETONATION	REMARKS
Low Explosives	Black Powder	Black, gray or cocoa brown	Safety fuze, Muzzle loaders	1,312 feet per second	very sensitive to friction heat and shock
	Smokeless Powder	Light brown to black	Small arms, mortars, rockets	Rapid burning	very sensitive to friction heat and shock
Primary Explosives	Lead Azide	White to buff gray	Detonators, priming compositions	17,500 feet per second	very sensitive to friction heat and shock
	Lead Styphnate	White to buff gray	Priming compositions	17,000 Feet per second	very sensitive to friction heat and shock
	Mercury Fulminate	Light orange to reddish brown	Detonators, priming compositions	13,940 feet per second	very sensitive to friction heat and shock
	Tetrazene	Pale yellow	Detonators, priming compositions	Less than 13,100 feet per second	sensitive to shock and heat. Used in combination with other explosives
Secondary Explosives	Amatol	Buff to yellow to dark brown	Main charge for bombs, projectiles	21,100 feet per second	Developed during WWII to conserve TNT
	Ammonal	Gray	Projectile filler	17,700 feet per second	water soluable

contd. ...

... contd.

CLASSIFICATION	EXPLOSIVE	COLOR	USES	RATE OF DETONATION	REMARKS
	Ammonium Nitrate	White but may be dyed other colors	Ingredient of many explosive mixtures	3,300 to 8,200 feet per second	Must be kept dry
	Ammonium Picrate	yellow to orange to red	Armor piercing projectiles and bombs	22,500 feet per second	Relatively insensitive to shock and friction
	Astrolite	White pellets	Demolition	25,000 to 26,200 feet per second	Inert until mixed. Do not use with Tetryl
	C-4	White to light brown	Plastic demolition explosive	26,400 feet per second	Insensitive to impact and friction
	Cyclotol	Buff to yellow to brown	Fragmentation bombs, projectiles	25,600 feet per second	Excellent for blast effects
	Flex-x	any color-- Usually olive drab or red	Cutting charges	22,300 feet per second	Flexible, waterproof, insensitve to shock
Secondary Explosives	HBX (Torpex)	Gray	Main charge filler for underwater bombs and torpedoes	22,700 to 23,700 feet per second	Excellent for blast effects
	HMX	White	Mixed with TNT in high blast warheads	29,900 feet per second	By-product of RDX manufacture
	Kinepak	Powder is white, the liquid is usually pink	Construction	20,700 feet per second	Inert until mixed
	Minol	gray	Filler for bombs and depth charges	19,100 to 19,700 feet per second	Comparable to TNT in sensitivity to initiation
	Nitro-Cellulose	White	Blasting, smokeless powder	21,900 feet per second	Used in flashless powder
	Nitroglycerin	Clear to amber. Red fumes mean "Beware"	Demolition, ingredient in dynamite	24,900 to 25,400 feet per second	Can be absorbed through skin causing headache
Secondary Explosives	Nitro-guanidine	White to yellow	Propellant and bursting charge ingredient	25,100 feet per second	One of the least sensitive military explosives

contd. ...

... contd.

CLASSIFICATION	EXPLOSIVE	COLOR	USES	RATE OF DETONATION	REMARKS
	Nitro-starch	white	Mortar shells, grenades	16,000 feet per second	Another form of Nitro-cellulose
	Octol	Buff	Projectile and bomb filler	27,500 to 28,350 feet per second	Excellent for blast effects
	Pentolite	White to yellow to gray	Shape charges, boosters	24,500 feet per second	Presence of grit increases impact sensitivity
	PETN	white unless dyed	Det cord, blasting caps, primer	27,500 feet per second	Presence of grit increases impact sensitivity
	Picratol	Yellow to brownish yellow	Armor piercing projectiles and bombs	22,900 feet per second	Insensitive to initiation
Secondary Explosives	Picric acid	Cream to yellow to red	Alternative filler	19,000 feet per second	Dangerous when it deteriorates
	RDX	White but may be dyed	Det cord, blasting caps, used to make C-4	26,800 to 28,000 feet per second	Not used much until WWII
	Tetryl	Clear to yellow to gray	Booster, blasting caps	25,800 feet per second	Colors skin reddish brown and causes rash
	Tetrytol	Light yellow to buff	Bursters, demolition blocks	24,000 to 24,200 feet per second	Similar to TNT and Tetryl
	TNT	Light yellow to brown to light gray	Bombs, projectiles, demolition	21,800 to 22,400 feet per second	Standard with which all other explosives are measured
	Torpex	Gray	Depth charges, mines	24,600 feet per second	Excellent for blast effects
	Tritonal	Silvergray	Bombs	21,200 to 22,000 feet per second	More powerful and more sensitive to shock than TNT

SELECTED BIBLIOGRAPHY

A Military Guide To Terrorism In The Twenty-First Century, United States Army Training And Doctrine Command, Office Of The Deputy Chief Of Staff For Intelligence. [Fort Leavenworth, KS] ,U.S. Army Training And Doctrine Command, Deputy Chief Of Staff For Intelligence, Ver. 3.0, 2005.

Akhavan, Jacqueline, The Chemistry Of Explosives, Royal Soc. Chem., Cambridge, UK, 1998.

Akhavan, Jacqueline, The Chemistry Of Explosives, Middle Eastern Terrorist Bomb Designs, Paladin Press, USA, 1990. (Formerly Titled: Bomb Squad: Defining And Defusing Terrorist Explosives).

An Improved Guide To Bridge Demolitions, Royal Engineers Training Notes No. 35 (Special), 1976, Ministry Of Defence, UK.

Assessment Of Technologies Deployed To Improve Aviation Security: First Report, Panel On Assessment Of Technologies Deployed To Improve Aviation Security, National Materials Advisory Board, Commission On Engineering And Technical Systems, National Research Council, Publication, National Academy Press, Washington, D.C. 1999.

Bailey, A., and Murray, S.G., Explosives, Propellants And Pyrotechnics, Brassey, London, 2000.

Baldwin, B., Ralph, The Deadly Fuze, (UK Edition Published By Janes), 1980.

Barrie, Alexander, War Underground: The Tunnellers Of The Great War, Tom Donovan, London.

Baytos, John F., Field Spot-Test Kit For Explosives, Los Alamos National Laboratory, USA.

Benson, Ragnar, Home-Built Claymore Mines: A Blueprint For Survival, Paladin Press, Boulder, Colorado, 1993.

Biasutti, G.S., History Of Accidents In The Explosives Industry, Eigenverlag, Vevey, Schweiz, 1985.

Black And Smokeless Powders: Technologies For Finding Bombs And The Bomb Makers, Committee On Smokeless And Black Powder, Board On Chemical Sciences And Technology, Commission On Physical Sciences, Mathematics, And Applications, National Research Council, National Academy Press, Washington, D.C. 1998.

Black Book Companion: State-Of-The-Art Improvised Munitions, Paladin Press, 1990.

Blaster's Handbook, E.I. Dupont De Nemours And Co., Wilmington, Delaware, 1980.

Bodurtha, F.T., Industrial Explosion: Prevention And Protection, McGraw-Hill, New York, 1980.

Border, Bill, Incendiaries: Advanced Improvised Explosives, Paladin Press, Boulder, Colo., U.S.A., 1988.

Bowden, F.P. and Yoffe, A.D., Initiation And Growth Of Explosion In Liquids And Solids, Cambridge University Press, 1952.

Brigitta, M. Dobratz, LLNL Explosives Handbook: Properties Of Chemical Explosives And Explosive Simulants, UCLR-52997, Livermore, California Nationai Technical Information Service, US Department Of Commerce, Springfield, VA, 1981.

Brodie, Thomas G., Bombs And Bombings: A Handbook To Protection, Security, Disposal, And Investigation For Industry, Police And Fire Departments, Springfield, Ill.: C.C. Thomas, 3rd Ed., 2005.

Brown, George I, The Big Bang: A History Of Explosives, Sutton Pub., Stroud, G.B., 1998.

Bruschini, Claudio , Commercial Systems For The Direct Detection Of Explosives (For Explosive Ordnance Disposal Tasks), Explostudy, Final Report , École Polytechnique Fédérale De Lausanne (Epfl) & Vrije Universiteit Brussel (Vub), 2001.

Budavari, Susan, The Merck Index, 11th Edition, Merck & Co Inc., NJ.

Bunker, Robert J, and Sullivan, John P, Suicide Bombings In Operation Iraqi Freedom, Land Warfare Paper No. 46w.Arlington: Institute Of Land Warfare, Association Of The United States Army, 2004.

Carleone, Joseph, (Ed.), Tactical Missile Warheads, Progress In Astronautics And Aeronautics Series (V-155), AIAA, 1993.

Chase, W. G., and Moore, H. K., Exploding Wires, Plenum Press, New York, 1959.

Chéret, R., Detonation Of Condensed Explosives, Springer Verlag, Berlin, 1993.

Chilton, Thomas. H, Strong Water, M. I. T. Press, Cambridge, MA, 1968.

Cole, R. H., Underwater Explosions, Dover Publ., New York, 1965.

Containing The Threat From Illegal Bombings: An Integrated National Strategy For Marking, Tagging, Rendering Inert, And Licensing Explosives And Their Precursors, Committee On Marking, Rendering Inert, And Licensing Of Explosive Materials, Board On Chemical Sciences And Technology, Commission On Physical Sciences, Mathematics, And Applications, National Research Council, National Academy Press, Washington, D.C. 1998.

Cook, James R., The Chemistry And Characteristics Of Explosive Materials, Vantage Press, New York, 2001.

Cook, M.A., The Science Of Industrial Explosives, Ireco Chemicals, Salt Lake City, Utah, U.S.A., 1974.

Cook, M.A., The Science Of High Explosives, Chapman & Hall, London, 1958.

Cooper, Paul W. and Kurowski, S. R., Introduction To The Technology Of Explosives, VCH Publishers, Inc., Weinheim, Germany, 1996.

Cooper, Paul W., Explosives Engineering, Wiley-VCH, New York, 1996.

Croll, Mike, History Of Landmines, Pen And Sword, 1998.

Cross, J. and Farrer, D., Dust Explosions, Plenum Press, New York, 1982.

Davis, Tenney L., The Chemistry Of Powder And Explosives, Angriff Press, Hollywood, CA, 1972. (also John Wiley & Sons, Inc., New York, 1956.)

Deadly Brew: Advanced Improvised Explosives, Paladin Press, Boulder, Colorado, U.S.A., 1987.

Dean, John A., Lang's Handbook Of Chemistry, McGraw-Hill Book Company, New York, 1987.

Deforest, M. J., Principles Of Improvised Explosive Devices, Paladin Press, Boulder, Colorado, U.S.A., 1984.

Doe, R., Zemlak, O. S., and Montreuil, M., Electronics For IEDs, Canadian Police College. Algonquin Pub. Centre, 2004.

Dussance, G., A Practical Treatise On The Fabrication Of Fulminating Powders, Philadelphia, 1864.

Electronic Noses & Sensors For The Detection Of Explosives, NATO Science Series II, Springer, 2004.

Elements Of Armament Engineering (Part One), Army Material Command, Army Research Office, Washington, D.C., 1964.

Elias, L. and Newdorfl, P., Laboratory Evaluation Of Portable And Walk-Through Explosive Vapor Detectors, Proceedings From The 3rd Symposium On The Analysis And Detection Of Explosives, Mannheim-Nenostheim, FRG, 1989.

Ellern, H., Military And Civilian Pyrotechnics, Chem. Publishing Co., 1968.

Encyclopedia Of Explosives And Related Items, PATR-2700, Picatinny Arsenal, USA.

EOD Improvised Explosives Manual, Paladin Press, Boulder, Colorado, 1990.

Existing And Potential Standoff Explosives Detection Techniques, National Research Council, National Academy Press, 2004.

Explosive Trains, Army Materiel Command Pamphlet 706-179, 1965.

Explosives And Blasting Procedures Manual, U.S. Bureau Of Mines, 1982.

Explosives And Propellants From Commonly Available Materials, Desert Publications, El Dorado, AR, USA.

Explosives And Rock Blasting, Atlas Powder Co., 1987.

Explosives For Sabotage, CIA Manual.

Explosives In Roadworks: User's Guide, Assoc. Of Australian State Road Authorities, 1982.

Federoff Basil T., et al, A Manual For Explosives Laboratories, Vol. IV, Lefax, Phila., Pa., 1946.

Fedoroff, Basil T., and Sheffield, Oliver E., Encyclopedia Of Explosives And Related Items, US Army Research And Development Command, Picatinny Arsenal, New Jersey, 1962.

Fester, Uncle, Home Workshop Explosives, Loompanics Unlimited, Port Townsend, WA, 1990

Feuer, Eigene, Complete Instructions On How To Build Undetectable Hand Grenades, Desert Publications, El Dorado, AR, USA.

Fickett, W. and Davis, W.C., Detonation, University Of California Press, 1979.

Fickett, W., Introduction To Detonation Theory, University Of California Press, 1985.

Field Expedient Incendiaries, CIA Manual.

Field, P., Dust Explosions, Elsevier, Amsterdam, 1982.

Fitts, Richard E., Fundamentals Of Electronic Warfare, U.S. Air Force Academy, Colorado Springs, 1972.

Flores, J., The Anarchist's Handbook, 1995.

FM 3-34.119, IED Defeat Operations, US Army, 2005.

FM 3-34.210, Explosive Hazards Operations, US Army.

FM 5-31, Boobytraps, US Army.

FM 5-250, Explosives And Demolitions, Headquarters, Department Of The Army, USA, 1992.

FM 23-30, Grenades And Pyrotechnic Signals, Headquarters, Department Of The Army, USA.

FM 9-6, Munitions Support In Theater Of Operations, US Army, 1989.

Fordham, S., High Explosives And Propellants, Pergamon Press, Oxford, 1980.

General Information Bulletin 97-1, FBI Explosives Unit-Bomb Data Center.

Gille, Bertrand, Engineers Of The Renaissance, The MIT Press, Cambridge, Massachusetts.

Glackin, James J., Elements Of Explosives Production, Paladin Press, Boulder, Colorado, 1976.

Goldman, Dirk, Improvised Primary Explosives, 1998.

Gonzales, Jo Jo, Deathtrap: Improvised Booby-Trap Devices, Paladin Press, Boulder, Colorado, 1989.

Gonzales, Jo Jo, Death By Deception: Advanced Improvised Booby Traps, Paladin Press, Boulder, Colorado, 1992.

Gonzales, Jo Jo, Deathtrap! Improvised Booby-Trap Devices, Paladin Press, Boulder, Colorado, 1989.

Gordon, Yefim, Soviet/Russian Aircraft Weapons Since World War II, Midland, 2005.

Grady, Sean M., Devices Of Controlled Destruction, Lucent Books, 1994.

Grupp, Larry, Claymore Mines: Their History And Development, Paladin Press, Boulder, Colorado, 1993.

GTA 90-01-001, Improvised Explosive Device (IED) and Vehicle Borne Improvised Explosive Device (VBIED), US Army.

GTA 90-10-044, MNC-I Iraq Theater IED & Explosive Hazards Awareness Guide, US Army.

GTA 90-10-045, MNC-I Improvised Explosive Device (IED) Components Smart Card, US Army.

GTA 90-11-001, MNC-I Warlock Family of Counter Radio Controlled IED Electronic Warfare, US Army.

Gustafson, R., Swedish Blasting Technique, Gothenburg, Sweden, 1981.

Handbook Of Electric Blasting, Atlas Powder Company, Dallas, Texas, 1985.

Hannum, D. W., and Parmeter, J. E., Survey Of Commercially Available Explosives Detection Technologies And Equipment, Written For The National Law Enforcement And Corrections Technology Center, A Program Of The National Institute Of Justice, National Institute Of Justice, USA, 1998.

Hannum, D.W., Performance Evaluations Of Commercial Explosives Vapor/Particulate Detectors, Sandia National Laboratories, USA, 1996.

Harber, David, Guerrilla's Arsenal: Advanced Techniques For Making Explosives And Time-Delay Bombs, Paladin Press, Boulder, Colorado.

Harber, David, Improvised Land Mines: Employment And Destructive Capabilities, Paladin Press, 1992.

Harber, David, The Advanced Anrachist Arsenal: Recipes For Improvised Incendiary And Explosives, Paladin Press, Boulder, Colorado, 1991.

Harber, David, The Anrachist Arsenal: Improvised Incendiary And Explosive Techniques, Paladin Press, Boulder, Colorado, 1990.

Hennessy, T. J., Development Of Techniques For Field Synthesis Of Potassium Nitrate And Nitric Acid, Frankford Arsenal, M67-17-1, Ammunition Development And Engineering Lab, Phila, Pa, 1967.

Hennessy, T. J., Field Expedient Preparation Of Black Powders, Frankford Arsenal, M67-16-1, Ammunition Development And Engineering Lab, Phila, Pa., 1967.

Hogg, Ian V., Ammunition: Small Arms, Grenades And Projected Munitions, Greenhill Books, 1998.

IED Trigger Recognition Guide, US Dept Of Homeland Security, Terrorist Threat Integration Center, Washington, DC, 2004.

IED Trigger Recognition Guide, US Dept Of Homeland Security, National Counterterrorism Center, Community Integration Group, IICT Secretariat, Washington, DC, 2005

Improvised Explosives: How To Make Your Own, Paladin Press, 1985.

Improvised Explosives For Use In Detonators, Desert Publications, El Dorado, AR, U.S.A., 1980.

Improvised Munitions Black Book, Vol. 1-3, Desert Publications, El Dorado, AR 1981.

Improvised Munitions From Ammonium Nitrate, Desert Publications, El Dorado, AR, 1980.

Improvised Munitions Handbook, Special Products Laboratory, Frankford Arsenal, Phila., Pa.

Improvised Sabotage Devices, CIA Manual.

Improvised Shaped Charges, Desert Publications, AR, USA.

Incendiary Weapons, A SIPRI Monograph, The MIT Press, Cambridge, Mass., 1975.

James, R., Propellants And Explosives, Noyes Data Corporation, Park Ridge, New Jersey, 1974.

Johnson, C. H., and Persson, P.A., The Detonics Of High Explosives, Academic Press, New York, NY 1970.

Joint Munitions Effectiveness Manual, Commander, Naval Air Systems Command, Washington, D.C., 1974.

Jones, Ian, Malice Afterthought: The History Of Booby Traps From World War One To Vietnam, Greenhill Books, 2004.

Jones, Tony L., Booby Trap Identification And Response Guide For Law Enforcement Personnel, Paladin Press, 1998.

Joseph, E. Zola, Chemical Safety Data Guide, The Bureau Of National Affairs, Inc., Washington, D.C.

Kaye, Seymour M., Encyclopedia Of Explosives And Related Items. PATR 2700, Vol. 1–10, US Army Armament Research And Development Command, Dover, NJ, 1980 (Available From National Technical Information Service, US Department Of Commerce, Springfield, Virginia)

Keller, J. J., OSHA Safety And Compliance Manual, 1992.

Kelly, Jack, Gunpowder: Alchemy, Bombards, & Pyrotechnics, Basic Books.

Khosrokhavar, Farhad, Suicide Bombers: Allah's New Martyrs, David Macey, Pluto Press, London, 2005.

Kit, B. and Evered, D.S., Rocket Propellant Handbook, Macmillan, New York 1960.

Kling, A., Report On An Yperite Mine, Report No. Z-741, Municipal Chemical Laboratory, Paris, France, DTIC No. AD499336, 1918.

Kohler, J. and R. Meyer, Explosives, VCH - Verlagsges., Weinheim, FRG, 1993.

Land Mine And Explosive Hazards (Iraq), Commander's Reference Guide: TC-20-32-5, 2003.

Langefors, U. and Kihlstrom, B., The Modern Technique Of Rock Blasting, Almquist & Wiksell, Stockholm, 1963.

Lecker, Seymor, Improvised Explosives: How To Make Your Own, Paladin Press, Boulder, Co, 1985.

Lecker, Seymour, Improvised Explosives: How To Make Your Own, Paladin Press, Boulder, Colo., U.S.A., 1985.

Lecker, Seymour, Professional Booby Traps: Special Devices And Techniques, Paladin Press, 1993.

Ledgard, Jared B., The Preparatory Manual Of Black Powder And Pyrotechnics, USA, 2006.

Lee, J. H., Knystautas, R. and Bach, G. G., Theory Of Explosion, McGill University Press, Montreal, 1969.

Lewis, Richard, Sr., Sax's Dangerous Properties Of Industrial Materials, 8th Ed. Van Norstrad Rienhold, 1993.

Luxon, S.G., Hazards In The Chemical Laboratory, Royal Society Of Chemistry, Cambridge, England.

Mader, Charles L. Numerical Modeling Of Explosives And Propellants. (2nd Edition), CRC Press, Boca Raton, Fla., U.S.A., 1998.

Mahan, D. H., An Elementary Course Of Military Engineering: Part I. Comprising Field Fortification, Military Mining, And Siege Operations, Wiley & Son, New York, 1867.

Manual Of Explosive Preparations & Formulas, EXP 5-100.

Marshall, A, Explosives, J & A Churchill, London, England, 1917.

McAdam, R., And Westwater, R., Mining Explosives, Oliver & Boyd, London, 1958.

Mcfee, J. E., and Das, Y., Advances In The Location And Identification Of Hidden Explosive Munitions, Defence Research Establishment Suffield, Report No. 548, 1991.

Mclean, Donald B, Medicine Chest Explosives: An Investigator's Guide To Chemicals Used In Home-Cooked Bombs, Paladin Press, Boulder, Colorado, 1995.

Mellor, J. W., A Comprehensive Treatise On Inorganic And Theoretical Chemistry, Vol. VI , Longman's Green & Co., New York, London, 1939.

Mellor, J. W., Modern Inorganic Chemistry, Revised Edn., Longman's Green & Co., New York, London, 1939.

Metamorphosis, The Terrorist Encyclopedia Vol. 1.02, ebook on the Web, The Phyco Department.

Meyer, Rudolf and Kohler, J., Explosives, 4th Edition, VCH-Verlagsges., Weinheim, New York, 1993.

Meyer, Rudolf, Kohler, J. And Homburg, A., Explosives, Wiley-VCH, 2002.

Mia Bloom, Dying To Kill: The Allure Of Suicide Terror, Columbia University Press, New York, 2005.

Military Explosives, Departments Of The Army And Air Force, Washington, D.C., 1967.

Military Training Pamphlet No. 30, Field Engineering (All Arms), Part VI-Demolitions, Chief Of The Imperial General Staff, The War Office, UK, 1945.

Moffatt, Thomas J, The Alchemist's Secrets Of Explosive Chemistry, Paladin Press, Boulder, Colorado, 1988.

Mordecai, Thomas, Professional Standards For Preparing, Handling, And Using Explosives, Paladin Press, Boulder, Colorado, 1995.

Murr, L. E., Shock Waves For Industrial Applications, Noyes Publications, Park Ridge, New York, 1989.

Myers, Lawrence W., Improvised Radio Detonation Techniques, Paladin Press, Boulder, Colorado, 1988.

Myers, Lawrence W., Counterbomb, Paladin Press, Boulder, Colorado, 1992.

Naoum, P., Nitroglycerine And Nitroglycerine Explosives, Williams And Wilkins Co., Baltimore, Md., 1928.

New And Improved C-4: Better-Than-Ever Recipes For Half The Money And Double The Fun, Paladin Press, Boulder, Colorado, 1995.

Newhouser, C. R., Introduction To Explosives, The National Bomb Data Center, Gaithersburg, U.S.A., 1973.

NGIC-116-9629-02, Land Mine and Explosive Hazards Reference Guide (Afghanistan), US Army, 2002.

Nobel, A., A Treatise On Blasting Agents, F. Soc. Arts, London, 1875.

Nobel, A, Tri-Nitro-Glycerine, 3rd ed. Rewritten, North And Adams, New York, 1874.

Noble, Andrew, Artillery And Explosives, E.P. Dutton And Co., New York, 1906.

Ordnance Technology, Indian Head Division, Naval Surface Warfare Center.

Partington, J.R., A History Of Greek Fire And Gunpowder.

Penner, S.S. and Mullins, B.P., Explosions, Detonations, Flammability And Ignition, Pergamon Press, London, 1959.

Penner, S.S., Chemical Rocket Propulsion And Combustion Research, Gordon & Breach, New York, 1962.

Penner, S.S., and Ducarme, J., The Chemistry Of Propellants And Combustion, Pergamon Press, London, 1960.

Persson, P. A., Holmberg, R., and Lee, J., Rock Blasting And Explosives Engineering, CRC Press Inc., Boca Raton, Florida, 1993.

Porter, Samuel J, Explosives And Blasting Agents, Fire Protection Handbook, 15th ed., 1981.

Powell, William, The Anarchist Cookbook, Stuart Lyle, New York, USA, 1971.

Powell, William, The Anarchist Cookbook, Barricade Books, 1990.

Principles Of Explosive Behavior, Army Materiel Command Pamphlet 706-180, 1972

Report On The Availability Of Bombmaking Information-1997, Prepared By The United States Department Of Justice As Required By Section 709(A) Of The Antiterrorism And Effective Death Penalty Act Of 1996 For The United States House Of Representatives And The United States Senate, April 1997 (Available On The Website Of The US Department Of Justice: www.usdoj.gov/criminal/cybercrime/bombmakinginfo.html)

Results Of Explosive Detection Research In Support Of Portable Explosive Detection, Naval Explosive Ordnance Disposal Technology Division, Indian Head, MD, USA (Unknown Date, Possibly 1997).

Rhykerd, Charles L., Hannum, David W., Murray, Dale W., and Parmeter, John E., Guide For The Selection Of Commercial Explosives Detection Systems For Law Enforcement Applications, NIJ Guide 100-99 Contraband Detection Technologies Department, Sandia National Laboratories, Albuquerque, (Prepared For the National Institute Of Justice, Office Of Science And Technology Washington,) 1999.

Roger, The Jolly, The Anarchist's Cookbook, ebook on the Web.

Rostberg, James I., Common Chemicals As Precursors Of Improvised Explosive Devices: The Challenges Of Controlling Domestic Terrorism, Naval Postgraduate School (U.S.), Monterey, Calif. : Naval Postgraduate School; Springfield, Va., 2005.

Rothman, Barry, Improvised Munitions Black Book, Vols. 1-3, Desert Publications, Cornville, USA.

Safety And Performance Tests For Qualification Of Explosives, Commander, Naval Ordnance Systems Command, Washington, D.C., 1972.

Safety Library Publication No. 20, Institute Of Makers Of Explosives, Washington D.C., 1988.

Samer, S.F., Propellant Chemistry, Reinhold, New York, 1966.

Saxon, Kurt, The Poor Man's James Bond, 12th Edition, Desert Publication, USA, 1991.

Schubert, Hiltmar, Detection And Disposal Of Improvised Explosives, Springer, 2006.

Scott, Lee, Car Bomb Recognition Guide: How They Are Made, How To Detect Them, Paladin Press, Boulder, Colorado, 1999.

Shock Sensitive Industrial Materials, (Advanced Improvised Explosives; 2), Paladin Press, Boulder, Colorado, 1988.

Sickler, Robert A, Explosive Principles: An Essential Guide To Understanding Explosives And Detonations, Paladin Press, 1992.

Sickler, Robert A.,Explosive Principles, Paladin Press, Boulder, Colorado, 1992.

Singer, et al, A History Of Technology Vol. V, Claredon Press, Oxford, 1958.

Sittig, Marshall, Handbook Of Toxic And Hazardous Chemicals, Noyes Publications, Park Ridge, NJ.

Smith, R.P., Gosselin, R. E. and Hodge, H.C., Clinical Toxicology Of Commercial Products, Williams & Wilkins, Baltimore, Md.

Smith, Jim, A Law Enforcement And Security Officers' Guide To Responding To Bomb Threats: Providing Working Knowledge Of Bombs, Preparing For Such Incidents, And Performing Basic Analysis Of Potential Threats, C.C. Thomas Publishers, Springfield, Ill. 2003.

Stiff, Peter, Taming The Landmine, Galago Publishing, Alberton, South Africa, 1986.

Stoffel, Joseph F., and Thomas, Charles C. Explosives And Homemade Bombs. Springfield, Ill., U.S.A., 1962 and 2nd ed. 1972.

Suceska, Muhamed, Test Methods For Explosives, Springer, New York, 1995.

Suicide Bombing In The COE, DCSINT Handbook No. 1.03, Army Training & Doctrine Command And Deputy Chief Of Staff For Intelligence, Fort Leavenworth, KS, 2005.

Summerfield, M., Solid Propellant Rocket Research, Academic Press, London, 1960.

Taylor, J., and Gay, P. F., British Coal Mining Explosives, George Newnes, London, 1958.

Taylor, J., Detonation In Condensed Explosives, Clarendon Press, Oxford, 1952.

Taylor, J., Solid Propellants And Exothermic Compositions, George Newnes, London, 1959

Taylor, W., Modern Explosives, The Royal Institute Of Chemistry, London, 1959.

TC 20-32-5, Commander's Reference Guide, Land Mine and Explosive Hazards (Iraq), US Army, 2003.

The Anarchist Cookbook Version 2000, ebook on the Web.

The Behavior Of Nitro-Glycerine, Tech Paper 12, U.S. Bureau Of Mines, Washington, D.C, 1912.

The Temple Of The Screaming Electron, How To Make Bombs Book 1, ebook Walnut Creek, CA.

Theisen, Lisa, Hannum, David W., Murray, Dale W., and Parmeter, John E., Survey Of Commercially Available Explosive Detection Technologies And Equipment 2004, The National Law Enforcement And Correction Technology Center, A Program Of The National Institute Of Justice, USA, 2005.

TM 5-223A, Soviet Mine Warfare Equipment, Department Of The Army, Washington, D.C., 1951.

TM 5-223C, German Mine Warfare Equipment, Department Of The Army, Washington, D.C., 1952.

TM 5-223D, British, French And Italian Mine Warfare Equipment, Department Of The Army, Washington, D.C., 1952.

TM 9-1300-206, Ammunition And Explosive Standards, US Army, 1973.

TM 9-1300-214, Military Explosives, Headquarters, Department Of The Army, USA, 1984.

TM 9-1985-2, German Explosive Ordnance (Bombs, Fuzes, Rockets, Land Mines, Grenades, And Igniters), Departments Of The Army And The Air Force, D.C., 1953.

TM 9-1985-6, Italian And French Explosive Ordnance, Departments Of The Army And The Air Force, 1953.

TM 31-201-1, Incendiary Devices & Techniques, US Army.

TM 31-210, Improvised Munitions Handbook, US Army.

TM 43-0001-38, Army Ammunition Data Sheets For Demolition Materials, US Army, 1981.

Tomlinson, W. R., Properties Of Explosives Of Military Interest, Picatinny Arsenal, Dover, N.J. 1971.

Turner, Stephen, Terrorist Explosive Sourcebook: Countering Terrorist Use Of Improvised Explosive Devices, Paladin Press, Boulder, Colorado, 2005.

Urbanski, Tadeusz, The Chemistry And Technology Of Explosives, Vol. 1-4, Pergamon Press, Oxford, England, 1985. (First ed. Polish Scientific Publishers, Warszawa, 1967.)

Viet Cong Boobytraps, Mines And Mine Warfare Techniques, United States Army, University Press Of The Pacific, 2004.

Vietnam, 1964-1969, Landmine And Countermine Warfare, Engineer Agency For Resources Inventories, Washington, D.C., 1972.

Wallace, William, FMX The Revised Black Book: A Guide To Field-Manufactured Explosives, Paladin Press, Boulder, Colorado, 1995.

Walters, W.P., and Zukas, J.A., Fundamentals Of Shaped Charges, John Wiley & Sons Inc., 1989.

Warheads Primer, Indian Head Division, Naval Surface Warfare Center.

Warren, F. A., Rocket Propellants, Reinhold, New York, 1960.

Weapons System Engineering, 6th ed., Weapons And Systems Engineering Department, U.S. Naval Academy, Annapolis, MD, 1984.

Weapons Systems Fundamentals, Commander, Naval Ordnance Systems Command, Washington, D.C., 1971.

Weast, Robert C., CRC Handbook Of Chemistry And Physics, CRC Press, Boca Raton, FL, 1988.

Western Europe, World War II, Landmine And Countermine Warfare, Engineer Agency For Resources Inventories, Washington, D.C., 1973.

Wiech, R. E. and Strauss, R.F., Fundamentals Of Rocket Propulsion, Reinhold, New York, 1960.

Winter, Ruth, Consumer's Dictionary Of Household, Yard, And Office Chemicals, Crown Publisher Inc., New York.

Yinon, J. and Zitrin, S., Modern Methods And Applications In Analysis Of Explosives, John Wiley & Sons, Chichester, England, 1993.

Yinon, J. and Zitrin, S., The Analysis Of Explosives, Pergamon Press, Oxford, 1984.

Yinon, J., Forensic And Environmental Detection Of Explosives, John Wiley & Sons, Chichester, UK, 1999.

Zahringer, A.J., Solid Propellant Rockets, Wyandotte, 1958.

Zedalis, Debra D., Female Suicide Bombers, Strategic Studies Institute Monograph, Carlisle, Strategic Studies Institute, US Army War College, 2004.

Zeldovich, J. B. and Raizer, J., Physics Of Shock Waves And High Temperature Hydrodynamic Phenomena (Translation), Academic Press, New York, 1966.

Zeldovich, J. B. and Kompaneets, A. S., Theory Of Detonation (Translation), Academic Press, New York, 1960.

Zukas, J., Walters, W.P. (eds.), Explosive Effects And Applications, Springer, New York, 1998.

❑❑❑

Subject Index